AUTO-CARTO 7

PROCEEDINGS

Digital Representations of Spatial Knowledge

Washington, D.C.
March 11–14, 1985

© 1985 by the AMERICAN SOCIETY OF PHOTOGRAMMETRY and the AMERICAN CONGRESS ON SURVEYING AND MAPPING. All rights reserved. This book, or any parts thereof, may not be reproduced in any form without permission of the publishers.

ISSN 0270-5133

ISBN 0-937294-65-9

Published by
AMERICAN SOCIETY OF PHOTOGRAMMETRY
and
AMERICAN CONGRESS ON SURVEYING AND MAPPING
210 Little Falls St.
Falls Church, Va. 22046

TECHNICAL PAPERS
Seventh International Symposium on Computer-Assisted Cartography
(Auto-Carto 7)
"Digital Representations of Spatial Knowledge"
Washington, D.C.
March 11-14, 1985

This volume contains the proceedings of all the Technical Program Sessions of the Seventh International Symposium on Computer-Assisted Cartography (Auto-Carto 7). This volume is produced by the Symposium Directorate as an aid to both the authors and the symposium registrants.

Only those papers or abstracts received in time for publication are contained in this volume. However, a complete alphabetical list of all authors whose papers were presented at the sessions also is included. Numbers in parentheses (starting with 85-501) indicate the order of presentation and listing in the final program for all papers in the volume.

The Auto-Carto 7 Technical Program Committee is deeply grateful to the many authors and their typists who made this volume possible. Without their effort, cooperation, understanding, and adherence to the required formats and scheduling, a project of this magnitude could not have been completed successfully in the time allotted. Special thanks go to Cliff Ludwig and his Publications and Distribution Committee, and particularly to Morris M. Thompson, Kim Peesel, and Don Memenway, for their overall scheduling and their working with the printer.

Many others gave a great deal of both their time and expertise in assisting the Technical Program Committee in arranging this selection of technical papers. Along with specific thanks to the individual members of the Technical Advisory Committee listed below, the Auto-Carto 7 Technical Program Committee especially is appreciative of all the ACSM and ASP members who helped put this program together.

The Auto-Carto 7 Technical Program Committee

Steven J. Vogel
Nicholas R. Chrisman
Mary G. Clawson
Donna M. Dixon
Bill Hezlep
Julie A. Prusky
Carlos A. Smith
Beverly L. Thompson
Sharon L. Tomlinson

The Auto-Carto 7 Technical Advisory Committee

Jack Foreman
Dennis P. Franklin
Lawrence W. Fritz
Armando Mancini
Robert W. Marx
Joel L. Morrison
Lowell E. Starr
Owen W. Williams

INDEX TO AUTHORS

AUTHOR		PAGE
ANDERSON, K. ERIC +	85-501	1
A Prospective Case for a National Land Data System: Ten Years Later		
ANSTAETT, MARK R. +	85-516	11
Area Calculations Using Pick's Theorem on Freeman-Encoded Polygons in Cartographic Systems		
ARMSTRONG, MARC P. +	85-515	283
Analytic and Cartographic Data Storage: A Two-Tiered Approach to Spatial Decision Support Systems		
ARONSON, PETER	85-517	23
Applying Software Engineering to a General Purpose Geographic Information System		
ARONSON, PETER +	85-571	313
Considerations in the Design and Maintenance of a Digital Geographic Library		
BAIN, STAN +	85-530	244
An Operational GIS for Flathead National Forest		
BENNETT, BRIAN S.	85-588	32
Applications of Producing Plots from DLG Data		
BICKINGS, CHARLENE K. +	85-589	38
Computer-Assisted Production of the 1980 Population Distribution Map		
BIRDSONG, R. B. +	85-589	38
Computer-Assisted Production of the 1980 Population Distribution Map		
BISHOP, DAVE +	85-529	228*
A Fuzzy and Heuristic Approach to Segment Intersection Detection and Reporting		
BLUMBERG, RICHARD E. +	85-557	43*
Application of Standard Linear Format to Digital Hydrographic Products		
BROOKS, WILLIAM F. +	85-558	44*
TLDB: A Global Data Base Designed for Rapid Digital Image Retrieval		
BROOKS, WILLIAM T.	85-559	46
Implementation of an Integrated Resource Information System and Its Application in the Management of Fish and Wildlife Resources in Alaska		

NOTES
+ : Coauthor
* : No paper was submitted or paper was not received in time for publication. A page number with an asterisk indicates the publication of an abstract.

INDEX TO AUTHORS

AUTHOR		PAGE
BROOME, FREDERICK R.	85-502	53*

TIGER Preliminary Design and Structure Overview: The Core of the Geographic Support System for 1990

BURR, DAVID J. +   85-509   399
  ACES: A Cartographic Expert System

CAMPBELL, WILLIAM C. +   85-568   254*
  The Map Analysis Package as the Core of an Integrated Water Resource Information System

CEBRIAN, JUAN ANTONIO +   85-518   55
  Analysis and Display of Digital Elevation Models Within A Quadtree-Based Geographic Information System

CHAPPUIS, ANNE +   85-560   66
  Info-Graphics: Its Applications to Planning in India

CHEN, ZI-TAN +   85-519   75
  Quad Tree Spatial Spectra Guides: A Fast Spatial Heuristic Search in a Large GIS

CHRISMAN, NICHOLAS R.   85-586   83*
  An Interim Proposed Standard for Digital Cartographic Data Quality

CHRISMAN, NICHOLAS R. +   85-552   84
  Alternative Routes to a Multipurpose Cadastre: Merging Institutional and Technical Reasoning

CHRISMAN, NICHOLAS R. +   85-577   512*
  Dane County Land Records Project

CHRISMAN, NICHOLAS R. +   85-555   552
  Tests to Establish the Quality of Digital Cartographic Data: Some Examples from the Dane County Land Records Project

CHUI WEIHONG +   85-561   95*
  Methods and Software Systems for Computer-aided Environmental Mapping

CLARKE, KEITH C.   85-520   97
  Strategies for Spatial Data Compression

COOKE, DONALD F.   85-579   108
  Vehicle Navigation Appliances

COOPER, JOHN R., JR. +   85-578   514
  The Automated Revision of Tennessee Valley Topographic 1:24,000 Quadrangles

NOTES
  + : Coauthor
  * : No paper was submitted or paper was not received in time for publication. A page number with an asterisk indicates the publication of an abstract.

INDEX TO AUTHORS

AUTHOR		PAGE

COWEN, DAVID J. +   85-590   116
   Alternative Approaches to Display of USGS Land Use/
   Land Cover Digital Data

CROMLEY, ROBERT G.   85-521   127
   An LP Relaxation Procedure for Annotating Point
   Features Using Interactive Graphics

CROSLEY, POWEL   85-522   133
   Creating User Friendly Geographic Information Systems
   Through User Friendly System Supports

DEARBAUGH, BRUCE +   85-562   141*
   The Conceptual Design for Products from the National
   Ocean Service's Automated Nautical Charting System II

DE GOLBÉRY, LUC +   85-560   66
   Info-Graphics: Its Applications to Planning in India

DE GREE, MELVIN +   85-591   142
   Digital Elevation Model Image Display and Editing

DEVEAU, TERRY J.   85-523   152
   Reducing the Number of Points in a Plane Curve
   Representation

DOYTSHER, Y. +   85-592   161*
   Terrain Analysis for Communication Purposes

DOYTSHER, Y. +   85-597   510*
   Automated Mapping

DRINNAN, CHARLES H.   85-563   162
   Military Base Planning Using Geographic Information
   Systems Technology

DUEKER, KENNETH J.   85-564   172
   Geographic Information Systems: Toward a Geo-Relational
   Structure

DWYER, MICHAEL F. +   85-567   237
   A Method for Comparing Data Entry Systems

DYE, ROBERT H. +   85-573   353*
   Development of Geographic Information Systems for Poorly
   Mapped Areas of the World

EDSON, DEAN T. +   85-584   178*
   An Interim Proposed Standard for Digital Cartographic
   Terms and Definitions

NOTES
    + : Coauthor
    * : No paper was submitted or paper was not received
       in time for publication. A page number with an
       asterisk indicates the publication of an abstract.

INDEX TO AUTHORS

AUTHORS		PAGE

EIDENSHINK, JEFFERY C. +   85-565   179*
　Merging Tabular and Spatial Data Sets in a Geographic
　Information System

FAGAN, JAMES +   85-524   180
　Geometric Algorithms for Curve-Fitting

FEGEAS, ROBIN G.   85-513   189*
　A Data Model for Spatial Data Processing

FORREST, DAVID   85-525   190
　Contouring with Small Computers: A Review

FOSNIGHT, EUGENE A. +   85-547   541
　A Relational Approach to Vector Data Structure Conversion

FRANK, ANDREW U. +   85-553   440
　About Different Kinds of Uncertainty in Collections of
　Spatial Data

GIBBONS, J. GLEN +   85-512   584
　An Interdepartmental Application of Spatial Data Bases--
　Building a Statistical Network File from a Topographic
　Feature File

GILLMAN, DANIEL W.   85-526   191
　Triangulations for Rubber-Sheeting

GLICK, BARRY J. +   85-527   *
　Hybrid Knowledge Representation for a Geographic
　Information System

GOEHLER, DAVID J.   85-505   200
　Laser Disc Technology Applications of Navigational
　Chart Compilation and Image Displays

GOULETTE, ANN M.   85-593   205
　Cartographic Use of Dithered Patterns on 8-color
　Computer Monitors

GOULETTE, ANN M. +   85-589   38
　Computer-Assisted Production of the 1980 Population
　Distribution Map

GRECZY, LASLO +   85-528   210*
　Terrain Analyst Work Station (TAWS)

GREENLAND, ARNOLD +   85-554   212
　Statistical Evaluation of Accuracy for Digital
　Cartographic Data Bases

NOTES
　+ : Coauthor
　* : No paper was submitted or paper was not received
　　　in time for publication. A page number with an
　　　asterisk indicates the publication of an abstract.

INDEX TO AUTHORS

AUTHOR		PAGE

GRELOT, JEAN-PHILIPPE 85-566 222
   Post-Census Cartography in France--Production Increase Despite Budgetary Restrictions

GRISWOLD, LORI ANN + 85-538 408
   A Step Towards Interactive Displays of Digital Elevation Models

GUEVARA, JOSE ARMANDO + 85-529 228*
   A Fuzzy and Heuristic Approach to Segment Intersection Detection and Reporting

GUPTILL, STEPHEN C. 85-506 229
   Functional Components of a Spatial Data Processor

HANSEN, DONALD B. + 85-567 237
   A Method for Comparing Data Entry Systems

HARRISON, DAVID A. + 85-509 399
   ACES: A Cartographic Expert System

HART, JUDY A. + 85-530 244
   An Operational GIS for Flathead National Forest

HECKMAN, BRADFORD K. + 85-509 399
   ACES: A Cartographic Expert System

HEIVLY, CHRISTOPHER + 85-568 254*
   The Map Analysis Package as the Core of an Integrated Water Resource Information System

HENDERSON, JAMES 85-556 257
   The Utility of a GIS in Evaluating the Accuracy of Classified LANDSAT Land Cover Type Maps on the Kenai Peninsula

HERRMANN, RICHARD A. 85-569 267
   Army Requirements for Digital Topographic Data

HIRSCH, STEPHEN A. + 85-527 *
   Hybrid Knowledge Representation for a Geographic Information System

HODGSON, MICHAEL E. 85-531 275
   Constructing Shaded Maps with the DIME Topological Structure: An Alternative to the Polygon Approach

HODGSON, MICHAEL E. + 85-590 116
   Alternative Approaches to Display of USGS Land Use/Land Cover Digital Data

---

NOTES
   + : Coauthor
   * : No paper was submitted or paper was not received in time for publication. A page number with an asterisk indicates the publication of an abstract.

INDEX TO AUTHORS

AUTHORS		PAGE
HOPKINS, LEWIS D. +	85-515	283
Analytic and Cartographic Data Storage: A Two-Tiered Approach to Spatial Decision Support Systems		
HUMPHREY, DAVID L. +	85-532	293
Vectorizing Surface Modeling Algorithms		
JACKSON, MICHAEL J. +	85-511	430
Expert Systems in Map Design		
JENSON, SUSAN K.	85-594	301
Automated Derivation of Hydrologic Basin Characteristics from Digital Elevation Model Data		
KEARNEY, JOHN A. +	85-570	311*
Capabilities for Source Assessment		
KEEGAN, HUGH +	85-571	313
Considerations in the Design and Maintenance of a Digital Geographic Library		
KEFFER, GERARD T.,	85-501	1
A Prospective Case for a National Land Data System: Ten Years Later		
LAI, POH-CHIN	85-595	322
Mis-Application of Automated Mapping, An Assessment		
LAM, ALVEN H. S.	85-572	327
Microcomputer Mapping Systems for Local Governments		
LAUZON, JEAN PAUL +	85-535	355
Approaches for Quadtree-Based Geographic Information Systems at Continental or Global Scales		
LICHTNER, WERNER	85-533	337
Pattern Recognition Procedures for Autmatic Digitizing of Cadastral Maps		
LIN HUAGIANG +	85-561	95*
Methods and Software Systems for Computer-aided Environmental Mapping		
LOON, JOSEPH C. +	85-534	342*
Evaluation Procedures for Testing Contour-to-Grid Interpolation Methods Using Synthetic Surfaces		
LYNCH, MAUREEN P. +	85-507	343
Conflation: Automated Map Compilation--A Video Game Approach		

NOTES
+ : Coauthor
* : No paper was submitted or paper was not received in time for publication. A page number with an asterisk indicates the publication of an abstract.

INDEX TO AUTHORS

AUTHOR		PAGE
MacRAE, BARRY D. +	85-573	353*

Development of Geographic Information Systems for Poorly Mapped Areas of the World

MANDEL, BETTY A. +   85-534   342*
Evaluation Procedures for Testing Contour-to-Grid Interpolation Methods Using Synthetic Surfaces

MARK, DAVID M. +   85-518   55
Analysis and Display of Digital Elevation Models Within A Quadtree-Based Geographic Information System

MARK, DAVID M. +   85-535   355
Approaches for Quadtree-Based Geographic Information Systems at Continental or Global Scales

MARTH, RICHARD +   85-528   210*
Terrain Analyst Work Station (TAWS)

MARX, ROBERT W. +   85-501   1
A Prospective Case for a National Land Data System: Ten Years Later

McCAUSLAND, ROBERT G. +   85-591   142
Digital Elevation Model Image Display and Editing

McCOLLOUGH, MAJOR C. R. +   85-578   514
The Automated Revision of Tennessee Valley Topographic 1:24,000 Quadrangles

McKEOWN, DAVID M., JR.   85-508   366*
Database Issues in Digital Mapping

McLAURIN, JOHN D.   85-503   368*
The U.S. Geological Survey 1:100,000-Scale Digital Cartographic Data Base

McQUARIE, DIANE S. +   85-504   582*
The Portable Map

MEIXLER, DAVID +   85-536   369
Storing, Retrieving and Maintaining Information on Geographic Structures: A Geographic Tabulation Unit Base (GTUB) Approach

MEZERA, DAVID F. +   85-577   512*
Dane County Land Records Project

MILLER, JAMES R. +   85-570   311*
Capabilities for Source Assessment

NOTES
+ : Coauthor
* : No paper was submitted or paper was not received in time for publication. A page number with an asterisk indicates the publication of an abstract.

INDEX TO AUTHORS

AUTHOR		PAGE
MITCHELL, ANDREW P.	85-574	377

An Approach to Microcomputer-based Cartographic Modeling

MOELLERING, HAROLD   85-583   387*
An Interim Proposed Standard for Digital Cartographic Data: Background and Review

MOELLERING, HAROLD +   85-584   178*
An Interim Proposed Standard for Digital Cartographic Terms and Definitions

MOELLERING, HAROLD +   85-516   11
Area Calculations Using Pick's Theorem on Freeman-Encoded Polygons in Cartographic Systems

MOGG, MARILYN A. +   85-567   770
A Method for Comparing Data Entry Systems

MOLNAR, RENNIE +   85-512   584
An Interdepartmental Application of Spatial Data Bases---Building a Statistical Network File from a Topographic Feature File

MOREHOUSE, SCOTT   85-514   388
ARC/INFO: A Geo-Relational Model for Spatial Information

MOWER, JAMES E. +   85-518   55
Analysis and Display of Digital Elevation Models Within A Quadtree-Based Geographic Information System

MOYER, D. DAVID +   85-577   512*
Dane County Land Records Project

NIEMANN, BERNARD J., JR. +   85-552   84
Alternative Routes to a Multipurpose Cadastre: Merging Institutional and Technical Reasoning

NIEMANN, BERNARD J., JR. +   85-577   512*
Dane County Land Records Project

NYERGES, TIMOTHY L.   85-587   398*
An Interim Proposed Standard for Digital Cartographic Data Organization

OHLEN, DONALD O. +   85-565   179*
Merging Tabular and Spatial Data Sets in a Geographic Information System

O'REAGAN, ROBERT T. +   85-537   491
Streamlining Curve-fitting Algorithms that Involve Vector-to-Raster Conversion

---

NOTES
+ : Coauthor
* : No paper was submitted or paper was not received in time for publication. A page number with an asterisk indicates the publication of an abstract.

INDEX TO AUTHORS

AUTHOR		PAGE
ORESKY, C. +	85-509	399

ACES: A Cartographic Expert System

PARKER, JEAN-PIERRE +   85-512   584
   An Interdepartmental Application of Spatial Data Bases--
   Building a Statistical Network File from a Topographic
   Feature File

PATIAS, PETROS G. +   85-534   342*
   Evaluation Procedures for Testing Contour-to-Grid
   Interpolation Methods Using Synthetic Surfaces

PEUQUET, DONNA +   85-519   75
   Quad Tree Spatial Spectra Guides: A Fast Spatial
   Heuristic Search in a Large GIS

PFEFFERKORN, CHARLES E. + 85-509   399
   ACES: A Cartographic Expert System

POIKER, THOMAS K. +   85-538   408
   A Step Towards Interactive Displays of Digital
   Elevation Models

POIKER, THOMAS K. +   85-539   472
   Automated Contour Labelling and the Contour Tree

PORTER, ELIZABETH +   85-528   210*
   Terrain Analyst Work Station (TAWS)

PRATT, STEPHEN H.   85-575   416*
   A Non-problematic Approach to Cartography within the
   Constructs of a Geographic Information System

PUTERSKI, ROBERT   85-576   417*
   Map Modelling as More than Weighting and Rating

RAYMOND, ROBERT H.   85-540   418
   Hyper-Isometry: n Dimensional Mapping in Two Projective
   Spaces

RIDD, MERRILL K. +   85-551   428*
   Building a Functional, Integrated GIS/Remote Sensing
   Resource Analysis and Planning System

ROBINSON, GARY +   85-511   430
   Expert Systems in Map Design

ROBINSON, VINCENT B. +   85-553   440
   About Different Kinds of Uncertainty in Collections of
   Spatial Data

NOTES
   + : Coauthor
   * : No paper was submitted or paper was not received
      in time for publication. A page number with an
      asterisk indicates the publication of an abstract.

INDEX TO AUTHORS

AUTHOR		PAGE
ROGOFF, MORTIMER	85-582	450
Navigating with Full-Color Electronic Charts: Differential Loran-C in Harbor Navigation		
ROSEN, BARBARA +	85-541	456
Match Criteria for Automatic Alignment		
ROSS, JOHN	85-596	463
Detecting Land Use Change on Omaha's Urban Fringe Using a Geographic Information System		
ROTHERMEL, JOHN G. +	85-509	399
ACES: A Cartographic Expert System		
ROUBAL, JOSEPH +	85-539	472
Automated Contour Labelling and the Contour Tree		
ROUSE, RICHARD E. +	85-568	254*
The Map Analysis Package as the Core of an Integrated Water Resource Information System		
SAALFELD, ALAN J.	85-542	482
Lattice Structures in Geography		
SAALFELD, ALAN J.	85-543	490*
Topics in Advanced Topology for Cartography		
SAALFELD, ALAN J. +	85-524	180
Geometric Algorithms for Curve-Fitting		
SAALFELD, ALAN J. +	85-507	343
Conflation: Automated Map Compilation--A Video Game Approach		
SAALFELD, ALAN J. +	85-536	369
Storing, Retrieving and Maintaining Information on Geographic Structures: A Geographic Tabulation Unit Base (GTUB) Approach		
SAALFELD, ALAN J. +	85-537	491
Streamlining Curve-Fitting Algorithms that Involve Vector-to-Raster Conversion		
SAALFELD, ALAN J. +	85-541	456
Match Criteria for Automatic Alignment		
SCHMIDT, WARREN E.	85-585	498*
An Interim Proposed Standard for Digital Cartographic Features		

NOTES
+ : Coauthor
* : No paper was submitted or paper was not received in time for publication. A page number with an asterisk indicates the publication of an abstract.

INDEX TO AUTHORS

AUTHOR		PAGE
SELDEN, DAVID D.	85-550	499
An Approach to Evaluation and Benchmark Testing of Cartographic Data Production Systems		
SHIRLEY, W. LYNN +	85-590	116
Alternative Approaches to Display of USGS Land Use/ Land Cover Digital Data		
SHMUTTER, BENJAMIN +	85-597	510*
Automated Mapping		
SHMUTTER, BENJAMIN +	85-592	161*
Terrain Analysis for Communication Purposes		
SIEFFERT, JAMES +	85-558	44*
TLDB: A Global Data Base Designed for Rapid Digital Image Retrieval		
SOCHER, ROBERT M. +	85-554	212
Statistical Evaluation of Accuracy for Digital Cartographic Data Bases		
SONZA-NOVERA, JOSEPH +	85-577	512*
Dane County Land Records Project		
STREETER, LYNN	85-581	*
Comparing Navigation Aids for Computer-Assisted Navigation		
STROUD, DONALD +	85-565	179*
Merging Tabular and Spatial Data Sets in a Geographic Information System		
TAMM-DANIELS, FREDERICK L. +	85-578	514
The Automated Revision of Tennessee Valley Topographic 1:24,000 Quadrangles		
TERENZETTI, JOAN R, +	85-544	523*
Cartographic Implications of Automated Multilayer Modification		
THOMPSON, DEREK	85-545	524*
Spatial Data Handling Software: A View from A University Geography Department		
THOMPSON, MICHAEL R. +	85-554	212
Statistical Evaluation of Accuracy for Digital Cartographic Data Bases		
TOBLER, WALDO R.	85-598	525*
Interactive Construction of Contiguous Cartograms		

NOTES
+ : Coauthor
* : No paper was submitted or paper was not received in time for publication. A page number with an asterisk indicates the publication of an abstract.

INDEX TO AUTHORS

AUTHOR		PAGE
TRAYLOR, CHARLES T. +	85-599	526

Map Symbols for Use in the Three Dimensional Graphic Display of Large Scale Digital Terrain Models Using Microcomputer Technology

VANG, ALFRED H. +   85-568   254*
   The Map_Analysis Package as the Core of an Integrated Water Resource Information System

VAN HORN, ERIC K.   85-546   532
   Generalizing Cartographic Data Bases

VAN ROESSEL, JAN W. +   85-547   541
   A Relational Approach to Vector Data Structure Conversion

VONDEROHE, ALAN P. +   85-577   512*
   Dane County Land Records Project

VONDEROHE, ALAN P. +   85-555   552
   Tests to Establish the Quality of Digital Cartographic Data: Some Examples from the Dane County Land Records Project

WAGNER, MELVIN L. +   85-557   43*
   Application of Standard Linear Format to Digital Hydrographic Products

WALLACE, THOMAS +   85-590   116
   Alternative Approaches to Display of USGS Land Use/ Land Cover Digital Data

WANG SHU-JIE +   85-561   95*
   Methods and Software Systems for Computer-aided Environmental Mapping

WASILENKO, MICHAEL T. + 85-544   523*
   Cartographic Implications of Automated Multilayer Modification

WATKINS, JAMES F. +   85-599   526
   Map Symbols for Use in the Three Dimensional Graphic Display of Large Scale Digital Terrain Models Using Microcomputer Technology

WEHDE, MICHAEL E. +   85-565   179*
   Merging Tabular and Spatial Data Sets in a Geographic Information System

WHEELER, DOUGLAS J. +   85-551   428*
   Building a Functional, Integrated GIS/Remote Sensing Resource Analysis and Planning System

NOTES
+ : Coauthor
* : No paper was submitted or paper was not received in time for publication. A page number with an asterisk indicates the publication of an abstract.

INDEX TO AUTHORS

AUTHOR		PAGE
WHERRY, DAVID B. +	85-530	244
An Operational GIS for Flathead National Forest		
WHITE, DENIS	85-548	560
A Generalized Cartographic Data Processing Program		
WHITE, MARVIN	85-580	570*
Building A Digital Map of the Nation for Automated Vehicle Navigation		
WHITE, TIMOTHY +	85-590	116
Alternative Approaches to Display of USGS Land Use/ Land Cover Digital Data		
WILLIAMS DONALD L.	85-549	571*
Suitability of Ada for Automated Cartography		
WILLIAMS, ROBERT J.	85-510	572
Enquiry Systems for the Interrogation of Infrastructure		
WILSON, CHARLES L. +	85-573	353*
Development of Geographic Information Systems for Poorly Mapped Areas of the World		
WILSON, PAUL M. +	85-504	582*
The Portable Map		
WILSON, ROBERT J. +	85-562	141*
The Conceptual Design for Products from the National Ocean Service's Automated Nautical Charting System II		
WINN, WALTER M., JR. +	85-562	141*
The Conceptual Design for Products from the National Ocean Service's Automated Nautical Charting System II		
WRAY, JAMES R.	85-600	583*
Potential Contributions of Digital Cartography and Spatial Analysis in Assessing Impacts of Acid Deposition		
YAN, JOEL Z. +	85-512	584
An Interdepartmental Application of Spatial Data Bases-- Building a Statistical Network File from a Topographic Feature File		
ZORASTER, STEVEN +	85-532	293
Vectorizing Surface Modeling Algorithms		

NOTES
+ : Coauthor
* : No paper was submitted or paper was not received in time for publication. A page number with an asterisk indicates the publication of an abstract.

A PROSPECTIVE CASE FOR A NATIONAL LAND DATA SYSTEM:
TEN YEARS LATER

K. Eric Anderson
U.S. Geological Survey
521 National Center
Reston, Virginia 22092
and
Robert W. Marx
Gerard T. Keffer
Bureau of the Census
Washington, D.C. 20233

ABSTRACT

Geographic information systems are emerging as powerful tools for the handling of spatial data. In the past, a geographic information system was often designed to meet the needs of a specific problem, and the data capture, management, and analysis functions were often restricted to the unique characteristic of specific data sets. More recently, systems have been designed for generic data types and functions to provide much greater flexibility and a wider range of applications.

The emergence of these systems, along with a new national data base being developed jointly by the Bureau of the Census and U.S. Geological Survey offer the opportunity to establish a solid foundation for a National Land Data System.

OVERVIEW

"There is a critical need for a better land information system in the United States to improve land conveyance procedures, furnish a basis for equitable taxation, and provide much-needed information for resource management and environmental planning." (National Research Council, 1980).

The need for a national file of land information is obvious. In fact, some describe the lack of such a data base as a glaring gap in our knowledge of the United States; probably resulting from the pioneer-inspired notion of an endless frontier and enough land for everyone. Several European countries, Canada, and Australia are much further along in developing a national data base of land parcels—but they also are far more involved in the development of administrative records systems. This paper presents an idea that might be viewed as the "ultimate" solution to a national land records system. Development of such a system would probably need to be handled in phases over many years or decades. It also recognizes "The Big Risk" involved in developing a national system. Specifically, there is a

---

Publication authorized by the Director, U.S. Geological Survey, on January 14, 1985.

serious question to be answered about the "big brother" aspects of a national system that allows easy access to records on land use and property holdings.

The idea of a national land data system is not new. The Greeks and Romans established elaborate land-record systems primarily in support of land taxation policies (Richeson, 1966). Approximately 10 years ago, the U.S. Geological Survey (USGS) and the National Aeronautics and Space Administration developed a report on the state of then current technology entitled "Toward a National Land Use Information System." (Ackerman and Alexander, 1975). They concluded, "In spite of striking progress made in the improvement of both the hardware and software needed for functioning geographic information systems ..., the ideal system for national use is not yet in sight."

Much has happened in the 10 years since that report was presented. While the ideal system still does not exist, it is in sight. The Governments Division of the U.S. Bureau of the Census conducts the Survey of Taxable Property Values every 5 years. While this is a limited data set, the Census Bureau has gained a considerable amount of expertise and data concerning the Nation's land inventory. The USGS is using high technology to produce digital data which serves as the basic foundation for many modern GIS activities. USGS also produces a series of maps compiled from remote sensing to portray general categories of land use at a scale of 1:250,000. Many State, county, and municipal governments are developing automated land records systems. Most recently, the Topologically Integrated Geographic Encoding and Referencing (TIGER) System being developed by the Geography Division at the U.S. Bureau of the Census uses the computer-readable 1:100,000-scale map base from the U.S. Geological Survey, to provide the potential mechanism for linking data sets at the Federal, State, and local levels. Finally, the Geological Survey is exploring the potential for linking Census Bureau data files, USGS cartographic, geographic, resource, minerals, water, and natural hazard files, as well as available geographically based files of other Federal, State, and local agencies to create a national land data system. This paper explores the potential for accomplishing this objective and the potential benefits.

INTRODUCTION

A modern nation, as a modern business, must have adequate information on the many complex, interrelated aspects of its activities in order to make decisions. Land use is only one such aspect, but knowledge about land use and land cover has become increasingly important as the Nation plans to achieve minerals independence and overcome the problems of haphazard, uncontrolled development, inequitable property tax assessment, deteriorating environmental quality, loss of prime agricultural lands, destruction of important wetlands, and loss of fish and wildlife habitat. Land use data are needed in the analysis of environmental processes and problems if living conditions and standards are to be improved or maintained at current levels.

One prerequisite for better use of land is information on existing land use patterns and changes in land use through time. The U.S. Department of Agriculture reported that during the decade of the 1960's, 730,000 acres (296,000 hectares) were urbanized each year, transportation land uses expanded by 130,000 acres (53,000 hectares) per year, and recreational area increased by about 1 million acres (409,000 hectares) per year. Knowledge of the present distribution and area of such agricultural, recreational, and urban lands, as well as information on their changing proportions, is needed by legislators, planners, and governmental officials at all levels to determine better land use policy, to protect transportation and utility demand, to identify future development pressure points, and to implement effective plans for regional development. As Clawson and Stewart (1965) stated: "In this dynamic situation, accurate, meaningful, current data on land use are essential. If public agencies and private organizations are to know what is happening, and are to make sound plans for their own future action, then reliable information is critical."

The variety of land use and land cover data needed is exceedingly broad. Current land use and land cover data are needed for equalization of tax assessments in many States. Land use and land cover data also are needed by Federal, State, and local agencies for water resource inventory, flood control, water supply planning, and waste water treatment. Many Federal agencies need current comprehensive inventories of existing activities on public lands combined with the existing and changing uses of adjacent private lands to improve the management of public lands. Federal agencies also need land use data to assess the environmental impact resulting from the development of energy resources, to manage wildlife resources and minimize man-wildlife ecosystem conflicts, to prepare national summaries of land use patterns for national policy formulation, and to prepare environmental impact statements and assess future impacts on environmental quality (J.R. Anderson and others, 1976).

Information on changes in land use is needed for water resource planning. As land is changed from agricultural or forestry uses to urban uses, surface water runoff increases in magnitude, flood peaks become sharper, surface-water and groundwater quality deteriorates, and water use increases, thereby reducing water availability. Statistics on the acres of agricultural, urban, and other types of land inundated by floodwaters would be invaluable in estimating damages, further crop losses, and consequent economic impacts. By monitoring and projecting land use trends, it will be possible to develop more effective plans for flood control, water supply, and waste water treatment.

THE PROPOSED LAND DATA SYSTEM

Many differing viewpoints exist about the role of land use planning and the need to regulate land use at a local, State, or Federal level. Regardless of these differences of opinion, there is a basic need to know how the Nation currently is using its land resources and what changes in land use are occurring. Unless an objective assessment of the current land use situation is made, and unless a process for measuring change is initiated, no one will have the facts necessary to evaluate trends and problems associated with the use of land resources.

The proposed land data system will be based on the computer-readable version of the Geological Survey's 1:100,000-scale maps. This file, once enhanced by the Census Bureau with the names of all mapped features, address ranges for each side of every addressable road between intersections, and the boundaries for all the tabulation units recognized in the decennial census, will provide a machine-readable geographic framework for the entire United States at the "city block" level. This nationally consistent geographic framework, is the key ingredient that has been missing in the past. The primary task then becomes developing the linkage mechanisms between systems.

Once the TIGER File is available in 1988, existing geographically based machine-readable land data files from other Federal agencies, State agencies, local governments, and private organizations can be related to one another using standard polygon overlay techniques. For example:

- The land use and land cover information contained in the USGS files can be related to the demographic information from the Census Bureau. Researchers can begin to derive relationships between the characteristics of people and the characteristics of the existing land information (for example, use, ownership, transfer, development potential, and so forth).

- The Environmental Protection Agency can assess proposed hazardous waste sites in terms of surrounding land uses and demographic characteristics.

- The Federal Emergency Management Agency can determine the extent of damage from natural disasters, such as floods or hurricanes, with detailed knowledge of the demographic characteristics of the people affected and the uses of the affected land.

- The extent of land in Federal or State ownership and the location of mineral and energy resources can be studied to develop comprehensive management strategies.

Many State and local data files use geographic area codes in common with the Census Bureau's scheme. The agencies can link their data files on this set of keys as follows:

- State and local agencies with data files organized using Census Bureau geographic codes can display those data graphically in relation to USGS land use and land cover information and Census Bureau demographic data.

- Private organizations can study the location of their facilities in relation to the characteristics of their service population and determine the nature of the surrounding land uses.

Using more sophisticated computer programs now being developed to match and merge two different, but similar computer-readable files:

- Local officials with parcel-based land records files can fit these files into the national file and analyze the characteristics of the people occupying the land parcels comprising each block.

- The coordinates of the national file can be upgraded in this process to the high-precision characteristic of parcel-based files and thus improve the overall utility and accuracy of the national file.

TESTING THE CONCEPT

Finishing the geographic base for the proposed land data system will take several years but the computer files for some areas will be ready much sooner. The USGS files for the State of Florida were part of the pilot project to test the use of 1:100,000-scale maps for the 1990 census. These Florida files will be among the early files processed through the remaining Census Bureau operations. Coincidentally, the State of Florida has a strong interest in a computer-based land data system and has given the task of coordinating developmental activities to Florida State University (Friedley, 1984). This offers the prospect for a State and Federal cooperative test. The suggested sequence of events for such a test is as follows:

1. The USGS captures the data from the 1:100,000-scale maps and edits/converts the information to a digital line graph (DLG) format.

2. The Census Bureau assigns classification codes to the road features in the file while the USGS assigns classification codes to the water, railroad, and other features.

3. The Census Bureau collects map update and feature name information using USGS 1:24,000-scale maps as the base.

4. The Census Bureau merges the road, water, railroad, and other features in the USGS separate files into a single file and enters the map update and feature name information from the 1:24,000-scale maps.

5. The Census Bureau collects the boundary and area name information for all geographic areas to be represented in the file.

6. The Census Bureau enters the boundaries and names for counties, townships, cities, census tracts, and so forth along with the names and geographic codes for these areas.

7. The Census Bureau collects address range information for areas and streets not already covered by existing files.

8. The Census Bureau enters the address range information for new areas.

9. Florida State University develops computer programs to match/merge State and local files with the geographic base.

10. The USGS develops computer programs to match/merge other Federal files with the geographic base.

11. All three agencies work together to test matching/merging State and Federal files with the geographic base.

POTENTIAL BENEFITS OF A NATIONAL LAND DATA SYSTEM

Despite the difficulty of implementing a national land data system, the potential benefits of such a system merit its further evaluation. This is true even if operationally such a system needs to be phased in over a period of 20 to 30 years, or even longer. The following is a partial list of potential benefits to various user groups of a multipurpose, standardized land data system.

- State Governments
    - --Provides accurate inventories of natural and manmade assets.
    - --Accurately locates State ownership or other interests in land.
    - --Provides a standardized data base for management of public land.
    - --Simplifies coordination among Federal, State, and local officials.

- Local Governments
    - --Improves accuracy of real property assessments.
    - --Provides a linkage to Federal and State maps for local planning and engineering base maps.
    - --Provides a standardized data base for neighborhood, municipal, county, or regional development plans.
    - --Avoids the cost of maintaining separate local map systems and land data files.
    - --Encourages coordination among public programs affecting land.

- Private Firms
    - --Provides accurate inventories of land parcels, available as a public record.
    - --Produces standard large-scale maps that can be used for planning, engineering, or routing studies.
    - --Speeds administration of public regulations.

- Individuals
    - --Provides faster access to records affecting individual rights, especially land titles.
    - --Clarifies the boundaries of areas restricted by zoning, wetlands regulations, pollution controls, or other use constraints.
    - --Reduces costs to public utilities by replacing present duplicative base-mapping programs.

- The Census Bureau
  - --Provides a data base so the decennial census can match its address control file of residential structures and select a sample of unmatched parcel records. This sample of supposedly nonresidential parcels can be used in a coverage evaluation survey to see what percentage actually has people living in commercial, industrial, public, or "vacant" space. (Some of the sample parcels might even be found to contain houses!)
  - --Provides a data base so the agriculture census can match its mailing list and select a sample of supposedly nonagricultural parcels for follow-on evaluation studies. In addition, over several census cycles, this will provide a record of the shift in land use from agricultural to nonagricultural pursuits.
  - --Provides a data base so the economic censuses can match the Standard Statistical Establishment List and Business Master File to select a sample of supposedly nonbusiness properties. The Bureau can survey this sample to help measure the significance of "cottage industries" in the American economy.
  - --Creates a file that the Governments Division can use as part of the periodic reporting of real property data: property uses, assessed values, parcel sizes, selling prices, property tax rates and so forth. In the short run, the birth of and support for a land data system can result in local governments moving toward a standard format for maintaining their land ownership/taxation records. This development can reduce the effort involved in collecting the assessed value and related information.
  - --Produces, ultimately, a data base that all of the censuses and surveys can use as a master control file for selective mailings and coverage evaluation studies. With this data base, the Census Bureau can begin to consider seriously the potential for an administrative records census--a dream of many in the Bureau and the nightmare of others.

- The Geological Survey
  - --Provides a geographic structure to integrate river basin data on stream flows, storage capacities, and water quality.
  - --Creates a file that can be used to analyze mineral deposits in relation to land use, demographic characteristics, and Federal and State land holdings.
  - --Provides a mechanism to organize and coordinate the land data records in Federal, State, and local agencies as part of the overall U.S. Department of the Interior stewardship of the Nation's lands.

- Other Federal Agencies
  - --Produces a data base that can be used to record Federal lands records now held in numerous district offices and separate agencies.

--Provides standardized data needed in updating Federal maps and statistics; for example, at the U.S. Departments of Agriculture and Housing and Urban Development, the Federal Emergency Management Agency, and the Environmental Protection Agency.
--Provides a data base for monitoring objects of national concern; for instance, agricultural land use and foreign ownership of United States real estate.
--Provides a reliable record of the locations of Federal ownership or other interests in land.
--Provides a standardized records system for managing Federal assistance to local programs such as housing, community development, and historic preservation.

UPDATING THE SYSTEM

Knowledge about existing land use, along with the rates and trends of change, is essential if the Nation is to overcome such problems as haphazard and uncontrolled growth, declining environmental quality, loss of prime agricultural, residential, and recreational development, and strip mining of coal.

The Geological Survey is developing a methodological framework for updating its land use and land cover baseline maps and data. Required elements of this framework include that it be timely, relatively inexpensive, and appropriate for widely varying needs at national, regional, State, multi-county, county, and city levels. The same technology used to compile the baseline data, that is, remote sensing, offers a timely and efficient approach to the collection and mapping of basic land use and land cover change data with some limitations. Remote sensing techniques span a wide range of capabilities with the promise of even more sophistication in the future.

Similarly, it is important to recognize that even in areas where remotely sensed data coverage exists, these data may not provide all the land use information that may be needed by a specific user. Supplemental source materials and fieldwork may be necessary for some purposes. In fact, the present range of data characteristics, combined with intermittent coverage dictates that the update methodology be designed to make maximum use of complementary sources of data. Even though some inconsistencies exist in the use of remotely sensed data, it is by far the most effective technology available for the timely and economical mapping of land use changes.

Another major component of a land use updating framework is a geographic information system employing a broad-based computer technology. For the user who is faced with developing a set of baseline maps and data pertaining to land use and land cover, updating such maps and data periodically, and relating several sets of associated data to information about the land and its use, the use of a national land data system with the capability for entering, storing, retrieving, and manipulating data geographically on a computer is an invaluable asset.

REFERENCES

Anderson, J.R., 1977, Land use and land cover changes--a framework for monitoring: U.S. Geological Survey Journal of Research, v. 5, no. 2, p. 143-153.

Anderson, J.R., Hardy, E.E., Roach, J.T., and Witmer, R.E., 1976, A land use and land cover classification system for use with remote sensor data: U.S. Geological Survey Professional Paper 964, 27 p.

Anderson, J.R., and Lins, H.F., Jr., 1978, Coastal applications of USGS land use data: American Society of Civil Engineers Coastal Zone '78 Symposium on Economic and Regulatory Aspects of Coastal Zone Management, San Francisco, Calif., March 1978, Proceedings, p. 943-964.

Ackerman, E.A., and Alexander, R.H., 1975, Toward a national land use information system: U.S. Geological Survey and National Aeronautics and Space Administration; prepared under Interagency Memorandum of Understanding S-70243-AG; Final Report, v. 3, 68 p. (National Technical Information Service Report No. E-77-10015. Report also referred to as CARETS).

Carr, J.H., and Duensing, E.E. (eds.) 1983, Land use issues of the 1980's: Center for Urban Policy Research, Rutgers University, New Brunswick, New Jersey, 304 p.

Clawson, Marion, and Stewart, Charles L., 1965, Land use information. A critical survey of U.S. statistics including possibilities for greater uniformity: Baltimore, Md., The John Hopkins Press for Resources for the Future, Inc., 402 p.

Friedley, Dale, 1984, Land boundary information systems: an implementation of a multipurpose cadastre in state governments: Florida State University, 17 p.

Kleckner, R.L., 1981, A national program of land use and land cover mapping and data compilation (chapter 2), in Planning Future Land Uses: American Society of Agronomy, CSSA, Madison, Wisconsin, p. 7-13.

Kleckner, R.L., 1982, Classification Systems for Natural Resource Management: Pecora VII Symposium, Sioux Falls, South Dakota, October 18-21, 1981, Proceedings, American Society of Photogrammetry, p. 65-70.

Milazzo, V.A., 1980, a review and evaluation of alternatives for updating U.S. Geological Survey land use and land cover maps: U.S. Geological Survey Circular 826, 19 p.

Milazzo, V.A., 1982, The Role of Change Data in a Land Use and Land Cover Map Updating Program: Pecora VII Symposium, Sioux Falls, South Dakota, October 18-21, 1981, Proceedings, American Society of Photogrammetry, p. 189-200.

National Research Council, Committee on Integrated Land Data Mapping, 1982. Modernization of the public land survey system: National Academy Press, Washington, D.C., 74 p.

National Research Council, Committee on Geodesy, 1980, Need for a multipurpose cadastre: National Academy Press, Washington, D.C., 112 p.

National Research Council, Committee on Geodesy, 1983, Procedures and standards for a multipurpose cadastre: National Academy Press, Washington, D.C., 173 p.

North American Institute for Modernization of Land Data Systems, 1979, Land data systems now: Second Molds Conference, Proceedings, Washington, D.C., October 5-7, 1978.

Reed, W.E., and Lewis, J.E., 1975, Land use information and air quality planning: U.S. Geological Survey Professional Paper 1099-B, 43 p.

Richeson, A.W., 1966, English land measuring to 1800: MIT Press, Cambridge, Massachusetts.

Soil Conservation Society of America, 1973, National land use policy: Soil Conservation Society of America, Proceedings, November 27-29, 1972, Des Moines, Iowa.

AREA CALCULATIONS USING PICK'S THEOREM

ON FREEMAN-ENCODED POLYGONS IN CARTOGRAPHIC SYSTEMS

Mark R. Anstaett
and
Harold Moellering
Department of Geography
The Ohio State University
Columbus, Ohio 43210

ABSTRACT

Pick's Theorem provides a means for calculating areas of closed figures formed by connecting points of a regular grid (Coxeter, 1961). The basic equation is

$$AREA = GI + 1/2*GB - 1$$

where GI is the count of grid points inside of the figure and GB is the count of points on the figure's perimeter. The application of Pick's Theorem on polygons represented by various chain codes is the primary topic of this paper. Two grid structures, those formed by corner points of rectangular and of hexagonal cells, and four Freeman-type encodings, 4 and 8-way on the rectangular grid structure and 6 and 12-way on the hexagonal grid structure, are those to which Pick's Theorem is specifically applied. A geometric proof of the proper area calculation through the original equation for 4, 6, and 8-way encodings is given. A revised Pick's Theorem for the grid formed by all 6 corner points of hexagonal cells allows application to 12-way encoded polygons. A proof is also given for this new equation. Through sequentially processing the Freeman codes, an algorithm produces the grid point values which are used in the area calculation. This algorithm is shown to correctly handle polygons with holes and pinched perimeter.

INTRODUCTION

The computation of polygonal areas is a common operation in cartographic systems. The method of area calculation employed is dependent to some extent on the data format. A formula which was proven long before the days of computer-assisted cartography, Pick's Theorem, calculates areas of polygons whose vertices are points in a regular grid. The basic equation is

$$AREA = GI + 1/2*GB - 1 \qquad (1)$$

where GI is the count of grid points inside of the polygon and GB is the count of grid points in the polygon's perimeter (Coxeter, 1961). Cartographic data originating from a variety of vector or raster formats, when quantized into a regular grid, can be efficiently stored as chains of neighboring grid points. Lines and polygonal boundaries are then represented as sets of integer codes which describe the line direction between points (Freeman, 1961). This representation has been called chain or starburst codes, but is commonly referred to as Freeman encoding. A particular Freeman encoding format is derived from the grid geometry and chosen number of angular directions from a grid point. Storage and processing efficiencies and computational and positional errors with use of Freeman encoding versus other data formats must be studied in building a case for or against use of this method with cartographic data. The purpose of this paper, however, is to show how Pick's Theorem can be applied for area calculation of polygons represented by four chain code formats. The four chosen formats are 4 and 8-way encoding in a rectangular grid and 6 and 12-way encoding in a hexagonal grid.

THEORY

Although Pick's Theorem is a simple equation for area calculation, its proof and application seem to be relatively obscure. The proof of the theorem given in (Coxeter, 1961) involves decomposing polygons into simpler ones. The area equation for a single polygon has been extended and proven to apply to a polygon with n holes (Sankar and Krishnamurthy, 1978). Applications of Pick's Theorem to 4 and 8-way encoded polygons have been shown without proofs (Rosen, 1980 and Kulpa, 1977). A simple proof of Equation 1 based on the geometry of Freeman encoding can be stated informally. Examples of 4, 8, and 6-way encoded polygons and their encoding schema appear in Figure 1. The 4 and 8-way encodings use a grid whose points are corners of rectangular cells, while grid points for 6-way encoding are 3 of the 6 points of hexagonal cells. Each grid point "owns" a Thiessen polygon which is identical in shape and area to the cells of the grid structure. To the right of each example is a grid point with its Thiessen polygon (dashed lines) and n-way chain codes. For grid points located within the polygon boundary, counted in GI of Pick's Theorem, an area contribution of one is made for its Thiessen cell. Considering points on the polygonal boundary, those counted by GB of Equation 1, a straight line of two equal chain codes crossing a grid point bisects that point's Thiessen cell. The area expectation for a GB point is then (1/2) cell. Where the chain code changes at a grid point, there is deviation from this expected area. As seen in the right half of each Figure 1 example, the n chain codes at a grid point cut its Thiessen cell into sectors with equal areas of (1/n) cell. For each unit change in chain code at a point, an area deviation of ($\pm$1/n) cell from the (1/2) cell expectation is produced. Assuming that Freeman chains circle polygons in a clockwise direction, polygon closure requires a net total of n clockwise code changes within the chain. The result of this is a subtracted area of n*(1/n) cells from the straight-line expectation of (1/2)*GB cells. This corner-correction value becomes the (-1) term of Equation 1, completing the informal proof.

The 12-way chain code is shown in Figure 2(a) in a polygon example, using the code scheme from (Scholten and Wilson, 1983) with X and Y grid orientations reversed. Figure 2(b)

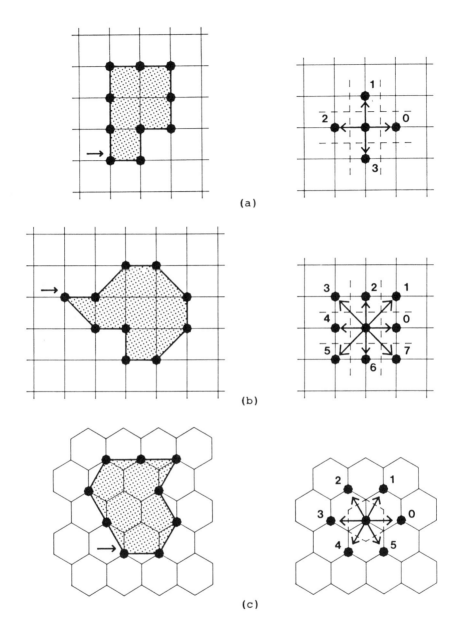

Figure 1. Examples of 4, 8, and 6-way encoded polygons.
Chain coding key is in right half of each figure.
(a) 4-way encoding:
polygon code chain=(1,1,1,0,0,3,3,2,3,2)
(b) 8-way encoding:
polygon code chain=(0,1,0,7,6,5,4,2,4,3)
(c) 6-way encoding:
polygon code chain=(2,2,1,0,0,4,5,4,3)

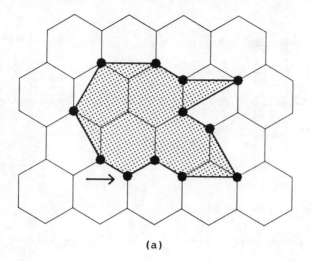
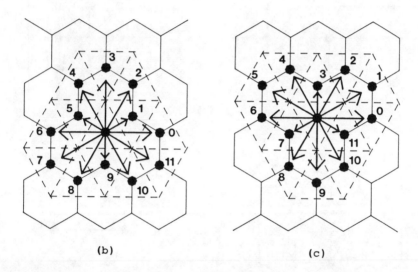

Figure 2. 12-way encoding.
    (a) 12-way encoded polygon:
        code chain=(5,4,2,0,11,0,7,11,10,6,5,7)
    (b) Chain code directions from upper-tier grid point. Thiessen triangles are shown in dotted lines.
    (c) Chain code directions from lower-tier grid point. Thiessen triangles are shown in dotted lines.

illustrates the chain codes at a grid point which is at the top of a vertical hex cell edge (upper-tier point), while Figure 2(c) shows the 12 codes at a lower-tier grid point. Thiessen polygons created through connecting perpendicular bisectors of a point with its 12 neighbors make overlapping tri-hexagonal shapes, then bisecting the area of overlap leaves the triangular Thiessen cells shown in dotted lines for each of Fig. 2(b) and (c). It is evident that a straight line of two equal chain codes crossing a grid point does not in all cases bisect the Thiessen triangle, and that the sector areas of deviation where codes change are not equal for all codes. For these reasons, 12-way encoding requires a different formulation of Pick's Theorem. The Pick's Theorem equation for 12-way Freeman encoding is

$$AREA = 1/2 * (GI + 1/2 * GB_{ODD} + 2/3 * GB_{EVEN,UPPER} + 1/3 * GB_{EVEN,LOWER} - 1) \quad (2)$$

where GI is the count of grid points inside of the polygon, GB(ODD) is the count of odd-valued 12-way codes in the chain, GB(EVEN,UPPER) is the count of even-valued codes in which the centroid of the hexagonal cell crossed by that code lies to the right of the code vector (inside of the polygon), and GB(EVEN,LOWER) is the count of even-valued codes in which the centroid of the hexagonal cell crossed by that code lies to the left of the code vector (outside of the polygon). For the proof of this equation, first consider that the equation counts Thiessen triangles, then gives area in terms of hex cells of the grid. The (1/2) factor scales the count of triangles to hex cells, as this is their area ratio. Each grid point located within the polygonal boundary, counted in GI, contributes one to the area. A straight line of two odd codes crossing a point bisects its Thiessen triangle, thus the 1/2*GB(ODD) term. Assuming polygon interior to the right of the chain vectors, straight-line GB(EVEN,UPPER) codes contribute 2/3 of a Thiessen triangle (codes 0,4,8 in Figure 2(b) and 2,6,10 in Figure 2(c)), while the remaining straight-line even code situations, counted in GB(EVEN,LOWER), contribute an area of 1/3. As in the proof of Equation 1, polygon closure requires a net total of 12 code changes in a clockwise direction, and each code change produces an area deviation from the expected straight-line GB area for the first of two codes. To prove that the total area deviation for a polygon equals one Thiessen triangle, one must consider an algorithm which marks sectors of the Thiessen triangle when that sector is the area deviation corresponding to a particular change of code. The starting grid point of a polygon sets the base triangle in either the Fig. 2(b) upper-tier or Fig. 2(c) lower-tier grid point situation. As the polygon chain is cycled, Thiessen triangle sectors are marked for clockwise code changes (GB area is less than straight-line expectation for the code) or unmarked for counterclockwise code change (GB area is more than the straight-line expectation). The grid point tier changes for each odd code, but the chain ends at the same tier as it begins. Area deviation sectors for a given odd code:odd code change are the same total area in either tier. Thus all code changes mark sectors in the base triangle or mark sectors between pairs of odd codes in the other tier's triangle, which equal the area of those same sectors in the base triangle. The result is that all sectors of the base triangle become marked, which is a total area deviation of (-1) from the straight-line area expectations for GB points. This is the term (-1) in Equation 2, completing the proof.

DESCRIPTION OF ALGORITHM

An algorithm has been written and implemented in a FORTRAN 77 program which computes areas of 4, 6, 8, or 12-way Freeman-encoded polygons by use of the Pick's Theorem Equations 1 and 2. Constraints on the input polygons are that they be simple closed figures, cycled clockwise (polygon interior to right), and that holes are linked to the perimeter by chain codes. Chain pre-processing removes zero-area peninsulas (detected by 180° change of direction in consecutive codes) and rotates the chain so that the last code has a nonzero X-component. A loop processes each chain code by breaking the code into X and Y grid coordinate components, updating the (X,Y) of the code vector destination, and incrementing GI and GB point counts. The function NC(Y) provides a count of grid points in a vertical column below current coordinate Y to a base level Y=0, and NCOFF(Y), used in the hexagonal grids, counts the column below (X$\pm$1,Y) for current grid point (X,Y). The actions taken for specific codes are shown in Table 1. After the last chain code has been processed, GB is set to the chain length, and GI is the chain length plus the number of internal grid points minus a correction factor for areal overcount by GB. This areal overcount may occur when GB counts every code while more than one code reaches a given grid point. The original Pick's Theorem works for some of these cases because such revisited grid points are visually counted only once, while in sequential processing of the codes no memory is kept of which grid points were reached. The equation

$$GI' = GI - GB \qquad (3)$$

produces the GI value for insertion into area Equations 1 and 2 which corrects for areal overcount in GB. The reader may wish to confirm from the situations given in Table 1 that when the chain revisits a point in such a way that the polygon area lies between the code vectors which meet the point, the algorithm only increments GI once while GB will count two, giving a net area correction of (-1) at this grid point by Equation 3. Holes in polygons are correctly handled by being cycled counterclockwise and linked to the polygon perimeter by chain codes; they are then made up of points counted in GB, and interior grid points within the hole become points external to the polygon, which are subtracted out in processing the top edge of the hole. For 12-way chains, the algorithm also maintains the three separate GB counts of Equation 2 by keeping track of upper/lower tier of grid points in the chain. The result of each polygonal calculation is the area in terms of grid cells, which can be directly converted to square user coordinate values by applying cell dimension information.

```
(situation and update of GI)        (description)
IF(DX.GT.0) THEN
   GI=GI+NC(Y)+1                    add column below and
                                       including current point
   IF(DX.EQ.2) GI=GI+NCOFF(Y)       add column passed over at
                                       (X-1) in hex grid
   IF(.NOT.XFLAG) THEN              change of X-direction found
      IF(...FORTRAN code to test for
           [diagrams]      cases)
                                    convex curve:
      GI=GI+NC(LY)+1                   add column below and
                                          including last point
      ELSE (must be one of
           [diagrams]      cases)
                                    concave curve:
      GI=GI+NC(LY)                     add column below
      ENDIF                               last point
   ENDIF
ELSE IF(DX.LT.0) THEN
   GI=GI-NC(Y)                      subtract column below
                                       current point
   IF(DX.EQ.-2) GI=GI-NCOFF(Y)      subtract column passed over
                                       at (X+1) in hex grid
   IF(XFLAG) THEN                   change of X-direction found
      IF(...FORTRAN code to test for
           [diagrams]      cases)
                                    concave curve:
      GI=GI-(NC(LY)+1)                 subtract column below and
                                          including last point
      ELSE (must be one of
           [diagrams]      cases)
                                    convex curve:
      GI=GI-NC(LY)                     subtract column below
      ENDIF                               last point
   ENDIF
ELSE IF(XFLAG.AND.DY.GT.0) THEN     ( [diagram] ) DX=0,
   GI=GI+1                                         add in current point

ELSE IF(.NOT.XFLAG.AND.DY.LT.0) THEN
   GI=GI+1                          ( [diagram] ) DX=0,
                                                  add in current point
ENDIF
```

Table 1.
Update of GI for each code situation, in which
(DX,DY) are X and Y grid components of current chain code,
(X,Y) are grid coordinates at end of current code vector,
LY is Y coordinate of origin of current code vector,
XFLAG=.TRUE. if last nonzero DX was greater than zero,
      =.FALSE. if last nonzero DX was less than zero,
NC(Y) is count of grid points in column below (X,Y),
NCOFF(Y) is count of grid points in column below X$\pm$1,Y).
Both NC and NCOFF use vertical column to and including
base level Y=0.

EXAMPLES

A set of four examples illustrate the algorithm's proper handling of 4, 6, 8, and 12-way encoded polygons. Figure 3 is a simple 4-way encoded polygon. Table 2 lists specific values computed during the sequential processing of the chain (see Table 1 for details) with the resulting values inserted into Equations 3 and 1 for the area in terms of cells. Figure 4 shows a 6-way encoded polygon containing a concavity, and Table 3 traces the area computation for it. An 8-way encoded polygon with a concavity and a pinched perimeter, that situation which Equation 3's modified GI value corrects for, is given in Figure 5 with the corresponding trace of area computation in Table 4. The last example, Figure 6, is a 12-way encoded polygon with a pinched perimeter and a hole. Table 5 lists the important values of the algorithm's execution, including the specific GB values which go into Equation 2.

SUMMARY AND CONCLUSIONS

The practical application of Pick's Theorem, an area calculation method for regular point grids in general, to specific grids used with Freeman encoding has been illustrated by this paper. The theorem was proven geometrically for 4, 6, and 8-way encodings. The modified area equation was introduced which allows a similar calculation on the irregular point grid of 12-way encoding, and this equation was also informally proven. The algorithm sketch describes an implementation of these area calculations on properly-encoded cartographic polygons, and the utility of the algorithm on holes and pinched perimeters has been illustrated.

While Freeman encoding may not be in widespread use with cartographic data, it does offer several advantages. The grid point-based structure allows easy integration of point, line, and area data into a single coordinate scheme. The chain codes themselves represent a vector format within a cellular structure, which permits data conversion between Freeman encoding and both vector and raster data. This quality of Freeman encoding, combined with its capacity for data volume compression, support its use in the new hybrid vaster data structure (Peuquet, 1983). Due to the versatility in dealing with various grid formats and the sequential nature and resolution-independence of the processing, the application of Pick's Theorem to Freeman-encoded polygons enhances the use of these data formats for cartographic data.

REFERENCES

Coxeter, H.S.M. 1961, <u>Introduction to Geometry</u>, pp. 208-210, John Wiley and Sons, Inc., New York

Freeman, H. 1961, On the Encoding of Arbitrary Geometric Configurations: <u>IRE Transactions on Electronic Computers</u>, June 1981, pp. 260-268

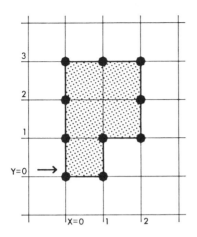
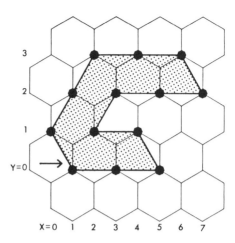

Figure 3.
4-way encoded polygon:
code chain=(1,1,1,0,0,3,3,
2,3,2), GB=10.

Figure 4.
6-way encoded polygon:
code chain=(2,1,1,0,0,5,3,
3,4,0,5,3,3), GB=13.

Table 2. Trace of algorithm execution for Figure 3 polygon.

```
CODE   1   1   1   0   0   3   3   2   3   2
DX     0   0   0   1   1   0   0  -1   0  -1
DY     1   1   1   0   0  -1  -1   0  -1   0
GI     0   0   0   8  12  12  12  10  11  11

GI' = (11) - (10) = 1                (Eq. 3)
AREA = (1) + 1/2*(10) - 1 = 5        (Eq. 1)
```

Table 3. Trace of algorithm execution for Figure 4 polygon.

```
CODE   2   1   1   0   0   5   3   3   4   0   5   3   3
DX    -1   1   1   2   2   1  -2  -2  -1   2   1  -2  -2
DY     1   1   1   0   0  -1   0   0  -1   0  -1   0   0
GI     0   3   5   9  13  15  12  10  10  12  13  13  13

GI' = (13) - (13) = 0                (Eq. 3)
AREA = (0) + 1/2*(13) - 1 = 11/2     (Eq. 1)
```

19

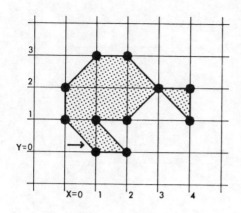

Figure 5.
8-way encoded polygon:
code chain=(3,2,1,0,7,0,6,
 3,5,4,7,4), GB=12.

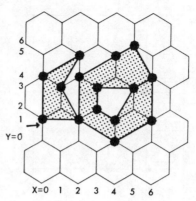

Figure 6.
12-way encoded polygon:
code chain=(3,1,8,10,3,1,1,
 10,7,6,9,11,2,1,9,7,5,6),
 GB=18.

Table 4. Trace of algorithm execution for Figure 5 polygon.

```
           CODE   3   2   1   0   7   0   6   3   5   4   7   4
           DX    -1   0   1   1   1   1   0  -1  -1  -1   1  -1
           DY     1   1   1   0  -1   0  -1   1  -1   0  -1   0
           GI    -1  -1   6  10  13  16  16  13  12  11  13  13

           GI' = (13) - (12) = 1            (Eq. 3)
           AREA = (1) + 1/2*(12) - 1 = 6    (Eq. 1)
```


Table 5. Trace of algorithm execution for Figure 6 polygon.

```
CODE   3   1   8  10   3   1   1  10   7   6   9  11   2   1   9   7   5   6
DX     0   2  -1   1   0   2   1   1  -1  -2   0   1   1   1   0  -2  -2  -2
DY     3   1  -2  -2   3   1   1  -2  -1   0  -1  -1   2   1  -3  -1   1   0
GI     0   9   5   8   9  15  18  21  18  15  16  18  20  23  23  22  21  20
GB_ODD
       1   2   2   2   3   4   5   5   6   6   7   8   8   9  10  11  12  12
GB_EVEN,UPPER
       0   0   1   1   1   1   1   2   2   2   2   2   2   2   2   2   2   2
GB_EVEN,LOWER
       0   0   0   1   1   1   1   1   1   2   2   2   3   3   3   3   3   4

GI' = (20) - (18) = 2                                        (Eq. 3)
AREA = 1/2*[(2) + 1/2*(12) + 2/3*(2) + 1/3*(4) - 1] = 29/6
                                                             (Eq. 2)
```

Kulpa, Z. 1977, Area and Perimeter Measurement of Blobs in Discrete Binary Pictures: *Computer Graphics and Image Processing*, Vol. 6, pp. 434-451

Peuquet, D.J. 1983, A Hybrid Structure for the Storage and Manipulation of Very Large Spatial Data Sets: *Computer Vision, Graphics, and Image Processing*, Vol. 24, pp. 14-27

Rosen, D. 1980, On the Areas and Boundaries of Quantized Objects: *Computer Graphics and Image Processing*, Vol. 13, pp. 94-98

Sankar, P.V. and Krishnamurthy, E.V. 1978, NOTE- On the Compactness of Subsets of Digital Pictures: *Computer Graphics and Image Processing*, Vol. 8, pp. 136-143

Scholten, D.K. and Wilson, S.G. 1983, Chain Coding with a Hexagonal Lattice: *IEEE Transactions on Pattern Analysis and Machine Intelligence*, Vol. PAMI-5, No. 5, pp. 526-533

Mark R. Anstaett

BIOGRAPHICAL SKETCH

December, 1984

Mark R. Anstaett received his B.A. in Computer and Information Science from The Ohio State University, where he is currently working on a master's degree in the Department of Geography with specialization in computer cartographics. He is employed as a programmer/analyst with the Ohio Department of Natural Resources, working on the Ohio Capability Analysis Program GIS. In the past he has worked as a research assistant for Dr. Harold Moellering and for the Department of Geography on various software and hardware projects. He is a member of ACSM and his research work currently focuses on error analysis of vector/raster data conversions.

Harold Moellering

BIOGRAPHICAL SKETCH

October, 1984

Harold Moellering received his Ph.D. from the University of Michigan and is currently Professor of Geography at Ohio State University. He is also Director of the Department's Numerical Cartography Laboratory. He is past Chairman of the ACSM Committee on Automation in Cartography and Surveying. He is Chairman of the National Committee for Digital Cartographic Data Standards in cooperation with ACSM, the U.S. Geological Survey and the U.S. National Bureau of Standards. He is a member of the U.S. National Committee for the ICA and travel chairman for the recent meetings in Australia. He has presented papers at the ICA Congresses in Maryland in 1978, Tokyo, 1980, Perth, 1984, and at the IGU Regional Congress in Rio de Janiero, in 1982. His research specialities include numerical, analytical and dynamic cartography. Professor Moellering is currently a member of the Committee on Cartography of the U.S. National Academy of Sciences/National Research Council.

APPLYING SOFTWARE ENGINEERING TO A GENERAL PURPOSE GEOGRAPHIC INFORMATION SYSTEM

Peter Aronson
Environmental Systems Research Institute
380 New York Street
Redlands, California 92373
(714) 793-2853

ABSTRACT

Over the past 14 years, ESRI has constructed three commercial general purpose Geographic Information Systems. The first two, PIOS and GRID, were constructed in the haphazard fashion typical of almost all past GIS development. The most recent, ARC/INFO, was from its initial design, constructed following modern software design and programming methodologies. In the course of the design, construction and maintenance of ARC/INFO, many lessons about the development of large software systems in general, and of GIS in particular, were learned.

INTRODUCTION

Most successful geographic data-processing software has not been designed by computer professionals (although there are exceptions, see Tomlinson, 1974). Such software has generally been designed by geographers or planners or foresters with limited programming background. This is due to the obscure nature of most geographic data-processing -- typically, no one outside those fields dealing with geographic data has had a sufficient grasp of its special requirements to produce useful tools to manipulate it.

ESRI's earlier GIS grew from such specific software packages (PIOS started as a simple package to digitize and to report overlay areas (Tomlinson, et al, 1976). They were not really planned as *systems*. The initial package was designed; then, whenever a new function was required, a new program or subroutine would be added to perform it. The systems simply grew. As long as a system fulfilled its current requirements, no matter how poorly, no overall system plan would be made.

In the winter of 1980-1981 the initial specifications for what was to become ARC/INFO were put together by Scott Morehouse. The original plan was for an arc/node digitizing system (Guevara, 1983), including an automatic topological "cleaning" program that would break line segments at intersections, snap closed undershoots and remove overshoots, and would feed into the PIOS system. This package was carefully designed, constructed and tested (and is currently still being employed after four years of heavy use).

At this time, dissatisfaction with PIOS prompted the next stage of ARC/INFO's development. A complete new design was performed (actually, the design had begun with the design of the digitizing system, however, for internal political reasons, the processes were officially separate). The ARC/INFO Geographic Toolbox concept was developed (see below), then the development of the ARC/INFO full geographic information system began.

ARC/INFO DESIGN REQUIREMENTS

ARC/INFO was planned as a general purpose geographic information system, including commands for data input, analysis, output and management. It needed to be efficient to compete with special purpose systems. Most important, it _had_ to be well designed and built.

That ARC/INFO be well engineered was vital for a number of reasons. First, because of practical limitations it would have to be built in stages, with capabilities being added as time went on. This would require that the system be carefully broken down into logical modules. Time constraints required that some problems be dealt with initially by simple, but non-optimum solutions, then later redone in a more efficient manner. For example, the original attribute handling after an overlay was done by a job control program, then recoded into FORTRAN a year later. This required that each module have well-defined inputs and outputs, as well as not generating side effects.

Second, the system would have to be easily expandable, both in the sense of adding new functions, and in increasing functional limits (such as the number of points in a polygon). When a system has a large and varied user base, there are sure to be additional capabilities required over time, not all of which can be anticipated at design time. For example, ARC/INFO did not have point-in-polygon overlay or point subsetting until Spring 1983 when they were requested by the State of Washington, Department of Natural Resources.

Third, it had to be portable. It was quite certain at the beginning that ESRI would at some point want to have ARC/INFO running on some machine other than PRIME, ESRI's current in-house development machine. PIOS and GRID were, at that time, available on PRIME, HP 3000 and IBM.

Finally, it had to be maintainable. Fixing problems in PIOS or GRID usually required either spending considerable time reading and comprehending obscure code, or writing a new program from scratch. This led to patched together systems where it was often easier to find some other way to solve a problem than to fix a bug.

ARC/INFO DESIGN PHILOSOPHY AND STRUCTURE

ARC/INFO was designed as a "toolbox" of geographic operators. ARC/INFO commands perform such operations as: overlay (Union, Intersection, Identity and Update), subsetting (Clip, Reselection, Sliver Elimination, Polygon Dissolve, Erase (reverse clip), and Multiway Split), combination (Mapjoin and Append), data input (Digitizing,

Generation, Grid-to-Polygon, COGO and Conversion programs),
output (Plotting, Conversion programs and Report Generation),
modeling (Theissen Polygon Generation, Network Analysis,
INFO (relational database manager) and Corridor Generation)
and data management (Librarian Map Library Manager).
ARC/INFO commands can be thought of as statements in a
geographic modeling language. Combined via system job control language (CPL on PRIME, DCL on VAX, CLI on Data
General and EXEC on IBM CMS/VM) with INFO's relational programming language, complex spatial models can be implemented
as relatively simple "programs".

A simple example of such a program could take as input a
map (coverage in ARC/INFO terminology) of an area within a
map library, and produce a report of the estimated total
value of state-owned forest stands and a map of those stands
shaded by value per acre. The procedure would go something
like this:

PROCEDURE STATE-FOREST-VALUE (AREA)
/*   */
/* This procedure produces an estimated value report and   */
/* an estimated value per acre plot for a specified area.   */
/*   */

EXTRACT from Map Library FOREST coverages: FOREST-STANDS,
SOILS and OWNERSHIP by AREA.

/*   */
/* Overlay FOREST-STANDS with SOILS and OWNERSHIP to   */
/* produce a coverage that contains all the information   */
/* required to calculate cost and owner. Remove those   */
/* polygons not owned by the state.   */
/*   */
IDENTITY FOREST-STANDS with SOILS to produce COST

INTERSECT COST and OWNERSHIP to produce VALUE

RESELECT VALUE so that OWNER = 'STATE' to produce
STATE-VALUE

/*   */
/* Create products.   */
/*   */
Model STATE-VALUE with INFO to produce Estimated Value and
Estimated Value per Acre (ESTVAL/AC), and produce a summary
report of Estimated Value and add ESTVAL/AC to STATE-VALUE.

Generate a plot of STATE-VALUE shaded by ESTVAL/AC.

Clean up (Delete FOREST-STANDS, SOILS, OWNERSHIP, COST,
VALUE and STATE-VALUE.

END

This adaptable structure is mirrored inside ARC/INFO as
well. ARC/INFO consists of about eighty FORTRAN 77 programs
linked in various fashions by job control language. The
CLIP command, for example, consists of nine FORTRAN programs
(not all of which would be used by any single execution)

linked together by about (on the PRIME) 60 lines of job
control language. Many times new ARC/INFO commands have been
created simply by linking these programs in new combinations
("As above, so below").

THE ARC/INFO SOFTWARE LIFE-CYCLE

Most commercial general purpose software systems are in a
continuous state of development. Operating systems, for
example, are re-released at a consistent rate, often with
fairly large changes in functionality coming every year or
so. A general purpose geographic information system, like
ARC/INFO, is also in such a state. There is typically a new
release every four to six months, and each release is sure
to contain one or more new commands, alteration of two or
three existing commands, and a reasonable number of fixed
bugs. As a result, ARC/INFO has its own internal development process.

The software life-cycle is one of the basic concepts in
software engineering. The waterfall life-cycle model shown
in Figure 1 is based on one by Barry W. Boehm (Boehm, 1981)
and modified for the ARC/INFO development process. This
model differs from the usual model in that it starts from
requirements (as a result of ESRI's operating procedure
discussed below) and the incorporation of phototyping as
separate steps.

Step by step the ARC/INFO life-cycle consists of:

   1) <u>Requirements</u>. Requirements for ARC/INFO come from
two sources: software modification request forms and the
systems group's section of the yearly company plan. The
software modification request forms (commonly called "bug
sheets") are filled out either by in-house personnel
(usually from the production group or from the systems
group) or by the software support group by user request.
These consist of reports of problems in the software ("bugs")
and of requests for enhancements. The yearly company plan
contains those enhancements that the management feels would
be useful in selling new systems or in satisfying an
important requirement of current users.

   2) <u>Feasibility</u>. This step is performed by the system
architect. The requirements are examined for practicality
and necessity, and those that can be performed acceptably
by existing software are usually rejected at this point.
There is a distinct effort made to minimize the amount of
code in ARC/INFO, when it can be done without sacrificing
significant functionality. Code that is not written does
not have to be maintained. Those requests that would
require excessive programmer time are usually tabled or
rejected at this point (however, occasional major system
additions are made, such as Network Analysis and Map Library
management packages added early in 1985). Those modifications deemed practical and desirable are then passed on to
the design stage.

   3) <u>Product Design</u>. This task is performed by one or two
development programmers, usually acting under the super-

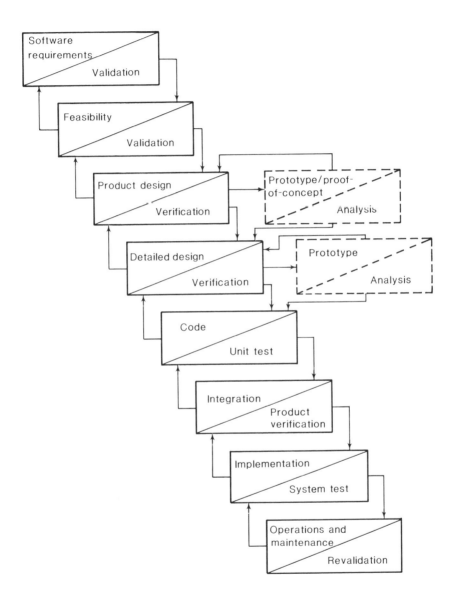

FIGURE 1 - The ARC/INFO Software Life-Cycle

(based on Boehm, 1981)

vision of the system architect. Performed at this time are the four major general design tasks: user interface design, initial algorithm selection, data flow design, and major module specification.

The user interface is usually described in terms of an entry in the command reference manual. Inputs, outputs, prompts and subcommands are fully described. This design also serves as a rough draft for the user documentation and on-line help files.

Boehm placed algorithm selection under detailed design, but in GIS development, where the algorithms are not always well understood, it has been found necessary to have at least an initial approach to an algorithm before proceding any further. At this stage, algorithms can often be compared via analysis using computational geometry (Guevara, 1983). If this proves inadequate, then prototyping may be required (see Step 4 below).

A data flow design shows each program or operational module in a command and each file/data structure in and out of each program. It looks at programs/modules as operations on files/data structures. Since most ARC/INFO commands are sets of programs linked operationally by files, a data flow design can be very helpful.

Defining subroutine modules might seem out of order at this stage, but in ARC/INFO the careful design of the low-level subroutine modules is considered as important or even more important than that of high level program sections. These subroutine modules, each based on a single data structure (such as variable record-length, random access files or coverage boundaries, or the user terminal), are the basic building blocks of ARC/INFO. As well as structuring the code and performing information hiding, these modules supply the primitives used in the construction of geographic operators. This approach has allowed ARC/INFO to be programmed in FORTRAN (selected for portability), without any problems due to the limited data structures available, as all major data structures are accessed via subroutine calls (70%+ of ARC/INFO code (exclusive of comments) consists of subprogram calls and control statements).

    4) _Prototyping/Proof-of-Concept_. This is an optional stage, used as an aid in algorithm selection. It often helps in choosing an algorithm to write a small program to perform comparative testing. This process can be invaluable for selecting among competing algorithms when mathematical analysis is inadequate. The current sets of sorts used in polygon overlay and the search algorithm used by the arc editor were both selected in this fashion.

    5) _Detailed Design_. A complete algorithm definition is produced. Pseudo-code is generated for all the main program sections. Both file formats and module contents are fully defined.

    6) _Prototype_. This optional stage is used when the detailed design will not, or cannot, be carried past a

certain point. This usually indicates that there is not
enough information available to complete the design. Often,
building a simple, quick version of the product will raise
all of the design questions, allowing them to be answered
before the final coding. This is particularly useful when
constructing a function that no one involved in the development has experience with, and for this there is little or no
published information available. The Map Librarian, for
example, went through this stage.

7) <u>Coding</u>. A complete ARC/INFO command is constructed
and tested by the development group. This is installed into
the active code set, along with any changes the installation
requires in other existing code.

8) <u>Integration</u>. The command reference manual, user
manual, training notes and programmers manual are updated
for changes. The software support group performs independent testing of the new/modified command(s). At this stage,
alterations in the operation of the software being tested
may be requested by the software support group. An alpha
release is made to inhouse users.

9) <u>Implementation</u>. The new software is officially
included into the next beta release, and listed in the
installation notes. After beta release, if no new problems
are reported it is included in the normal release. Finally,
it is then installed on all systems and trained.

10) <u>Operation and Maintenance</u>. As the software is
operated, problems or shortcomings are discovered and transcribed onto software modification request forms. This
cycle continues until either the software is entirely replaced or the function it performs is no longer needed (the
former has happened to the original line-overlay command
OVERLINE, the latter will happen to the ARC to PIOS conversion command ARCIDC).

PROGRAMMING PROVERBS FOR GIS PROGRAMMERS

This section is a collection of small tidbits of advice for
prospective programmers of geographic information systems
(as well as being my tribute to Henry F. Ledgard). For
the most part, these simply reiterate the material discussed
above, but in a more succinct form.

<u>Always minimalize code</u>. The more code you have, the
greater the maintenance load. Whenever possible, use
existing code in preference to writing new code. Deprogram
(reduce the amount of code without reducing functionality
or readibility) whenever the opportunity arises.

<u>Construct useful subroutine packages</u>. Low level functions will be used again and again. Putting them in
coherent packages allows simple reimplementation, isolation
of system dependent code, and information hiding.

<u>Any special case that can occur, will occur</u>. Because of
the vast quantities of information involved in geographic
and cartographic data processing, GIS procedure algorithms

must be particularly robust, as any possible special case will arise in short order (and typically in a vital data set).

<u>When in doubt, test</u>. If the choice between two algorithms is unclear, submit your case to the computer. Write test programs for each case, and run them through a variety of typical and extreme cases to learn their behavior. Then choose the algorithm on the basis of experimental results.

<u>GIS procedure algorithms will have unexpected characteristics when coded</u>. GIS procedures are often not well understood. Formal analysis can only tell so much. A host of practical issues involving machine precision, special cases, and so forth, are sure to arise in the coding. This characteristic, combined with the above lesson, leads to the requirement that GIS require extensive prototyping before acceptance.

<u>A general purpose GIS needs to be made of flexible building blocks</u>, to allow new functions to be easily added. Since a general purpose GIS is not locked into a single set of operations, from time to time a user will require some capability that has not been hitherto present in the system. It is necessary that such capabilities not be difficult to add.

CONCLUSIONS

The design and development methodology of ARC/INFO presented above was the result of a combination of the deliberate application of software engineering methodologies and of trial and error. Even given a complete set of methodologies and techniques, it is not a simple matter to apply them. Long ingrained procedures must be removed, and new habits learned in their place. The procedures described here, simple as they are, can seem unwieldly and overly slow to an overworked programming staff. There is often a strong temptation to skip methodology and "just fix the problem". However, the consequences of ignoring methodology in differing versions and bad releases will drive home the need to follow a well defined procedure.

Geographic information systems are, by their nature, large and complex systems. If they are not well built, they will simply not work well. Software engineering is the logical application of methodologies such that software products and systems are well constructed. Therefore, usable geographic information systems need to be constructed following the principles of software engineering.

REFERENCES

Boehm, B.W. 1981, <u>Software Engineering Economics</u>, Prentice-Hall, Inc., Englewood Cliffs, New Jersey.

Guevara, J.A. 1983, <u>A Framework for the Analysis of Geographic Information System Procedures: The Polygon Overlay Problem, Computational Complexity and Polyline Intersection</u>, Unpublished Ph.D Dissertation, SUNY at Buffalo.

Tomlinson, R.F. 1974, *Geographic Information Systems, Spatial Data Analysis, and Decision Making in Government*, Unpublished Ph.D. Thesis, University of London.

Tomlinson, R.F., Calkins, H.W., Marble, D.F. 1976, *Computer Handling of Geographic Data*, The Unesco Press.

APPLICATIONS OF PRODUCING PLOTS FROM DLG DATA

Brian S. Bennett
U.S. Geological Survey
345 Middlefield Road
Menlo Park, CA. 94025

ABSTRACT

The National Mapping Division of the U.S. Geological Survey is producing digital data representing boundary and Public Land Survey information derived from standard Geological Survey quadrangle maps. The digital data are formatted into digital line graphs (DLG's) and archived in a National Digital Cartographic Data Base. This paper describes the use of this data to produce publication-quality boundary and land net manuscripts.

BACKGROUND

The Geological Survey has developed the capability to produce publication-quality boundary and Public Land Survey (land net) overlay plots using digital line graph data. The plots will meet National Map Accuracy Standards and will use symbology that conforms to cartographic standards for provisional mapping.

Current procedures call for Public Land Survey information to be manually plotted on a manuscript and then digitized as a DLG. A film positive of the land net information, symbolized according to provisional map standards, is then plotted using software written for that purpose by the Western Mapping Center. This digitally produced land net manuscript is then used as the basis for a buildup of the boundary information. An incomplete DLG is produced by this procedure because the Public Land Survey information is not coincidence-referenced to the boundary information.

A significant amount of opaquing and rescribing of manuscripts is required to produce overlays which meet publication requirements. Additional editing of the DLG is required before it can be archived in the National Digital Cartographic Data Base (NDCDB).

NEXT GENERATION SOFTWARE

A programming effort is underway to address the problems and shortcomings of the first-generation land net plotting program and, at the same time, provide the ability to plot boundary information. The program will allow coincident features to be plotted--or not plotted--based on a predefined order of precedence. It is assumed that fully annotated land net and boundary source material is available for DLG digitizing so the information for both overlays is spatially correct and the attribute codes indicating coincidence can be included in the digital data. The software is being written in standard FORTRAN and many of the routines will be useful for future DLG plotting.

PROVISIONAL MAPPING

As part of the National Mapping Program, the U.S. Geological Survey produces a series of Provisional Edition maps containing essentially the same information as a standard quarangle map but with a provisional rather than a finished map appearance. This reduced amount of map finishing results in a significant
cost savings in map production and a shorter map completion cycle. Provisional mapping symbology for Public Land Survey and boundary overlays can be more easily computer generated than corresponding symbology for traditional 1:24,000scale quadrangle mapping.

DLG STRUCTURE

A DLG is structured so that each graph is defined by nodes, lines, and areas. The nodes are the points that define the ends of each line. The line elements contain coordinate pairs that define the position of the line. Additionally, each line element contains spatial information that topologically describes the areas to the left and right of the line. Theoretically, there is no limit to the number of coordinate pairs which can be used to represent a line. Two coordinate pairs are sufficient, for example, to define a straight line between two section corners, but a complex boundary requires many pairs to accurately represent it. In addition to coordinate information, each node, area, and line element contains attribute codes that further describe properties of the feature. An attribute code is composed of two numeric fields: a major code that identifies the major catagory to which the element belongs, and a minor code that specifically describes the element. In the case of the Public Land Survey and boundary overlays, most of the information that describes that type of feature is contained in the area attributes.

FEATURE DETERMINATION

For the Public Land Survey and boundary overlays, each area has attributes associated with it that identify the feature codes for that particular polygon. For a boundary overlay, most areas will have the attributes of a State and a county. The major code for a State is 091 and for a county is 092. In both cases, the minor code is the equivalent Federal Information Processing Standards (FIPS) code for that particular State or county. Other area attribute codes designate civil townships, national parks, military reservations, and so forth. Line types are determined by sorting through the attributes on both sides of any line to determine the symbology that should be used to represent the feature. The standards for 1:24,000-scale quadrangle maps establish the order of precedence for representation of land net and boundary features.

Figure 1 illustrates two examples using a city boundary and a park boundary. In part A the park is adjacent to the city but outside the city limits. The common line between the two has a city attribute (90.101) on one side and a park attribute (90.151) on

33

the other. The city boundary symbol will take precedence. Part B illustrates a variation of the same situation. In this case the park is contained entirely within the city limits. The lines representing the park have a city attribute on both sides and a park attribute on one side. Although the city attribute takes precedence, the program assumes that since 90.101 appears on both sides of the line, there is no city boundary, so the park boundary symbology will be plotted.

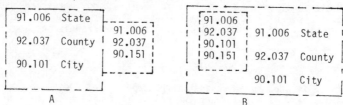

FIGURE 1--Example of symbology precedence.

LINE ATTRIBUTES

Additional information which may affect line symbology is provided by the line attributes. For boundaries, line attributes are used to define approximate, indefinite, disputed, or historical lines. For land net, the attributes define approximate or protracted lines, arbitrary closure lines, and base lines. For plotting purposes, lines with one or more of these attributes may require a reduction in line weight, line annotation, or perhaps a change in line symbology. An indefinite boundary will have normal symbolization but will be labeled as indefinite. An approximate section line will be plotted as a dashed line, rather than as a solid line.

NODE ATTRIBUTES

The node attributes supply specific information describing the feature positioned at the node location. For the boundary overlay, the nodes might be coded as boundary monuments or turning points. For the Public Land Survey data, the attributes would identify U.S. Public Land Survey section corners, closing corners, meander corners, witness marks, reference marks, and other appropriate marks. An additional attribute would designate corners which were identified in the field and corners with horizontal coordinates or elevation values. Again, for plotting purposes, these attributes would determine the symbology for the monumented features.

COINCIDENT FEATURES

Attribute codes are applied to lines or nodes to designate coincidence for features which occupy the same spatial position on more than one overlay. This code determines if a line should be plotted or not based on a hierarchical scheme. For example, a section line or boundary may be coincident with a road. For a double line road, major boundarys are plotted with a reduced line weight and placed between the two road casings. If a boundary is coincident with a section line, the boundary takes precedence and the section line is omitted from the land net plot.

SYMBOLOGY

In the program being developed, the symbology will conform to standards for publication symbols for 1:24,000-scale primary quadrangle mapping. A flatbed plotter capable of making photo plots will produce line weights and dashed line symbology which will meet these standards. Because pen plotters or electrostatic plotters do not have the same aperture selection capability as a film plotter, line weights of plots produced on paper will not meet standards in

PROBLEMS

There are cases where deficiencies in attribute coding results in incorrect line type determination. This often happens in complex urban areas. Unique situations have surfaced in southern California because of the combination of land grants intermixed with public land surveys. Two identical feature types which join at a common line frequently cause problems. Figure 2 uses two examples to illustrate how attribute codes may not always uniquely identify line symbology.

This problem can also cause misplaced labels for sections and tracts. A section or tract intersected by a line is treated as two polygons, and under the present programming scheme, each polygon is labeled separately. The placement of labels in irregularly shaped polygons may not be correct. Although plots produced by the present software are positionally correct, some capability needs to be developed which will allow editing of line symbology and label placement. Editing could be accomplished by an interactive interface or it could be in the form of correction overlays.

ADVANTAGES

There are many advantages to working with data in digital form. Plots can be made at any reasonable scale. Any combination of feature types can be plotted. Editing and replotting of overlays can be done easily and quickly. One very important feature that surfaced from symbolized plots is the ability to use the plots as a quality control check on the attribute coding. When an attribute code is incorrect or missing, the line symbology will often be displayed incorrectly.

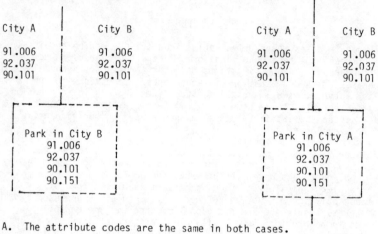

A. The attribute codes are the same in both cases.

B. Attribute coding for adjacent tracts in one land grant is identical to coding for adjacent tracts in different land grants.

FIGURE 2

In these examples, the line symbology is not uniquely defined by the attribute codes.

CONCLUSION

DLG data can be used to produce symbolized land net and boundary manuscripts that meet publication requirements for provisional mapping with a minimum amount of editing. Techniques need to be developed which can be used to correct line symbology and label placement. A very important fallout of this programming effort will be the use of symbolized plots to quality control DLG attribute data.

SELECTED REFERENCES

Allder, W. R., and Elassal, A. A., 1983, Digital Line Graphs from 1:24,000-Scale Maps: U.S. Geological Survey Circular 895-C, 79p.

Allder, W. R., Sziede, A. J., McEwen, R. B., and Beck, F. J., 1983, Digital line Graph Attribute Coding Standards: U.S. Geological Survey Circular 895-G, 31p.

COMPUTER-ASSISTED PRODUCTION OF THE 1980 POPULATION DISTRIBUTION MAP

R. B. Birdsong
Charlene K. Bickings
Ann M. Goulette

Geography Division
U.S. Bureau of the Census
Washington, D.C. 20233

ABSTRACT

The Census Bureau is completing production of a map showing the 1980 population distribution of the United States. This product of the Geography Division represents a transition in Census Bureau mapping operations; it is the first wall-sized map produced largely by computer-assisted cartographic methods. This paper describes the problems encountered and solved by computer-assisted production of the map.

INTRODUCTION

The Geography Division traditionally publishes a map showing the distribution of the population as reported in each decennial census; the map displays the urban and rural character of the distribution using multiple symbolization techniques. For example, the 1970 Population Distribution Map used three types of symbols: point symbols to represent rural population and urban place population outside of urbanized areas, graduated circles to represent urbanized area population, and an areal fill pattern to represent the extent of urbanized areas.

In the past, all wall-sized Census Bureau maps were produced using manual cartographic techniques; however, for the 1980 Population Distribution Map, an effort has been made to automate production as much as possible. This has allowed and sometimes has necessitated many changes in the design of the map.

Computer-assisted cartography allows the cartographer a great deal of flexibility in the design process. With automation, several different versions of a single map can be made quickly and compared before final production. Map parameters such as line width, symbol size and symbol placement can be adjusted by minor changes to computer programs and the results can be seen almost immediately. In addition, computers can provide accurate and repeatable linework with a quality only previously produced by the most skilled technicians.

Computer-assisted cartography is not without its shortcomings. Automation creates a set of problems not encountered in manual production. The use of computers can place constraints on design considerations. The hardware available to the cartographer (in the case of the Census Bureau, Tektronix 4014 and 4115 graphic terminals for research and design work, and a Gerber 4477 photo head plotter for production of overlays) limits what can be plotted on the map. By using the computer as a mapping tool, the cartographer also is confronted with handling massive data files. In turn, the information on the files limits the area and subject matter to be mapped.

This paper outlines the various stages encountered in the automated production of the 1980 Population Distribution Map. In particular, the predesign considerations, the design of the test map, and final design stage are discussed.

PREPROCESSING DATA FILES

The structure and content of the data files took precedence among the design considerations. Considerable preprocessing was necessary long before a mapping product could appear or a program could be written. The Census Bureau file used to produce the map was the Master Area Reference File 2 (MARF 2)*. The file contains much more information than is necessary to produce the map. The MARF 2 contains population and locational data for all levels of census geography for each state or state equivalent. The file was stripped of extraneous information (for example, unneeded variables) before being used as input to the mapping programs.

INITIAL DESIGN PROBLEMS

The preliminary design problems centered on programming to plot the map elements. A map program was needed to convert latitude-longitude coordinates of the centroids to the map projection. In addition, algorithms were required for placement of symbols for three classes of population: rural, non-rural place outside urbanized areas, and urbanized area.

PROJECTION

A new projection program was written for the 1980 Population Distribution Map. The projection chosen was the Albers equal-area polyconic based upon the Clarke 1866 ellipsoid. The coterminous United States was shown at a scale of 1:5,000,000 while the insets were shown at various scales.

---

*This file is available for public use through the U.S. Bureau of the Census, Data User Services Division, Washington, DC. 20233.

RURAL POPULATION SYMBOLS

The rural population symbols were placed using data available in MARF 2. A program was written to sum the population of rural enumeration districts. When the population exceeded the value to be represented by one symbol on the map, a symbol was placed at the centroid of the last enumeration district added to the total. The population value per symbol was then subtracted from the total and the remainder carried over to be used in the next cumulative sum.

Placement of the symbol at the exact centroid, however, created an undesirable gridlike pattern especially apparent in the midwestern states, as seen on the 1970 map. A routine was, therefore, developed to randomly scatter the placement of rural symbols within approximately two symbol diameters from the centroid to reduce this problem.

PLACE POPULATION SYMBOLS

The positioning of rural and urban place population symbols was determined from a file containing the centroids and population class code of all "place" enumeration districts. Most places are made up of more than one enumeration district, so placement of the symbols became problematic. Simply placing the symbol at the average location of the centroids for a given place could, in some cases, cause a symbol to be placed inaccurately (for example, in a body of water or outside the areal extent of the place). To solve this problem, symbol placement was determined by an iterative process which moved the symbol location close to the densest concentration of enumeration districts for each place.

The symbols chosen to represent place population (filled circles, squares and triangles) were not included in the Gerber plotter software repertoire; therefore, programmed instructions were written to draw the symbols. Although these symbols appear solid on the map, they are actually plotted as successively larger symbol outlines, each of which uses the same center point.

URBANIZED AREA POPULATION

Previous population distribution maps have represented the extent of urbanized areas by a solid fill pattern bounded by each area's defined limits. Because these boundaries were not available in the MARF 2, a new method of depiction was necessary for the 1980 map. Using a file similar to that used for place population, a dot was placed at the centroid of each block group in urbanized areas. These dots coalesce to delineate the areal extent of the population.

A further change to the 1970 design involved the mapping method used to depict the size of urbanized area population. The graduated circle representation was dropped and a choroplethic technique using redefined class intervals was substituted.

THE DESIGN TEST

The northeastern United States was chosen as a test area. Since this area is the most densely populated area of the United States, problems in design would likely surface here.

Two versions of the map were tested. On one version, six classes of urbanized areas were shown (each in a different hue) and four classes of place symbols were plotted. The other version had three classes of urbanized areas (shown in a progression of hue from light purple to dark purple) and three classes of place population. In addition, two versions of the rural population symbols overlay were produced: one dot per 500 people and one dot per 1000 people. Color keys were produced for each version; these were then circulated within the Census Bureau for review.

As a result of the review, the six-class urbanized area, four-class place population version was rejected. The colors used in that version were thought to have a "fruit-salad" appearance. Furthermore, the class intervals chosen for this version did not represent a sufficiently smooth progression. Bureau tradition prevailed in retaining rural population symbolization of one dot per 500 people.

The three-class urbanized area, three-class place population version was modified, however, before acceptance as the final design. The size of the dot representing rural population was reduced. The dots used in the test had overwhelmed the other data on the map as a result of their large size and solid black color. In addition, the program to scatter the rural dots was adjusted to reduce the amount of scatter, thereby reducing the number of dots being erroneously placed within urbanized areas.

TRADITIONAL CARTOGRAPHIC INPUT

As a result of time constraints and the unavailability of certain data, the production of some elements of the 1980 Population Distribution Map will not be automated. Although the Census Bureau continues its efforts in computer-assisted cartography, as this paper goes to press, all type, coastlines, state and county boundaries are planned to be created using traditional methods. If time allows before publication of the map, all linework (coastlines, state and county boundaries, graticule) will be through computer-assisted means. The production of all publication negatives will be achieved through photo and photo-mechanical techniques.

SUMMARY

The 1980 Population Distribution Map is another step for the Census Bureau toward producing all thematic mapping products through computer-assisted means. With minor adaptations, the software and data files created for this project will be used in future automated projects. Although the time spent in research and design was considerable, the lessons learned from this experiment in mapping methodology will benefit the Census Bureau's mapping operations in the future.

APPLICATION OF STANDARD LINEAR FORMAT TO
DIGITAL HYDROGRAPHIC PRODUCTS

CDR Richard E. Blumberg
Mr. Melvin L. Wagner

DMA Hydrographic/Topographic Center

SUMMARY

With an ever increasing number of new navigation, weapon, and training systems being developed which use digital hydrographic information, the Defense Mapping Agency (DMA) can no longer provide system specific, digital products. DMA has adopted an internal digital data exchange format, called Standard Linear Format (SLF), to permit the exchange of digital data between various production elements. The DMA Feature File (DMAFF), a dictionary of cartographic features and attributes to describe those features, has been developed to standardize digital cartographic features.

To test these standards, a digital chart of Portsmouth, Virginia has been produced using the DMAFF features encoded in SLF. This digital prototype is being evaluated for its ability to serve as a multiproduct data base to satisfy both digital and paper chart requirements.

Title

TLDB: A Global Data Base Designed for Rapid Digital Image Retrieval

Abstract

   The TLDB (Target Location Data Base) was developed by the personnel at Rome Air Development Center (Griffiss AFB, NY) to meet the challenge of managing and utilizing a global digital imagery data base capable of displaying monoscopic digital imagery to the interpreter/analyst to support the target identification and location functions. The data base is composed of cells encompassing data within a one degree geographic square, which are further subdivided into 36 subsectors. Each ten minute square subsector contains its own set of doubly-linked lists, keyed on decreasing latitude and increasing longitude of the center of each digital image.

   Experience with the system has shown that the display of a 1024 by 1024 digital image, centered about an operator-entered coordinate pair, takes apporoximately ten seconds, and includes the necessary ancilliary data needed to perform the monoscopic target location function.

Authors

Mr. William Brooks earned a B.S. degree in Forest Engineering from the State University of New York (SUNY) College of Environmental Science and Forestry in 1977. He earned an M.S. degree in Photogrammetry from SUNY in 1978, and an M.S. degree in Remote Sensing from Cornell University in 1984. Mr. Brooks developed the TLDB data base structure while an employee of RADC.

Mr. James Sieffert earned an A.A.S. degree in Forestry from Paul Smiths College in 1976. He earned a B.S. degree in Forest Engineering from SUNY in 1978, and a M.S. degree in Photogrammetry from SUNY in 1980. Mr. Sieffert is currently employed by RADC as a Physical Scientist where he is engaged in developing a softcopy target mensuration system.

Addresses/Telephones

Mr. Brooks can be reached at the following address:
63-G Cowperthwait St.,
Danbury, Connecticut. 06810
Tel (work): 203-797-6663
Tel (home): 203-744-5161

Mr. Sieffert can be reached at the following address:
RR 6, Box 295
5101 Rt. 69
Rome, NY. 13440
Tel (work): 315-330-4378

IMPLEMENTATION OF AN INTEGRATED RESOURCE INFORMATION SYSTEM AND ITS APPLICATION IN THE MANAGEMENT OF FISH AND WILDLIFE RESOURCES IN ALASKA

William T. Brooks
U.S. Fish and Wildlife Service
1011 E. Tudor Road
Anchorage, Alaska 99503

ABSTRACT

The U.S. Fish and Wildlife Service in Alaska is composed of several resource management and research components involved in the collection and analysis of data. The specific mission objectives of these component organizations vary from the definition and inventory of fish and wildlife resources to the assessment of environmental impacts and consequences given the implementation of proposed management actions. The data acquired through direct observation and monitoring of species, habitat surveys, remote sensing, and traditional mapping modes have been indexed, stored, referenced, and analyzed in both automated and non-automated "systems" for many years.

Increasingly complex issues and a significantly enlarged management role in a spatial sense have forced the evaluation and development of more expeditious and efficient means of managing these diverse data bases and providing a facility for information integration and analysis. Fiscal constraints have also created an environment condusive to the sharing of technical resources within the agency and among other natural resource management entities within Alaska. The design of the Integrated Resource Information System (IRIS) focuses on the use of remote micro-computers networked with a central mini-computer facility, utilizing hierarchical and relational data base Management Systems (DBMS), Geographic Information Systems (GIS), and customized applications. Hardware systems and software packages supporting local (tactical) reporting and data management needs are employed in field offices, while facilities and technical personnel required for the planning, design, and development of these systems andthe accomodation of regional management and policy (strategic) requirements are centrally located. Wherever possible, data entry andediting are accomodated using tactical facilities and abstracted or summarized data are incorporated into strategic data bases as required. System designs under the provisions of the IRIS concept make use of existing commercial software components (DBMS and GIS) to maximize standardization and minimize the necessity to develop software "in house".

Formal and informal contact is maintained with other resource management agencies (U.S. Bureau of Land Management, U.S. Forest Service, U.S. Geological Survey, the State of Alaska Department of Fish and Game, and the State of Alaska Department of Natural Resources for the purposes of sharing data, facilities, and expertise, reducing costs by minimizing redundant systems and facilities, and jointly developing systems to accomplish common mission objectives.

INTRODUCTION

The organization of government agencies is conceived out of political necessity and a recognition of professional, administrative, and technical disciplines. These organizational constraints often promote a parochial approach to the accomplishment of objectives, even in areas of mutual interest with other organizations. In 1980, with the passage of the Alaska National Interest Lands Conservation Act (ANILCA , jurisdiction over federal lands in Alaska was allocated to those Federal agencies in the Departments of Interior and Agriculture according to the traditional rules and conventions applied to the "lower 48". Lands conveyed to the State of Alaska under the provisions of the Alaska Statehood Act of 1959 are administered by similarly organized state agencies. This spatial approach to resource and land management and administvation has resulted in a complex array of invisible "boundaries" where the emphasis in management is based upon the interest, expertise, and political priorities of the administering agency. While the areas of jurisdiction reflect the nature and content of the resources they contain in a general sense, the potential to overlook causes and effects of actions taken in one area on another area exists. One example of this is the navigable waters issue, where state ownership of navigable waters has been mandated in areas otherwise administered by a federal agency.

Within the individual agencies themselves, organization reflects technical and scientific specializations. Within the U.S. Fish and Wildlife Service in Alaska, a multi-dimensional organization exists. Vertically, the Service is comprised of Wildlife Resources, Fishery Resources, Habitat Resources, and Research and Development programs. Horizontally, these programs have both regional and field station offices in Alaska. At the field station level, there are 16 Wildlife Refuges, 3 Fishery Field Stations, 3 Ecological Services Offices, and several Research Field Stations. While the mission objectives of these components vary according to program and locale, the overall goal of the Fish and Wildlife Service in Alaska is the effective management of Fish and Wildlife Resources. This includes the protection of endangered species and critical habitat, enforcement of federal fish and wildlife regulations, assessment of environmental impact and land use planning within the National Wildlife Refuge System in Alaska, and monitoring of various species. The most important tool in the attainment of that goal is information. This information must be objective, contextual, accurate, and timely. It must also be available to the decision makers.

BACKGROUND

With the passage of ANILCA by Congress in 1980, the U.S. Fish and Wildlife Service in Alaska was required to prepare a "Comprehensive Conservation Plan" for each of the 16 National Wildlife Refuges created at that time. Each of these plans must inventory and describe the natural resources and values within the refuge, the management programs to conserve those resources and values, the uses of the natural resources that are compatible with the purposes of the refuge, and the opportunities for fish and wildlife oriented recreation, research, and education within the refuge. In 1981, a regional computer facility was established. A Data General MV8000 minicomputer was acquired to support the data collection, storage, analysis, and display requirments of the Refuge Planning Group. A Geographic Information System (GIS), consisting of the Analytical Mapping System (AMS), Map Overlay and Statistical System (MOSS), and GRID was implemented. AMS and MOSS were developed for the U.S. Fish and Wildlife Service by Autometric, Inc. several years earlier. GRID was acquired from the Environmental Systems Research Institute jointly with the State of Alaska Department of Fish and Game and modified for use on this system. Since the original development of this facility, other program needs in the collection, storage, analysis and evaluation of data have been addressed as well. In the process of evaluating these needs, it became evident that despite the diversity of data capture methods and objectives, common needs which transcended organizational constraints existed. With the evolution of computer technology, manifested in the availability of relatively inexpensive yet powerful microcomputers and the concept of distributed processing, the advantages of networking and systems integration became obvious.

In 1983, the Office of Information Resources Management was created in the Alaska Region of the U.S. Fish and Wildlife Service. The objectives of this office were to (1) provide operational support of the information needs of the Service in Alaska, (2) to manage and coordinate the use of the data processing, remote sensing, telecommunications, office automation, and library services, (3) to develop policies, standards and procedures for the application of these technologies to program needs, and (4) to promote further cost-effective means through networking, integration, and resource sharing with other agencies. Subsequent to the creation of that office, consideration was given to the variations in needs between programs and their hierarchical levels. The concept of "strategic" and "tactical" systems was used to provide a simple basis for the design of both software and hardware systems to meet these needs.

PHILOSOPHY

The Integrated Resource Information System (IRIS) is an invocation of a managerial philosophy in the design and implementation of systems rather than a design in itself. IRIS focuses on the use of (1) remote systems to support local data capture, edit, and reporting needs (2) a central facility for the support of data consolidation, integration, and (3) telecommunications networking between and among the local systems and the central system. It is recognized that this is not a necessarily unique approach. Centralized systems have long been criticized for insulating users from control over their processing environment, while decentralized systems create concern among higher level managers regarding security, data veracity and availability, and control of individual productivity. Distributed systems are advantageous in that, while users have direct control and access to local resources and data, data sharing and integration are facilitated through the use of telecommunications networks. Variations in the approach to networking are numerous. The most crucial elements in the design and development of a distributed system are (1) the standardization of data base elements, (2) quality control of local data, (3) minimization of unnecessary redundancy in hardware resources, (4) maintenance of software, and (5) systems compatability. The resolution of potential problems in these areas must be complete and absolute in order to assure a totally successful networking scheme.

In addition, networking provides for the maximization of resource utilization and economy of scale. In a time-sharing environment, common utilities may be simultaneously accessed by a wide variety of users on a single system. While there will be an increase in hardware utilization and requirements (processor cycles, storage, etc.), it must be realized that hardware components are becoming less expensive and software licensing and devlopment costs are rising rapidly. Unfortunately, any cost savings resulting from the use of timesharing in Alaska are more than offset by prohibitive communications costs. As a result, under the IRIS concept, data entry and editing are accomplished on remote facilities, or microcomputers. Interactive processes with the central facility using switched, or "dial-up", telecommunications are discouraged except for the transfer of data and information.

Another important consideration in the design of IRIS and its subsystems is the "contextuality" of information. Analysis of data for the purpose of obtaining information has as an inherent limitation the natural bias of the individual performing the analysis. A recognition and acceptance of this bias is an important design criterion. Unfortunately, the natural propensity for subjective evaluation cannot be easily documented or codified. Limiting the user's access to data for the purpose of controlling or limiting the subjective aspects of evaluation typically tend to only exacerbate the situation. For these reasons, the IRIS approach to data availability must be liberal. IRIS will provide raw data and tools for the analysis of that data, evaluation of the results, and depiction and reporting of information with minimum constraint upon the end user. If one were to enter a workshop and find piles of wood, woodworking tools, and instructions for the use of those tools, one would have the flexibility to build whatever one needed. Obviously, the quality and applicability of the finished product will be directly proportional to one's experience and expertise, but these limitations exist regardless of the approach taken. Care must be taken in this area not to create an environment where the use of the tools becomes more important than the quality of the results obtained. Therefore, significant emphasis will be placed on training and periodic evaluation of system resource utilization. The alternative is user dependance upon the limited resources and time constraints of a centralized support organization and facility. This facility must exist to accomodate the needs of strategic reporting and overall system design and maintenance, but the user must have the flexibility at the local, or tactical, level to provide for their own daily operational information needs.

Strategic information needs typically differ from tactical needs in both scope and detail. IRIS assumes that strategic information is a composite of the results of tactical evaluations. In order for the composite information to have credibility, certain constraints must be placed upon the tactical evaluation approach and format. This is, however, a management rather than a technical consideration. For this reason, IRIS employs a User Needs Analysis Document in the initial design and implementation of any system that ultimately affects more than one tactical location. In the performance of this User Needs Analysis, a representative of the user group is designated as the Project Manager and the Data Base Administrator. This individual must have the authority to make all decisions regarding the design of the tactical and strategic data bases, the rules which will comprise the supporting algorithms, and system

access and security. This individual must also accept the
responsibility for the success or failure of the completed system.
The duties of the Project Manager include consolidation and
documentation of user design criteria and documentation of the
completed system. The Office of Information Resources Management
performs the analysis of user needs and recommends alternative
approaches in terms of system development, aquisition of
commercially available software, or use of existing facilities. The
user makes the selection of the preferred alternative based upon
time and fiscal constraints.

The availability of the composite data in the strategic data base is
not a constraint upon the design and implementation of new systems.
The primary reason for this is that there is not a single strategic
data base, but several. These data bases will be linked indirectly
by an internal bibliographic index system, or directory. Access to
individual data bases is controlled by the intrinsic file access
controls system of the regional computer, which can provide levels
of access ranging from the ability to modify records to the ability
only to read indirectly through intermediate processes. The use of
temporary data bases consisting of the results of inquiries of
several permanent data bases can be accomodated through the use of a
commercially available Data Base Management System (DBMS). This
DBMS, "INFO", also provides the necessary interface to the
Geographic Information System for analysis and graphic
representation.

In summary, the IRIS philosophy is that strategic reporting
requirements may support the development or acquisition of tactical
systems, but must not be the basis for their design. The tactical
user typically has the responsibility for data entry and validation
and must see direct results from their efforts in the accomplishment
of their objectives.

IMPLEMENTATION

Fisheries Information Network

A subsystem of IRIS, the Fisheries Information Network (FIN), was
selected as the pilot project using these design criteria. In 1984,
three Data General 10SP microcomputer systems were acquired for the
three Fishery Field stations. In general, the design of FIN
requires a significant amount of detailed biologically oriented
data, collected through sampling surveys, to be collected by the
field stations. Analysis of the survey data will be performed at
the local, or "tactical", level. Resultant statistics will be
integrated from the three stations into a composite data base
located on the regional, or "strategic", system. All data
collection and editing will be done on the tactical systems. Once
the data has been collected and edited and the local analysis is
completed, the resultant information will be "uploaded" into the
strategic data base. The strategic system will accomodate regional
reporting and mapping needs while the more detailed data remains
available to the field stations for local reporting and analysis
requirements. The Project Manager for FIN determined that the
regional, or strategic, data needs would be met as a result of the
data collection and analysis performed at the field, or tactical,
level. While all three of the field stations had some differences
in their initial data elements definitions, subsequent discussions

resulted in a common design which met all of their needs. In Phase I
of the implementation of FIN, the data collection and editing
applications were written and installed on the tactical systems.
Phase II will consist of the development and installation of the
data analysis and reporting requirements of the tactical users.
Phase III will consist of the implementation of the strategic data
base and incorporation of the Geographic Information System
capabilities into FIN at the strategic level. In this system, it
was determined that it would not be cost effective to provide direct
access to GIS capabilities at the tactical level at this time. This
is due primarily to limitations in data communications facilities
and the cost of GIS support hardware such as graphics terminals and
plotters for those sites.

Wildlife and Habitat Information Systems
Over a period of several years, a number of small systems were
developed to support the collection, editing, and reporting of
information relative to the observations of a number of wildlife and
waterfowl species. These systems were developed on an ad hoc basis
with little or no consideration given to the desire to relate data
from one system to another. This has resulted in redundancy,
unnecessary modification to accomodate changes in minute reporting
detail, and a lack of compatability. As previously discussed, a
parochial approach to system development and data management will
evolve without a mandate for standardization. In addition, this
standard must result from a recognition on the part of the user that
such an approach is the desireable alternative. Given limited
resources for the development of new applications or modifications
to existing systems, IRIS has provided a means for the consolidation
of these smaller data bases using data base management technology
that may not have been available when the original systems were
implemented. As a result of this assimilation, certain analysis
techniques have been made available to users and relationships
between causes and effects which were not previously considered will
result. For example, a consolidation of data regarding the location
of marine birds and mammmals and other species would provide a basis
for easily determining the potential impact of an oil spill or other
contaminants if a common element for establishing that relationship
existed. This would provide an opportunity to document relationships
that were previously only speculative. While differences will exist
in certain elements, it became obvious in the analysis process that
there are many similarities among these smaller systems. The common
element throughout these data bases is location, expressed in
latitude and longitude, of observations and sitings. At present, it
is planned to assimilate the data contained in these smaller systems
into the Wildlife and Habitat Information Management System (WHIMS),
which will be supported by an ARC/INFO, an integrated GIS/DBMS
package developed by the Environmental Systems Research Institute
(ESRI). While this system will remain resident on the regional
computer system as a strategic application, data entry and editing
will be accomodated tactically using portable microcomputers.
Observation data will be recorded in the field and uploaded in
either larger microcomputer systems or directly into the regional
computer. Verification reports and plots will be generated by the
strategic system and returned to the field biologists for review and
correction.

Strategic Support System

The regional computer system is located in Anchorage, Alaska. It consists of a Data General MV8000 minicomputer. At present, this system is comprised of 6 megabytes of memory, 2 gigabytes of disk storage, and supports a network of 64 dedicated terminals and 8 switched telecommunications ports. Peripheral equipment includes a CALCOMP digital drum plotter, a Tektronix 4115B Color Graphics System, two digitizing stations, and several Tektronix graphics work stations. Supporting software packages include AMS/MOSS, ARC/INFO, DBMSII and INFOS (Data General Data Base Management Systems), and a number of custom applications developed for regional users to support administrative and resource management programs. Telecommunications applications for networking with tactical systems and the facilities of other agencies are also maintained.

Tactical Support Systems

In October, 1984, an additional 17 Data General 10SP microcomputer systems were acquired for the Wildlife Refuge Offices, Ecologocal Services Offices, and Research Stations. These systems will intially be used to provide word processing and generic spreadsheet and statistical capabilities. Support systems similar in concept to FIN will be implemented as they are developed. The Office of Information Resources Management in the Regional Office is in the process of acquiring additional commercially available software to support GIS needs on the tactical systems. As stated earlier, the primary obstacle to the effective implementation of GIS at the tactical level is the high cost of telecommunications. At present, Fish and Wildlife Service telecommunications facilities consist of the use of switched, or "dial-up" access, using 1200 Baud Racal-Vadic 3400 Series modems and commercial phone lines. Where available, the use of TYMENET and TELENET are encouraged for electronic mail and transfer of text. At present, these services are available only in Fairbanks, Juneau, and Anchorage. Stations located in other areas must use commercial long-distance carriers. By early 1985, those stations located in Fairbanks will be able to access the regional facility in Anchorage using a statistical multiplexer and high-speed data circuit under an arrangement with the Alaska State Office of the Bureau of Land Management. Other plans include development of similar resource sharing agreements with Federal and State agencies as well as the use of commercially available networks where cost-effective.

CONCLUSION

In the absence of a long term plan and design strategy, small, seemingly unrelated systems develop on an ad hoc basis. The concept of Information Resources Management imposes a requirement for system planning that transcends the traditional approaches to the acquisition of hardware to support "data" processing needs. There must be a recognition that data is a raw resource, expensive in its acquisition and, in some cases, non-renewable. As with any resource, consideration must be given to the maximization of its use and availability while protecting it from waste or loss. While there are obvious exceptions to a concept that all natural resource information must be available to anyone expressing an interest and having the resources to exploit it, there must be a recognition that we in government can no longer afford an approach that allows duplicity, redundancy, and parochialism in the management of natural resource information.

TIGER PRELIMINARY DESIGN AND STRUCTURE OVERVIEW:
The Core of the Geographic Support System for 1990

Frederick R. Broome
Geography Division
U.S. Bureau of the Census
Washington, D.C. 20233

The U.S. Bureau of the Census has embarked upon an effort to automate the geographic support process in time to meet the needs of the 1990 Decennial Census. The effort will result in an integrated system capable of performing the following three major geographical functions.

1. Assignment of residential and business addresses to a geographic location for data collection.

2. Provision of the geographic structure for tabulation and publication of the collected data.

3. Production of cartographic products to support data collection and publication operations.

The system will utilize the power of the computer to assure coordination of the geographic content between the various control lists and maps. At the core of this system will be the Topologically Integrated Geographic Encoding and Referencing (TIGER) System.

The theoretical foundation of the TIGER file structure is drawn from principles of topology and associated mathematics. The implementation of the TIGER System uses some of the latest concepts in computer science.

The term TIGER file refers to the computer file that contains all the data representing the position of roads, rivers, railroads, political and statistical boundaries, and other census-required features along with their attributes. The TIGER file, plus the specifications, procedures, computer programs, and related materials required to build and use the file, constitute the TIGER System. The Geographic Support System consists of the TIGER System and all other activities undertaken by the Geography Division to support the census and survey missions of the Bureau.

The TIGER file structure is a series of interlocked files and programs accessible through a master control program. The system is made up of a number of subsystems which handle the major functions. The TIGER file currently resides on a Sperry Univac mainframe computer. However, because future computer equipment cannot be determined, every effort has been made to produce software independent of the hardware. This included the mandate that all programming be done in Fortran-77. The mandate even extends to all programming for the microcomputer-based peripherals used during data capture.

The spatial framework for the TIGER file is being provided through a
cooperative program between the U.S. Geological Survey (USGS) and the
Bureau of the Census. Each USGS 1:100,000-scale map is being raster-scanned
and vectorized by USGS to produce four logical computer files: a roads file,
a hydrography file, a railroads file, and an "other transportation" file
containing among other things, pipelines and powerlines. The Census Bureau
is tagging the roads file with the seven-character USGS attribute codes.
USGS is tagging the other files with similar codes. Then the files are
combined at USGS to form their National Digital Cartographic Data Base. From
this combined, tagged data base, files of all transportation and hydrography
are provided to the Bureau for use as the TIGER spatial framework, an extremely
complex data and file structure whose projected final size will make it one
of the largest geographic data bases in the nation.

ANALYSIS AND DISPLAY OF DIGITAL ELEVATION MODELS WITHIN A
QUADTREE-BASED GEOGRAPHIC INFORMATION SYSTEM

Juan A. Cebrian, James E. Mower, and David M. Mark
State University of New York at Buffalo
Department of Geography
Buffalo, New York 14260

ABSTRACT

Digital elevation models (DEM) are an essential component of many geographic information systems (GIS), especially in natural resources applications. Recently, a number of prototype geographic information systems based on quadtrees have been developed. Digital elevation data create special problems for quadtree systems, since neighboring grid cells seldom have identical values. In this paper, we present a strategy for integrating DEM data into a quadtree-based GIS. The Morton sequence, which is the basis of most linear quadtree systems, is used to provide cell addresses in the DEM file.

INTRODUCTION

A *Geographic Information System* (GIS) may be defined as a computerized, spatially-referenced data base organized in such a way that spatial data input, analysis, and output may be accomplished. As is the case for any computer application, the issue of *data structures* is a critical one. Once a data structure for a GIS has been adopted, it is very difficult to change it; data structures are also a major factor in determining the efficiency with which queries can be answered within a GIS. Geographic information systems frequently involve natural resource data. In most such applications, elevation data (*digital elevation models*, or DEM) represent an important component, both directly and in the form of derived measures such as slope. Thus, any data structure adopted for a resources GIS must be able to efficiently integrate DEM data with other geographic data.

Recently, *quadtrees* have received considerable attention as a data structure for GIS applications (see below). Quadtrees appear to have many advantages for handling coherent ('blocky') spatial data, but are inefficient for continuous surfaces such as topography. However, if quadtrees are to be used in natural resources and related applications, it is essential to develop strategies for efficient integration of DEM data into the quadtree environment. This paper presents such a strategy.

Quadtrees
The quadtree is a data structure which is based on a regular decomposition of an image into quadrants and subquadrants. The region quadtree is relatively new, having been first suggested in 1971 (Klinger, 1971), and

fully developed for the first time in 1976 (Klinger and Dyer, 1976). As an example, we will consider the construction of a quadtree of a soils map. We assume that the soils map exists as a grid cell representation, that is, we begin with an array of square cells, each of which has a soil type. The map is assumed to be enclosed by a square region of side length 2^n cells. The cells are nodes of level 0, and the entire image is a node of level n (note that a node of level L has side length 2^L, and contains 4^L cells).

The 'top-down' construction of the quadtree begins by examining the entire image. If it contains only a single soil type, then the process stops, and the quadtree is a single node representing the whole image. If, however, the region contains more than one value (soil type), it is divided into four quadrants (level n-1 nodes), each of which is referred to as a child of the image square. Each of these quadrants is then examined for uniformity of value, and the process is applied recursively until all subquadrants are uniform. Note that since single cells are of uniform value and are valid subquadrants, the process must stop at the cells if not before. For blocky images, the quadtree typically will occupy much less space than the grid cell representation.

Linear Quadtrees for GIS. Early work on quadtrees used explicit pointers to represent the tree. Recently, however, alternative structures termed linear quadtrees have been adopted by many researchers (Gargantini, 1982; Lauzon, 1983, Abel, 1984; Mark and Lauzon, 1984; Samet and others, 1984; Lauzon and others, in press). In a linear quadtree, each leaf node is given a unique key number (or key and level combination), based on an ordered list of the node's ancestors. Then, the quadtree is simply a list of all the leaves, sorted by key. Neighbor-finding and other tree traversals are accomplished through the use of modular arithmetic or by bit addressing individual bits in the keys. Gray nodes can be omitted. Our research group has adopted the 'Morton matrix' sequence (Morton, 1966) as an appropriate key system for linear quadtrees. Briefly, the Morton number of a cell can be obtained from the row and column coordinates simply by interleaving the bits from the coordinates (coordinates and cell keys begin from 0).

Two-dimensional run-encoding (Lauzon, 1983; Lauzon and others, in press) is a technique for compacting a linear quadtree. Whenever the linear quadtree has consecutive leaves of the same color, only the key of the last leaf in such a 'run' is stored. Lauzon discovered that such runs could readily be decoded into maximal leaves; run-encoding usually reduces the number of records in a file by more than fifty percent (Lauzon, 1983; Lauzon and others, in press; Mark and Lauzon, in prep.). Furthermore, since any valid quadtree leaf of level m represents a sequence of 4^m consecutive Morton numbers all of the same color, runs of leaves are simply runs of pixels. A 2DRE file could be constructed simply by visiting the cells of a digital image in Morton sequence, saving the Morton key and color of the last cell in any run.

Digital Elevation Models

A *digital elevation model* (DEM) can be defined as any machine-readable representation of topographic elevation data. Digital elavtion models are widely used in a variety of fields, including cartography, engineering, mining, land use planning, and defense. A major issue in DEM research is the selection of an appropriate *data structure* (Mark, 1979). The most common data structure for digital elvation models is a regular square grid; the geographic space is partitioned into square cells of constant size, and an elevation value is associated with each cell. One problem with regular grids is that they are often highly redundant. The grid size must be sufficiently small to sample the smallest feature of interest in the entire study area, and to define the boundaries of larger features to some required level of precision. Then, because a constant cell size is used throughout the area, cells in most of the region will be smaller than needed (i.e., there will be too many cells).

In an attempt to address this problem, alternative data structures for digital elvation models have been designed. The most widely used of these is the triangulated irregular network (TIN), which represents the terrain by a triangulation on a set of points chosen to represent the surface (see Peucker and others, 1978). While this approach appears to have many advantages, it has not made a major inroad into the use of regular grids for digital elevation models. In large part, this is because devices are now available to produce very dense regular grids directly from aerial photographs (Noma and Spencer, 1978; Swann and others, 1978; Allam, 1978).

The quadtree would seem to hold considerable potential here, since it is an adaptive structure with variable resolution. There is, however, a major problem with the strict application of quadtree concepts to topographic data: it is unusual to find sets of four mutually adjacent cells which are of identical height. Of course, this depends on the height resolution of the data, but a typical case is the DEM data distributed by the U.S. Geological Survey (USGS). These data sets report elevations of 30 by 30 meter cells as integers in meters (Elassal and Caruso, 1983). Ground slope has to be less than 3 percent before adjacent points will have the same value.

Digital Elevation Models and Quadtrees

In existing quadtree-based geographic information systems at the University of Maryland (Samet and others, 1984) and at the University of California at Santa Barbara (Pequet, 1984; Chen, 1984), topographic data has been included by classifying the DEM data and producing quadtrees of the resulting coverage files. Both of the above systems have included in their published examples only maps of inter-contour areas, presented as quadtree files. The topographic component thus consists of as many binary quadtrees as there are inter-contour areas. We assume that slope data would be included in these systems in a similar way, as quadtrees of images of classified slope maps.

Inclusion of such secondary or derived data in a GIS is undesirable. In the slope example, it is unlikely that various users of a GIS will require maps at the same critical slope values. Once a set of critical slopes has been adopted and included in the data base, the flexibility of this aspect of the GIS has been greatly reduced. We propose that it is highly desirable to represent the topography as precisely as possible within the GIS; then, classified images can be derived to fit the user's exact requirements. In this case, there seems to be little alternative to the storing of all cells in the DEM grid.

We propose to do exactly that. However, in our system, the grid is not stored row by row, but rather, the grid cells are stored in Morton sequence. Lauzon (1983) and Lauzon and others (in press) have proposed that digital elevation models and other full matrix data be integrated into the linear quadtree GIS by using the Morton numbers as <u>addresses</u> for cell elevation data in a one-dimensional array or file. The use of the Morton key as a position within the DEM file allows for easy and efficient integration of DEM data into a GIS based on a linear quadtree structure such as 2DRE.

DEM DISPLAY WITHIN A LINEAR QUADTREE GIS

DEM display algorithms for grid data can be easily modified to process data in the Morton sequence. Quadtrees of two-dimensional (coverage) data, such as land-use or geology, can then be used to select portions of the DEM for analysis or display. In this section, we describe QDEM, a system based on this approach. Algorithms for DEM display using this strategy will be presented, and examples will be shown. The data display techniques discussed include: elevation band maps; slope steepness maps; slope aspect maps; and analytical hill shading.

Hardware and Operating System

QDEM is written in the C programming language, under Venix, a UNIX-like operating system, on a Terak microcomputer. The hardware configuration includes a Terak 8510/b Graphics Computer System and an 8600 Color System. The 8510/b uses the DEC LSI-11/23 16-bit microprocessor (and thus shares the architecture of the DEC PDP-11), and has 256k bytes of random access memory. The 8600 is based on the Intel 8086 16-bit microprocessor, and accesses up to 32k of display processor memory as well as 256k of frame buffer memory. The frame buffer is logically divided into four banks, each controlling 64k of RAM and three bit planes. As configured, the user can control two visually interlaced 640 by 240 pixel displays and 6 bit planes, providing a virtual resolution of 640 by 480 pixels and a palette of 64 colors. The 8600 provides a default palette, and also allows the user to define customized palettes. The red, green, and blue color guns each have 8 possible values (0-7), giving 512 possible colors.

Within a program, the user selects a color for drawing via a color index (0-63). Specifying a positive color causes the new drawing index to alter all bit planes, setting them to '0' or '1' according to the binary representation of the color index. Specifying a negative color index causes a bitwise OR to be performed on the current color index (ignoring the sign) and the color index already present in the pixel. In effect, only the bit planes equivalent to the bits which are '1' in the current color index will be changed to '1's. Thus, users may overlay and manipulate up to six distinct binary images (or, as described below, two distinct 8-value images).

The Terak Color Graphics Software package provides both ACM-SIGGRAPH compatible routines and Terak extensions. Many of the Terak extensions are primitives which allow display addressing at the physical device coordinate level. Since quadtrees are based on pixels, we have found that the primitive physical device routines are much more useful than the SIGGRAPH ones, and have written our own quadtree-based higher level graphics procedures.

Data Preparation
The first procedure developed reads a grid DEM, such as one of the U.S. Geological Survey's 7-1/2 minute Quadrangle series of digital data sets (Elassal and Caruso, 1983), and creates two files. The first one, which is called the 'header' file, contains such information as the map title, location, maximum and minimum elevations, grid cell size, and the parameters necessary for the conversion of the data to screen coordinates. The second file (termed the 'Morton' file) contains only the altitudes of the grid cells (30 by 30 meter squares in the case of the USGS data), which will be represented by single pixels in the display device. As the heights in the original file are usually processed in row-column order, the pre-processing routine calls the primitive 'morton' to generate the Morton key of the cell, which can be easily matched with several 2DRE thematic files by the system. Morton is a procedure which converts x-y (column, row) coordinates to a Morton key, using an algorithm based on modular arithmetic (see Lauzon, 1983). Then, procedure 'put_height' stores the height of the point in the appropriate location in the Morton file. Given the Morton number of a pixel, its height can then be directly accesed by key, using procedure 'get_height'. The procedures get_height and put_height have been implemented in C , and use the procedure 'lseek' (Kernighan and Ritchie, 1978, p. 164), which goes to a position in a file which is a specified offset (number of bytes) from the beginning of the file. Each integer height occupies two bytes, and thus the offset is just twice the Morton key.

In a 2DRE file of some type of coverage data, two Morton numbers K_1 and K_2, with $K_1 < K_2$, define a homogeneous region. As a consequence of the organization of the Morton file containing the heights, the current system can very quickly retrieve all the topographic information concerning such a uniform area. A program can retrieve the heights of all points which are on, say, a certain soil type simply by reading the soil type file and examining the soil value. Whenever this corresponds to the type being examined, the program can read the heights of all points with keys from the previous (in the 2DRE file) record's key, plus one, to the current record's Morton key.

Software Configuration

As noted above, QDEM is a package of procedures written in C. The program is interactive and menu driven, and manages a number of other routines which perform user-requested graphic and other analytical tasks. Dialogue between the system and users is provided by means of an alphanumeric keyboard and a graphic cursor. Within this dialogue, users select appropriate DEM and thematic files, and define analysis and/or display tasks and associated parameters. The data base is not yet accessible through queries based on coordinates.

The principal characteristics of each user query are resolved in a three stage dialogue. The first selects the type of analysis to be performed, the second determines whether or not thematic data are to be used to window the DEM, and the third defines the type of output (graphic or other).

Immediately after this stage, a second level of dialogue is established to determine the parameters which will drive the chosen process; this will depend on the type of analysis and/or display selected. There are two different sets of parameters: those which refer to the classification of the pixels (number of intervals, amplitude of the intervals, and so on) and those which will characterize the graphic output, if required. The first ones are introduced from the keyboard.

Users can select colors for the graphic representation from any of several palettes. With the aid of the graphic cursor, users can easily pick colors by moving over the palette menu. If none of the palettes available on the system are suitable, users can build their own palettes by calling the appropriate routines.

Types of Analysis and Display Procedures

The basic DEM displays described by Kikuchi and others (1982) in their DEMGS system have been implemented here; we refer the reader to that paper for further details of the algorithms. From each of these types of analysis, users may wish to produce graphic output, to compute summary statistics, or to output a linear quadtree (2DRE) file. We have decided that at present only the first and last option will be available, and that statistical analyses will be performed on the 2DRE files rather than on the DEM directly.

Contour Band Map. This procedure simply classifies the pixels according to their heights. The parameters of this classification are requested from the user; all parameters do, however, have default values. The maximum and minimum elevations are already recorded in the corresponding header file. After users determine the number of class intervals, the uniform height range for all classes is then derived from the other parameters. Users can also define varying height intervals. Elevation maps would seldom use varying intervals, but in GIS queries some users may be interested in a specific partition of the height dimension in order to match this result with some other spatial phenomenon.

Slope Steepness Map and Slope Aspect Map. These images are produced using simple linear algebra operations (see Kikuchi and others, 1982). Users can determine the number of slope classes, their limits, and the colors to be used on graphic output. The pixels are processed in Morton order. The elevations of two non-adjacent neighbors (in this case, arbitrarily the cells to the east and south) are used to establish a cross-product vector orthogonal to the surface. The length of the z-component of the unit vector orthogonal to the surface equals the cosine of the slope angle; the slope aspect can be obtained from the x- and y-components of the orthogonal vector (Kikuchi and others, 1982). Note that, for every cell, two neighbors must be examined. Thus, in the current implementation, each elevation is actually read three times, reducing the speed of the algorithm. We intend to remedy this by modifying the program to read contiguous blocks of data, 512 bytes at a time.

Analytical Hill Shading. This procedure is based on the same linear algebra operations as the slope maps. However, in this display, the classification is done automatically. The cosine of the angle between the surface normal vector (see above) and the light source is determined as the dot product between those two unit vectors. All the pixels are then classified into one of 8 colors: black for surfaces directed away from the light source, white for surfaces perfectly orthogonal to the light source, and 6 gray shades for the intermediate cases. The user can, however, select the location (azimuth and altitude) for the light source.

Relating Thematic Attributes to the DEM
The system has been designed in order to optimize the time response of special queries about the topographic characteristic of specific zones inside the area covered by the DEM. As the areas under consideration are described by linear quadtrees, and the DEM is sorted in Morton order, the windowing of the DEM is straightforward. Only the pixels in the area of interest are examined and classified. There is no need to deal first with the DEM within a rectangular area which encloses the search polygon and then later clip the rectangular area following the edges of such polygons.

Two different overlay procedures have been devised and implemented. The first one displays topographic properties (heights, slopes, slope aspects, etc.) within a definite zone (such as a certain land cover or soil type). It is also possible to prepare statistical reports about these properties. The search area is always described by a 2DRE file. Algorithms are available to convert vector descriptions of polygons into linear quadtree representations (see Mark and Abel, 1984).

The second overlay procedure produces graphic output. We

HUE

Lightness		Red	Blue	Magenta	Green	Yellow	Cyan	White		Lower Three Bit Planes
1.00	(7,7,7)	(7,0,0)	(0,0,7)	(7,0,7)	(0,7,0)	(7,7,0)	(0,7,7)	(7,7,7)	111	
0.85	(6,6,6)	(6,0,0)	(0,0,6)	(6,0,6)	(0,6,0)	(6,6,0)	(0,6,6)	(6,6,6)	110	
0.71	(5,5,5)	(5,0,0)	(0,0,5)	(5,0,5)	(0,5,0)	(5,5,0)	(0,5,5)	(5,5,5)	101	
0.57	(4,4,4)	(4,0,0)	(0,0,4)	(4,0,4)	(0,4,0)	(4,4,0)	(0,4,4)	(4,4,4)	100	
0.42	(3,3,3)	(3,0,0)	(0,0,3)	(3,0,3)	(0,3,0)	(3,3,0)	(0,3,3)	(3,3,3)	011	
0.28	(2,2,2)	(2,0,0)	(0,0,2)	(2,0,2)	(0,2,0)	(2,2,0)	(0,2,2)	(2,2,2)	010	
0.14	(1,1,1)	(1,0,0)	(0,0,1)	(1,0,1)	(0,1,0)	(1,1,0)	(0,1,1)	(1,1,1)	001	
0.00	(0,0,0)	(0,0,0)	(0,0,0)	(0,0,0)	(0,0,0)	(0,0,0)	(0,0,0)	(0,0,0)	000	
	000	001	010	011	100	101	110	111		

Upper Three Bit Planes

Figure 1: Color gun values (in the order: red, green, blue) for the video look-up table used form combining thematic data with analytical hill shading.

have defined a special palette consisting of several series of color shades; each series has a particular hue, but with different _values_ (see Figure 1). Using this color palette, we have devised and implemented a procedure which produces a hill shading image in many colors, with each hue corresponding to a different area type (soil, rock type, etc.). The eight gray levels of the hill-shading image can be drawn using colors 0 to 7. This is equivalent to drawing that image in the lower three bit-planes of the image buffer. Then, a thematic map with up to 7 classes (from a decoded 2DRE file) is drawn in the upper three bit planes without altering the contents of the lower bit planes. On the Terak, this is done by using negative color indices of the form -8*i, i=1,7, where i represents the thematic class. It is then possible to overlay a series of thematic coverages, without any need to re-compute the hill-shaded 'base-map'. The six bit-planes of the Terak limit the display to at most 8 thematic classes; the 8 discrete levels for each color gun further limit the display, so that only 7 thematic classes are practical. The method can be extended trivially on more powerful systems.

SUMMARY

As a DEM display system, QDEM is little more than an implementation of DEMGS (Kikuchi and others, 1982) in C. However, the change from row-and-column to Morton sequencing for the grid DEM means that DEM data can be efficiently interfaced with other geographic data within a GIS. The use of a full DEM matrix, referenced by Morton numbers, appears to be the most appropriate way to include topographic data in a quadtree-based GIS.

ACKNOWLEDGEMENTS

We wish to thank the Fulbright program for their assistance in allowing Juan A. Cebrian to conduct research at SUNY at Buffalo as a visiting scholar from Spain. Also, J.P. Lauzon provided useful comments on the work reported in this paper.

REFERENCES

Abel, D.J., 1984, A B^+-tree structure for large quadtrees: Computer Vision, Graphics, and Image Processing, 27: 19-31.

Allam, M.M., 1978, DTM's application in topographic mapping: Proceedings, Digital Terrain Models (DTM) Symposium, ASP-ACSM, St.Louis, Missouri, May 9-11, 1978, p. 1-15.

Chen, Z.-T., 1984, Quad tree spatial spectrum: Its generation and application: Proceedings, International Symposium on Spatial Data Handling, Zurich, Switzerland, August 1984, vol. 1, pp. 218-237.

Elassal, A.A., and Caruso, V.M., 1983, Digital elevation models: U.S. Geological Survey Circular 895-B, 40pp.

Gargantini, I., 1982, An effective way to represent quadtrees: Communications of the ACM, 25:905-10.

Kernighan, B.W., and Ritchie, D.M., 1978, The C Programming Language: Englewood Cliffs, New Jersey: Prentice-Hall, Inc., 228pp.

Kikuchi, L., Guevara, J.A., Mark, D.M., and Marble, D.F., 1982, Rapid display of digital elevation models in a minicomputer environment: Proceedings, ISPSR Commission IV Symposium (Auto-Carto V), pp. 297-307.

Klinger, A., 1971, Patterns and search statistics: in Optimizing Methods in Statistics, J.S. Rustagi, ed., New York: Academic Press.

Klinger, A. and Dyer, C.R., 1976, Experiments on picture representation using regular decomposition: Computer Graphics and Image Processing 5, 69-105.

Lauzon, J.P., 1983, Two-dimensional Run-encoding for Spatially Referenced Data: unpublished M.A. project, Department of Geography, State University of New York at Buffalo.

Lauzon, J.P., Mark, D.M., Kikuchi, L., and Guevara, J.A., in press, Two-dimensional run-encoding for quadtree representation: Computer Vision, Graphics, and Image Processing, forthcoming.

Mark, D.M., 1979, Phenomenon-based data-structuring and digital terrain modelling, Geo-Processing 1: 27-36.

Mark, D.M., and Abel, D.J., 1984, Linear Quadtrees from Vector Representations: Polygon to Quadtree Conversion: CSIRONET Technical Report #18, Canberra, Australia (also submitted to IEEE Transactions on Pattern Analysis and Machine Intelligence, February 1984; revised October 1984).

Mark, D.M., and Lauzon, J.P., 1984, Linear quadtrees for geographic information systems: Proceedings, International Symposium on Spatial Data Handling, Zurich, Switzerland, August 1984, vol. 2, pp. 412-430.

Mark, D.M., and Lauzon, J.P., in preparation, The space efficiency of quadtrees: An empirical examination including the effects of two-dimensional run-encoding. Submitted to Geo-Processing, July 1984.

Morton, G., 1966, A computer oriented geodetic data base, and a new technique in file sequencing: IBM Canada Limited, unpublished report, March 1, 1966.

Noma, A.A., and Spencer, N.S., 1978, Development of a DMATC digital terrain data base system, Proceedings, Digital Terrain Models (DTM) Symposium, ASP-ACSM, St.Louis, Missouri, May 9-11, 1978, p. 493-505.

Pequet, D.J., 1984, Data structures for a knowledge-based geographic information system: Proceedings, International Symposium on Spatial Data Handling, Zurich, Switzerland, August 1984, vol. 2, pp. 372-391.

Peucker, T.K., Fowler, R.F., Little, J.J., and Mark, D.M., 1978, The triangulated irregular network, Proceedings, Digital Terrain Models (DTM) Symposium, ASP-ACSM, St.Louis, Missouri, May 9-11, 1978, p. 516-540.

Samet, H., Rosenfeld, A., Shaffer, C.A., and Webber, R.E., 1984, Use of hierarchical data structures in geographic information systems: Proceedings, International Symposium on Spatial Data Handling, Zurich, Switzerland, August 1984, vol. 2, pp. 392-411.

Swann, R., Thompson, J., and Daykin, S.E., 1978, Application of low cost dense digital terrain models, Proceedings, Digital Terrain Models (DTM) Symposium, ASP-ACSM, St.Louis, Missouri, May 9-11, 1978, p. 141-155.

ANALYSIS AND DISPLAY OF DIGITAL ELEVATION MODELS WITHIN A
QUADTREE-BASED GEOGRAPHIC INFORMATION SYSTEM

Juan A. Cebrian, James E. Mower, and David M. Mark
State University of New York at Buffalo
Department of Geography
Buffalo, New York 14260

BIOGRAPHICAL SKETCHES

Juan A. Cebrian received his Ph.D. from the Complutense University of Madrid in 1983. He was an Assistant Professor in the Department of Human Geography at the Complutense University of Madrid from 1978 to 1983, and an Associate Professor in the Department of Cartography at the Polytechnic University of Madrid in 1984. He is currently a Fulbright Scholar in the Department of Geography, State University of New York at Buffalo. His current research interests include computer assisted cartography, geographic information systems, and quadtrees.

James E. Mower is a Doctoral candidate in Geography at the State University of New York at Buffalo, where he is studying cartography and geographic information systems. He holds the degrees of B.A. (1977) in English from the State University of New York, College at Geneseo, and M.A. (1981) in Linguistics from Indiana University (1981). His current research interests include line generalization algorithms based on feature recognition.

David M. Mark received the degrees of B.A. (1970) and Ph.D. (1977) from Simon Fraser University, and M.A. (1974) from the University of British Columbia. Since 1981 he has been a member of the Department of Geography, State University of New York at Buffalo, where he is currently Associate Professor and Director of the Cartographic Laboratory. His current research interests include quadtrees, digital elevation models, artificial intelligence, and theoretical geomorphology.

INFO-GRAPHICS :
ITS APPLICATIONS TO
PLANNING IN INDIA

Anne Chappuis, Luc de Golbéry
Indo-French Compu-Graphic and Planning Project
Bureau of Economics & Statistics
Planning Department, Government of Andhra Pradesh
Khairatabad P.O.   - HYDERABAD 500004 - INDIA

ABSTRACT

India's only hope of fighting poverty successfully is to use effective methods. It is essential to provide appropriate tools for analysis and aid to decision. Info-Graphics, the micro-computerisation of Bertin's Graphics Information Processing, is a cheap, reliable, easy to use, extremely powerfull and universal tool for decision making. If it can be considered a luxury in developped countries, it is an absolute necessity for the developing countries if we really want to reduce the inequalities. Even more so if we plan transportable units.

INTRODUCTION

In India the Planning Department plays a key role at all levels of development. With most major infrastructures in place, the problem faced by planners is that of "unbalanced development", that is unequal development of regions and/or social groups. In such conditions the challenges for development are to analyse and communicate more and more numerous, detailed and complex information.

It is consequently essential to supply Planners with apropriate tools for analysis and aid to decision. The vast amount of data required to be processed and analysed and their timely output advocates for the use of computers. However the decentralisation process, which has already started, and the difficult environment call for a network of micro-computers rather than a big centralised main-frame.

On another hand the necessity to provide meaningfull informations in an unambiguous and quickly digestible form pleads for the use of Bertin's Graphics Information Processing *. The foundations of Graphics Science have been laid down twenty years ago, but it is only now, with the advent of "computer graphics", telematics and other videotex, with the universal acceptation that "a small picture tells more than a long talk" that the image

---
Bertin J., 1981, "Graphics and Graphic Information Processing", Berlin - New York, de Gruyter

specialists realise to what extent Bertin was an extraordinary precursor. So much a precursor in fact that Graphics had to wait for micro-computers with graphic facilities to really find its full application.

The unfortunate fact that Graphics Science remained unknown among computer specialists led to the development of poor graphics softwares. They grossly under utilize the computer graphic hardware capabilities as well as the extraordinary powers of the human visual perception successfully tapped by Graphics Science.

INFO-GRAPHICS METHODS AND TECHNICS

Info-Graphics (Info for informatics as well as information) methods and technics try to bridge this gap by implanting Graphics Science logic into micro-computers through specific softwares. This will lead in a first phase to the development of a complete Expert System for Automatic Cartography and later to a more general Information Processing Expert System designed as a decision aid.

Graphics like Mathematics is a science and a rational language whose vocation is to be applied in the vast domains of information processing and communication. This includes cartography, graphs, diagrams and goes far beyond by proposing methods and technics similar to multi-dimensional analysis in statistics. In other words Graphics Science transforms radically the empirical and artistical approach of traditional visual presentation into rigorous and scientific methods which multiply manyfold the efficiency of their outputs and simultaneously minimises its cost by cutting down the empirical component.

Info-Graphics applied to development is usefull in two different domains:

1) <u>Over worked decision-makers have to analyse and synthetise an enormous amount of information to rapidly diagnose a problem.</u> They also need easily communicable information to be able to argue about the choices with the politicians: For example during an administrative reform which consists of replacing the old 316 units by a thousand new ones, the new potential headquarters have to be identified amongst 27,000 villages. It is obvious that politicians will try to influence decisions to include their village in the list even if it is a very small unqualified one. If the decision-maker has:

   a) a list of potential headquarters with their population and infrastructures

   b) a map which locates them he can:

      1) verify whether the proposed village is on the list
      2) show to the politician that it is not and that it is a very small dot on the map

3) argue that the next village having a big dot should be selected instead.

To prepare such a list (27,000 X 15) and map the results in three months it is obvious that a computer is needed.

2) <u>Communication of information is essential in the democratic process</u>. In a mostly illiterate environment, graphic communication is a necessity. The following example will demonstrate this: The Gujarat Planning Secretary visited our laboratory three years back and was convinced that visual translation of information is important. He built up a cell in his department. He mapped the available village infrastructure (a black dot for presence, a blank for absence). These maps were displayed in the local administration. The reaction of an illiterate elected panchayat woman member illustrates the universality of graphics: She said "How is it that my village has only two "bindies" (the dot that women draw on their forehead) while the next village has so many". Privileges and inequalities become immediately obvious and democratic control by citizen voters becomes efficient.

The same maps also created an awareness of some of the development constraints among villagers: some villagers asked the Government to construct a power line to their village. It became very easy to show them on the map that their village being the furthest it needed a consensus of four villages for the department to construct the line. What would have been difficult to explain through speech became obvious on the map.

Through regional comparison people became conscious of their relative backwardness. This helped them setting up higher goals for their development programmes to try to bridge the regional unbalances.

In third world countries graphic translation of information is the most efficient and perhaps the only mean of democratisation provided:

- it is fast, cheap and reliable, which means micro-computerised
- it follows strict procedures to produce efficient and true pictures, which means the use of Bertin's "Semiology of Graphics" rules*.

Bertin J., 1983, "Semiology of Graphics", Madison, The University of Wisconsin Press

TEST OF THE METHODS

To test the methods on a real scale a relatively cheap micro-computer graphic laboratory was installed in 1981, in the Bureau of Economics and Statistics, Planning Department, Government of Andhra Pradesh. It started functioning in a difficult environment (high fluctuation in voltage, frequent power cuts, heat, humidity, dust...). It is composed of the following equipment:

- micro-computer :   Hewlett-Packard 9825, 64 K
- memory   :   cassets of 280 K or
                     diskets of 500 K (single face,
                     double density)
- digitizer   :   Summagraphics (750 X 1000 mm)
- plotters   :   H.P 9872 A, 4 pens, 285 X 400 mm
                     Benson 1332, 4 pens, 92cm width
- graphic printers: Logabax with programmable
                     characters
                     Epson with graphic funtion
- ondulator/de-ondulator with battery back up for power control.

The system which has been developped links data banks which can be geographiccal (state, district, taluk, village, hamlet) or not (household, individual, company etc...) to statistical and graphic processing and output. It works under the following constraints:

- easy use by non specialists
- fast retrieval of information
- security

Recorded information is documented on a key-word library system which allows fast and easy retrieval.

Base maps are digitised and recorded using

- polygon centers to produce plotter punctual maps
- vectors to produce plotter linear maps
- polygons to produce plotter zonal maps
- matrices to produce printer grid maps (punctual/zonal)

Once the user has identified his data he is faced with the following choices:

- direct listing
- statistical processing : uni-variate
                             multi-variate
- graphical processing   : direct
                             after statistical processing

At any given time he can go back to any one of the steps to make new calculations and/or new graphic outputs.

The ease of use can be demonstrated by the fact that some of our Indian assistants who are villagers with a level of education varying between 10th grade and bachelor degree and without any computer training are able to use easily part of the system (data entry, printing, calculations, classification and mapping). Trainees who were sent from other states to our laboratory were able to start using the system the very first day and could use it fully after three weeks. Besides they were able to go back to their home states with about thirty graphic documents (maps, graphs, bar diagrams, visual matrices).

THE APPLICATIONS

They are at present at two levels:

1) <u>the State of Andhra Pradesh (60 million people):</u>

   - analysis, visualisation and communication of development data:

   a) <u>on a regular basis:</u>

   * Documents at different scales (state, district, taluk) on population, climate, agriculture, industry, finance etc...
   are printed in the yearly publication of the Bureau of Economics and Statistics, (Season and Crop Report, Statistical Abstract, Quarterly Bulletin) meant for administrators and the public.

   * Rainfall: daily rainfall from 360 raingauges scattered throughout the state is recorded with one week delay to set up an alarm system which will inform the decision-makers that a drought is setting in and it is time to launch the special aid programmes for the daily agricultural labourers and marginal farmers. This system will later be linked to crop data for crop insurance.

   * Yields: sample seasonal crop harvest are recorded and analysed.

   * Timely crop report: a 20% village sample of the areas under different crops is recorded season-wise to arrive at an early estimation of the total areas. Linked to the yields it allows estimates of production.

   * A regularly up-dated data bank is available for decision-makers.

   b) <u>occasionally</u>

   * Population: 1981 census data have been analysed for the 316 taluks of the state and a hundred thematic maps were prepared in a month. Multi-variate analysis is in progress especially on weaker sections.

DISTRICT MAPS

Figure 1 : VILLAGE-WISE (PRINTER)

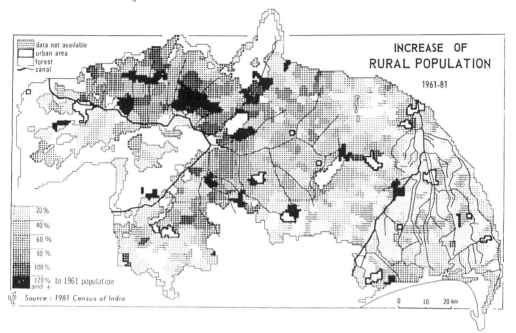

Figure 2 : VILLAGE-WISE (PLOTTER)

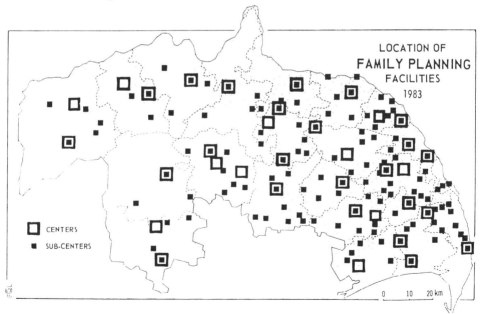

* Finances: an inter-states comparative study of the evolution of state income heads over 25 years was prepared for the finance department.

* Administrative reform: the Government of Andhra Pradesh has decided to change its smallest administrative units and create smaller units to reach the people more efficiently. They had to identify the thousand potential headquaters among the 27,000 villages of the state. Fifteen parameters on population and available infrastructure and services were recorded. A score was calculated and lists of reclassified villages were output and mapped for each district. These documents were sent to the District Collectors (heads of the administration) to allow them to prepare the final list without too much political interference. This was done in three months.

2) at the regional level: a district planning atlas

One of the districts, Guntur (three million people), was selected to build a village data base (for 730 villages) with data on population, agriculture, infrastructure, amenities, administration, non agricultural activities (about 300 parameters). Data is up-dated seasonally whenever possible. A planning atlas with a hundred thematic maps can now be prepared in less than two months for any one of the districts. Output on plastic sheets alows cheap and fast reproduction. This atlas is more a dictionary which the Collector can carry along with him and consult easily any time on his tour programmes. The format (22 X 34 cm) is that of the usual administrative files (figure 7 to 10). This document helped in the preparation of the seventh plan.

RESPONSE

1) The local government has decided:

* to strenghen the state laboratory

* to put info-graphic laboratories equipped with Indian micros in each of the 22 rural districts in order to:

  - keep and up-date data bases at different levels:
      village
      hamlet
      household
    analyse them and make the necessary graphical presentation.

  - identify the backward zones and/or groups qualified for special development programmes.

  - follow and monitor these programmes and control their correct execution

- prepare a new system of agricultural data
  collection since the past system of "Village
  Officer" through whom the data was collected has
  been recently abolished.

2) <u>The Land Record Department</u> wants to micro-
computerise its cadastral survey and clear their 18
years backlog in three phases:

- village skeletons
- plots with related owners
- link these with crop information

3) <u>Seven other States</u> (Kerala, Karnataka,
Maharashtra, Gujarat, Jammu and Kashmir, Sikkim,
Mizoram) and Sri Lanka have contacted us to help them
set up similar laboratories.

FUTURE

Info-Graphics opens up new areas of applications to
visual processing :

- decision aid for :
  * economic and social development
  * business management
  * government administration
  * regional planning

- communication :
  * mass media
  * road signs and other visual signals
  * advertising

- education
  * structuring of knowledge
  * development of powers of analysis and synthesis

- real time :
  * instrument panels in vehicles, scientific
    processes, industry etc...

In order to meet these demands it also make possible the
creation of :

- Cartographic robots which are :
  * inexpensive
  * capable of assisting the cartographer and
    increasing his productivity, particularly in the
    area of "Business Graphics"

- Expert Graphics systems which may
  * introduce Graphics in areas where it has not
    hitherto been used (government, business)
  * replace the cartographer in areas where he is a
    luxury (e.g. the third world).

- Expert Information Processing System which will be
    able to
    * assist any decision maker (administrator,
      businessman but also pilot) in his task
    * assist scientists in their research process
    * help everybody get a meaningfull and non
      distorted understanding of the reality through
      telematic media
    * help the mass media have a more powerfull
      analysis and presentation of the news
    * etc...

- Info-Graphics Education System which will
    * help the teachers in their work
    * help pupils in acquiring a much better power of
      analysis and synthesis

CONCLUSION

Bertin's "Semiology of Graphics", a universal tool which functions everywhere, particularly well with illiterates, applied on micro-computers gives a cheap, reliable, easy to use and extremely powerfull tool for decision making. If it is considered a luxury in developed countries, it is an absolute necessity for the developing countries if we really want to reduce the inequalities. Even more so if we plan transportable units which can be installed anywhere for the required time.

BIBLIOGRAPHY

BERTIN J., 1967, "La semiologie Graphique", Paris, Gauthier-Villars.

1977, "La Graphique et le Traitement Graphique de l'information, Paris, Flammarion.

1981, "Graphics and Graphic Information Processing", Berlin-New York, de Gruyter.

1983, "Semiology of Graphics", Madison, University of Wisconsin Press.

GIMENO R., 1980, "Apprendre a l'ecole par la Graphique", Paris, Retz.

de GOLBERY L., 1979, "Micro-informatique et traitement graphique de l'information", in Cahiers Geographiques de Rouen No 10-11, Rouen, pp157-183.

CHAPPUIS A., de GOLBERY L., PINCHON C., TAN CHoNG CHIN, VERLUT X., 1982, "Cartographie automatique et info-graphique, essai d'application de la semiologie graphique a la sortie de cartes sur imprimante a caracteres programmables", in 2eme Colloque de Micro-Info-Graphiquem pp L1-23.

QUAD TREE SPATIAL SPECTRA GUIDE:
A FAST SPATIAL HEURISTIC SEARCH IN A LARGE GIS

Zi-Tan Chen
Donna Peuquet

Geography Dept., Univ. of California,
Santa Barbara, CA 93106, U.S.A

ABSTRACT

An approach for improving spatial search efficiency in large-scale GIS is presented in this paper. This improvement is achieved by a spatial heuristic search instead of the conventional blind search in GIS. This approach is a practical application of artificial intelligence for solving the difficulties caused by large-scale GIS, which spends excessively long CPU time for many search operations on huge volume data. The traditional spatial search can be categorized as blind search, which wastes many efforts on those impossible regions. This research increases the spatial search efficiency by a heuristic search strategy, which finds the candidate areas by utilizing spatial knowledges. The spatial knowledges, required by the heuristic search, are the quad tree spatial spectra(QTSS) trees of spatial data and goal objects. A forward up-down strategy controls the search procedure. The interpretation algorithms among the QTSS are developed as the search rules. The test results are shown in this paper.

INTRODUCTION

The requirements for geographic information systems are increasing rapidly. Likewise, the contrast between huge volume data and limited computer sources is also increasing. For large countries such as the USA, China and Canada, size is a real problem. One of the causes of this problem is the inherent limitation of algorithm efficiency. Any practical GIS has to use some 'slow' algorithms because there is not always a better alternative.

After these expensive large computer GIS systems are established, their manipulation efficiency is a serious problem. For practically applying GIS, it is better that the computer system not only can do geographic data storage and simple manipulations, but also that the system is smarter --- we hope GIS has some degree of artificial intelligence(Smith, 1984).

An important concept of artificial intelligence is the utility of human being's knowledges, especially those professional knowledges of the experts(Nilsson, 1980). The heuristic search is a way to use these knowledges(Pearl, 1984). A GIS is a spatial data base. For improving the spatial search efficiency, the heuristic search must be implemented not only on symbolic data, but also on spatial data.

This approach focuses on the spatial location heuristic search based on spatial knowledge extraction and its applications. The geographic data are represented as quad trees. The spatial knowledges are extracted as quad tree spatial spectra (QTSS). (Chen, 1984). Then, a spatial heuristic search control strategy is implemented based on the interpretation of QTSS knowledges.

Those GIS with the capabilities of location heuristic search are more intelligent. The GIS can pick up the higher potential candidate regions just using these compact knowledges. The complex algorithms are only executed on the candidates. Thus, much of the CPU time can be avoided.

Suppose the CPU time for implementing an expensive algorithm for the whole area S is TA. Then the CPU time for implementing at a partial area S/N is assumed to be TA/N. The CPU time for finding the candidate area is TC, while a heuristic search and implement time TH is:

$$TH = TC + TA/N \qquad (1)$$

If TC is much less than TA, and N is not small, TH will be much less than TA.

QTSS TO REPRESENT SPATIAL DISTRIBUTION KNOWLEDGES

The principles for evaluating spatial knowledges

The selection from different kind of spatial distribution knowledges to be used in GIS should consider the following principles:
(1) These knowledges include richer spatial distribution information;
(2) the storage of these knowledges takes small space;
(3) various interpretation algorithms can be easily made from these spatial knowledges and they are efficient;
(4) the generation of these knowledge is efficient.

QTSS definition

The concepts of quad tree spatial spectra QTSS have been proposed(Chen, 1984). First, a raster image converts its quad tree representation. A $2^N * 2^N$ binary raster image can be represented by its quad tree by successively subdividing it into four quadrants until all pixels within a quadrant are homogeneous(Klinger and Dyer, 1976). The number of black nodes at each tree level constitutes a set. The length of this set is the level number N. For instance, the evergreen forest at Black Hill area, South Dakota, has this QTSS:

(0 0 0 7 117 452 1354 3299 7765 14997 29893)

This QTSS has a length of 11. Such a set is defined as a quad tree spatial spectrum(QTSS) of this spatial phenomenon for this area(Chen, 1984). To judge QTSS based on the previous principles, the QTSS can be generated much faster than those conventional transformations such as FFT or FWT, and they store more compactly. QTSS do not contain complete spatial information, but, nevertheless, they do include a great deal of statistical spatial information, as well as some contextual information. For example, an interpretative function, called as "QAREA", calculates the area of this phenomenon at a quad tree node by the formula:

$$QAREA(QTSS) = \sum_{l=0}^{L} \frac{QTSS(l)}{4^l} \qquad (2)$$

where l is the level sequence and L is the length of QTSS. Many interpretations of the spatial distribution properties based on QTSS are possible.

Elementary phenomena represented by QTSS

There are two kind of geographic data in GIS. One is elementary phenomena, which has been input directly from data sources, such as lakes or railways. They can be stored in GIS as one simple binary image. Another kind of data is complex phenomena, which consists of several elementary phenomena. A phenomenon can be either elementary or complex depending on the GIS in question. For example, the evergreen forest in a rough GIS of land-use might be an elementary data, but it would be a complex phenomenon in a thematic forest GIS.

The QTSS can describe the spatial distribution properties of an elementary phenomenon, such as density, clustering and approximate location.

Figure 1 shows the quad tree image of the residential areas at Black Hill. The Rapid City is at the middle east of the image. It is an elementary phenomenon in the GIS. Table 1 shows its QTSS at the root (whole image) and its four quadrants. The QTSS at the root shows that some residential areas exist, but the total size of these residential areas only is small fiction of the whole area. Also, they are dispersed. From the QTSSs for four quadrants, it can be seen that most of the residential areas are in quadrant 2 and 4, while a few are in quadrant 1 and none is in quadrant 3.

Figure 1. Quad tree representation of
the residential area at the Black Hill

THE ARCHITECTURE OF GIS

The GIS includes three parts: a data base, a knowledge base, and a search module.

A spatial data base

All geographic data are stored in this data base. They are original spatial data, inputed from maps, classified images, etc.. The data structure representing these spatial data is quad trees. All data are registered at a same square area. Each elementary geographic phenomenon is converted into quad tree file. (Peuquet, 1984).

A spatial knowledge base

The spatial distribution knowledges are stored in this part. These knowledges are the QTSS trees, extracted from these original spatial data. First, for each node at the top several levels of a quad tree, one QTSS is generated from the original data. Table 1 shows five QTSS of residential area, one is in the root, other four are on four quadrants, respectively.

Table 1. An example of five QTSS
of residential area (land use code ld11)

$QTSS_{ld11}(0, 0) = (0\ 0\ 0\ 0\ 0\ 0\ 2\ 82\ 472\ 1456\ 3311)$
$QTSS_{ld11}(1, 1) = (0\ 0\ 0\ 0\ 0\ 0\ 6\ 64\ 252\ 525)$
$QTSS_{ld11}(1, 2) = (0\ 0\ 0\ 0\ 0\ 1\ 42\ 214\ 692\ 1729)$
$QTSS_{ld11}(1, 3) = (0\ 0\ 0\ 0\ 0\ 0\ 0\ 0\ 0)$
$QTSS_{ld11}(1, 4) = (0\ 0\ 0\ 0\ 0\ 1\ 34\ 194\ 512\ 1057)$

Where the $QTSS_p(L, N)$ is the QTSS of phenomenon P of the N_{th} node at the L level. These separate QTSS can be organized into a tree structure called as QTSS tree. Figure 2 shows a QTSS tree of the same data.

Figure 2. The structure of a QTSS tree
(only the top two levels of the QTSS tree is showing here)

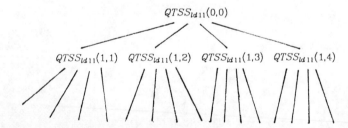

In the test GIS, an image has 2048*2048 pixels. Only the top four levels of the QTSS tree have been stored in this GIS. The QTSS at root has a length of 11. It means that the smallest candidate region can be turned out by the heuristic search has 1/64 size of the whole area.

A spatial heuristic search module

In artificial intelligence terminology, expert system is a type of software that can solve user's queries based on utilization of knowledges. For implementing a spatial heuristic search in GIS, a simple search module has be build in the GIS. The search module is essentially a simple analogue of expert systems.

When the module receives an user's query, it selects corresponding algorithms called 'rules' in terminology of AI. Then, the module executes these rules in the knowledge base. The results of the search module are candidate region(s), where the user's query may be satisfied with great possibility.

SPATIAL HEURISTIC SEARCH MODULE

The search module, similar to an expert system, includes three components: a control strategy, a set of roles, and a knowledge base. The knowledge base is the spatial knowledge base, mentioned above.

Control strategy

The heuristic search is implemented on the QTSS tree. only one kind of user's queries is dealing with, it is called as "EXTREME" type. User asks that find out the region(s) with the extreme value of a certain geographic measure, such as the highest density region, or the largest connective area, etc.. Different queries ask for different search methods. Most of them are following up-down breath-first strategy(Nilsson, 1983).

Searching algorithms (rules)

For example, two sets of rules are described in more detail.

Largest connective block finding --- An up-down breath-first search procedure is executed. This is a query of "EXTREME" type, so only one candidate node at each level is enough. The first candidate node is the root of the QTSS tree. The evaluation function QAREA (Eqn.2) is implemented on all sons of the candidate node to find the node with the maximum value. This node is the new candidate node at this level. Recurring this procedure until arriving a given level L_g. The candidate node at L_g is the final region. If a candidate node has nothing, the search procedure stops because there is not any the phenomenon at all in the area. The set of these rules is listed here in a formal representation.

(1). The search procedure is implemented in the QUAD tree of the phenomenon.
(2). At the beginning of the search procedure, the root node is the first candidate node.
(3). One candidate node at the level L has four son nodes at level L+1. the new candidate node at level L+1 is the one who has the maximum value of the area function "QAREA(QTSS)".

$$QAREA(QTSS_p(L+1,I)) = MAXIMUM(QAREA(QTSS_p(L+1,i))) \quad where \ i=1,2,3,4.$$

where the quadrant I is the new candidate node at the level L+1. This rule can be used recursively.
(4). At the given level L_g, the search procedure stops. The candidate node at this level is the final candidate region.
(5). If the QTSS of a candidate node have zero value from "QAREA" function, The search procedure stops. The result is that there is not any such phenomenon in the whole area.
(6). At the level L, if two or more than two nodes have same value from "QAREA(QTSS)". They are juxtaposed as candidate node at level L. All

their sons of these candidate nodes have to be evaluated for chose the new candidate node at next level. This rule can be used recurse.

Highest density area --- This is also a query of "EXTREME" type. Its search procedure is more simple. From the QTSS tree, all nodes at the given level L_g are directly compared each other. The node who has the maximum value from "QAREA($QTSS_p(L_g,I)$)" is the final candidate node. A formal representation of these rules is listed here.
(1). The search is executed in the QTSS tree of this phenomenon. The candidate node is among the all nodes of the QTSS tree at the given level L_g.
(2). The candidate node has the maximum value from "QAREA".

$$QAREA(QTSS_p(L_g,I)) = MAXIMUM\ (QAREA(QTSS_p(L_g,N))\ \ where\ N\ is\ all$$

(3). If the candidate node has zero value from "QAREA", the result is that There is not any such phenomenon at the whole area.

RESULTS ANALYSIS

The test GIS covers an area about half size of LANDSAT images. It is at Black Hill, South Dakota, USA. There are 2048*2048 pixels covering this area with 50*50 meter resolution. The GIS is working at the environment of VAX750/UNIX4.2.

For test time efficiency, the whole test area is cut into 8*8 grids. The user's query is to find the highest density block among 64 blocks. This procedure has only a few of steps for each pixel in a raster image. However, it has to calculate 64 areas from 64 block, and compare them to find the largest one. The CPU time spent to calculate the area of each block is about 0.0275 seconds. The whole procedure takes about CPU time 1.760 seconds(TA). As a comparison, the heuristic search to find out the candidate block needs 0.17 seconds(TC). Plus the CPU time to calculate the candidate block, the whole procedure takes 0.195 seconds(TH). The ratio of 0.195/1.760 shows that the improvement for time efficiency is obvious.

Second, the correction of the heuristic search is tested. Heuristic search does not always guarantee a correct result, But it often turns out better results(Pearl, 1984). Thus, besides the CPU time advantage, the correct probability of its results is tested. Among 23 types of land-use in the test area, the heuristic search is used to find the block with a highest density all types have correct results. While the heuristic search is used to find the largest connect block, 22 types have correct results. It means that this heuristic search usually gives right answers but not always.

Figure 2 shows the finding steps of the highest density block on the crops area. Figure 3 shows the finding steps of the largest connective block on the mixed forest area. Those candidate blocks at top several levels have been plotted on both figures. They show the procedure from finding rough approximation candidate blocks to point out more precise candidate blocks.

CONCLUSIONS

This research has provided a fast spatial heuristic search for large scale geographic information systems. It shows the potential feature of utilization of artificial intelligence in GIS. The quad tree spatial spectra(QTSS) is shown its capability of compact way to extract spatial information which can guide spatial

Figure 2. The candidates of the highest density blocks generated by the heuristic search at level 1, 2, and 3 of crops area at Black Hill.

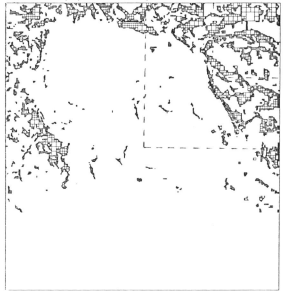

Figure 3. The candidates of the largest connective blocks generated by the heuristic search at level 1, 2, and 3 of mixed forest area at Black Hill.

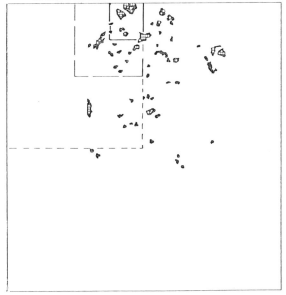

heuristic search for several kind of queries. The spatial heuristic search avoids the disadvantages of the conventional blind search. It trys finding the higher potential candidate area more directly based the knowledges.

From the above test results, the sum of the time spend for finding the candidate block and the time executing on the block is much less than the conventional blind search time. The correction probability is also acceptable.

The further research will pay more attention on the algorithms of using spatial knowledges to find different kind of candidate blocks.

REFERENCES

Chen, Z.T., 1984, "Quad Tree Spatial Spectra --- Its generation and applications", Proceedings of the International Symposium on spatial data handling, Zurich, Vol.1, pp.208-237.

Klinger, A. and Dyer, C.D., 1976, "Experiments in picture representation using regular decomposition", Computer Graphics and Image Processing, Vol.5, pp.68-105.

Nilsson, N.J., 1980, "Principles of Artificial Intelligence", Tioga Publication Co., California.

Pearl, J., 1984, "Heuristics - Intelligent search strategies for computer problem solving", Addison-Wesley Publs. Co., London.

Peuquet, D., 1984, "Data structure for a knowledge-based Geographic Information System", Proceedings of the International Symposium on spatial data handling, Zurich, Vol.2, pp.372-391.

Smith, T.R. and Pazner, M., 1984, "Knowledge-based control of search and learning in a large-scale GIS", Proceedings of the International Symposium on spatial data handling, Zurich, Vol.2, pp.498-519.

AN INTERIM PROPOSED STANDARD
FOR DIGITAL CARTOGRAPHIC DATA QUALITY

National Committee for Digital Cartographic Data Standards
Working Group II on Data Set Quality:

Nicholas Chrisman (chair)	University of Wisconsin
Charles Poeppelmeier (vice-chair)	Defense Mapping Agency
Wallace Crisco	Bureau of Land Management
John Davis	Kansas Geological Survey
Gunther Greulich	Survey Engineering of Boston
George Johnson	National Ocean Service
David Meixler	Bureau of the Census
Dean Merchant	The Ohio State University
George Rosenfield	U.S. Geological Survey
John Stout	Petroleum Information Corp.

ABSTRACT

The Working Group on Data Set Quality of the National Committee for Digital Cartographic Data Quality has progressed through cycles of defining the issues and evaluating alternatives. The group has now completed an Interim Proposed Standard. This standard specifies the contents of a quality report pertaining to digital cartographic information.

The purpose of the quality report is to provide detailed information for a user to evaluate fitness for a particular use. The general philosophy is characterized as "truth-in-labelling", because the goal is to communicate information on fitness for use, rather than fixing arbitrary numerical thresholds of quality. The subsections of the standard permit different levels of testing for the basic components of the quality report: lineage, positional accuracy, attribute accuracy, logical consistency and completeness. While testing will be at the option of the producer, the purpose of the quality report is best served by the most rigorous and quantitative information available. Each section of the report must also deal with issues of currency and other temporal information.

ALTERNATIVE ROUTES TO A MULTIPURPOSE CADASTRE:
MERGING INSTITUTIONAL AND TECHNICAL REASONING

Nicholas R. Chrisman and Bernard J. Niemann, Jr.
Land Information and Computer Graphics Facility
Department of Landscape Architecture
School of Natural Resources
College of Agricultural and Life Sciences
University of Wisconsin - Madison
Madison, WI 53706

BIOGRAPHICAL SKETCH

Nicholas Chrisman is Assistant Professor at the University of Wisconsin-Madison in the Department of Landscape Architecture and the Institute of Environmental Studies. He holds a BA in Geography from the University of Massachusetts Amherst and a PhD in Geography for Bristol University (UK). Before coming to Wisconsin, he was a programmer for ten years at the Harvard Laboratory for Computer Graphics and Spatial Analysis where he participated in the development of the ODYSSEY system and other software.

Bernard (Ben) J. Niemann Jr. is Professor at the University of Wisconsin-Madison in the Department of Landscape Architecture, the Institute for Environmental Studies and Cooperative Extension. He holds a BFA in Landscape Architecture from the University of Illinois and a MLA from Harvard University. Since 1970, he has been applying computer technology to land management and location issues. In 1982, he served as a member of the National Research Council Committee on Integrated Land Data Mapping.

ABSTRACT

The National Research Council report "Procedures and Standards for a Multipurpose Cadastre" sets admirable goals, but may not provide the most workable means to the ends. In the search for modernization, technical and institutional reasoning should be merged. Institutions can act as barriers to modernization, but they can also act in a more positive manner, shaping the technical needs. This paper considers three topics related to construction of a multipurpose cadastre: incrementalism, basic unit, and compilation procedures. In each case, we present some dilemmas that deserve further interdisciplinary discussion.

THE MULTIPURPOSE CADASTRE

To date, the development of automated spatial information systems has progressed dramatically but with limited planning and almost no sense of vision. Systems are purchased by agencies and companies to assist in their preestablished roles. This behavior is perfectly rational, but only in the short term.

In contrast to the general drift of computerizing current functions, there is a growing interest tied to the term "multipurpose cadastre". Starting from a group of committed professionals, activity has expanded through three reports of the National Research Council (occasionally abbreviated as NRC below). The Committee on Geodesy Panel on a Multipurpose Cadastre first published <u>Need for a Multipurpose Cadastre</u> (NRC, 1980). This panel, somewhat reconstituted, continued with <u>Procedures and Standards for a Multipurpose Cadastre</u> (NRC, 1983). During this period, the Committee on Integrated Land Data Mapping produced a related report <u>Modernization of the Public Land Survey System</u> (NRC, 1982).

Importantly, these NRC reports have brought attention to a factor missing in many past attempts at implementing and automated information system. This missing factor has been the cadastral or land ownership record (Clapp and Niemann, 1980). The need to define, spatially and legally, our rights in property has resulted in the surveying profession as we know it today in North America. Forward-looking surveyors and their international counterparts, have been instrumental in formulating the concept of a multipurpose cadastre.

In contrast, most geographers and planners involved in geographic information systems have overlooked or seen no value in cadastral information.
"The many persons whose job it is to regulate land apparently don't know or care about this most basic political and economic fact of the resource they guard." (Popper, 1978, p. 5)
The concept of a multipurpose cadastre offers a spatially based integration of property rights with the uses, values and distribution of natural and cultural resources. The introduction of this concept and its associated terminology has resulted in some confusion as to which term best describes a system which explicitly integrates ownership with other land-based information. Some argue (NRC, 1983; Marble, 1984) that a multipurpose cadastre is a subset of a land information system which is a subset of a geographic information system. Others argue that it is not quite that simple (Hamilton and Williams, 1984). Since most terminology is dependent on professional training and occupational outlook, the debate over terminology cannot be resolved without interdisciplinary negotiation.

For the purposes of this paper, we define a multipurpose cadastre to be interchangeable with multipurpose land information system. "Multipurpose" requires use and access to a mix of routinely maintained records which explicitly includes the cadastral or ownership records as well as some other, independently and spatially defined environmental records.

In order to create a public debate on the conclusions of the National Research Council panels, we offer some comments targetted on a few specific topics. Our intention is not to defeat the overall objectives, which we share, but to explore some technical alternatives which may be more attuned to institutional concerns. Our specific concerns cover three topics: incrementalism, the nature of the basic unit, and compilation procedures. In each section we will explore some dilemmas posed by the proposed "procedures and standards".

MERGING INSTITUTIONAL AND TECHNICAL REASONING

AUTO-CARTO symposia have concentrated almost exclusively on technology without considering institutional concerns. A few authors in related literature have emphasized the negative effects of human institutions on the new technology (Cook, 1969; Moyer, 1977; McLaughlin, 1975; Larsen and others, 1978). Others, particularly at the Ottawa symposium, have suggested that the wrong technologies may be being developed (Bie, 1983; Wellar, 1983). While recognizing that institutions can and do act as barriers to modernization, new technologies can be developed so that the institutional issues reenforce the technical ones.

INCREMENTALISM

Any information system development falls somewhere on a continuum of incrementalism. At one extreme, new procedures are brought in gradually through the effects of routine record-keeping. In the maintenance of land records, it is common to see innovations, such a a parcel index to deeds, established on a "day-forward" basis. A similar incremental approach applies to cartography, where scales of topographic maps shifted from 1:63360 to 1:62500 then to 1:24000, without immediately throwing out the old series. An incremental approach is often the cheapest to implement because it does not require reworking old information. Institutionally, it also produces least disruption. Extreme incrementalism has the problem that it may take a long time (if ever) for the system to be completely modernized.

At the other extreme, modernization can be performed by a wholesale replacement. We term this approach a "parachutted system" because a wholly functional system appears out of the blue to supplant the existing one. While parachutted systems may make sense on technical criteria, and even on economic grounds, wholesale replacement can create institutional friction. Often the reasons advanced for developing a system separately are that skills in the existing agencies are not sufficient (Hanigan, 1979). The experience of other large digital data base projects points towards the opposite conclusion (for example, Huxold and others, 1982). Even in a retrospective of the METROCOM project, Hanigan suggests in a section on "lessons learned" that "data collection, analysis and preparation tasks .. might better be done by the client whose data are being processed" (Hanigan, 1983, p. 147).

The NRC panel does not fit easily on the incrementalism continuum. When it comes to funding, the group is quite pragmatic and expects a twenty year process of implementation. Their fiscal incrementalism corresponds with a hesitance to require digital mapping. The "procedures and standards" talk of the promise of automated systems, but seem to leave substantial room for manual procedures as well. On certain other topics, the NRC panel rejects the incremental approach. This rejection is based on technical reasons explored below.

Computerization

The adoption of computer technology in too many cases has resembled a parachutted model. A whole new technology is acquired without preparing the staff and the rest of the institution. Many parachutted systems fail; for example many of the state natural resource inventory systems are now moribund (Mead, 1981). Such events are probably inevitable in the early years of a technology, but they should not continue.

Computerization is not automatically disruptive. If the initiative arises from existing agency staff who develop skills naturally and gradually, implementation can be positive for all concerned. One example of this process is the City of Milwaukee which rejected a centralized service bureau approach in favor of a model based on upgrading existing agencies (Huxold and others, 1982).

Computers may have been exotic and threatening in the past, but there seems to be much less resistance to them now. Our experience in dealing with local government officials is that we are on the verge of explosive growth in the adoption of automation in land records functions. Even the person making property maps in a county of 16,000 population is considering acquiring a computer drafting system. Rapid adoption of computer technology, if it is unmanaged, could result in each agency going its own way. The promise of a true multipurpose system could be lost.

Geodetic Reference

Adopting the adage for the Public Land Survey System, the NRC panel seems to expect survey and monumentation before anything else. There is little mention of continual improvements in geodetic control. In fact, the report explicitly cautions against underinvesting in the geodetic control network (NRC, 1983, p. 22). Considering the rapid introduction of satellite positioning (Bossler and Hanson, 1984), the panel's conclusions on new technology seem conservative.

We believe that the panel's approach has technical and institutional drawbacks. On the technical side, computer-based maps can be incrementally tied to control in ways that traditional graphic products cannot be. Survey control has logically preceded the rest of mapping efforts due to the nature of manual mapping technology. Even if a massive survey is done first, continual improvement in geodetic information is inevitable. Eventually, any system must be readjusted to fit new observations. On the institutional side, building a geodetic framework could create a dilemma in the competition for funds. The geodetic framework might have long-term benefits, but it might not be able to compete with more immediate needs. One way to diminish the conflict is to begin the implementation of the whole information system, expecting to upgrade the geometry as new control is added. Proponents of accurate positional information should not misinterpret an incremental approach. Descriptions of data quality (Chrisman, 1983) and tests for accuracy (Vonderohe and Chrisman, 1985) should be used to prevent uninformed use. There is a dilemma between quickly satisfying less stringent needs and the possibility that such a process may actually slow down more accurate surveys.

Cadastral Overlay
The report seems ambiguous on the issue of incremental development of the cadastral maps. In one section it recognizes the possibility of an incremental approach (NRC, 1983, p. 57). However, the report also makes the strong statement: "it is particularly important to resist the temptation to use only paper records of mapped locations as a basis for the development of a land-data system in order to save initial costs." (NRC, 1983, p. 22) Data quality is an important concern, which should not be minimized. Still, there may be alternative paths to achieve high quality. For example, in the West German Land of Hesse, existing parcel maps are being digitized for reasons of economy and speed (Eichhorn, 1984, p. 5). Institutionally, the existing paper records have their defenders in the bureaucratic structure. It may be more effective to have these groups participate in the modernization rather than having them fight the process.

NATURE OF BASIC UNIT

The multipurpose cadastre concept, as represented in the 1980 NRC report, uses the land ownership parcel as the basic building block (see Figure 1).

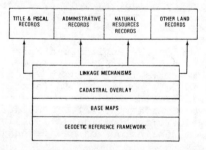

Figure 1: Components of a multipurpose cadastre (1980)

In the 1983 report, the diagram changed to recognize that some forms of land information are difficult to attach to parcels (see Figure 2). The 1983 report still contains the same basic ingredients: geodetic reference, base maps, cadastral overlays and registers of attributes associated with parcel identifiers. The report tries to explain the differences by assigning all non-parcel information to other components of a more broadly defined land information system, not the multipurpose cadastre.

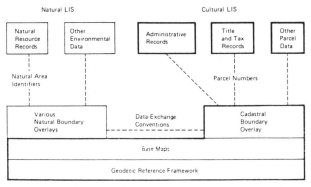

Components of a multipurpose cadastre (in heavy outline) as the foundation for Land-Information Systems (LIS's).

Figure 2: Components of a multipurpose cadastre (1983)

The definition of a cadastral parcel as "a continuous area within which unique, homogeneous interests are recognized" (NRC, 1983, p. 14) runs into institutional trouble in defining those interests. In our experience of building cadastral information for the Dane County Land Records Project (Chrisman and others, 1984), the assessor, the surveyor and the zoning administrator defined parcels somewhat differently. This situation of autonomous behavior, with each actor behaving rationally within a limited horizon, is a part of the land records problem (Portner and Niemann, 1983). The NRC seems to favor the immediate imposition of a unified system. It may be more reasonable to have the divergent approaches merge over time, with the information system as a positive catalyst.

Beyond these institutional difficulties, parcels are hard to define because property rights are affected by a patchwork of environmental concerns with a spatial expression. In Wisconsin, for example, counties perform zoning including floodplains, protect farmland through protective zoning, and will produce soil erosion control plans; the state regulates water pollution, shoreland use, and wetland preservation. This array of public interests in private lands complicates the cadastral concept. Similarly, Eichhorn (1984, p. 9) finds "the parcel as the smallest reference unit loses importance" in rural areas. Environmental information is absolutely crucial to describe the actual interests in the land.

Given that no particular object is indivisible, it may seem to make technical sense to create the currently undivided units and manage them. This approach characterizes the thinking in the early development of topological information systems (Chrisman, 1975) and in the Integrated Terrain Unit phase (Robinove, 1979; Dangermond, 1979). However, managing the overlaid polygons creates difficulties for recording data quality and for maintenance (Chrisman, 1983). A new alternative is needed.

Placing the cadastral parcel as the central focus of an information system also projects the wrong kind of institutional arrangement. A multipurpose system should accept different views of what exists. These different views may be fundamental to keeping each group contributing to the overall system.

Instead of a parcel-based system, a system can be based on handling the features identified by each contributing agency. Each agency is then responsible for their data. Responsibility has to be carefully examined to remove duplication and to ensure cooperation. Such a system could be termed "layer-based". It would have no single permanent basic unit, merely the amalgamation of the units distinguished by the current participants. Individual layers would be integrated as needed (see Figure 3). Of course, this approach depends on reliable integration through geodetic control, but so does any multipurpose mapping system.

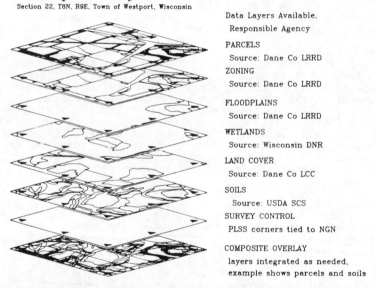

Figure 3: The process of integration through overlay

Another crucial requirement is reliable, robust software for polygon overlay. While this may have been unavailable in the past, recent software systems have conquered this limitation (Dougenik, 1980; Dangermond, 1983). The lesson here is that technical problems, such as integrating layers, can be solved. Systems should not be designed for our imperfect technology, but should reflect the inherent structure of the problem.

Organizationally, a layer-based system is more modular than the scheme proposed on the parcel base. There would be less need for centralization and reorganization. In the long-term a layer-based system would rely on "horizontal" cooperation rather than "vertical" authority.

COMPILATION PROCEDURES

The final issue in this discussion concerns the role of base maps. The NRC report strongly advocates the creation of detailed planimetric base maps as a necessary preliminary to cadastral mapping. The reasoning seems to be mixed.
"Good planning and engineering practice dictate the preparation of large-scale maps as a basis for sound community development and redevelopment." (NRC, 1983, p. 37)
This important multiple purpose intention may not be applicable in the extensive rural areas with limited engineering needs. In other passages in the report, the cadastral map is seen as an "overlay" related to the geodetic control through the base map. While this approach may have been necessary with previous mapping technology, we believe that this approach misses some of the power of modern digital systems. It also may create institutional conflict.

No mapping professional is opposed to accurate large-scale maps. Greater detail and higher accuracy are assumed to be useful eventually. However, it is easy to place one's own biases into decisions about the features required on a base map. The features on a traditional base map are often selective, not an exhaustive inventory useful in multipurpose analytical situations. In a truly multipurpose system, the needs for base information will vary enormously. An assessor needs an inventory of buildings and thus wants to see through the suburban forest canopy. A wetland expert needs infrared images of a different time of year. Agricultural programs require coverage during the crop cycle. All these needs for "base information" are valid, but they conflict. From reading the NRC report, a rural county in the US might be scared off. We agree that accurate engineering maps form a possible layer in a multipurpose system. However, we do not think that it is required as a "base" to compile the cadastral maps.

Kjerne and Dueker (1984) contrast the methods of constructing a cadastre directly from legal descriptions and surveys with the planimetric base map approach. Although they support the base map approach, the registered survey approach is the common basis for the property systems in some of the countries with functional cadastres (for example, South Africa, and Australia). Kjerne and Dueker (1984) place our research project in an ill-defined category as "implicit base map", but we disagree. Every layer in a multipurpose system must be referenced to the geodetic control network through its coordinate system. No layers are more central than others, although some may be compiled with indirect reference to the geodetic measurements. In a fully operational multipurpose system (which we suspect exists nowhere yet), features in one layer intended to be identical to features in another can be enforced through software, not simply by relying on compilation steps.

In any event, there are substantial legal problems with compilation of property boundaries from physical features. Although fences and other visible features can be used as evidence for boundaries, they are not automatically correct. Other information, only found in deeds and surveys, along with monuments take precedence.

Our conclusion is that the components of a multipurpose cadastre should be revised. The geodetically defined coordinate system plays a central role, serving as the basis for integration of diverse layers through polygon overlay (the ill-defined "data exchange conventions" in the NRC report, see Figure 2). Layers will be compiled by different approaches, such as direct photogrammetry based on geodetic control, or compilation on existing map products.

One important method has particular application to the Public Land Survey portion of the United States. The treatment of PLSS section corners is a major technical issue, because the NRC "procedures and standards" insist on legal remonumentation and accurate survey as the basis of the whole process. We believe that they underestimate the power of digital technology to transform coordinates and warp data to fit new control. In one passage, the report talks about wasting all of the investment on inaccurate information (NRC, 1983, p. 22). With traditional mapping, this may be true, because distortions would not be removed. In a digital environment, however, it is possible to remove systematic errors at a later time, once they are detected. Although it runs counter to the "procedures and standards", the Panel on Integrated Land Data Mapping (NRC, 1982) suggested a focus on the PLSS corners, beginning with digitizing information from the 1:24000 quadrangles. While this may seem unacceptably crude for urban engineering projects, it may prove sufficient to ensure that resource information and other data for rural areas are at least put into a spatially referenced framework leading towards a multipurpose system. In our experience with tying existing soils maps to a geodetic framework, we found the PLSS corners to be more useful than the other "well-defined" features normally shown on a base map. This result is specific to a particular landscape, but it indicates a role for the PLSS corners in a future digital multipurpose system.

SUMMARY

A multipurpose cadastre will be achieved more smoothly and operate more effectively if the technical and institutional components are in tune. We suggest that data sources must be seen as incremental. Computerization, by contrast, should be expected to be rapid. A true multipurpose system will have no simple basic unit; it must be based on separate layers maintained be cooperating agencies. The layers must rely on the coordinate system, not a base map, for integration.

ACKNOWLEGMENT

These ideas evolved through discussions with students and visitors in the Seminar on the Multipurpose Cadastre, and particularly with those in Landscape Archtecture 875 in Spring 1984. Partial support comes from USDA Hatch. The opinions are strictly those of the authors.

REFERENCES

Bie, S.W. 1983, Organizational Needs for Technological Advancement - The Parenthood of Autocartography Revisted, paper presented at AUTO-CARTO 6

Bossler, J.D. and Hanson, R.H. 1984, The Impact of VLBI and GPS on Geodesy, Surveying and Mapping, vol. 44, p. 105-113

Chrisman, N.R. 1975, Topological Data Structures for Geographic Representation: Proc. AUTO-CARTO 2, p. 346-351

Chrisman, N.R. 1983, The Role of Quality Information in the Long-Term Functioning of a Geographic Information System: Proc. AUTO-CARTO 6, vol. 1, p. 303-317

Chrisman, N.R., Mezera, D.F., Moyer, D.D., Niemann, B.J., Vonderohe, A.P. 1984, Modernization of Routine Land Records in Dane County Wisconsin: Implications to Rural Landscape Assessment and Planning: URISA Prof. Paper 84-1

Clapp, J.L. and Niemmann, B.J. 1980, Land Records: Costs, Issues and Requirements: p. 1-26 in The Planning and Engineering Interface with a Modernized Land Data System, Am. Soc. Civil Eng., New York

Cook, R.N. 1969, Land Law Reform: A Modern Computerized System of Land Records: University of Cincinatti Law Review, vol. 38, p. 385-448.

Dangermond, J. 1979, A Case Study of the Zulia Regional Planning Study, Describing the Work Completed: Harvard Library of Computer Graphics, vol. 3, p. 35-62

Dangermond, J. 1983, ARC/INFO, A Modern GIS System for Large Spatial Data Bases: Proc. ACSM Fall Meeting, p. 81-89

Dougenik, J.A. 1980, WHIRLPOOL: A Geometric Processor for Polygon Coverage Data, Proc. AUTO-CARTO 4, 304-311

Eichhorn, G. 1984, Interaction of Different Land Information Systems, Proc. FIG Symp. on LIS, Edmonton

Hamilton, A.C. and Williamson, I.P. 1984, Critique of the FIG Definition of a "Land Information System", Proc. FIG Symp. on Land Information Systems, Edmonton

Hanigan, F.L. 1979, METROCOM: Houston's Metropolitan Common Digital Data Base: Surveying and Mapping, 39, p. 215-222

Hanigan, F.L. 1983, Houston's Metropolitan Common Data Base: Four Years Later: Surveying and Mapping, vol. 43, p. 141-151

Huxold, W.E., Allen, R.K., Gshwind, R.A. 1982, An Evaluation of the City of Milwaukee Automated Geographic Information and Cartographic System in Retrospect, paper presented at Harvard Computer Graphics Week

Kjerne, D. and Dueker, K.J. 1984, Two Approaches to Building the Base Map for a Computer-Aided Land Records System, Proc. URISA 84, p. 233-244

Larsen, B.J., Clapp, J.L., Miller, A., Niemann, B.J., Ziegler, A. 1978, Land Records: The Cost to the Citizen to Maintain the Present Information Base, A Case Study of Wisconsin, Dept. of Admin., Madison

Marble, D. 1984, Geographic Information Systems and Land Information Systems: Differences and Similarities: paper presented at FIG Symp. on Land Information Systems, Edmonton

Mead, D.A. 1981, Statewide Natural-Resource Information Systems - A Status Report: J. Forestry, vol. 79, p. 369-372

McLaughlin, J.D. 1975, The Nature, Design and Development of Multipurpose Cadastres, Ph.D. thesis, U. Wisconsin - Madison

Moyer, D.D. 1977, An Economic Analysis of the Land Title Record System, Ph.D. thesis, U. Wisconsin - Madison

National Research Council, 1980, Need for a Multipurpose Cadastre, National Academy Press, Washington DC

National Research Council, 1982, Modernization of the Public Land Survey System, National Academy Press, Washington DC

National Research Council, 1983, Procedures and Standards for a Multipurpose Cadastre, National Academy Press, Washington DC

Popper, F.J. 1978, What's the Hidden Factor in Land Regulation: Urban Land, p. 4-6

Portner, J. and Niemann, B.J. 1983, Belief Differences among Land Records Officials and Users: Implications for Land Records Modernization, paper presented at URISA, Atlanta

Robinove, C.J. 1979, Integrated Terrain Unit Mapping with Digital Landsat Images in Queensland, Australia: USGS Prof. Paper 1102, Reston

Vonderohe, A.P. and Chrisman, N.R. 1985, Tests to Establish the Quality of Digital Cartographic Data: Some Examples from the Dane County Land Records Project: Proc. AUTO-CARTO 7

Wellar, B.S. 1983, Achievements and Challenges in Automated Cartography: Their Societal Significance: Proc. AUTO-CARTO 6, vol. 1, p. 2-13

Methods and Software Systems for
Computer-aided Environmental Mapping

by
Chui Weihong, Lin Huagiang, and Wang Shu-jie

Institute of Remote Sensing Application
Academia Sinica
China

Local address: % Prof. Robert D. Rugg
Department of Urban Studies and Planning
Virginia Commonwealth University
Richmond, VA 23284 USA

phone: 804-257-1134

(through May, 1985)

Computer aided mapping or the urban environment differs from other environmental mapping. It was set up and developed according to the characteristics of the spatial distribution of each environmental factor in the artificial urban environment, and according to the needs of planning, management and evaluation of that environment. Based on the practical experience of producing the <u>Atlas of Environmental Quality</u> in Tianjin in recent years, this paper summarizes the methods and software systems used in computer aided mapping of the urban environment.

The primary methods of urban environmental mapping include locating environmental factors, data processing, urban environmental dynamic analyses, computer aided mapping, urban environmental system analyses and evaluation; also multielement multilayers and multistructure mapping methods for contract analyses of urban areas. The six types of map structures we have used are grid cell, isodensity, scatter (distracted) point, polygon, network and three dimensional map. These six structures can be linked or transformed by urban environmental mapping systems software.

This paper introduces the urban environmental mapping software system. The system was set up depending on the needs and characteristics of urban environmental mapping. It is written in FORTRAN IV and Assembler languages, and is compatible with minicomputers such as PDP 11/23-24, NOVA-3D and DJS-130,140. The system is able to deal with environmental monitoring data, investigative and single statistic maps, human air photo interpretation, environmental system analysis and evaluation data. It is able to position and quantify environmental factors and brings data into the urban environmental data system-- to establish data files, to organize and manage data, to classify and analyze data, and to draw the six structural types of map by computer. It provides three kinds of graphic output: plotter output, line print output, and display in a VT-100 terminal.

The basic structure and main contents of the software system include four parts: (1) data treatment and analysis software, for instance, polygon digitizing data treatment program, data testing and correcting program, data organization management program, map digital analysis program; (2) basic mapping software, which can offer address information, control information and mapping information to control and drive a plotter; (3) mapping function software, offering basic functions including drawing map symbols, writing numerals and characters, drawing different lines; and (4) urban environmental mapping software, which can be used not only to draw the six structureal types of map described above, but also to draw map projection grids, map scale and other essential map elements.

This paper also goes further into the process of urban environmental mapping in connection with the work of the <u>Atlas of Environmental Quality in Tianjin</u>.

STRATEGIES FOR SPATIAL DATA COMPRESSION *

Keith C. Clarke
Hunter College
New York, NY 10021

ABSTRACT

As part of the research on remote workstations and networking for NASA´s Land Data Pilot Information System the issue of data compression is being investigated in the context of spatial data. Spatial data compression can be logical, physical, information retaining or information reducing, and often requires different strategies than non-spatial data compression. Several different compression methods are considered in the context of polygonal land use data, and a test system is described which uses these methods to send data between a mini and a microcomputer. Spatial data compression is shown to have a number of potentially important applications in automated cartography.

INTRODUCTION

Digital data bases have increased in size at a remarkable rate since the emergence of the Geographic Information System as a management and research tool. This increase in data volume is the result of two forces; the conversion of the immense reserve of analog map data into digital form, and the new, usually higher resolution sensors and mapping instruments now coming into use. The transition of remotely sensed data, for example, from Multispectral scanners to the Thematic mapper to the French SPOT sattelite, and in the future to the Shuttle Imaging Spectrometer, represents a data density increase from 6.25 to 77.78 to 100.00 to 1422.2 bytes per hectare on the ground, a 227 fold increase.

Faster computers and new mass storage devices have alleviated the problem to some extent, but at many stages in the processing of cartographic information bottlenecks exist where the rate of data flow through communications links can fall to 1200 baud and less. Examples of these bottlenecks are magnetic tape and floppy disk input/output operations and communications between nodes in a network. One solution to the problem of low speed data transmission is the use of data compression. Data compression methods have long been used in the handling of textual and other non-spatial data, but few spatial data compression techniques are used on a regular basis. One notable exception is in image processing, where compression of array-type data is well understood and frequently used.

---

* Funds for the support of this study have been allocated by the NASA-Ames Research Center, Moffett Field, California, under Interchange No. NCA2-1R305-401

This paper examines the potential for data compression in spatial data processing, and considers the suitability of different data structures for compression. Finally, a test system for compression of polygonal data is described, as well as some of the algorithms behind the compression methods implemented in the system.

THE WORKSTATIONS CONCEPT

The National Aeronautics and Space Administration, as part of the Land Data Pilot Information System, is investigating the development of a computer network to provide data links between the NASA Technical centers, NASA headquarters, other NASA units, and principal investigators on land-based NASA research projects. Similar projects are already under way for the climatic, planetary, and oceanographic data from the manned and unmanned space programs. One of the problems specific to the Land Data Pilot Information system is the broad variety of needs for system users, and the huge volume of land-related data available.

The network has no shortage of computing power, but the major problem is seen as one of access at the local level, since investigators are geographically dispersed and may have a limited computing environment. At the users end, the working arrangement is assumed to be a microcomputer-based workstation, with limited storage capability (less than 12 megabytes) perhaps on a hard-disk, minimal graphics capability, and other assorted peripherals. As a standard at the node end, the VAX 11/780 has been suggested, with VAX VMS or Unix as the operating system. This favors microcomputers with Unix-like operating systems, or VAX VMS communications capabilities.

The communications link for the workstation would normally consist of microcomputer software which is commercially available and uses telephone lines at 300 or 1200 baud. Several packages exist, including the KERMIT system (DaCruz and Catchings, 1984). The standard communications link is asynchronous serial, using the RS-232C EIA standard, and may or may not involve timing, packet switching, buffering, or error-checking within the protocol.

Given these low transmission speeds, two data problems exist. First, users require the ability to browse through directories and get previews of selected data and data subsets. Secondly, when data have been selected, they have to be transmitted from the main node to the workstation, preferably with minimal error. It is the first problem to which the current research is addressed. However, this problem necessarily involves considering spatial data structures and how data volume reduction and data compression can be used to facilitate a preview capability for workstations.

DATA VOLUME REDUCTION AND DATA COMPRESSION

In computer science, data compression is subdivided into logical and physical. Logical data compression using spatial data involves changing the geocodes to minimize redundancy. An example might be leaving off the first few digits of a U.T.M. map reference to limit the number of bytes necessary to represent the eastings and northings. The U.S.G.S., in its Digital Land Use GIRAS files, uses an arbitrary coordinate system based on a local origin to achieve the same goal. Into this category of logical compression also might fit the actual way that the digital map data are converted to computer readable files, perhaps using different file structures (random-access, sequential, etc.), or digital representations (ASCII, EBCDIC, binary).

Physical data compression involves an alteration in the logical data structure to reduce the total number of bytes needed to represent the data set. An example might be the conversion between vector and raster data structures, or the conversion of spatial data to a volume reducing transform.

Cartographic data does not fit neatly into these categories due to the added complications of data resolution and data dimensionality. In the context of textual data, where sequence is critical, leaving out the redundant blanks, punctuation and every n-th letter would reduce the information content to zero. Yet in the spatial context, point elimination is a necessary and valid way of reducing the volume of digitized line data, and a large number of point elimination methods have been devised (McMaster, 1983). Data volume reduction in a spatial context is equivalent to cartographic generalization. Generalization normally occurs when either of two processes take place, either data are sampled and/or data are reduced in dimensionality or scaling. Dimensionality involves the traditional division of cartographic data into point, line, area, and volume data. Scaling involves the type of measurement, either nominal, ordinal, interval, or ratio.

In terms of data transmission, generalization can be thought of as information loss. In the spatial context we can, and indeed must, generalize to clarify map information, so generalization may aid in the communication of cartographic information. This gives an increased flexibility to spatial data compaction, especially since we need only be concerned at this stage with a "preview" or "quick look" capability, which can be achieved with a high degree of generalization.

For this purpose, we will categorize spatial data compression as information loss compression and information retention compression. In many cases, similar techniques may be used for each, and therefore may be of use in both preview and full data transmission cases. The distinction is that information retention compression allows full reconstruction of spatial data after compression while information loss compression may preserve only spatial relations or generalizations.

The volume of spatial data may be considerably reduced by information loss compaction. However, this remains as physical compaction since no change has taken place in the actual data structure. Data structure is taken to mean the abstract representation of the cartographic objects and their attributes, which may or may not have an impact on the physical storage characteristics of the data. Physical spatial data compression, therefore, may be defined as an alteration in the abstract representation of cartographic objects and their associated attributes, with the intent of reducing the total number of bytes needed for their physical digital representation.

POLYGONAL DATA COMPACTION

To investigate the problems associated with these various types of compaction, a single type of spatial data was considered in isolation. Since large quantities of data are encoded in the polygon, arc, node system (Fegeas et al., 1983) the basic unit of a plane polygon was selected. The polygon in this context is a nominal areal feature and is assumed to be a vector chain representation of a set of non-overlapping, contiguous, closed polygons which may or may not contain holes.

To contrast information loss/retention and physical/logical data reduction, consider a polygon consisting of three chains (figure 1). To logically compact this file, we could rewrite the file using Huffman coding (Rosenfeld and Kak, 1982; Held, 1983). This technique uses an adaptive compaction method to reduce the file size. The ASCII file containing the chain and chain relational (linkage) data is analyzed in terms of its content. If the file contained 50% ASCII code 32 (blank), the probability associated with encountering this code in the file would be 0.5. Similarly, if we had only one negative sign (-) in the file, perhaps representing a counter-clockwise chain link, and the file was one kilobyte in size, the probability of finding a negative sign (ASCII code 45) would be 1/1024.

Huffman coding starts by determining the lowest of these probabilities, and links this with the next lowest, summing the two probabilities to form a hierarchical link for each step (figure 2). This process continues until all possible values are linked and the total probability is unity. The hierarchical tree is then followed, coding a left turn as a one and a right turn as a zero. The string of digits leading to each ASCII code is then its Huffman code. The codes are arranged so that the lowest number of bits are associated with the most frequent ASCII codes and vice versa. A key property of the Huffman code is that the codes can be instantaneously translated upon reception by matching with the possible codes. This process uses the hierarchical tree backwards, with each bit leading the decoding closer to the actual ASCII code.

Some operating systems contain system utilities for adaptive Huffman coding, for example the "compact" command in Unix. Regular use of Huffman coding can greatly reduce the storage needed for files, and when both a transmitting and receiving network node have the compact and decompact utilities the data transmission bottleneck can be avoided to some extent. It may be that a more spatial Huffman coding could be developed, to take advantages of particular spatial data properties for example. Freeman codes for land use polygons in a north-south oriented township and range system, where fields follow the lines, may be overweighted in the orthogonal rather than the diagonal axes, and therefore could be compacted efficiently with Huffman codes. It is notable that

Figure 1: Types of Spatial Data Compression

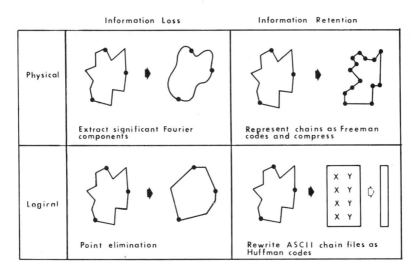

Figure 2: Huffman Coding

ASCII Code	Huffman Code	#times in file	Probability	Links [code 1 for left branch, 0 for right]
32 ( )	0	512	0.500	
50 (2)	1000	57	0.056	
55 (7)	1001	60	0.059	
52 (4)	1010	50	0.049	
54 (6)	1011	45	0.044	
56 (8)	11000	43	0.042	
57 (9)	11001	44	0.043	
53 (5)	110110	39	0.038	
45 (-)	110111	1	0.001	
49 (1)	11011	40	0.039	
51 (3)	1110	62	0.061	
43 (0)	1111	72	0.070	

the less random the data, the more useful the Huffman coding is for data reduction, since most spatial data are far from random.

An alternative (figure 3) is to convert the chains to Freeman codes (Freeman, 1974). To some extent this represents a resolution change, and is equivalent to a vector to raster conversion, since the Freeman code has a finite resolution. The Freeman codes can be represented in three bits, and so may produce a volume reduction. However, if the fourth bit is used as a flag, a zero flag can be used to mean a regular Freeman code, while a one flag can be used to signify a repetition of the previous code by the number of times represented in the next three bits. This means that sequences of seven identical Freeman codes can be reduced to eight bits, as opposed to the twenty-one bits necessary without compaction. Huffman and Freeman coding are information retention compression methods, though by decreasing the resolution (increasing the length of the vectors) Freeman coding can be made information reducing.

Logical information loss methods include line smoothing. It is possible to use a variety of methods, each of which uses different criteria for point elimination. A crude method is to leave out every n-th point. More sophisticated techniques consider points as "significant", (i.e. high information content) on the basis of the angle between successive segments or the displacement from a straight line between chain end points. McMaster (1983) contains a review and empirical evaluation of these methods.

Reducing the number of points in a chain is equivalent to generalization, just as the cartographer smooths a coastline by eye to reduce minor intricacies. It is also equivalent to the generalization produced by changing to a smaller map scale. If this scale change retains self-similarity, i.e. if the basic shape of the chains is repeated in smaller and smaller sub-segments as we go to larger scales, then it may be possible to invert the point elimination process by fractal enhancement. Dutton (1981) has proposed such a method. Although point elimination and inversion by fractal enhancement does represent data compression, and may indeed produce maps which are perfectly adequate for preview purposes, the final information is not the same as the original, and this method would have to be classified as information-loss compaction. In addition, not all types of lines on maps show self similarity at different scales.

An important point to note is that when the data are chain encoded, the end points of chains should not be included in the elimination process. This means that for a chain containing N points and N-1 segments, elimination of every n-th point will eliminate $(N-1)/n$ points. For large numbers of points, the figure of merit for the compression (defined as the ratio of the length of the compressed to the length of the original data string) is approximately $1 - 1/n$.

Physical information loss compaction techniques are numerous, but not all of them are invertible. Shape parameterization, such as the computation of form ratios, attempts to derive descriptive values for

Figure 3: Freeman Codes

Vector	Freeman Code decimal	Freeman Code binary	Index in array[x][y]
→	0	000	x+1, y
↗	1	001	x+1, y+1
↑	2	010	x, y+1
↖	3	011	x-1, y+1
←	4	100	x-1, y
↙	5	101	x-1, y-1
↓	6	110	x, y-1
↘	7	111	x+1, y-1

Note: Code has resolution of one unit in x,y but $\sqrt{2}$ units diagonally.

Table 1: Compaction Estimates
Sample Land Use Data

	GIRAS format	Polygon List
Full Data Set	137,579 bytes	211,680 bytes
Point Elimination n=2	95,239 (69.2%)	118,520 (56.0%)
Huffman Coding	81,227 (59.0%)	124,976 (59.0%)
Freeman Code Compaction	86,140 (62.7%)	107,163 (50.6%)
Relations only + nodes	43,158 (31.4%)	-------
Fourier Abstraction	-------	83,700 (39.5%)

polygons, which are often single scalar values. Boots and Lamoureaux (1972) provided an annotated bibliography on these measurement methods. None of these methods involves inversion, so that although a polygon is reduced to a single real value, there is an incomplete mapping of this value back onto a particular shape. In addition, the parameters are designed to be independent of the size, location and orientation of the polygon, and these are all properties which a reconstruction of the polygon needs to preserve.

Considerably more information is contained in the end point and chain linkage information for the polygons. Regeneration of simple polygon networks from this data alone, equivalent to point elimination with n = 1, often can yield a satisfactory preview map. Topologically complex sets of polygons, however, are not well represented. Islands, for example, close onto points, and concave polygons may have crossed segments. One experiment, using the boundaries of the Departements of France, was able to reproduce a polygon set from contiguity and spacing information alone (Kendall, 1971) and may have potential as a method for spatial data compression.

One reversible shape-based technique was adapted from the work of Moellering and Rayner (1979), who considered a series of digitized points in terms of two real series and their Fourier transform. The real series constitutes the x and y values of the polygon outline, and should be of even spacing. In the transform, it is possible to extract the harmonics which contribute most to a polygon's shape, and to save the Fourier coefficients for these harmonics. An approximation of the polygon outline can then be reconstructed from the Fourier coefficients. Depending of the complexity of the polygon, comparatively few, sometimes as few as four or five, harmonics can produce a very good approximation of the polygon.

While the complex Fourier series representation is more elegant, the working approach taken here was to perform two separate Fourier transforms, one for the x and one for the y axis. The discrete Fourier transform was used, since the limitations on the string length imposed by the Fast Fourier transform algorithm was found to produce overflow into the padded part of the transform. The inverse transform involves summing over a reconstructed series, with zeros for non-significant Fourier coefficients, and the series length being passed as part of the compressed data.

TEST DATA SET

The above mentioned data compression methods are currently being tested on a U.S.G.S. digital land use data file, at the scale of 1 to 250,000. Section 4, a 20 minutes of latitude by 36 minutes of longitude area in south-western Massachusetts was used, since this area contained the smallest number of land use polygons of the 15 for the whole one degree by two degree quadrangle. The GIRAS data file for this area contains 465 land use polygons, consisting of 1,260 arcs or chains and 10,161 coordinate pairs.

The GIRAS files contain several non-essential attributes, such as the x and y ranges of each arc. Without these attributes, the GIRAS file in ASCII random access format, blocked to 80 characters per line with line feed characters, constitutes 137,579 bytes. Based on a simple topology, and leaving out the double-counted interior island polygons used within GIRAS, this file can be rewritten as closed polygon lists in 211,680 bytes.

Some preliminary and expected compactions and their figures of merits are listed in table 1. The figures of merit are stated as percentages of original data set size in bytes. For the GIRAS data, a low of 31.4% derives from the elimination of all points along each arc. The implications of this structure for data display make this a preview option only. Similarly, the 39.5% expected figure of merit for the Fourier abstraction method may produce gaps and slivers along the edges of the reconstructed polygons.

In the case of Freeman code compaction, a byte oriented approach was taken. A C-language program named SQUEEZE reads Freeman code files and writes sequential ASCII files. This strategy loses some of compaction of the bit-based methods, but allows a code to be repeated nine times instead of seven. Preliminary tests have given figures of merit of about 50%, and the sizes of the Freeman coded data sets have been approximately the same as the polygon list sets at the same resolution.

The Fourier abstraction method is still undergoing testing to determine the optimal number of spatial harmonics necessary for regeneration of polygon shapes. A single harmonic in x and y requires the storage of at least two pointers to the harmonics, and two real values. Including identifier and attribute information takes the number of bytes necessary to store a polygon with five harmonics up to 180. Five harmonics were used to generate the estimates, and no more than ten may be necessary to rebuild the polygon quite precisely. Finally, no algorithm has yet been devised to build polygons from relational information.

TEST CONFIGURATION

The test data sets reside on a VAX 11/730, runnning the Unix operating system. To simulate a network, the VAX is connected to a SAGE IV microcomputer running the IDRIS operating system, a Unix lookalike. Considerable differences in the two Unix versions have made the Unix "cu" (call Unix) command inoperative. Alternate communications software is currently being installed. The link will be by 1200 baud modem and telephone line, although the two computers are actually in adjoining rooms. Final displays will be produced on a pen plotter, with the polygon list as a destination format. Software for both the VAX and the SAGE is being written in the C-language, using portable calls so that transportation problems to other systems are minimized.

The figures of merit for the data compressions are only one of three sets of test figures being compiled. The other two are communications errors and compaction/ reconstruction processing time. File transfer time will be assumed proportional to file size in bytes.

DISCUSSION

For networking, a preview capability may favor information loss compression, with the benefit of not having to transform between data structures. How effective a highly generalized cartographic image is is likely to determine the method chosen for the Land Data Pilot or any other compaction system. Fourier abstraction and reconstruction from relational data offer high degrees of compaction, but the "fuzzy" images they produce should be regarded as preview or throw-away only, and will not satisfy cartographic purposes.

Certainly, Huffman coding offers important savings in transmission time and storage at all levels. However, this method depends upon the compatibility of different operating systems, something which only occasionally exists. A double compaction, perhaps a combination of Freeman code and then Huffman compaction may be a highly desirable combination. Also of importance in the communications link is processing time. A suitable system should concentrate processing at the VAX end of the transmission, minimizing the work load on the slower workstation.

CONCLUSION

Clearly much work remains to be done, and the Land Data Pilot research is in its infancy. Already apparent, however, is the potential for applications of data compaction in the spatial sciences, where data sets are typically very large. In analytical cartography, the emphasis has often been upon the elegance of a data structure rather than its storage requirements. With such a rapid increase in the volume of spatial data, and the need to integrate these data into Geographic Information Systems, cartographers and spatial scientists must begin to incorporate this thinking into new systems. In particular, as the volume of archived, general purpose spatial data begins to expand, the need to compress spatial data will become self evident. Rather than waiting for the day when the sheer volume of data is overwhelming, cartographers should anticipate the need, and become involved in investigating the relative merits of the different strategies for spatial data compression.

REFERENCES

Boots, B. and Lamoureaux, L. (1972) Working Notes and Bibliography on the Study of Shape in Human Geography and Planning , Exchange Bibliography 346, Council of Planning Librarians.

DaCruz, F. and Catchings, B. (1984) "KERMIT: A file transfer protocol for universities", Byte , June 1984, 255-278.

Dutton, G. H. (1981) "Fractal enhancement of cartographic line detail", The American Cartographer , 8, 23-40.

Fegeas, R. G. at al. (1983) Land Use and Land Cover Digital Data Geological Survey Circular 895-E, Reston, Va., U.S.G.S.

Freeman, H. (1974) "Computer processing of line-drawing images", <u>Computing Surveys</u>, 6, 1, 57-97.
Held, G. (1983) <u>Data Compression</u>, New York, J. Wiley.
Kendall, D. (1971) "Construction of maps from odd bits of information", <u>Nature</u>, 231, 158-159.
McMaster, R. B. (1983) "A mathematical evaluation of simplification algorithms", <u>Proceedings</u>, <u>AUTOCARTO 6</u>, Ottawa, Canada, 2, 267-276.
Moellering, H. and Rayner, J. N. (1979) <u>Measurement of Shape in Geography and Cartography</u>, Numerical Cartography Laboratory, Ohio State University, Columbus, Ohio.
Rosenfeld, A. and Kak, A. C. (1982) <u>Digital Picture Processing</u>, 2, Computer Science and Applied Mathematics, New York, Academic Press.

VEHICLE NAVIGATION APPLIANCES

Donald F. Cooke, Chairman
Geographic Data Technology, Inc.
13 Dartmouth College Highway
Lyme, New Hampshire 03768

BIOGRAPHICAL SKETCH

Donald Cooke was a member of the Census Use Study team that developed the DIME method of map encoding in 1967. He was a founder of Urban Data Processing, Inc, a company that sells geographic data analysis services and software to commercial clients. Since 1980 his present company has served as an author and publisher of digital maps supporting thematic data display, vehicle routing and in-car navigation. His publications include a series of technical comic books written for the Census Bureau. He is a member of ACSM, URISA and the Wild Goose Association.

ABSTRACT

A Vehicle Navigation Appliance (VNA) is a device installed in a car which displays information such as location of the car in relation to a chosen destination. No longer a "James Bond" novelty, VNAs are currently on the market and could be showroom options before 1990. Consumer acceptance will depend on having a great body of useful and entertaining information stored in the vehicle and indexed to an on-board digital map. Probable use of compact audio disks for map distribution presents an unprecedented map publishing opportunity because of the enormous amount of data which can accompany the map.

INTRODUCTION

Vehicle Navigation Appliances -- devices that keep track of the location of a vehicle and display the location in map form -- are no longer the stuff of James Bond movies. Honda has had it's "Gyrocator" system on the market in Japan since 1982. ETAK's "Navigator" is for sale as an after-market auto accessory in California. Most automobile manufacturers have navigation systems in their concept cars and promise systems as showroom options by the end of the decade.

The technology required to build navigation appliances stems from personal computing, high-fidelity music, administering the decennial census and guiding airplanes and missles. The most probable VNA hardware designs and the most promising marketing strategy for VNAs together present an unprecedented map publishing challenge to cartographers and a unique opportunity for electronic publishing of large volumes of information. This paper briefly surveys VNA technology and concentrates on the map publishing challenge it presents.

DEFINITION

Two functions define a vehicle navigation appliance: 1) it

must be able to locate the vehicle on the ground and 2) it must be able to display the location in a familiar and useful form.

VNAs now on the market provide these functions which satisfy gadget buffs and commercial fleet operations. However to reach a broad consumer market a VNA must also: 3) provide useful and entertaining information related to the location of the car or the desired destination.

VNA HARDWARE

This functional definition of a VNA requires six hardware sub-systems for implementation: 1) a location device, 2) a computer, 3) an output facility, 4) an input facility, 5) map storage and 6) storage for "other" information.

Location Devices

Three location techniques seem promising for locating land vehicles: Loran, Satellite, and Dead Reckoning. Although "electronic signposts" serve to monitor buses along fixed routes, they are dismissed for supporting a broad consumer market because of the cost of the infrastructure required to work nationwide.

Loran. A Loran set receives coordinated radio signals from three fixed ground stations. It measures the time differences in the arrival of the signals. The time differences serve as coordinates, which can be looked up on special maps with time-difference coordinate overlays or converted to latitude and longitude.

About 250,000 ships and boats have Loran sets costing between $700 and $2500 for navigation. Loran transmitters cover not only the coastal and offshore waters but also the Great Lakes and the majority of the continental land area. Loran sets adapted for general aviation use are being well received; some units have thousands of locations of airports and navigation aids stored internally for recall and use by the pilot.

Loran manufacturers are also investigating land vehicle installations as another market. Loran accuracy is affected by a seasonal temperature-related drift and more drastically by local electrical disturbances. On land, the best-case accuracy will discriminate between most adjacent street intersections; use of a locally-broadcast differential signal reportedly improves accuracy to 15 feet.

Satellite. Three proposed satellite systems may be able to support vehicle navigation: Geostar, Startech and the Navstar Global Positioning System (GPS). Geostar and Startech are both private ventures which would have to charge users a fee for system access. Both have demonstrated their concepts, Geostar with simulated satellites on mountaintops and Startech with a facility currently in orbit. Each system appears to yield sufficient accuracy for land vehicle location.

On the other hand, GPS/Navstar is a Defense Department

system planned for operational deployment between 1986 and 1988. The mature system will consist of 21 satellites; six test satellites are currently available for testing. GPS is really two systems: a highly accurate military one with coded signals unusable by civilians. The open channel will be deliberately degraded for national security reasons to yield a location accuracy that is marginal for land vehicle location. Ironically, it appears possible to determine and broadcast a differential correction signal, or even generate a "pseudolite" that would look to a GPS receiver like another satellite, further increasing accuracy. Either of these corrections would make GPS adequate for VNA use. The issue of user fees remains unresolved, with sentiment for a free system generated in part by the KAL flight 007 disaster.

Dead Reckoning. Dead reckoning is a "position keeping" rather than "position finding" technology; a DR device keeps track of how far and in what direction a vehicle has gone to compute location relative to a known starting point. Dead reckoning hardware can be quite simple and inexpensive: an odometer and a heading indicator. Heading can be sensed by a magnetic compass, a differential odometer or inertial sensor. A combination of compass and differential odometer is used by most current systems.

Simple open-loop DR yields a position measurement that degrades as the vehicle moves because of errors in distance and heading readings. At least four firms have overcome this deficiency by map correlation/matching or closed-loop DR. In closed-loop DR the navigation computer compares sensed evidence of distinct turns to bends or intersections in a digital map. The computer choses the most probable turn point and updates the vehicle's location, eliminating accumulated errors. Even closed-loop DR can get lost and require re-initializing if turns that allow position updating are too far apart; otherwise, closed-loop DR yields excellent accuracy.

Computer
Computing functions in a VNA include retrieving and displaying the appropriate map, updating the car's location on the display, managing queries from the operator, searching the directory of destinations, and supporting the dead-reckoning function, if used, including map correlation in the closed-loop case. Other navigation-related functions might include calculation of a shortest path and voice synthesis of directions. These requirements are typical of today's home and business computers; a car navigation computer could be a general device available also for entertaining children or business use, especially if linked to an office computer through a mobile cellular telephone.

Output Facility
The VNA output facility must have a screen to display maps and to present various directories of streets and destinations. The 1985 Buick Riviera has an optional touch-screen CRT; other cars such as recent Corvettes have full-color LCD instrument displays. Another appealing output device would be a voice-synthesis unit to read directions, instructions

or descriptive text to the driver while the car is moving.

Input Facility
The VNA user must be able to request a function, specify a destination, or search through information stored in the system. A touch-screen or several function buttons around the display screen should suffice.

Map Storage
Navigation systems to date have used three different methods of map storage: photographic, digital image and digital encoded.

Photographic. Honda's Gyrocator and Omni Devices' Navigator both use photographic map storage. Honda requires the user to slip a transparent map overlay over the display CRT. Omni stores maps in a film cartridge and uses a complex fiber-optical/LCD system to generate the display. Although Omni's system provides more flexible panning and zooming than Honda's it suffers from relatively low resolution: ETAK's CRT, for example, displays 50 times as many pixels.

Digital Image. The most publicized concept car, Chrysler's Stealth, uses videodisk images of paper maps in its "CLASS" (Chrysler Laser Atlas Satellite System). Others use videotape to achieve a similar effect. Chrysler reports having over 13,000 maps stored on its demonstration disk, covering the USA at several scale levels to permit zooming.

Digital Encoded. A digitally encoded map is a data file describing the street network by connectivity and coordinates, augmented by street and city names and address ranges. This flexible and compact recording form supports many functions: DR map correlation, map display, specification of origin for dead reckoning, and selection of destination by street address or intersection. The digitally encoded map provides the most flexibility in windowing and zooming, and can support shortest-path route-finding computations.

All three of these options presume that the map is stored on board the car, although Boeing's FLAIR (Fleet Location and Information Reporting) system was supporting closed loop DR for 200 cars with central computing and map storage eight years ago. ETAK stores maps on audio cassette, taking advantage of a familiar medium and an inexpensive read mechanism. This has two disadvantages: limited storage capacity and sequential access. Current "floppy" disks have about the same capacity as cassettes, but their virtue of random access is offset by a ten-fold cost disadvantage for the hardware unit.

A panacea for map storage in vehicle navigation systems appears to be the compact audio disk used as a read-only memory for digital data. "CD-ROM" players are on the market for automotive music systems; only a modification to the error detection and correction circuitry is needed for data storage. Philips -- a promulgator of the CD standard -- is proposing automotive CD players wired for both in-car entertainment and information. Each CD-ROM disk holds 500 million bytes of data, more than enough for a digital map of all streets in the USA.

Storage of "Other" Information

Only the CD-ROM or other laserdisk like Chrysler's holds out the possibility of storing much information beyond the basic map. The point to establish here is that one CD-ROM disk can store a complete digital map of every street in New England plus 300 unabridged copies of "Moby Dick".

HARDWARE SUMMARY

Anticipating the hardware environment of 1988 or 1989 (dates usually mentioned in the popular press) one can project vehicle navigation appliances built around a computer with roughly the power of a 256K IBM PC, using a color CRT screen and sharing a CD player for storage of both map information and music. Although it is tempting to predict that GPS will be the predominant location technology, there are too many successes with inexpensive DR and too much uncertainty with GPS cost, accuracy and timing to do so. All location techniques have drawbacks: Loran and GPS won't work in tunnels; DR only yields a relative position and needs initializing and occasional resetting. A reliable and trouble-free system would ideally use two complementary techniques: either GPS and DR or Loran and DR.

MAPPING CONSIDERATIONS

Although it is easy to build digital map images by scanning existing paper maps at various scales, image storage is inferior to digital encoding for three main reasons: images do not support closed-loop DR, they are inflexible for zooming and panning and they require a separate index for looking up addresses of destinations. On the last point, note that the index required to support an image database is almost equivalent to the full content of a digitally encoded map.

An ideal digitally encoded map would have all streets represented with accurate and visually pleasing coordinates, and sufficient housenumber and intersection information indexed to permit lookup of a precise street address or a street intersection. Additional useful information includes flagging one-way streets and turn restrictions and identifying highways and major arteries. These additions permit highlighting of expressways on the map display and improve performance of route-finding algorithms.

This basic map information, judiciously compressed, amounts to 120 to 150 bytes per street. Sixty percent of the USA population lives on about one million streets represented in the Census Bureau's "GBF/DIME" files, a series of digital maps prepared for administering the 1980 census. A simple extrapolation, allowing for rural streets that wiggle more than their urban counterparts, yields a nationwide digital map that fits easily on one 500 megabyte CD-ROM disk.

Though enticing as a resource for vehicle navigation, the GBF/DIME files have numerous errors in segment connectivity; they represent the street network of 1977; they cover only one percent of the USA land area; the coordinate digitizing has a systematic clipping that causes straight streets to appear to wander back and forth and some new streets added just before the last census were scribed inaccurately enough to guarantee malfunction of a closed-loop DR system. The Census Bureau has acknowledged these deficiencies, but ironically its solution, the TIGER (Topologically Integrated Geographic Encoding and Referencing) system will not be in place in time to have an impact on the first wave of VNAs. Nevertheless, the GBF/DIME files are useful enough resources to serve as the starting point for proprietary vehicle navigation digital maps prepared by both ETAK and Geographic Data Technology, Inc.

From a marketing point of view, does it make sense to sell the whole country on one CD-ROM? The answer is no; split the map up into states, regions or metropolitan areas, and use the saved disk space to generate many custom editions and make multiple sales to VNA owners.

For example, target one edition for business travellers in rental cars or their own vehicles, containing a complete directory of all business establishments in the region. There is room for a lot more data on each firm than name and address (just delete 299 of the copies of Moby Dick); Dun and Bradstreet -- a candidate for map publisher -- could put its directory of business establishments for any state on one disk, along with the digital map. Augmented with appropriate indexes, this disk could display to a salesperson all businesses of a certain size, in a particular industry within five minutes driving time, including the names of officers and phone number for use with the cellular phone.

For the family market, substitute the kind of data found in the Yellow Pages and supplemented with more detail about each entry: hours of operation, product lines carried, or sample restaurant menus. This information would help in many family emergencies where a fast-food restaurant or a children's museum is needed in a hurry, or to take advantage of unexpected free time by querying for nearby antique book stores or those elusive outlets selling Liz Claiborn garments. The obvious publishers, the Yellow Pages operations, are restricted from publishing an "electronic yellow pages" until 1989. Whether distribution of an "annotated" map on CD-ROM is also restricted is an interesting question; if so, there may be an excellent opportunity for another company to compete in this highly profitable service.

The Yellow Pages is profitable because of the same mechanism that supports publication of most paper maps: both are really advertising media. The CD-ROM map with its huge capacity presents an unprecedented opportunity for advertisers.

The dynamic nature of the map display screen permits display of commercial logos at the location of facilities of various kinds. For example, Hyatt could buy the capability to display its logo at individual hotel locations as any car drove within range. Marketing executives for competing hotels would be hard-pressed not to pay for similar advertising of their facilities.

A final consideration regarding functions of paper maps concerns their potential for relieving boredom on a long trip. The physical limitations of a thirty inch square of paper restrict its entertainment and educational potential to listings of state birds and insects. Contrast this with the ability of a CD-ROM to store an immense amount of localized information, indexed to the map: schedules of summer playhouses, town histories, agriculture, or geology (sample: "In 1958, when this highway was under construction, dinosaur bones were unearthed right where you're driving"). Making the CD-ROM map into a tour guide would enhance its attractiveness and promote sales.

SUMMARY -- UNRESOLVED ISSUES

Vehicle Navigation Appliances are no longer a curiosity in James Bond movies. Potential technical, cost, and regulatory barriers are identified; none appear to prevent a robust competitive market developing by the end of this decade.

There is clearly technical groundwork still to be done such as defining the standards for broadcasting a GPS differential correction signal, or establishing formats and accuracy standards for the on-board map. Manufacturers have to settle on methodologies, build and test prototypes and set up production and marketing.

If CD-ROM is chosen for map distribution, the major unresolved issues of publisher and content of the "other" data remain. One must ask if a Magnavox or Texas Instruments -- motivated to sell GPS receivers -- is willing to enter the digital map/yellow pages business to effect the GPS sales? Might a General Motors or Ford take the lead and compete with the telephone companies? How does this opportunity sit with McGraw-Hill and its aggressive movement into electronic publishing, or Rand McNally and its century-plus experience in map publishing, or RR Donnelley -- a map maker that also prints the Yellow Pages, or Philips and Sony, both of which already sell CD's for music storage?

The answers to these questions -- which we practitioners of automated cartography can affect -- will change forever the largest use of maps: navigating private automobiles. The challenge we face is an unprecedented opportunity for innovative publishing of a new kind of map.

FURTHER INFORMATION BY SUBJECT

Popular Press
Popular Science, August 1984, Cover story
Radio-Electronics, July 1983, Cover story
Business Week, June 18, 1984, Page 82
Science Digest, December 1984, Page 34
Venture, October 1984, Page 12
Computerworld, June 25, 1984 Page SR-3

Loran
McGillem, C.D, et al, Feb 1982, Experimentally Determined

Accuracy and Stability of Loran C Signals for Land Vehicle Location: IEEE Transactions on Vehicular Technology, Vol VT-31, No 1, page 15.
Also: Proceedings of Wild Goose Association (Loran users)
    WGA, 118 Quaint Acres Dr, Silver Spring, Md 20904
and: Proceedings of The Institute of Navigation
    815 15th Street, Suite 832, Washington DC 20005
Note: Differential Loran: Navigation Sciences, Inc, 6900 Wisconsin Ave, Bethesda, Md 20815

Satellite
GPS: see papers by Parkinson & Gilbert and by Stansell in Proceedings of the IEEE Volume E71, Number 10, October 1983; entire issue is on Global Navigation Systems.
Geostar Corp: P.O. Box 82, Princeton, NJ 08540
Startech: P.O. Box 1015, Humble, Texas 77347
Differential GPS Standards: By summer, 1985; R.T.C.M., 655 Fifteenth St NW, Suite 300, Washington, DC 20005

Dead Reckoning
Lezniak, T.W., et al, Feb 1977, A Dead Reckoning/Map Correlation System for Automatic Vehicle Tracking: IEEE Transactions on Vehicular Technology, Vol VT-26, No 1, Page 47.
French, R, Sept 1984, Autonomous Route Guidance Using Electronic Differential Odometer Dead Reckoning with Vectorized Map Matching: US Army Artificial Intelligence and Robotics Symposium, Indianapolis, Indiana
Also: ETAK, 1287 Lawrence Station Rd, Sunnyvale, CA 94089

Electronic Signposts
"Automatic Vehicle Monitoring Program Digest", April 1981, Department of Transportation report # DOT-TSC-UMTA-81-11

Computers in Cars
Whitmore, S, Buick Engineers Experimenting with Dashboard PCs: PC Week, Vol 1 No 43 October 30, 1984, Page 5

Compact Audio Disks
Monforte, J, The Digital Reproduction of Sound: Scientific American, December 1984, Page 78.

CD-ROM
Schipper, J. 1984, In-Car Entertainment and Information Systems, Part 2: Application of the Compact Disc in Car Information and Navigation Systems: Nederlanse Philips Bedrijven B.V.

Digital Map Standards for Vehicle Navigation
Transportation Research Board Research Problem Statement:
R. French, Radiation Research Associates, Inc. 3550 Hulen St Fort Worth, Texas 76107

GBF/DIME & TIGER
Robert Marx, Chief, Geography Division, U.S. Bureau of the Census, Washington, DC 20233: various publications.

Route Finding
Elliott, R and Lesk, M, Let Your Fingers Do the Driving: Maps, Yellow Pages, and Shortest Path Algorithms: Bell Laboratories, Murray Hill, NJ 07974, October 1982

ALTERNATIVE APPROACHES TO DISPLAY OF USGS
LAND USE/LAND COVER DIGITAL DATA

David J. Cowen, Michael Hodgson, W. Lynn Shirley,
Thomas Wallace and Timothy White
Department of Geography
and
Social and Behavioral Sciences Lab
University of South Carolina
Columbia, SC 29208

ABSTRACT

This paper outlines a procedure for converting the USGS GEODATA Land Use and Land Cover files into polygon and raster formats that are compatible with several widely used software systems. Examples are included for SAS/GRAPH®, GIMMS and MAP. There is also a discussion of an approach to handling the grid format on IBM-PCs and color frame buffer systems.

INTRODUCTION

Through the recent advertisements of GEODATA in commercial magazines, the US Geological Survey has made a major statement concerning digital cartographic products (Computer Graphics World, 1984). The "Quiet Revolution" is ready to evolve into the accepted standard for future map products (USGS, 1983). While the suppliers of digital cartographic data appear eager to disseminate their products, the ultimate success of digital cartography will be measured only by its adoption within the user community. The value of all data is determined by its utility in problem solving environments. The utility of digital data is dependent upon its integration with powerful, inexpensive and easy to use hardware and software systems. While the geographic equivalent of the financial spreadsheet may not yet exist there are several alternatives available which can greatly facilitate the use of the GEODATA files. The twofold purpose of this paper is to examine the basic file structure of the land use/land cover component of the GEODATA files and to demonstrate a procedure that converts the file structure into formats compatible with the widely used vector and raster based processing systems. Furthermore, this paper will demonstrate how this type of data now can be handled on the new generation of personal computers and color frame buffer systems.

THE LAND USE/LAND COVER FILES AND GIRAS

In 1973 the USGS began creating maps and digital files of land use and land cover at the scale of 1:250,000 (Fegeas et al, 1983). At this scale, each land use polygon is classified into the 37 USGS level II codes (Anderson et al, 1976). Urban, or built-up land and water are compiled and mapped at 4 ha. minimum polygon sizes, while other polygons must exceed 16 ha. in area. At these minimum mapping units the $1°$ by $2°$ quadrangles contain an average of almost 3000 polygons. The decision to map the Nation in digital form at this scale dictated that the USGS develop both an efficient topological data structure and a sophisticated geographical data handling system. The data structure, based on arcs, is a direct descendant of DIME segments and POLYVRT chains (Fig. 1) (Peucker and Christman, 1975). These arcs can be created, edited, mapped and converted into polygons and grid cells with the Geographical Information Retrieval and Analysis System (GIRAS) (Mitchell et al, 1977). Although it was not designed to be a mapping system, GIRAS does include two map generating programs: SHADE, a sophisticated shading routine that efficiently handles complex islands (Fig. 2): and PTG-MAP which generates crude lineprinter displays of the grid cell data. Neither of these routines allows the type of flexibility needed in a comprehensive mapping system.

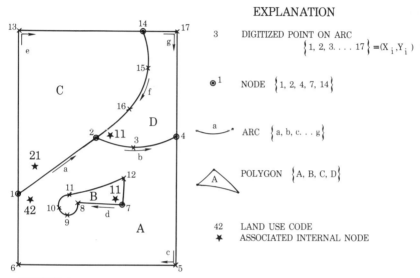

FIGURE 1
Simplified Version of Giras Topological Structure after Fegeas et al. 1983, p.5.

While GIRAS incorporated several approaches to geographical data processing which were state-of-the-art during the mid 1970's it is cumbersome to operate and, was never widely distributed by the USGS. Clearly a public organization such as the USGS is not staffed to provide user support for software as complex as GIRAS. More importantly, it can be argued that such developmental activities should be left to the private sector. The consequence of this decision has limited applications of digital land use data to organizations with sufficient resources to support a full scale integrated GIS system. Unfortunately, a large number of groups that could potentially benefit from the data collection efforts of the USGS have been left without appropriate software alternatives.

POLYGON CONVERSION ROUTINE

One approach to increasing the use of any database is to ensure that it is compatible with widely used software systems. Based on this premise an effort was undertaken to develop a computer program to convert the efficient GIRAS data structure into a more redundant but versatile format. The program creates a file compatible with any standard choropleth mapping routine, requiring a file of coordinate pairs (X_i, Y_i) to define the perimeter of a closed polygon. Usually this ring around the polygon (RAP) file circumscribes the polygon in a

clockwise direction and jumps inside the polygon to trace any embedded
islands in a counterclockwise direction.

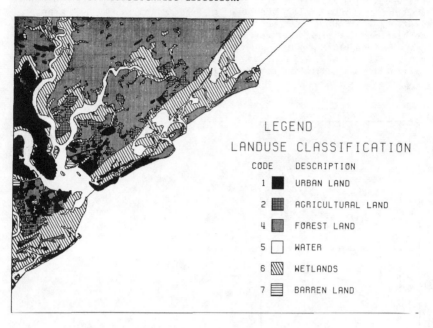

Fig. 2   GIRAS Output - SHADE Program - James Island, SC Quad.

The procedure developing a RAP file from four GIRAS subfiles is
demonstrated in Table 1 which corresponds to the simple topological
structure portrayed in Figure 1. The four subfiles consist of a
sequential list of polygon labels, or two digit land use codes (PID), a
list of ordered arcs that are chained to form a polygon (FAP), a file
of pointers to the last coordinate contained in each arc (PLC), and a
file of the arc coordinates arranged sequentially by arc (ARC).   It
should be noted that the **nodes** identified by the USGS on Figure 1 are
actually end points of arcs that are computed during the GIRAS
digitizing and editing steps. These **nodes** should not be confused with
the points which locate internal labels for each polygon during this
process. Furthermore, since the nodes in Figure 1 are common to at
least two arcs they must be duplicated in the ARC file.

The conversion process combines each polygon in sequential order
and writes out a file with land use codes and the corresponding
coordinate points needed for mapping. For example, the first polygon
(A) is processed by reading its land use code (42) from the PID file
and writing it to a data file. The topology of the polygon is then
read from the FAP subfile. This file contains the number of arcs (5)
and the correct order (a, b, c, 0, -d) in which the arcs are to be
linked to complete the polygon. In this example the first three arcs
(a, b, c) define the perimeter of the polygon, while the zero "arc"
acts as a flag to indicate the inclusion of an island. Such islands
are always followed by the arcs that define the island in a
counterclockwise direction (-d). The next polygon processed is the
island B which consists solely of one arc (d). Similarly the FAP entry

for polygons C and D in Figure 1 would be 3, e, f, -a and 3, -f, g, -b respectively. Once the topology of the polygon is determined, the location of the coordinates in the ARC subfile are read from the PLC subfile. For example, the first arc (a) ends at the location of the second coordinate pair in the ARC file. Therefore, we know that it consists of X_1, Y_1 and X_2, Y_2. The example in Table 1 demonstrates the manner in which these coordinates are ordered in the RAP file.

TABLE 1. CREATION OF RAP FILE FROM GIRAS

Polygon Sequence	GIRAS FILES					OUTPUT FILE	
	PID Polygon ID	FAP ARCS of Polygon	PLC Pointer Last Coord.	ARC Coordinates		RAP (Ring Around Polygon)	
				Order	Xi,Yi	Polygon	Xi,Yi
A	42	5 (# of ARCS)					
(Ever. Forest)							
		a (1st ARC)	2 ------->	1	1	1	1
				2	2	1	2
		b (2nd ARC)	5 ------->	3	2		
				4	3	1	3
				5	4	1	4
		c (3rd ARC)	9 ------->	6	4		
				7	5	1	5
				8	6	1	6
				9	1		
		0 (Flag)				1	0,0
		-d (Island)	16 ------>	16	7	1	7
				15	12	1	12
				14	11	1	11
				13	10	1	10
				12	9	1	9
				11	8	1	8
				10	7	1	7
--- End of 1st Polygon ---							
B	11	1 (Single ARC Polygon - Must be an Island)					
(Residential)							
		d	16 ------>	10	7	2	7
				11	8	2	8
				12	9	2	9
				13	10	2	10
				14	11	2	11
				15	12	2	12
				16	7		

SAS Environment

The format of the RAP file described in Table 1 is directly compatible with SAS/GRAPH® polygon mapping files (SAS Institute, 1981). The PROC GMAP choropleth mapping procedure is capable of handling the lengthy polygon files and complex islands encountered within the land use/land cover files (Fig. 3). Although it lacks flexibility in terms of placement of legends and does not provide extensive shading options it does conveniently eliminates polygons which do not have corresponding data values (Fig. 4).

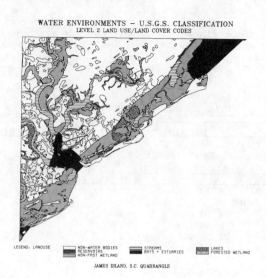

Fig. 3. SAS/GRAPH® Output - PROC GMAP

Fig. 4. SAS/GRAPH® Output - Drop Line Feature

Alternative Choropleth Programs

SYMAP: Once the polygon data are linked to SAS in RAP format they can easily be reformatted into basemap formats compatible with most other choropleth mapping routines. For example, even SYMAP A-Conformolines can be created from SAS/GRAPH® by explicitly closing the polygon, writing out appropriate labels, shifting the output order to

Y, X and translating the origin to the upper left. While limited experiments indicate that SYMAP has some difficulty handling the large number points contained in very complex land use polygons the logic of the conversion is straight forward.

GIMMS provides an excellent set of options for handling the GEODATA land use data. It actually serves as a useful combination of mapping procedures and geographical information system functions. In addition to extensive options for legends and symbolization (Fig. 5) GIMMS also provides functions for area calculation, centroid labeling and even statistical graphics. The SAS to GIMMS link simply formats the SAS file by removing the ID variable and closing each polygon with a slash.

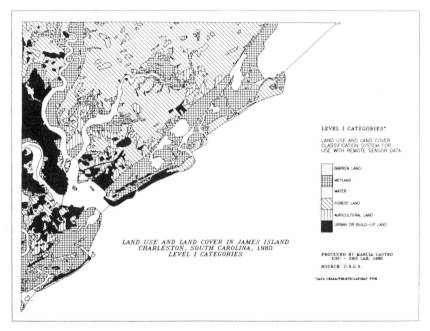

Fig. 5. GIMMS Output - Various Text and Legend Options

RASTER PROCESSING WITH MAP

Except for a polygon to grid conversion routine and some area calculation procedures, GIRAS remains a data base creation system. The GIRAS polygon to grid (PTG) procedure, which generates a single theme grid file in metric units registered to a UTM projection, provides a useful interface into grid oriented GIS functions. The Map Analysis Package (MAP) developed at Yale University provides an excellent set of procedures to analyze, integrate and display the USGS land use data in grid format. Since a MAP formatted database consists of rectangular matrices of map overlays, the output from PTG can be converted into a MAP dataset. Within the MAP environment line printer maps can easily be generated with the DISPLAY command (Fig. 6). Although the variation in lineprinter spacing results in distorted output of cellular data, MAP's ability to locate each cell and to associate it with its neighbors or other overlays more than offset the poor quality of the output for many applications.

Fig. 6. MAP Lineprinter Output - 500 Meter Grid Cells

Data stored in a grid cell format data are ideally suited for raster display devices, such as electrostatic and ink jet plotters, and color frame buffer systems. With such output devices a direct association can be made between row and column cell locations in the data base and pixel locations on the output mechanisms. Based on this linkage a plotting routine was developed which displays any cellular data structure. The routine includes options to shade each cell, display cell values, or generate a plot of boundaries between cells of different values (figs. 7 and 8). Furthermore, by not advancing the plotter, the user can easily overlay different MAP datasets and combine various options onto one plot. While this routine is actually written for any plotting device the large number of pen up and down commands are most suitable for raster output devices.

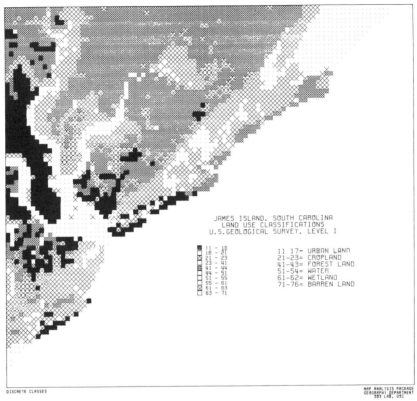

Fig. 7. Electrostatic BITMAP Output - 500 m. Grid Cells

IBM-PC COLOR FRAME BUFFER

Current technology has made it cost effective for most serious geographic data processing organizations to begin using high resolution color frame buffer systems. For example, several frame buffer systems are available for the IBM PC which provide high resolution and extensive color capabilities for less than $5,000 (including monitor). Although MAP is still a mainframe based system several recent developments make it easy to download data from such systems to the IBM-PC. The IBM-3270 PC allows the user to run concurrent sessions of MAP using CMS on the mainframe as well as PC-DOS. MAP data sets can be quickly transferred (9600 BAUD) onto floppy or hard disk for additional processing. Using the XT-370 the entire MAP session can be downloaded and run locally. Using either device a MAP dataset can be converted into a bit plane format that can be converted easily and displayed by the frame buffer system.

A system with 9 bit planes and a resolution of 672 x 480 was used to display, at a scale of 200 m, the USGS land use data previously stored as a MAP data base (Fig. 8). Within the system the data are represented as color palette locations which can be assigned any of 16,000,000 different colors by indicating the 512 intensity levels of the RGB guns. Since color change does not involve any additional data transfer the resultant image can be changed quickly based on user criteria. For example, by a simple color assignment step any of the land use categories can be viewed individually or in combination of any of the colors. In addition, the user can interactively add legends and text to the image (Fig. 9). Presently there are several color ink jet and thermal transfer devices available that should provide inexpensive means of generating color hard copy from the frame buffer system. Such devices will have an enormous impact on the practicality of using color raster processing of such land use data and, probably, will further accelerate raster processing of land use data. In such a processing environment there could be wide spread use made of the composite theme grid format for the USGS GEODATA digital data.

CONCLUSION

This paper has demonstrated a set of procedures which can greatly increase the use of the USGS GEODATA Land Use and Land Cover digital data by converting them into more compatible polygon and grid formats. In these formats the data can be easily processed by a number of widely available software systems. Furthermore, the paper has demonstrated that the data can also take advantage of recent advancements in personal computers and raster display technology.

ACKNOWLEDGMENTS

The original algorithm to create the RAP file was written by Peter Oppenheimer. The GIMMS maps were produced by Marcia Castro. Their contributions are greatly appreciated.

REFERENCES

Anderson, J., E. Hardy, J. Roach and R. Witmer 1976, A Land Use and Land Cover Classification System for Use With Remote Sensing Data, U.S.G.S. Professional Paper 964, GPO, Washington, DC

Computer Graphics World, 1984, Vol 7, No. 10, p. 85

Dougenik, J. A., and D. E. Sheehan 1975, SYMAP Users Reference Manual, Harvard Lab for Computer Graphics and Spatial Analysis, Cambridge, MA

Fegeas, R., R. Claire, S. Guptill, K. E. Anderson and C. Hallams 1983, Land Use and Land Cover Digital Data, U.S.G.S. Circular 895-E, GPO, Washington, DC

Mitchel, W., S. Guptill, K. E. Anderson, R. Fegeas and C. Hallam 1977, GIRAS: A Geographic Information Retrieval and Analysis System for Handling Land Use and Land Cover Data, U.S.G.S. Professional Paper 1059, GPO, Washington, DC

Peucker, T. and N. Christman 1975, Cartographic Data Structures: American Cartographer, Vol 2, No. 1, pp. 55-69

SAS Institute 1981, SAS/GRAPH Users Guide, SAS Institute Inc., Cary, NC

Tomlin, C. D. 1983, Digital Cartographic Modelling Techniques in Environmental Planning, Doctoral Dissertation, Yale University School of Forestry and Environmental Studies, New Haven, CT

U.S. Geological Survey 1983, "The Quiet Revolution in Mapping", Reprint from 1979 Annual Report, GPO, Washington, DC

Waugh, T. C. and J. McCalden 1982, <u>GIMMS Reference Manual</u>, GIMMS LTD., Edinburgh, Scotland

Fig. 8. IBM-PC/Color Frame Buffer Configuration

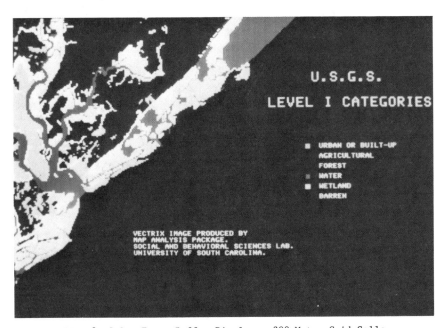

Fig. 9. Color Frame Buffer Display - 200 Meter Grid Cells

UNIVERSITY OF SOUTH CAROLINA

COLUMBIA, S. C. 29208

SOCIAL AND BEHAVIORAL SCIENCES LABORATORY

(803) 777-7840

Biographical Sketch

David J. Cowen is a professor of Geography and Director of the Social and Behavioral Sciences Laboratory at the University of South Carolina.

W. Lynn Shirley is a program manager and instructor in computer mapping.

Timothy White is a systems analyst for the Department of Geography.

Michael Hodgson and Thomas Wallace are graduate students in the Department of Geography.

AN LP RELAXATION PROCEDURE FOR ANNOTATING POINT FEATURES
USING INTERACTIVE GRAPHICS

Robert G. Cromley
University of Connecticut
Storrs, CT 06268

ABSTRACT

The development of computer-assisted mapping procedures has made the design of maps a normative problem in the processing and communication of information. However, mathematical programming techniques have rarely been used by cartographers in the optimal design of maps. Instead, heuristics have been employed. A modified linear programming technique is used here to operationalize the annotation of point features. An interactive relaxation procedure is utilized to avoid the problem of having to define algebraically the set of cartographic restrictions on the placement of labels. Because all constraints in the point annotation problem are graphic in nature, any violated constraint in the relaxed problem can be visually detected on a screen image. The viewer interactively adds violated constraints to the algebraic problem until no violations appear on the screen image. This approach is promising as many cartographic rules are partially qualitative and human interaction is an appropriate method to resolve conflicts.

INTRODUCTION

As cartographers have defined mapping procedures for the accurate and efficient graphic display of information, the concept of designing an optimal map by machine has been proposed by various researchers. The map as a normative model for processing information has been used in determining class intervals (Jenks and Caspall, 1971; Monmonier, 1973), line generalization (Douglas and Peucker, 1973), gray-tone selection (Smith, 1980), and more recently annotation (Ahn and Freeman, 1983; Freeman and Ahn, 1984). With few exceptions, though, cartographers have not used techniques of optimization from mathematical programming but instead have opted for heuristics as cartographic problems have been difficult to define as a linear or nonlinear program. This paper is an initial investigation into the integration of linear programming procedures with interactive graphics to overcome some of the difficulties of operationalizing the optimal annotation of point features.

The interface between mathematical programming and computer graphics holds promise for the solution of many optimization problems in cartography and spatial analysis that may be more difficult to solve using either technique in isolation of the other. Many algebraic constraints of linear and nonlinear programs used in spatial analysis can be transformed into a graphic display. Thus, constraints can be modelled graphically and the set of feasible solutions identified visually. In this manner, algebraic programs have the potential to be solved by inspection on a graphics terminal. Composite mapping is one such graphic technique that has been applied to site location problems. Conversely, mathematical programs can find solutions that minimize or maximize some utility function over space that are not recognizable by the human eye. By combining these two techniques, it is possible to synthesize the positive attributes of each. The point annotation problem is amenable to this approach because it has the properties of a mathematical program although the final solution has a graphic form. First, a discussion of the point feature annotation problem is presented and then a linear programming (LP) relaxation procedure is defined to solve the problem.

MACHINE ANNOTATION OF POINT FEATURES

The annotation of point features is just one of three name placement problems in mapping. The positions of names for area features, line features, and point features must be determined in unison although a feature name with a smaller degree of freedom with respect to its placement is usually placed before names with a larger latitude in their positioning. The study is restricted here to just the point feature problem for ease of exposition and will be expanded in the future.

While the positioning of names on a map is somewhat subjective, cartographers have attempted to standardize the rules for an aesthetically pleasing yet informative map. Imhof (1975) has formulated a set of guidelines for general map annotation that has been refined and expanded by Freeman and Ahn for the point feature problem (Ahn and Freeman, 1983; Freeman and Ahn, 1984). Their rules are summarized here as: 1) names should be horizontal in an east-west orientation, 2) each name should be near but not too close to the point feature that it annotates, and 3) name placement above and to the right of the point is generally preferred. Although numerous name positions are possible with respect to each point, Figure 1 presents six of the more preferred positions. While the rank ordering of positions in this figure is subjective, it is illustrative of a potential preference function. This utility function could be determined by a panel of experts.

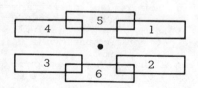

Figure 1. An example set of name positions

The crux of annotating point features is not where to position the name with respect to its point but where it should be positioned relative to the location of other point labels (Freeman and Ahn, 1984, p. 556). Names may need to be repositioned until a final placement pattern is determined. For example, in Figure 2 the preferred name placement for point four conflicts with the name placement for point three. However, moving point four's name to another position may conflict with the placement of names for other points. Therefore, the positions of several names may need to be readjusted before the final name pattern for all four points is determined.

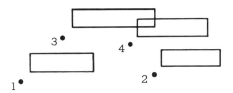

Figure 2. An example positioning conflict

This adjustment process has been resolved by constructing a graph of possible name conflicts (Freemand and Ahn, 1984). Each node in the graph represents a point; two nodes are connected if their name placement areas overlap. Each connected component of the graph is processed independently to resolve any potential conflicts within the area represented by the component. As each node is processed, its branches to other points are examined to determine the position of its name. If no name can be placed, the procedure backtracks to a previous node and the positions of placed labels are altered to accommodate the new point.

This resolution of overlap problems may be indeterminant with respect to the preference for name placement. The resolution to the positioning conflict in Figure 2 may have required the relocation of point three's name or alternatively the relocation of names for points one, two, and four. Therefore, the prefernce function for individual points should be extended to a utility function for the entire distribution of names. One pattern should be preferred over another; this problem can be overcome within the context of linear programming.

AN LP RELAXATION PROCEDURE

The problem of annotating point features is amenable to solution by linear programming techniques because preferences for the possible positioning of a point label have been outlined and a general set of rules (i.e. no names should overlap) that constrain the final distribution of of names has been identified. Position preferences can be modelled as the objective function of the program and the general set of rules forms the constraint set. The range of potential name positions, as previously discussed, is limited to one of six places for each point (see Figure 1). Each position is assigned a priority weight from one to six based on the position's preference ranking where one is the most desirable position and six is the least. The objective of the point feature annotation problem is to produce a map that minimizes the sum of the priority weights for the entire label distribution. This goal is expressed algebraically as:

$$\text{Minimize} \quad \sum_{i=1}^{n} \sum_{j=1}^{6} W_{ij} X_{ij} ; \quad (1)$$

where, n is the number of points to be annotated;
W_{ij} is the priority weight of the jth position for the ith point;
and, X_{ij} is a decision variable that determines whether the jth position is chosen for the ith point.

The constraint set for this program is defined by the following two rules: 1) each point must be assigned a label, and 2) no label can overlap or intersect another point or label. The first rule can be modelled algebraically by the following inequality:

$$\sum_{j=1}^{6} X_{ij} \geq 1 \quad \text{for all i.} \quad (2)$$

This inequality forces at least one of the positions to be selected for each point and no more than one would be chosen given the minimizing goal of the objective function. Using just equations (1)-(2) as an LP, the optimal solution would produce a map where the label for each point would be assigned to its most preferable position. While finding the solution to this problem is trivial, it could satisfy the broader annotation problem if no labels overlapped. In a clustered or dense point distribution, however, this would be an unlikely occurance.

The modelling of the overlap rule is more difficult than the assignment requirement. Usually, linear programs are executed on a computer in batch mode because as a normative model once the parameters have been set, the solution is deterministic. Unfortunately, the point annotation problem does not lend itself to solution by this approach. The overlap contraints for labelling point features define a set of geographic relationships that cannot be determined without analyzing a graphic display of all possible placement combinations. The final constraint set may contain too many inequalities for efficient solution and some violations of the overlap rule may be accidentally overlooked and thus omitted from the LP model.

The model presented here utilizes a relaxation procedure to avoid these problems. The general philosophy of relaxation techniques is to solve initially a simpler problem with fewer constraints than the original one. If the optimal solution to the relaxed problem does not violate any of the omitted constraints, then it is also the optimal solution to the original problem (Lasdon, 1970, p. 268). Whenever constraints are violated, they are explicitly included and the problem is iteratively solved until no relaxed constraint is violated. Normally, this iterative process is transparent to the user as violated constraints are identified algebraically by the program. Because overlap constraints in the point annotation problem are graphic in nature, each violated relaxed constraint in the program has a graphic manifestation on a map image. This suggests that an interactive graphic approach can be used where violated constraints are detected visually by the user and denoted for inclusion as an explicit constraint in the next iteration. For this model, the following algebraic inequality is used to implement each overlap restriction:

$$X_{ir} + X_{jk} \leq 1 ; \qquad (3)$$

where, X_{ir} denotes the rth label position of the ith point;
and, X_{jk} is the kth label position of the jth point.

This inequality only permits one of the two overlapping positions to be present in the next iteration while the objective function searches for the next best preference pattern of point labels. This process continues until the user cannot detect any more overlap violations on the screen image.

This LP relaxation procedure was written in FORTRAN for implementation on a Tektronix 4012 graphics terminal at the University of Connecticut. To improve the overall aesthetic appearance of the final map display, the user is allowed to make minor adjustments to any of the map parameters used by the program. Additionally, the user has the option to modify the system objective function for choosing the optimal annotation distribution. The utility function discussed above minimizes total priority weight; although this goal will try to place names in an overall system design, individual points may have their worst label position selected. An alternative utility function would be to place names such that the worst placement for any single point is avoided. This minimax criterion is used frequently in public facility location and can easily be implemented by powering the priority weights for each point.

SUMMARY

The interface between mathematical programming and computer graphics has the potential to solve problems in cartographic design and spatial analysis. Whenever system constraints are graphic in nature, they can be modelled as such and solved using a relaxation procedure. If any constraints are violated in the relaxed problem, they can be interactively included and the problem solved again until no relaxed constraints are detected. While this approach was applied here to the point annotation problem, this man-display-program has appeal whenever system rules are partially qualitative by nature and human interaction is an appropriate method for resolving any conflicts. The system also has the flexibility to use different utility functions to find the most aesthetically pleasing name distribution.

ACKNOWLEDGMENTS

The author wishes to thank Jeanne Mengwasser for final preparation of all figures.

REFERENCES

Ahn, J. and Freeman, H. 1983, A Program for Automatic Name Placement: Proceedings, AUTO-CARTO VI, Vol. 2, pp. 444-453.

Douglas, D. and Peucker, T. 1973, Algorithms for the Reduction of the Number of Points Required to Represent a Digitized Line or its Caricature: The Canadian Cartographer, Vol. 10, pp. 112-122.

Freeman, H. and Ahn, J. 1984, AUTONAP - An Expert System for Automatic Map Name Placement: Proceedings, International Symposium on Spatial Data Handling, Vol. 2, pp. 544-569.

Imhof, E. 1975, Positioning Names on Maps: The American Cartographer, Vol. 2, pp. 128-144.

Jenks, G. and Caspall, F. 1971, Error on Choroplethic Maps: Definition, Measurement, Reduction: Annals, the Association of American Geographers, Vol. 61, pp. 217-244.

Lasdon, L. Optimization Theory for Large Systems. 1970. New York: The Macmillan Company.

Monmonier, M. 1973, Analogs Between Class-Interval Selection and Location-Allocation Models: The Canadian Cartographer, Vol. 10, pp. 123-131.

Smith, R. 1980, Improved Areal Symbols for Computer Line-Printed Maps: The American Cartographer, Vol. 7, pp. 51-57.

CREATING USER FRIENDLY GEOGRAPHIC INFORMATION SYSTEMS THROUGH USER FRIENDLY SYSTEM SUPPORTS

Powel Crosley
Data Research and Applications, Inc.
412 Cedar Bluff Rd. Suite 203
Knoxville, Tennessee 37923

ABSTRACT

For over twenty years, geographic information systems have been growing in stature as more and more firms and governmental agencies realize the need for automated methods to retrieve, analyze, and display geographic data. This has been paralleled by a growth in the number of information systems available, and consultants to develop and/or manage them. Technical applications have evolved to the point that it is sometimes user friendliness which sells one system over another, rather than what can be produced by either system. This paper will explore the concept of user friendliness in geographic information systems from the standpoint that a truly friendly system has to have friendly 'supports' on which to build. These 'supports' include a database management system, a device independent graphics package, a human-machine interface, and a highly available computer system. Following a discussion of these supports, a brief description of a prototype system utilizing these 'supports'.

GEOGRAPHIC INFORMATION SYSTEMS

Many professions have developed automated methods for assisting in data analysis and decision support. Terms such as 'decision support system' (DSS), 'management information system' (MIS), and 'interactive information system' (IIS) are commonly discussed and closely related. So too are geographic information systems (GIS), which will be defined here as a set of procedures that support the acquisition, manipulation and use of spatial data. Until recently there was no interaction between these fields to try and utilize or develop concepts which are in fact common to all the above systems [Johnson, 81].

There are obvious differences in the nature, purpose, and requirements of different forms of information systems. In geographic information systems the primary difference is its use of spatial data. Spatial data can be defined as any attribute, entity, or identifier which can be referenced to a geographic place. It must be recognized, though, that geographic information systems are just another flavor of decision support systems, and face many problems and considerations common to its non-geographic counterparts.

USER FRIENDLINESS IN INFORMATION SYSTEMS

Regardless of the type of information system, all decision support systems have one common criteria; a well designed, "friendly" user interface. It is not uncommon to see over 50% of generated code devoted to the user interface in non-geographic systems. In contrast, geographic information systems often devote 30 - 35% of their code to a user interface [Nicholson, 83]. Does this mean that geographic information systems are inherently less user friendly than their non-geographic counterparts? No! What this indicates is that "user friendly" is one of the most ambiguous criteria faced by system designers.

What Is User Friendliness?

While concrete definition may be difficult to arrive at, there are several attributes of user friendly systems that can be generally accepted. First, system directions or prompts should be understandable by non-technically oriented personnel. Second, the system should be flexible enough to provide shortcuts for experienced or technical personnel. Third, error messages should be clear in their meaning, and provide some direction on how to correct the problem that occurred. Fourth, the system should be able to handle different approaches to the same query. Fifth, there should be verbose directions under a 'help' umbrella, to fully illustrate what is expected of the user. Sixth, the system should be available when the user wants to use it.

These attributes are not all inclusive, but illustrate some primary areas where systems commonly have problems. Designing and developing user friendly systems is an evolutionary process. Obviously today's interactive systems are much more friendly than the batch oriented systems of days gone by. Even today, though, the person accessing the system is often not the decision maker but rather a middleman, possibly because of the decision makers concern over his ability to extract what he wants from the system. This will probably change in the future because today's middlemen will become tomorrow's decision makers, but that in and of itself will not mean that systems are more friendly. The continued evolution of user friendly systems depends upon incorporation of new hardware technologies, and software methodologies.

DEVICE INDEPENDENT COMPUTER GRAPHICS

A common occurrence in earlier geographic information systems was the constraint of what graphics device could be utilized to view system output products. While this particular constraint helps the turnkey vendor who offers a total system, it negatively impacts a buyer who has similar existing equipment and limited budgets. Dependence on specific devices or protocols also leads to major software rewrites when display devices become obsolete, or new display technology calls for upgrading the existing system.

Since the early seventies, this problem has led to attempts to develop and standardize device independent computer graphics software [Scott, 84].

What is Device Independence?

Device independence is the characteristic of a graphics support package which allows application programs to run on different types of display devices [Warner, 81]. For instance, a program which generates a map utilizing solid fill colors on a Tektronix (or any other) raster display device can create the same image on a Hewlett-Packard (or any other) pen plotter without any change to the source code. This is accompilished by the graphics support package, which maps the application program graphics requirements into the protocol of the intended display device. This protocol mapping is handled by individual device drivers.

Device independence decreases the impact that device obsolescence has on geographic information system development. It provides the developer with a set of graphics tools which allows the system to address a virtual, universal display device [Scott, 84]. Buyers of device independent GIS technology also benefit from the abilty to incorporate any hardware they may already have, as an integral part of their system.

Standards in Device Independence

Standardization of device independent graphics support packages is an obvious necessity to insure transportability of developed software, and programmer skills. Currently, several standards either exist, have been proposed, or are in development. These standards include the SIGGRAPH Core, Graphics Kernel System (GKS), North American Presentation Level Protocol Syntax (NAPLPS), and Programmer Hierarchical Interface to Graphics Standards (PHIGS), as well as the Initial Graphics Exchange Specification (IGES) [McLeod, 84; Scott, 84; Warren, 84]. Soon it should be possible for entire geographic base files and graphic databases to be transferred from one vendor system to another. Imagine the time and cost savings that could be realized if your entire graphic database which was developed on system 'X' can be directly installed on system 'Z'.

FAULT TOLERANT PROCESSING

Throughout the vast array of automated information systems, some can be seen as 'heartbeat' applications. In other words some systems have to be constantly running, highly available, or some serious consequences will result. Obviously these types of operations can not afford the slightest amount of down time. In less critical systems down time can also pose serious consequences if the system is not available when you need it. How many times has a demonstration been scheduled only to be scrubbed or postponed because the 'computer is down'? How much programming time has been lost due to down time on development systems? The need for fault tolerance then not only applies to 'heartbeat' systems, where continuous processing is critical, but also to any system where availability is important at any point in time.

What is Fault Tolerance?

A fault tolerant computer system is one in which any single failure is transparent to the user. To be considered fault tolerant the following characteristics must be present. First, each critical component must be replicated to allow replacement after failure. Second, the system must be able to identify a failed component automatically. Third, the failed component must be isolated electrically and logically from the rest of the system. Fourth, the system must reconfigure itself to continue processing uninterrupted. Fifth, the faulty component must be repairable without disrupting ongoing system operation. Sixth, once repaired the component must be reintroduced to the system. Finally, no component failure should allow the database to be corrupted [Highleyman, 84].

Until recently there was only one fault tolerant computer system on the market. In the past five years, though, over a dozen new systems have become available utilizing different means to the same end [Serlin, 84]. To illustrate how fault tolerant systems operate, consider yourself working away on a machine that appears to be operating normally. Perhaps what you don't see is that one of the duplexed CPU boards malfunctioned, which triggered an automated call from your system to the central support center across the country. There the trouble was logged and a spare part placed in overnight mail to your office. When the part arrives the operator simply opens the cabinet pulls out the failed board and slides in the replacement, while your important demonstration continues uninterrupted.

DATABASE MANAGEMENT SYSTEMS

The heart and soul of any geographic information system is the database management system (DBMS). In general, all database management systems serve to separate the user's logical view of the data from the physical organization of the data. The differences between hierarchical, network, and relational systems are mainly the way they allow the user to view the data, and the way they map that view into the physical organization [Martin, 77]. Within the past few years relational database management systems have become more widely accepted as vendors and buyers see some inherent benefits in the relational model over the network or hierarchical model.

What is a Relational Database Management System?

There are several criteria for labeling a DBMS as relational. First, the database is represented in the form of two-dimensional tables. Second, there are no navigational requirements imposed on the user. Third, any number of tables may be projected or joined [Venkatakrishnan, 84]. In terms of applications, the benefits of a relational DBMS can be summarized as follows:

application programs are smaller, and easier to develop; fewer application errors; shortened development time; ability to handle ad hoc queries; and ease of expanding data elements and tables. The relational DBMS is ideally suited for applications where queries are not well known ahead of time, because retrieval schemes and logic do not need to be understood by the user [James, 84; Wood, 84].

Neither traditional nor relational data models are going to be able to efficiently handle all the requirements of a geographic information system. However the wide range of queries that might be asked of a GIS, and the need to be able to handle new layers of information, or modify existing ones, clearly indicate a relational model should be the basis of geographic information systems.

HUMAN - MACHINE INTERFACES

The most visible aspect of any information system is its user interface, how the machine communicates with the human user. Not surprisingly, it is also the most common area where users have problems with a system. Users generally have a good idea of what they want from the system, and a general understanding of how to instruct the system to perform the task. Everyone, though, occasionally commits errors, needs extra information, or has a lapse in memory. When these situations occur the system should be a forgiving friend to the user, helping them back on track. What too often happens is the messages or actions taken by the system are unclear in meaning, and further confuse the user raising their frustration level.

Forms Management Systems in Geographic Information Systems

A contributing factor to user confusion can stem from the fact that different application modules were developed by different programmers with different levels of sympathy for the user. One module may have well designed, explicitly worded prompts and error messages, another may have a tersely worded prompt or error message with ambiguous meaning [Norman, 83]. The incorporation of a forms management sytem (FMS) can limit some of these problems because it provides uniformity in what the user sees on the screen, usually in the form of menus. A robust FMS will include capabilities for explicit help screens, error messages, look back features, and terminal independence. In addition to assisting the novice user, FMS packages typically allow random traversing of the menus. This is a benefit for the saavy users who know exactly what they want and what menus they need to answer [Mason, 83].

Another feature of robust FMS packages was briefly noted above, terminal independence. Similar to device independent graphics software, terminal independence allows a user to work with the system utilizing any terminal type he may have available. The application program is developed to utilize a virtual terminal, and the FMS translates the virtual commands into the volcabulary of the target device. This feature is terrific for information sytems with time sharing users in remote locations. If the user has a terminal that uses TTY protocol, they can access the system just as well as an onsite user who may have VT100 protocol.

BUILDING A USER FRIENDLY INFORMATION SYSTEM

Utilizing the previously discussed 'supports'; device independent graphics; relational database management system; forms management system, and fault tolerant processing; it is possible to develop a more totally friendly geographic information system. A system which can minimize initial costs for display peripherals; a system that can handle ad hoc queries with no a priori knowledge by the user; and a system that is available when it is needed. To illustrate, the following description details the purpose, requirements, and solutions for a geographic information system with parcel level resolution both for the geographic base file and attribute information, for which the prototype currently exists.

Development of the GIS was commissioned to provide one immediate capability, and capacity for some specific enhancements to enable broadening the user base at a later time. The immediate capability dealt with allowing a distributed user network to query a geographic data base for prior title policies on specific parcels of land. Expanded capacity was specified as including capability for demographic analysis, tax appraisal, title exceptions and property liens, and structural attributes for realty evaluation.

Immediately, it is apparent that the database will require extension and modification in the future. Queries will be ad hoc in nature and may need to aggregate parcel data to block or census tract level, or retrieve census data based on title policy numbers. After reviewing several DBMS packages based on network and relational models, it was clear that the relational model was the one that could most efficiently manage the data to be resident in the system.

In terms of peripherals, most users or potential users would already have some type of terminal or personal computer. The need to minimize additional hardware cost was an explicit requirement of the finished system. Even though most of those existing terminals are non-graphic, a large majority of them can be retrofit with a graphics board to emulate one of several graphics devices. Device independent graphics software is the only sensible alternative in this case, because it reduces system development time, and allows the user to maintain his familiar equipment.

Menu driven prompts was another requirement for the user interface. Most existing users had experience on a time sharing non-geographic system which was to interface the finished GIS for providing title information. This existing system was menu driven, but not by a FMS. A design decision was made to utilize a FMS package to aestheticaly emulate the structure of the existing menu system, provide terminal independence, and reduce development costs.

Finally, the system is expected to be heavily utilized and system availability is a must. The prototype system has therefore been developed on a fault tolerant processor to eliminate system down time. While it appears at first glance that fault tolerance would be prohibitively expensive, in fact the difference between a DEC VAX processor and some equally powerful fault tolerant systems entails only about a ten percent premium for fault tolerance. How often would you have been willing to pay ten percent more for a system that would not go down at all the wrong times?

CONCLUSIONS

User friendly geographic information systems will never be 100% user friendly, because everyone has a different definition of user friendly. It is currently possible, though, to develop systems that:

o   Do not require a priori knowledge of database navigation paths,

o   Do not require purchasing of peripherals for which a user already has similar equipment in place,

o   Do not require cryptographic translation of error messages or responses to 'help',

o   Do not require users to use the excuse, or operators to invent explanations why, 'the computer is down'.

A geographic information system with these features is surely a positive development in the continued evolution of user friendly systems.

REFERENCES

Highleyman, W. H., 1984, "The coming day of fault tolerance":Computerworld, September 10, pp. ID/1 - ID/11.

James, B., 1984, "Network vs. relational: Fit the solution to the need": Computerworld, September 24, page SR/38.

Johnson, T. R., 1981, "Evaluation and Improvement of the Geographic Information System Design Model": an unpublished Masters Project, Department of Geography, State University of New York at Buffalo.

Martin, J., 1977, Computer Data-Base Organization: Prentice-Hall, Inc., Englwood Cliffs, New Jersey.

Mason, R.E.A. and Carey, T.T, 1983, "Prototyping Interactive Information Systems": Communications of the ACM, Volume 26, No. 5, pp. 347 - 354.

McLeod, J., 1984, "Graphics Capabilites Abound": <u>Systems & Software</u>, Volume 3, No. 8, pp. 82 - 94.

Nicholson, R. L., 1983, "Mapping for Decision Makers": <u>Computerworld</u>, November 28, pp. ID/43 - ID/48.

Norman, D. A., 1983, "Design Rules Based on Analyses of Human Error": <u>Communications of the ACM</u>, Volume 26, No. 4, pp. 254 - 258.

Scott, G. M., 1984, "Practical Implementation of GKS": <u>Computer Graphics World</u>, Volume 7, No. 12, pp. 35 - 38.

Serlin, O., 1984, "Fault-Tolerant Systems in Commercial Applications":<u>Computer</u>, Volume 17, No. 8, pp. 19 - 30.

Venkatakrishnan, V., 1984, "Handling data integrity in relational DBMS": <u>Computerworld</u>, October 8, pp. ID/25 - ID/30.

Warner, J. R., 1981, "Principles of Device-Independent Computer Graphics Software": <u>IEEE Computer Graphics and Applications</u>, Reprint from October Issue, pp. 1 - 10.

Warren, C., 1984, "Graphics Software Schemes Enhance Peripheral Interfacing": <u>Mini-Micro Systems</u>, Volume 17, No. 9, pp. 163 - 178.

Wood, D., 1984, "Emphasis upon physical data structures can distract from accurate analysis of relational DBMS makeup": <u>Computerworld</u>, September 24, pp. SR/12 - SR/13.

The Conceptual Design for Products from the National
Ocean Service's Automated Nautical Charting System II

Bruce Dearbaugh
Robert J. Wilson
Walter M. Winn, Jr.
National Charting Research and
Development Laboratory
Nautical Charting Division
National Ocean Service, NOAA
Rockville, Maryland 20852

ABSTRACT

The National Ocean Service (NOS) is currently upgrading the Automated Information System to more effectively serve the needs for both digital cartographic data and digital marine navigation information. The needs for data and information with respect to qualities, quantities, product types, and distribution schemes in support of NOS nautical products are well defined; however, the needs of those users outside NOS are unclear.

The intent of this paper is to make the marine information community aware of the current status of the Automated Information System, the conceptual design for the next generation Automated Nautical Charting System II, and the current NOS perception of the community's needs for digital marine navigation information.

DIGITAL ELEVATION MODEL IMAGE DISPLAY AND EDITING

Melvin De Gree
Robert G. McCausland
U.S. Geological Survey
526 National Center
Reston, Virginia 22092

BIOGRAPHICAL SKETCHES

Melvin De Gree is Chief, Branch of Applied Sciences, Office of Systems and Techniques Development, National Mapping Division, U.S. Geological Survey. He received a B.S. degree in electrical engineering from North Carolina A. & T. State University in 1965, and an M.S. degree in computer science from Howard University in 1975. He joined the USGS from private industry in 1981, where he gained the majority of his career experience in large-scale software systems engineering, design, and development.

Robert G. McCausland is a Senior Electronics Engineer in the Branch of Applied Sciences, Office of Systems and Techniques Development, U.S. Geological Survey. He received a B.S. degree in electronics engineering from Virginia Polytechnic Institute in 1960. Prior to joining the USGS in 1979, he served in several electronics design and development capacities with the Harry Diamond Laboratory, U.S. Department of the Army.

ABSTRACT

The display, manipulation, and editing of digital elevation models, prior to their entry into the National Digital Cartographic Data Base, using an Image Display and Editing System are addressed. Visual inspection and verification are readily performed by computer processing the DEM data to create color-banded-elevation displays, shaded-relief displays, and anaglyphic stereodisplays, and observing the displays on an image display cathode ray tube. Histogram equalization and pseudocolor enhancement techniques can be applied to the elevation points.

The imagery is derived from DEM data on the host computer and transferred to an interactive display system. Areas needing editing are outlined into graphic image overlays, which are then used with DEM editing programs. For individual point editing, the overlay degenerates into a single point. The display techniques are implemented in a combination of FORTRAN and Assembler languages on an LSI 11/23 computer.

Publication authorized by the Director, U.S. Geological Survey, on December 18, 1984. Any use of trade names and trademarks in this publication is for descriptive purposes only and does not constitute endorsement by the U.S. Geological Survey.

INTRODUCTION

The U.S. Geological Survey has undertaken the creation of a spatial topographic data base for the United States, and the task of defining, generating, and maintaining this data base has been assigned to the National Mapping Division (NMD). In response to this assignment, the NMD has created the National Digital Cartographic Data Base (NDCDB). The NDCDB will contain thousands of digital elevation models (DEM), one of the basic categories of the data base and the need for comprehensive interactive display and editing of DEM data has been recognized.

The display, manipulation, and editing of DEMs before and after their entry into the NDCDB provides a powerful and efficient tool for error detection, error correction, and quality control. This capability is provided by the DEM Image Display and Editing System (IDES) described herein. The system provides for visual inspection and verification by computer processing the DEM data to create color-banded-elevation displays, shaded relief and slope displays, and anaglyphic stereo displays, which can be viewed on an image display cathode ray tube (CRT). Image enhancement techniques such as pseudocolor and histogram equalization can be applied to the elevation data.

Interactive editing techniques can also be applied to the elevation data, including single point, editing based on either simple point replacement or sophisticated neighborhood averaging, and area editing, based on either a priori information or on data collected from an online map digitizing capability.

REQUIREMENTS

Design of the IDES was determined by consideration of seven basic factors:

o Low cost,
o Effectiveness in a production environment,
o Compatibility with the NDCDB data structure,
o Capability to display data from a standard 7 1/2-minute quadrangle,
o Capability to display data from a standard 7 1/2-minute quadrangle in the Gestalt Photomapper (GPM) format,
o Capability to interactively edit individual points, and
o Capability to interactively edit areas, i.e., collections of points.

Applications at several points in the production process were envisioned for the IDES, and it was determined that in order to be cost effective, the system cost should not exceed $100,000. The production environment dictated relatively unskilled operators, high throughput, and an easy-to-use system. The National Mapping Division DEM tape specifications determined the software format for the system, while the need to display standard 7 1/2-minute quadrangles was the controlling factor on system display memory requirements.

The need to perform effective editing was basically driven by three factors: an interactive technique with an efficient and easy-to-use operator-machine interface; editing of isolated points; and editing of water body errors.

The average Gestalt Photomapper DEM contains approximately 614,400 points, requiring a memory of 1,228,800 bytes for full storage. The GPM data may contain errors which must be corrected before further processing is applied. After the errors have been resolved, the GPM data are processed with a resampling algorithm which produces a relatively sparse DEM in the NDCDB format. After the resampling process, the average DEM contains approximately 164,000 elevations, which require a memory of 328,000 bytes for full storage. DEM data which are currently in the NDCDB may also contain both isolated point elevation errors and water body elevation errors (areas) which require further processing for correction. Therefore, the system must handle data in both GPM and NDCDB formats.

SYSTEM DESIGN

Design tradeoffs among cost, user training, speed, and memory requirements led to a system based on the Digital Equipment Corporation (DEC) LSI 11/23 microcomputer system. The system configuration is shown in figure 1. The LSI 11/23 is an MOS/LSI technology machine which brings the functional power of the PDP 11/34 minicomputer to the microcomputer level.

The Comtal Vision ONE/10 Color Graphic Image Display System forms the nucleus of the DEM viewing capability. It is an interactive system which provides a wide variety of image manipulation and display functions in a combination of well-designed firmware/software capabilities. Chief among these are such functions as a four-to-one hardware zoom factor; full image roaming; complete random addressing to a single pixel element; nondestructive superimposition of images, and combinations of image addition, subtraction, division, and multiplication.

In addition, a 24-bit color representation can be obtained in a configuration which provides address space for 8 image planes of 24-bit color. The Comtal is linked to the LSI 11/23 via a Direct Memory Access (DMA) channel to provide high image transfer rates, which enhances operational speed. Standard DEMs are input to the IDES via the magnetic tape drive, and a 30-megabyte (MB), Winchester-type magnetic disk drive provides ample storage for the various display files, applications programs, and the DEC RT-11 operating system programs.

Operator input to the Comtal is through use of an alphanumeric keyboard and a track-ball cursor control unit. Operator input to the LSI 11/23 is provided through a

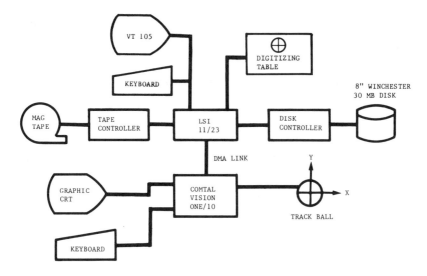

Figure 1.—DEM image display and editing system.

standard CRT terminal. Digitized map point and line data are entered directly into the LSI 11/23 via a manual digitizing table.

DISPLAY DESIGN

The main purpose of the IDES is to detect errors and anomalies in the DEM data. These errors may be classified as substantive or cosmetic and may be due to a number of causes, e.g., missing or inaccurate data, datum shifts, correlation failures in the case of automatically scanned imagery, overlooked data in the case of manually scanned maps, abberations due to software logic, software faults, and others. The following displays were selected as having the best potential for highlighting these errors:

- o Color-banded-elevation,
- o Shaded relief, and
- o Anaglyph stereo.

Finite memory size is the driving element of display software design. The LSI 11/23 has an 18-bit extended memory space, which allows for only 262,144 bytes of real memory space. Conversely, the set of DEM images on the Comtal consists of four 512 x 512-pixel arrays of 8-bit points, which translates into 524,288 16-bit words (generally written as 512K words). This means that two Comtal images would require four times the entire address space of an LSI 11/23 if it were processed at one time.

To overcome this obstacle, each DEM tape data file of approximately 328,000 bytes (representing a standard 7 1/2-

minute quadrangle) is reformatted into a series of 512 x 512-pixel arrays or image planes, with separate image planes provided for the upper and lower bytes. As shown in figure 2, each array is partitioned into 16 smaller arrays, each consisting of a 32 x 512 x 2-byte array. Each of these smaller arrays is manipulated rather nicely within the LSI 11/23 address space. For DEMs which do not fill a particular subarray, the remaining points are zero filled. The successive 32-kilobyte subarrays are then stored on disk as a partitioned subimage matrix. This disk

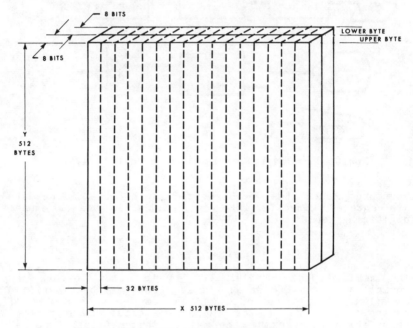

Figure 2.--DEM data partition array.

file is used by the display generation applications programs to process the various displays. Successive subimages are then linked together to form the 512 x 512-pixel image planes for the IDES, thereby allowing display of a full 7 1/2-minute quadrangle. The display image planes are stored back on the same disk, where they can be transferred to the Comtal Image Display System.

Color-Banded Elevation Display

The color-banded-elevation display uses point elevations to modulate a color intensity function. The function has a 24-bit range, with 8 bits assigned to each color. Point elevations are contained in the DEM so the color banding is straightforward. The major advantage of this display is to show slope irregularities, since contours of equal height should be of the same color. In addition, the display gives the experienced observer a feeling for the real appearance of the terrain.

Shaded-Relief Display

The shaded-relief display uses the slope between points to modulate an intensity function of a monochrome image. The intensity function is further modulated by an illuminance function, which represents sunlight coming from the north-west. Figure 3 shows the typical shaded-relief computational model. The amount of light striking a unit surface, S, on a horizontal map plane is related to the slope angle, Θ_s, of the terrain, and the elevation angle, Φ, of the light source (Monmonier, 1982). A surface sloping away from the light source will receive and reflect less light than one facing the light source. The amount of light, L, impinging on this surface is a function of the inclination angle, Θ_s, of the surface facing the light source and the elevation angle, Φ, of the light source above the horizon. L can be computed as L = S SIN $(\Phi + \Theta_s)$. In this manner, an intensity value is

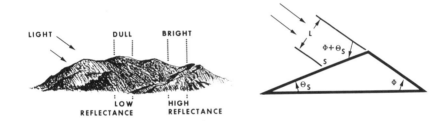

Figure 3.--Shaded-Relief computational model.

calculated for each point in the DEM. The intensity value is then translated to a display brightness value via a software table lookup function.

Figure 4 shows a typical DEM for which a shaded relief is computed. The shaded-relief display overcomes the limitation of the color-banded-elevation display by providing an approximate view of the terrain. However, it still provides no direct elevation information. Despite this rather undesirable result, it does provide the major advantage of displaying the terrain shape. The GPM correlation algorithm is deficient in water bodies and other poor contrast areas. The computer program which corrects these deficiencies rounds off all elevations to within plus or minus one meter. The resulting mismatch tends to show as a series of lines on the display.

Anaglyph Stero Display

The anaglyph stereo display is one which uses two super-imposed images, each with a slightly different view of the same surface. Each image is generated in a different color, one red and one cyan. The viewer wears a pair of

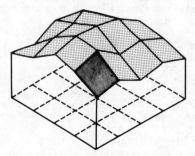

Figure 4.--DEM for shaded-relief computation.

filter glasses, of which each lens transmits only one color, and the viewer thereby perceives a three-dimensional view.

The geometry of the anaglyph stereo display is based on well-known photogrammetric principles. In order to generate a stereopair which is perceived to have elevation, one image is shifted away from the position where the eye would normally find the unelevated image to the intersection of a line of sight through the elevated image and the ground surface (Monmonier, 1982; Slama, 1980). Figure 5 illustrates this principle.

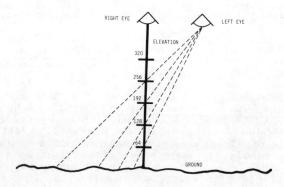

Figure 5.--Anaglyph stereo-display principle.

Since a CRT is composed of fixed-interval pixels, the anaglyph principle previously stated must be modified. This is accomplished by three simplifying procedures.

First, a shaded-relief image is calculated and held constant. Second, parallax is introduced into the first image in order to create a second image. Each corresponding pixel in the second image, relative to the first, is shifted one pixel for each 64-meter increase in elevation. This produces a somewhat less exaggerated three-dimensional image than a comparable true stereoscopic model, but one which, nevertheless, is quite acceptable to the eye. Finally, the intensity values for the shaded-relief image are assigned to the three 8-bit color planes precisely as if each image were a standard shaded relief. The left image is assigned to red, while the right image is assigned to both blue and green to yield cyan. The resulting display has the appearance of a full anaglyph stereo terrain model. The major advantage of this display is that an operator can rather easily identify elevation and slope errors.

EDITING

The IDES brings the power and simplicity of interactive processing to the task of editing. Editing needs can be essentially divided into two classes, cosmetic and substantive.

Cosmetic editing may include such functions as eliminating systematic tilt in GPM patch-derived data, smoothing of shoreline data around water bodies, and edge-matching of GPM patches. Substantive editing may include adding missing profile points based on nearest neighbor algorithms and correcting water-body elevation errors through interactive graphic polygon-fill methods. The latter problem is particularly prevalent in GPM data due to its inability to correlate images of areas which have poor contrast or are homogeneous in reflectance values, such as water, snow, and certain crops. Substantive editing also includes elimination of gross errors at isolated points, which singly would have little effect on the root-mean-square error of the DEM, but are troublesome to users during data applications.

Three basic editing capabilities have been provided for both point and area elevations:

    o Averaging,
    o Replacement, and
    o Increment/Decrement.

Averaging replaces the value of a point with a weighted average computed using the eight nearest neighbor points. The weighting factors are interactively entered into a 3 x 3 matrix. Figure 6 illustrates the weighting matrix.

W1	W2	W3
W4	W5	W6
W7	W8	W9

Figure 6.--Weighting Matrix

The replacement capability simply replaces the value of a point with an operator-selected value, which is entered interactively. The Increment/Decrement capability incrementally adjusts a point value by plus or minus 1. The system operator interactively selects which capability is to be used for a given editing task.

Single point editing is accomplished interactively using the Comtal display screen and the system console CRT. Using the track ball, the operator selects a point to be edited by moving the Comtal screen cursor to the desired point on the elevation display. The software uses the Comtal cursor position to define a 64 x 64-pixel processing window and moves the window data from disk to memory. An independent 10 x 10-pixel display window, initially centered around the selected point, is created for display on the system console CRT. The elevation of each point within the 10 x 10-pixel window is displayed on the console CRT, with the current cursor position being highlighted on the display. In effect, this window displays a 10 x 10-pixel matrix of elevations about a point. Figure 7 illustrates the matrix of elevations display.

2149	2161	2175	2190	2204	2215	2225	2238	2248	2258
2144	2157	2172	2184	2195	2204	2214	2224	2236	2247
2135	2149	2163	2175	2183	2192	2201	2211	2223	2234
2127	2137	2149	2161	2169	2178	2189	2200	2211	2221
2115	2125	2134	2143	2155	2165	2176	2187	2198	2208
2100	2111	2120	2129	2138	2150	2162	2173	2185	2195
2085	2095	2104	2113	2123	2136	2149	2160	2172	2181
2069	2077	2086	2096	2107	2122	2135	2146	2158	2168
2057	2058	2066	2078	2093	2108	2121	2133	2145	2155
2045	2046	2052	2063	2079	2094	2106	2118	2131	2141

X POSITION   13   Y POSITION   254
ENTER COMMAND (AVERAGE, INCRMENT, REPLACE, OR EXIT)

Figure 7.—Matrix of elevations display.

The result of the editing is shown on the 10 x 10 matrix of elevations display and on the Comtal elevation display. As the operator moves the cursor around the display, a new 64 x 64-pixel window and the corresponding 10 x 10-pixel windows are calculated and stored in memory, as appropriate. Each modified 64 x 64-pixel block is then written to disk, as necessary.

Area editing is accomplished by having the system operator use the Comtal track ball to outline an area on the Comtal display screen. The display may be a color-banded-elevation or a shaded relief. The software transforms the operator-defined area into a graphic overlay. The graphic overlay is scanned and the operator-selected algorithm is performed on each point within the overlay. After the scan is completed, the display is updated with the new information.

The location of map features from published maps, map separates, or orthophotos can be transferred to the IDES using the digitizer table. The outline of the feature is traced on the digitizer and transferred to the IDES, where it is loaded into a graphic plane. A map registration program is used to align the graphic plane boundaries with the stored DEM data. After the editing is completed, the display is again updated with the new data.

CONCLUSION

The IDES provides a practical and cost-effective means for viewing and editing DEMs formatted as 7 1/2-minute quadrangles. A powerful image processing system, coupled with interactive software, provides a significant error detection and correction capability. These capabilities can be used for both substantive and cosmetic editing. In addition, its wide color range affords a significant ability to detect slope irregularities. The IDES also provides anaglyph stereo and shaded-relief displays comparable in quality to those generated on high-powered and more expensive minicomputers. A configuration comparable to that shown in figure 1 can be reproduced for approximately $65,000. In all, the combination of a low-cost microcomputer, a state-of-the-art display system, and an innovative, interactive software design provides powerful image display, manipulation, and editing capabilities for use in a production environment for error detection and correction, and quality control of digital elevation models.

BIBLIOGRAPHY

Monmonier, Mark S., 1982, Computer-Assisted Cartography: Prentice-Hall

Slama, Chester C., Editor-in-Chief, 1980, Manual of Photogrammetry: American Society of Photogrammetry.

REDUCING THE NUMBER OF POINTS IN A PLANE CURVE REPRESENTATION

Terry J. Deveau

Halifax Fisheries Research Laboratory
Department of Fisheries and Oceans
Scotia-Fundy Region
P.O. Box 550
Halifax, N.S., Canada, B3J 2S7

Telephone: (902) 426-4224

ABSTRACT

The most common representation of a plane curve in a digital computer system consists of an ordered set of points implicitly joined by straight line segments. An algorithm is presented which attempts to reduce the number of points in this representation while retaining the curve's salient features. The algorithm is compared with two other published procedures.

INTRODUCTION

General Problem
When a plane curve is represented by a subset of its points, two requirements arise immediately:
(a) the number of points included in the subset should be minimized to lessen the resources required in storing and processing the curve,
(b) the subset selected should be such that the polygon formed by joining the member points with straight line segments matches the original curve within some acceptable tolerance.

Note that in other contexts the 'straight line segments' of (b) could be replaced with 'circular arcs', 'cubic splines', 'conic segments', or some other curve family; this work deals only with the straight line segment case.

Clearly, (b) can be satisfied by increasing the size of the subset; we need some way to find the smallest sufficient subset. In developing such an algorithm we must be mindful of a third constraint:
(c) the processing required to select the reduced points should be kept to a minimum.

Application to Cartography
Consider the problem of rendering the same coastline repeatedly on a number of separate maps as an example of the practical aspects of this problem. The coastline may have been digitized at high resolution and stored in a computer data base. The required maps could be at various scales, requiring that only a subset of the data base be used for each map, to provide the proper generalization of coastline features and effective use of the plotting hardware.

To avoid the cost of having several copies of the coastline at different resolutions stored in the data base, an efficient reduction algorithm is required which can produce the desired subset at only a small additional cost over that of a straight sequential read.

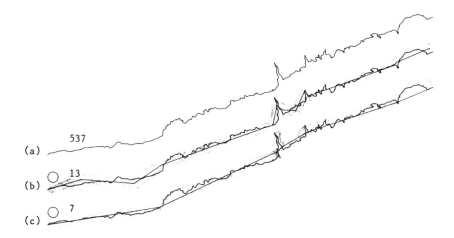

Figure 1. (a) A section of the 500 fathom depth contour off the Scotian Shelf. The number at the left is the number of points in the line. (b) The same contour overlaid with a reduced line fit using a centred band (NW=1), the tolerance used (TOL) is indicated by the diameter of the circle at the left and the width of the shaded band; (c) shows a second fit using a floating band (NW=10).

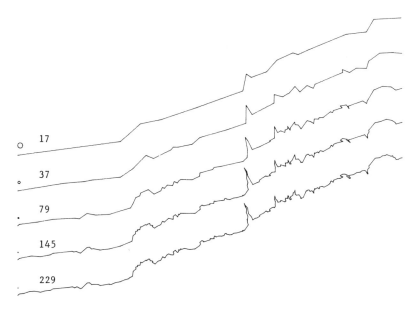

Figure 2. These lines have been fit to the same depth contour shown in figure 1. The same parameters were used as in line (c) of figure 1 except for TOL which was halved for each successive fit.

Additional Requirements
After consideration of the practical problems associated with automatic cartography we find that there are additional features we would like to see in the reduction algorithm:
(d) works with open as well as closed curves;
(e) avoids systematic deviations from original curve (even within tolerance);
(f) prefers members of original set for inclusion in the reduced set (i.e. interpolation minimized);
(g) promontories not blunted indiscriminately;
(h) eliminates small closed curves in the furtherance of cartographic generalization (must use a test which produces results in line with the decisions a cartographer would make).

With the exception of (h), these further requirements need not compromise the algorithm's ability to be used for non-cartographic work (as long as attention to them does not make the algorithm significantly more expensive to use).

DEVELOPMENT

General Solution
The first step in developing the required algorithm is to quantify what is meant in (b) by 'match within some acceptable tolerance'. One way is to require that a band of width TOL, centred on the reduced polygon, include within it the original curve. The value TOL then becomes a parameter driving the reduction algorithm. Algorithms which use this fit criterion will be called 'centred band' algorithms.

For cartographic applications however, a second (looser) definition is often useful: for each segment of the reduced polygon, some band of uniform width TOL must exist which contains both the polygon segment and that section of the original curve. This will be referred to as the 'floating band' criterion; it allows very narrow (but long and straight) anomalies to be removed when reducing the curve (e.g. a coastline broken by a small river).

The floating band criterion also allows fewer polygon segments to represent the curve to the same (numerical) tolerance than is allowed by the centred band approach. An illustration of the two methods is shown in figure 1 and the effect of various values of TOL in figure 2.

In consideration of (c) we want to process the points in sequential order and compute the reduced set during a single pass. In general terms, the solution will be obtained by finding a band which contains two consecutive points in the original curve, then adjusting this band for each subsequent point so that it includes all points under consideration. When a point is found for which such an adjustment is not possible, the first and last points which did fit are used as vertices of the reduced polygon. The process is then repeated starting with the latest polygon vertex.

Practical Algorithm
The algorithm presented here uses the first point to fix one degree of freedom of a line through the centre of the tolerance band, leaving a single degree of freedom for fitting the band to subsequent points. That freedom can be viewed as the angular orientation of the band.

Since the band's orientation will be constrained within two extreme angular positions, the centre lines of these two limiting positions are maintained by the algorithm. Both lines pass through the starting point and form the edges of an **acceptability wedge**. Each subsequent point must come within TOL/2 of this wedge, otherwise the last point to fit will become a new vertex in the reduced polygon. One or both edges of the wedge will usually be adjusted with each point until a new vertex is required.

This approach allows the new vertex to fall near one edge of the tolerance band or anywhere inside it. The initial vertex however, has been restricted to the centre of the band. This is halfway between the criteria defined above because the tolerance band is centred at one end and floating at the other.

If instead of using the last point to fit as the new vertex, we use the nearest point to it which is in or on the wedge, the new algorithm becomes a strictly centred band procedure.

A floating band algorithm can also be derived from the above by keeping track of several wedges simultaneously, in addition to the wedge starting at the last polygon vertex. These additional wedges can start around the circumference of a circle of diameter TOL centred on the last vertex. When a point is found which is further than TOL/2 from one of the wedges, that wedge is dropped from the active set. Only when the last wedge is dropped does the last point which did fit become the new vertex.

The number of wedges (NW) is a parameter characterizing the algorithm. The author has implemented an algorithm which uses a centred band criterion when NW=1 and a floating band criterion for NW>1; this allows the best method to be selected for each application. Figure 5 shows several reductions of the same curve with different values of NW.

The centred band method outlined above can be improved slightly by using several extra wedges on the first segment and using the starting point of the last surviving wedge as the first vertex instead of the initial point of the curve. Credit for this idea belongs to Ivan Tomek who embodied something similar in his 'Algorithim F' (Tomek 1974).

Delays in Assignment of Initial Points

Features of a curve which have dimensions less than TOL are removed during reduction. If the entire curve (whether open or closed) has no dimension greater than TOL it seems reasonable to eliminate it completely. This consideration is independent of (h) above; here it is simply a matter of removing those curves which are so small that they would otherwise be reduced to just two points separated by less than TOL (or two identical points in the case of closed curves).

A practical algorithm can, in consideration of the above, delay treating the initial point of the curve until a point is found which cannot fit in the same circle of diameter TOL as all the preceding points. When this happens, the centre of the circle which has fit all points thus far can be used as the initial point.

If the final point in the curve is reached before the initial circle is exceeded, the entire curve can be dropped. Otherwise if the final point is intended to close the curve it should be replaced with the initial vertex of the reduced polygon.

Closed curves which exceed TOL in one dimension only would be reduced to degenerate curves of zero area and only two distinct points; these

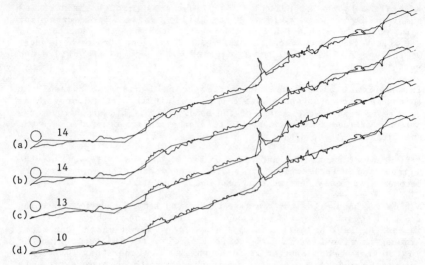

Figure 3. The 500 fathom depth contour is again fit with the same parameters as in figure 1 (b) except for bump angle (ANG). The effects of varying ANG can be seen here. Note that for a feature to be considered as a "bump" it must usually exceed TOL in both height and width. The values of ANG used here are (a) 0, (b) 60, (c) 120, and (d) 180.

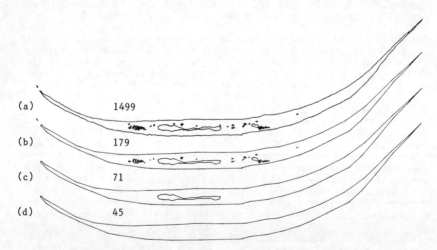

Figure 4. The coastline of Sable Island including some of its ponds. (a) Prior to reduction. (b) After reduction with FACTOR = 0.0; notice that the smallest features have been removed (no dimension > TOL). (c) After reduction with FACTOR = 1.0 and (d) with FACTOR = 2.0; use of FACTOR = 3.0 would completely remove the island itself. Other reduction parameters were held fixed in (b) through (d). As elsewhere, the number above each curve gives the number of points in the curve.

can also be removed (the next section will discuss other features which exceed TOL in only one dimension). For this reason (and to effect the improvement mentioned in the last paragraph of the preceding section) assigment of the initial polygon vertex can be delayed beyond the acceptance of the initial curve point to when the second polygon vertex is obtained.

Meeting Cartographic Requirements

The procedure outlined thus far will largely satisfy (a) through (f). Requirement (g) demands that we define another parameter to quantify what is meant by 'indiscriminatly blunting promontories'. Lets say that if the vertex angle of a bump on the curve is less than ANG and its height and width are greater than TOL then the bump is a significant feature which should not be smoothed out in the process of finding the minimal polygon.

Figure 3 shows several reductions of a curve with different values of the parameter ANG. A value of $110°$ seems to produce good results for cartographic applications. Theo Pavlidis has also found a bimodality in vertex angles of text characters with a trough between $110°$ and $130°$ (Pavlidis 1983); this suggests that vertex angles on either side of this range are subjectively perceived as different features.

Because it is applicable only to cartography (h) can be looked at separately. Under this requirement we want to drop those closed curves which would appear too small on a map (in one or both dimensions). A simple test of the maximum and minimum extents of the curve would not do the trick however; a thin but curved feature (e.g. a horseshoe lake) would pass the test even if it was so thin on the map that both banks overlapped.

One test that was found to do the job fairly well is:

$$\frac{\text{area of closed curve}}{\text{perimeter}} > \text{TOL} \times \text{FACTOR} \qquad (1)$$

where FACTOR is set to about one or two. Curves which fail the test are dropped. Figure 4 illustrates the control over auto-generalization provided by the adjustment of FACTOR in (1).

COMPARISONS

Implimentation

The author has implemented the algorithm outlined above as a FORTRAN subroutine suitable for use with a variety of application programs. A listing of this subroutine (which is too long to be included here) is available from the author.

This subroutine has been used for removal of redundant points from line-followed digitizer output, reduction of map data from a high-resolution data base to an appropriate level of generalization for larger scale maps, and matching general graphics output to the hardware resolution of a particular plotter.

The four parameters TOL, ANG, NW, and FACTOR provide the facility for tuning the method to the particular application without requiring maintenance of separate subprograms.

Douglas and Peucker Algorithm

This is basically a centred band algorithm (Douglas and Peucker 1973)

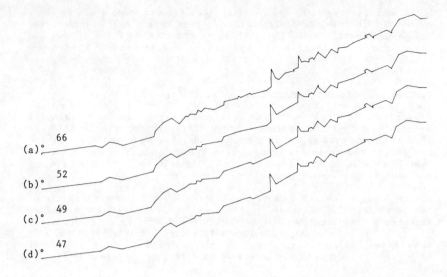

Figure 5. The effects of varying the number of wedges (NW) can be seen in these reductions of the depth contour shown in figure 1 (a). The first reduction (a) was done with NW=1 and uses a centred tolerance band; (b) NW=3, (c) NW=5, and (d) NW=10 use a floating band.

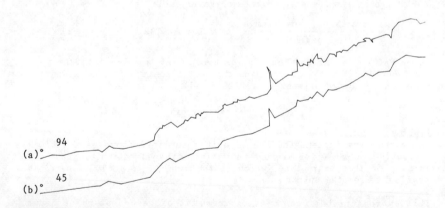

Figure 6. These reductions of the 500 fathom contour were done using the same tolerance as in figure 5; (a) was done using the Douglas and Peucker algorithm which uses a centred band. The Dettori and Falcidieno algorithm was used to produce (b), it uses a floating band.

with their 'offset tolerance' corresponding to half the band width. It has received wide recognition among cartographers as it was the first published procedure which attempted to address their specific needs.

Figure 6 (a) shows a reduced line produced by this procedure, it can be compared with figure 5 (a) which is also a centred band reduction. Note that although the same tolerance was used in both reductions, the method described here was able to find a smaller polygon.

Dettori and Falcidieno Algorithm

Dettori and Falcidieno (1982) describe a floating band algorithm which employs the convex hull of a subset to decide if a new polygon vertex is needed. Unfortunately, the example FORTRAN subroutine included in their paper contains several critical logic errors. Substantial debugging was necessary before the example subprogram accurately implemented the algorithm they presented.

This done, the method worked very well; as can be seen in figure 6 (b). This reduction was done at the same tolerance as that of figure 5 (d) and their method found the smaller polygon. Of course our use of the bump angle test (controled by ANG) means that extra (cartographically significant) points will be included beyond those required simply to satisfy the tolerance criterion.

Objective Comparison

To provide an objective basis for comparsion, a coastline file of 13,216 points was processed by these algorithms with equivalent settings of their control parameters. The input data included 794 coastal islands as well as a single 7437 point coastline curve.

The auto-generalization feature was not used for this comparison (FACTOR = 0.0) but 750 islands were eliminated by our algorithm because they were smaller than TOL in one or both dimensions.

The results are listed in table I. A CPU second is a measure of the computing effort used to perform the fit. The last two columns show the results of a second comparison run on the coastline segment only (no islands).

TABLE I - Comparison Data

Method	Points Output	CPU sec.	No Islands	CPU sec.
NW = 1	667	57.859	566	25.119
NW = 10	481	53.692	394	31.032
Douglas and Peucker	2766	58.936	909	25.339
Dettori and Falcidieno	2015	61.198	369	26.561

A more thorough and detailed comparison of these and other reduction procedures is in preparation and will be published separately.

CONCLUSIONS

The algorithm presented here provides extended control over the reduction process via four control parameters. It has several features of particular relevance to automatic cartography and provides a useful degree of auto-generalization. The flexiblity afforded by this procedure allows it to serve diverse applications where a single algorithm may have been considered inadequate.

The techniques used here can also be used with non-linear piecewise curve fitting. Circular arc fitting could be especially useful as some plotting hardware now accepts circular arc input.

The problem of auto-generalization also needs much more work with such things as merging nearby islands, deleting branching rivulets, and either widening or eliminating narrow channels offering a particularly tough challenge.

ACKNOWLEGEMENTS

The author is indebted to Dr. Robert Mohn for his valuable suggestions and comments.

REFERENCES

Dettori, G. and B. Falcidieno 1982, An Algorithm for Selecting Main Points on a Line, Computers and Geosciences, Vol. 8, No. 1, pp 3-10

Douglas, D. and T. Peucker 1973, Algorithms for the Reduction of the Number of Points Required to Represent a Digitized Line or its Caricature: The Canadian Cartographer, Vol. 10, No. 2, pp. 112-122

Pavlidis, T. 1983, Curve Fitting with Conic Splines, ACM Transactions on Graphics, Vol. 2, No. 1, pp 1-31

Tomek, I. 1974, Two Algorithms for Piecewise-Linear Continuous Approximation of Functions of One Variable, IEEE Transactions on Computers, Vol. C-23, pp 445-448

TERRAIN ANALYSIS FOR COMMUNICATION PURPOSES

Y. Doytsher & B. Shmutter
Technion I.I.T. , Haifa , Israel

Summary

Terrain analysis embraces various problems, one of these being the demarcation of areas regarded as dead ground with respect to certain stations. This problem is frequently encountered when analysing the topography in the course of planning a communication network. A solution to that problem is presented in the paper. It comprises the following steps:

- Determination of dead ground in relation to a single station.
- Determination of dead ground with respect to several stations.
- Superposition of dead ground regions.
- Definition of dead ground with respect to a moving station.

The above problems are solved on the basis of a DTM consisting of two layers of information - a grid of elevations and a layer of topographic features, break lines and salient points. When defining the boundaries of the dead ground various factors, such as the nature of the transmitting device expressing itself in the refraction of the propagating waves and the assumed accuracy of the DTM, are taken into account.

The procedure can be executed either off line, the stations being selected on existing maps, or interactively on a graphical station. In the latter case a topographic map is produced from the DTM and displayed on the CRT. In both cases, the established boundaries of the dead ground are displayed on the background of the topographic map.

Examples of dead ground determinations illustrate the discussed procedure.

MILITARY BASE PLANNING USING GEOGRAPHIC INFORMATION SYSTEMS TECHNOLOGY

Charles H. Drinnan
Manager Advanced Projects
Federal Systems Operation
Synercom Technology, Inc.
10405 Corporate Drive
Sugar Land, Texas 77478

BIOGRAPHICAL SKETCH

Mr. Drinnan has designed, developed, implemented and marketed Mapping Information Management Systems for more than ten years. He is currently Manager, Advanced Projects for Synercom Technology, Inc. He is SIG Co-chairman for Geoprocessing for URISA, Member of NCGA's Technical, Research and Standards Committee and co-author of the APWA CAMRAS reports. He is a member of ACSM.

ABSTRACT

Planning for a military base is done in a regulated command environment over a contained geographical area. Contingency plans which radically change the base in a very short time are also developed and maintained. Over 65 different maps may be in one master plan. The effect of change is posted on all the affected maps and incorporated within the contingency plans.

Geographic Information System (GIS) technology provides the planner tools to manage both the spatial and temporal aspects of plans and changes. By developing integrated parametric units, the data may be maintained in an integrated manner and many of the maps generated automatically by aggregating the parametric units. The temporal aspect of the system may be studied by developing the union of integrated parametric unit coverages from past years and future plans. Changes may be made to the collection of data rather than each map.

By incorporating specialized reporting capability within the system capability, much of the regulation paperwork may be generated automatically. This paper explores the use of GIS within the particular environment of military base planning.

INTRODUCTION

Military base planning includes the recording and analysis of spatial data. Geographical Information System (GIS) technology provides the automated tools to effectively manage the spatial data. By combining spatial data with data that is naturally stored as lines and symbols, a complete digital plan of a military base may be stored and maintained on a Mapping Information Management System (MIMS). MIMS is a broad system including GIS technology as a subset. The integration of spatial, inventory data, utility, planimetric and topographic data into a single digital plan offers the ability to relate these data quickly in ways that would be impractical in a manual drafting system.

This paper proposes a technical approach to the conversion, use and maintenance of spatial data within the framework of a totally integrated data base. Geographic Information System (GIS) is defined and basic (not all inclusive) requirements are described.

To relate different coverages of spatial data an integrated parametric unit approach is proposed. Examples of the use of the basic GIS algorithms and integrated parametric units are developed for data conversion, analysis, temporal data management and maintenance. Special reporting requirements are discussed briefly.

Examples have been drawn from data captured and tested on the Synercom INFORMAP II and Environmental Management Information System (EMIS) software products. INFORMAP II is a proven MIMS system with many different installations throughout the world. EMIS is the GIS subsystem of INFORMAP II and is derived from the Harvard Laboratory for Computer Graphics and Spatial Analysis Odyssey program.

MILITARY BASE PLANNING MAPS

Planning for a militray base is done in a regulated command environment over a contained geographical area. A regulated command environment requires thut procedures and regulations be strictly followed and that extensive reports and forms accompany each change. The contained area restricts the size of the problem although the size may be large for the scale of maps produced (Vandenburg Air Force Base is 100,000 acres, China Lake Naval Weapons Center is 1,100,000 acres, for example). The constrained, usually crowded, area makes the planning more difficult.

A master plan may include over 65 different maps at several scales and all with different content. The maps are developed by composite techniques utilizing the planimetric and topographic maps as the base. Many of the maps are a geographically oriented inventory of facility elements rather than classified spatial data. The integration of facility maps with spatial maps provides unique analysis opportunities that are difficult to accomplish in a manual drafting system.

GEOGRAPHIC INFORMATION SYSTEMS - A DEFINITION

The term Geographic Information System (GIS) could be used to describe any system which processes geographically oriented data. However the term was first used to describe systems which analyze data that is spatially organized and topologically defined. This data is generally represented by collective polygons (or grid cells) forming a coverage. A coverage is a map with each area classified and the boundaries delineated. Output results are portrayed in two or three dimensions with the data classified and thematically presented. GIS includes the ability to determine the interrelationship between several coverages. This definition has been widely used in the literature and will be followed in this paper. The scope of this paper is vector oriented data and not grid cell data.

The term Mapping Information Management Systems (MIMS) represents a broader system including planimetric, topographic, facilities mapping, GIS and other mapping technologies in an automated computer graphics system. Both GIS and MIMS associate attributes with geographical location entities. GIS is concerned with the relationship between homogenous areas defined topologically while MIMS includes attributes for specific elements such as the facilities making up a public works system as well as GIS, spatial type entities. MIMS manage both line and symbol entities as well as spatial entities. MIMS systems produce high quality cartographic products. See Reference 1.

The analysis capabilities within a GIS include the ability to:

1. Determine and manage polygons from arbitrarily digitized line strings.
2. Associate attribute data with each polygon.
3. Determine zones and areas about line and symbol entities.
4. Perform a union of two coverages to form a new coverage with all the informational content of the original two coverages.
5. Thematically represent the classified results of coverages.
6. Determine areas by classification.
7. Aggregate polygons of similar classifications into a single polygon.

These are basic capabilities and numerous additional capabilities are commercially available in GIS products. This paper will show examples of how each of these analysis capabilities may be used within the military base planning application.

INTEGRATED PARAMETRIC UNIT

An integrated parametric unit is the result from unioning two or more coverages. Each parametric unit has homogenous attributes from the original coverages. An integrated parametric unit means that small inconsistencies in the representation of common boundaries such as roads, hydrology, soil changes, etc. have been removed by the digitizer and by the system. Otherwise the integrated coverages are dominated by small meaningless sliver polygons. The advantage of consistent development of integrated parametric units from highly correlated data is that the inconsistencies are systematically removed and analysis can usually be done by examining the attribute data associated with each parametric unit rather than reprocessing the polygon data.

DATA CONVERSION

The military base plans are derived from aerial photography, existing map sources, inventory records, field surveys and other sources. The source documents have different scales, accuracy levels, and confidence levels.

Spatial data is a set of lines forming polygons with some notation of the classification of each polygon. The conversion of buildings into topological correct data, so that a GIS can analyze the data, illustrates the data conversion technique:

1. Using stereo photography and classical stereo compilation techniques determine the boundaries of the buildings. The user is not required to digitize the building outline in any particular order.

2. Using the GIS technology determine the building polygons. The GIS system will recognize each line segment that makes up a polygon and will cause end points that are close to each other (within a tolerance) to coincide. Doubly digitized lines, dangling lines and other error conditions are recognized and recorded.

3. The GIS technology determines a logical sequence number and centroid.

4. The user, from existing records, corresponds the centroid sequence number and the attribute information. This can be done at an alphanumeric CRT.

5. Map products may then be produced by classifying the buildings based on the attributes (building use for example) and thematically presented the buildings using crosshatching, gray scale, halftone and color representation capability. Figure I provides an example of the final product.

Modern GIS does not require the user to directly define the polygon boundaries, digitize the polygons in any particular order or associate left and right attributes directly with each polygon boundary segment. Removing these requirements reduces the data conversion effort substantially.

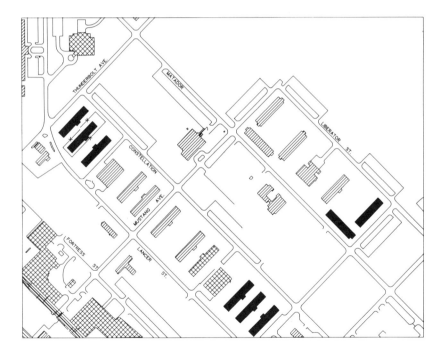

Figure I

Buildings Converted to Topological Polygons,
Classified, and Crosshatched

OTHER POLYGON DETERMINATION

Zones such as approach-departure zones, obstructions, habitat limits, and historical sites are determined geometrically based on data that has already been entered into the digital master plan. A GIS system can determine polygons in the following manners:

1. Fixed distance about a specific point such as a historical site.
2. Fixed distance about a line. Suitability zone about a road for example.
3. Fixed distance about a facility or structures such as hazard zones about a munition dump.

The integration of inventory, planimetric, topographic and special data into a single digital data base combined with the GIS technology provides for the generation of these polygons automatically.

Topographic data are used to generate slope, aspect ratios, viewing angles, etc. These values may then be classified and polygons derived to form a coverage.

HIGHLY CORRELATED DATA

Each type of spatial data forms a coverage. The buildings, hazard zones, soil, etc. form separate coverages. Many of these coverages are highly correlated. For example, roads, hydrology, soils, vegetation, and land use often have common boundaries. Some of these coverages are more accurately captured than others. The coverages may be unioned together to produce integrated parametric units (sometimes called integrated terrain units). An integrated parametric unit is a coverage itself with the data classified by the attributes of the contributing coverages. The generation of integrated parametric units usually proceeds from the most accurate data to the less accurate data. For the example above, the roads and hydrology will be captured accurately from a stereo photography. To develop the land use coverage, the accurate location of the roads and hydrology should be considered. There is a wide spectrum of techniques to produce integrated parametric units from essentially manual drafting techniques to the GIS doing most of the resolution of the data. The manual techniques are manpower intensive; the completely automated techniques require careful determination of tolerances and precise digitizing. In a digitizing environment utilizing interactive graphics feedback a compromise between the extremes is most effective. One procedure is:

1. Include the roads and hydrology as potential boundaries in the coverages.
2. Assuming a separate source for the land use, place the land use source on the digitizer and determine the non-linear transformation from the land use to the accurate compiled map by recognizing common points. If the system has cursor tracking, then digitizer location may be reflected on a display of the digitized data.
3. Digitize the land use boundaries. When a natural boundary exists digitize up to it or slightly over it. There is no need to snap to the natural boundary since the polygon building logic will do that subsequently. The natural boundary is also not redigitized or otherwise traced. The user can always view the interrelationships on the graphics CRT.

4. Execute the GIS polygon building algorithm to determine the polygons and automatically snap the boundaries to the natural boundaries.

5. Resolve any inconsistancies that the GIS has found using the graphics CRT.

6. Assign land use codes to the computed centroids. An alternative is to digitize and classify the centroids at the time the boundaries are digitized.

7. If a number of polygons are homogenous across the boundary, then the polygons can be aggregated to form a new set of polygons which do not include the natural boundary when the land use does not change.

Producing high cartographic quality maps requires specialized display techniques. This is not only the classical thematic tones that GIS usually supports but also the display of lines that form mutual boundaries. For example the hydrology may be displayed in blue while the land use in green. For mutual boundaries the hydrology will have precedence. If a land use map is required that does not include hydrology, the same line in the data base would be green. This is easily accomplished with a system whose displays are driven by representational tables external to the converted data base. It is even easier if the system supports multiple display modes. See reference 2 and 3.

ANALYSIS USING INTEGRATED PARAMETRIC UNITS

The results of analysis of spatial data is either a tabular report listing access by classification or some form of thematic map where each classification is displayed uniquely in either quantized steps or continuous graduations. If all the related data has been integrated together, then each polygon attribute record captures each polygon it belongs to within each coverage. Thus any form of analysis merely involves operations on the attribute data and does not require processing of the polygon data. If a thematic result is required the classification as an output of the analysis is stored in the attribute record for the polygons and the display logic formats the output referencing this classification. Complex classification procedures may be easily developed.

The land use suitability results (Figure 2) is an example of the analysis that can be done using GIS algorithms and integrated parametric units. The map was derived as follows:

1. Develop 1984 Land Use Coverage by digitizing the boundaries and programatically determining topological polygons.

2. Develop Historical site and Woodpecker Habitat restricted areas using bounding radius about the historical sites.

3. Develop Military Land Use Map coverage including a buffer about the roads.

4. Develop Equipment Limits and Erosion Factor of Soils. This coverage was derived from other coverages.

5. Union the 1984 Land Use coverage with the Historical Site coverage.

6. Union Military Land Use with Equipment Limits.

7. Union the coverages determined in step 5 and step 6. Derive a classification based on the classifications in the original coverages. At this step a full integrated parametric unit coverage has been derived.

8. Aggregate the polygons based on the classification levels.

9. Assign symbolization to each classification and produce the final product (Figure 2).

This example exercises all the basic GIS technology requirements.

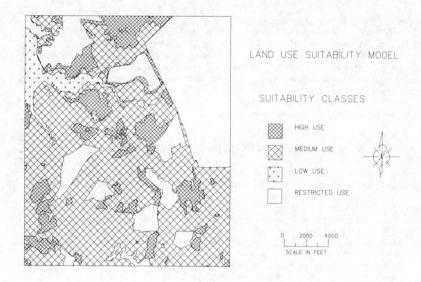

Figure 2
Land Use Suitability Analysis Final Product

TEMPORAL ANALYSIS

Military base plans require historical as well as planned improvements and contingency plans. These plans become an integral portion of the data base and thus proposed changes can be considered in relation to other plans and contingencies without additional effort. One approach is to define a coverage of the planned changes. For example, the five year plan for building changes could be developed as follows:

1. Copy the existing buildings and attribute data to a new coverage.

2. Set the change flag for each building to no change.

3. For each change modify the attribute file to indicate the change and include the expected date of the change.

4. For new construction add the building to the new coverage.

To produce a map of the status at any given time select from the plan coverage the changes to be completed by that time and then overlay it with the existing condition. The data that has changed (torn down, built, modified) can be highlighted by color or other symbolization. As a change is implemented the as built condition is placed in the plan coverage, selected separately from the other changes in the plan, overlayed with the existing condition, and the existing condition updated. Yearly the existing conditions are archived in coverage form so a historical record is kept. A change map may be produced at any time.

REPORTING REQUIREMENTS

In a regulated command environment the reporting requirements are extensive. By utilizing an integrated parametric unit approach including proposed changes and past usage, many of the reports may be automated. For example the Strategic Air Command Form 246 (Figure 3) documents the area presently used, existing usable facilities and facilities planned for future use by use category for a particular Air Force base. This may be developed by examining the integrated data for each building.

CONCLUSION

The Geographic Information System (GIS) technology has numerous applications for military base planning. It offers rapid data conversion, extensive analysis capabilities and graphic as well as narrative reports.

Developing the base plan utilizing integrated parametric units recognizes the interrelationships of some coverages and resolves data conversion inconsistencies as the data is captured. The result is a more accurate data base supporting many analysis requirements quickly. Specialized reports and temporal analysis are readily automated.

By integrating data derived in point or line format with the GIS data, new coverages may be easily defined. By displaying these other maps in conjunction with the GIS output, a higher quality cartographic product may be produced.

The Synercom products, EMIS and INFORMAP II, provides an integrated system technology to support mapping as well as GIS requirements for the militray base planner. The integration of this technology broadens the use of the digital military base plan.

```
                           S.A.C.   246
SUPPORTING DATA PROGRAMMED INSTALLATION    DATE      REPORTS CONTROL
   END MISSION REQUIREMENTS JACKSON AFB, CTG 19-Oct-84 HAF-PRE(AR)  7115
SECTION A
1 CATEGORY  2 NOMENCLATURE        3 AGENCY   4 UNITS    5 QNTY. REQ.
   721312      DORM AM PP/PCS-S      SAC        SF         146665
6 EXT. UNUSABLE    7 EXISTING USABLE 8 NOT IN INVENT 9 CURRENT DEF.
     118001 SF        152664   SF       0              5999  SF
 10 PROGRAMMED AMT.     11 PROG. DEFICIENCY   12 PROG DEFICIENCY COST
           0                     0                     0
SECTION B                 COMPUTATIONS AND SOURCE DOCUMENTS
 1 COMPUTATION OF REQUIREMENTS           2 AUTHORIZATION OR SOUR.
   COMPLIMENT 18 MEN EQUALS 5700 SF          AFM 86-2, PARA 16-10C

   MOBILITY EQUIPEMENT STORAGE 3150 SF




SECTION C
EXISTING FACILITIES PRESENTLY USABLE OR PROGRAMMED TO BECOME USABLE
                                                     DISPOSAL ACTION
FACILITY           SCOPE              STATUS          PROCESS
NUMBER                              VAC    OCC       YES       NO
0301       17159  SF                 X                X
0302       17159  SF                 X                          X
0304       16116  SF                        X                   X
0305       17159  SF                        X                   X
0307       17159  SF                        X         X
0309       16090  SF                 X                          X
0311       17159  SF                        X                   X
          118001  SF

2            EXISTING USABLE FACILITIES
FACILITY           SCOPE             STATUS        EXISTING  PROGRAMMED
NUMBER                            VAC    OCC        CODE        CODE
0327       26035  SF               X                  0
0328       25185  SF                      X           2
0329       26035  SF                      X           2
0331       25039  SF                      X           1
0332       25185  SF                      X           1
0333       25185  SF                      X           1
          152664  SF
SECTION D     FACILITIES PROGRAMMED FOR ANOTHER USE
FACILITY           SCOPE          FUTURE NOMENCLATURE   FUTURE   FUTURE
NUMBER                                                 CATEG.   CODE


REMARKS                                    COORDINATOR
                                           UNIT COMMANDER

                                           ORGANIZATION
                                           305 CSG/SP
                                           CATEGORY CODE
                                              721312
SAC      246                SYNERCOM INFORMAP II SYSTEM
                                  Figure 3
                         Example Report Derived From
                         Integrated Building Data
```

REFERENCE

1. Drinnan, Charles H., Vertical integration of small and large scale mapping systems, Third Annual Alaskan Conference for Computer Graphics and Geoprocessing, September 1982.

2. Drinnan, Charles H., Design considerations for mapping information management systems to support multipurpose cadastres, Computers Environment Urban Systems, Vol. 9, Pergamon Press 1985.

3. Drinnan, Charles H., Data base considerations in small scale mapping, Electronic Imaging, June 1984.

GEOGRAPHIC INFORMATION SYSTEMS:
TOWARD A GEO-RELATIONAL STRUCTURE

Kenneth J. Dueker
Portland State University
School of Urban and Public Affairs
P.O. Box 751
Portland, OR 97207

ABSTRACT

The paper distinguishes features of computer-aided mapping and geographic information systems and identifies an emerging link. Common to both systems is the need to link graphic data and attribute data. This geo-relational structure requires a one-to-one relationship of the graphic record with the record of the table of attributes. Combining distance-referenced line data with a topological data structure is the key to a powerful geographical information system.

INFORMATION

The similarities and differences between a system for computer-aided mapping (CAM)[1] and a geographic information system (GIS) are difficult to articulate, but important. The following definitions are offered as a means of initiating this presentation of distinguishing features:

    1. A GIS is defined as a specialized information system in which locational identifiers are attached to data for spatial analysis and/or mapping. Thus a GIS provides spatial information to users for decision making in management, planning, and research. Importantly a GIS allows the spatial collation of separately collected data. Cartographic modeling or overlaying is employed for analysis across layers of data to express the spatial relationship among the variables.

    2. On the other hand, computer-aided mapping is a more limited display of layers of data with the ability to select layers, window, scale, and display, but without the ability for analysis across layers.

These definitions ignore a feature that both kinds of systems are stretching to meet -- a data base management system that relates objects on the map to records in data files containing attributes of those map objects. This portends that the distinctions identified in the definitions will diminish. This will be illustrated by the subsequent development of a geo-relational structured GIS.

The purpose of the paper is to distinguish better between computer-aided mapping and a GIS and to identify an emerging link. Although the distinctions are becoming less in actuality as systems evolve and mature, it is important to understand and to be able to describe clearly the difference. Belaboring the distinctions draws attention to the required properties and power of the GIS, while exploring ways by which

---
[1] Not to be confused with Computer-Aided Manufacturing (CAM) systems.

computer-aided mapping systems can be augmented for greater intelligence.[2]

The Urban and Regional Information Systems Association (URISA) supports this on-going process of promulgating consistent terminology and in clarifying concepts. A draft of the paper was prepared for discussion at the 1984 URISA conference in a special session on geoprocessing and the present version has benefited from feedback received.

COMPUTER-AIDED MAPPING

Computer-aided drafting systems when applied to mapping problems are hereinafter referred to as computer-aided mapping (CAM). The map is the product and the information to draw the map is not used for computation or analysis. Although, more current models of CAM have analytical capabilities, early versions merely overplotted layers of data with no analytical capability to relate data across layers. Streets, railroads, streams, boundaries, water mains, etc. were digitized as separate layers and the relationship among layers was only discernable visually. This inability to relate data across layers was due to a lack of data structure. The strings of data that made up a layer were in effect cartographic spaghetti. The strings were not structure to correspond to logical entities, i.e., points, lines, or areas. Windowing was about the only logical operation that could be performed. Within a layer no further selection by attribute could be performed. The roads, sewer lines, or streams were not segmented logically by width or diameter, or capacity. Even if they had been, selection by another attribute, say slope, would not have been possible.

More current CAM systems utilize within a layer a distance-based system of recording location on a line, such as a milepost, stream mile, or station, to select graphically from a layer and to provide linkage to attribute data about segments between stations or of stations. This linkage between graphic data and attribute data is extremely important. Before developing this importance, the emergence of GIS technology is described.

GIS APPROACHES: TOPOLOGICAL STRUCTURE

Early approaches to dealing with the limitations of CAM for analysis of spatial data as well as mapping data recognized the need to structure graphic data as point, line, and area, or grid cell elements.

Thematic mapping programs were built around a one-time specification of the graphic structure of points or areas, which would be related to values or value ranges for a number of attributes. In this way, a number of maps depicting a variety of data for the same graphic structure could be mapped efficiently.

Relating attributes or data across layers was more difficult. Grid data

---

[2]Dangermond and Freedman (1984), p. 4, draw a distinction between automated geographic information systems and computer-aided design-type system. The latter type are based on relationships visible to the user, resulting in "implicit" understandings about how data interrelates, whereas the topological relationships of GIS result in "explicit" data relating.

structures were adopted for this reason. The imposition of a regular lattice on layers of data allowed comparison of or computations in corresponding grid cells.

For vector formatted data, point-in-polygon, line-in-polygon, or polygon overlay were used to relate layers of data. Encoding data layers for polygon processing poses quality control problems, especially in natural resource applications. Polygon digitizing necessitates line segments of a network of areas to be digitized twice, once for each polygon it bounds. This process yields small gaps and overlaps as the same line yields two sets of coordinates when digitized. Overlays of polygons so derived results in many spurious polygons.

To avoid the spurious polygon problem, many systems adopted a chain encoding scheme wherein a line between junctions or nodes was digitized once and polygons formed by the computer, either using digitized center identifiers or by use of right and left polygon identifiers. This topological data structure was useful in editing to ensure the logical consistency of the network of areas.

Segmenting graphic records to correspond to chains connecting nodes in a topological data structure is useful for relating data across layers. It is a limited approach though in relating graphic data of line features to records in data files used by agencies responsible for operating and managing a resource or facility type characterized by a single graphic layer, say sewers or streets. Consequently, GIS technology also sought ways to relate graphic records of linear features to corresponding attribute records.

GIS EXTENSIONS: GEO-RELATIONAL STRUCTURE

The application of relational data base concepts to GIS yields a geo-relational structure. This geo-relational structure requires a one-to-one relationship of the graphic record with the record of the table of attributes. This means segmenting chains between nodes of a topological GIS data structure. Segmenting is necessary to produce graphic elements that are homogeneous with respect to attributes that may be used for selection.

Whereas a topological structure only requires a record pertaining to a section of road between intersections, a geo-relational structure requires segments that are homogeneous with respect to roadway width, depth of pavement, traffic volume, conditions of surface, etc. As these attributes change in value, segmentation would be necessary and a corresponding record in the table of attributes would be added. This geo-relational data structure allows selection and analysis by both geographic and attribute criteria. Roads or sewer lines of specific slope, age and/or condition can be selected.

In much the same way as drop-line algorithms can be used to delete boundaries that are redundant in a topological structure, segment-aggregation algorithms are needed to collapse networks into more generalized forms, when an attribute is common in value, dropped or ignored.

However, it is difficult to forsee all possible segmentations imposed by combinations of attributes. Consequently, a combined topological and distance-based system (station, milepost, stream mile, etc.) has the greatest potential for implementation of a geo-relational structured

GIS. Thus, segmentation of topological entities is not needed. The distance-based system (stations, road miles, stream miles) constitutes means by which attribute data can be used to select graphic elements.

With distance referenced line data it is possible to both select lines within arbitrary polygons and to select attribute records for distance units for lines contained in the polygon that possess specific attribute levels. Thus we could select roads of a given width, functional classification, and volume-to-capacity ration range within specified counties and between specified elevations. Or all two-mile sections of rivers downstream of specified discharge points having specified low flow rates could be selected. Property owners within 200 yards of the stream with stream withdrawal permits could be selected.

Distance referenced line data with associated attribute data that is coordinate based and topologically structured constitutes a basis for a powerful knowledge-based system. The addition of topologically structured and coordinate-based networks of areas, and coordinated point data add greater intelligence to the digital map data for inclusion within a geographic information system.

CONCLUSION

An intentially brief articulation of GIS concepts have been presented. An extreme distinction between CAM and GIS is stressed to draw attention to the importance of data structure for computer-assisted analysis. These distinctions are summarized and illustrated in the attached figures.

A distinction was then drawn within the GIS field that identifies a stand-alone topological data structure used in natural resource application. This topological GIS approach was compared to a geo-relational GIS that has great potential for facilities management applications.

This paper results from observations by the author that the concepts contained herein are understood by those working in the field, but poorly articulated. This is an attempt to clarify and explain the concepts and by so doing, hopefully encouraging standard terminology for communicating to potential users of GIS.

REFERENCES

Dangermond, J. and C. Freedman 1984, "Findings Regarding A Conceptual Model of a Municipal Data Base and Implications for Software Design," Environmental Systems Research Institute, Redlands, CA. (unpublished).

Strings, or "cartographic spaghetti"

Overplotting map layers

CAM (Computer-Aided Mapping)

Figure 1

Network of areas; digital line graphs

Topological data structures
(lines only intersect at ends)

Manipulation:
Overlaying

Line in polygon

Point in polygon

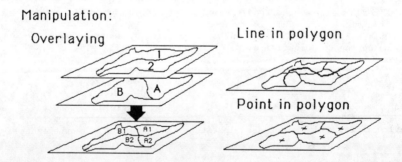

GIS (Geographic Information System)

Topological Structure

Figure 2

Correspondence of graphic records to non-graphic records

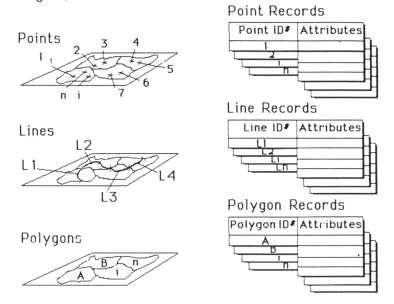

GIS (Geographic Information System)

Geo-Relational Structures

Figure 3

Station	Row Width	Pavement Width	Pavement Condition	Traffic Count	Culvert Diameter
0+00	100	26	3		
10+33					24
50+00			2		
63+21					48
100+00		24			
121+11					24
148+50				500	

Figure 4

AN INTERIM PROPOSED STANDARD FOR

DIGITAL CARTOGRAPHIC TERMS AND DEFINITIONS

Prepared and Edited by:

Dean T. Edson, Co-chairman, Working Group IV

and

Harold Moellering, Co-chairman, Working Group IV

National Committee for Digital
Cartographic Data Standards
Numerical Cartographic Laboratory
158 Derby Hall
Ohio State University
Columbus, Ohio 43210
(614) 422-0468

Working Group on Terms and Definitions is composed of:

Dean T. Edson (Co-chairman)   E-Quad Associates
Harold Moellering (Co-chairman)   Ohio State University
Eric Frey (vice-chairman)   National Ocean Survey
Frank Beck   U.S. Geological Survey
Mark Monmonier   Syracuse University
Hugh Calkins   State University of
                                    New York

Abstract

The National Committee for Digital Cartographic Data Standards has made considerable progress in establishing uniform definitions for approximately fifty key terms which are used in the proposed Interim Standards for Digital Cartographic Data. The approval process, specified in the National Standards Institute guidelines, requires a formal ballot vote and comment by members of the NCDCDS Steering Committee. This paper deals with the "push and shove" debate to resolve the formidable body of sometimes ambiguous input collected from committee members, from recent literature and from public comment, and our procedure to state these key terms and definitions in clear notation consistent with our goal of maintaining majority opinion in the Glossary.

The paper will be presented by the Co-chairmen of the Working Group on Terms and Definitions in two parts: (1) General Terms and (2) Cartographic Objects.

MERGING TABULAR AND SPATIAL DATA SETS
IN A GEOGRAPHIC INFORMATION SYSTEM

by J. C. Eidenshink, D. O. Ohlen, M. E. Wehde
Technicolor Government Services, Inc.
U.S. Geological Survey
EROS Data Center
Sioux Falls, South Dakota 57198
(605) 594-6114

and

Donald Stroud
Bureau of Indian Affairs

ABSTRACT

Tabular attribute information within existing data bases of the Bureau of Indian Affairs was merged with digitized spatial information within a geographic information system. The procedure enables incorporation of attribute data into spatial analysis processes and enables the processing results to be used for updating and refining both spatial and tabular data bases.

GEOMETRIC ALGORITHMS FOR CURVE-FITTING

James Fagan and Alan Saalfeld
Statistical Research Division
Bureau of the Census
Washington, DC 20233

ABSTRACT

Some curve-fitting procedures may be described entirely in geometric terms, and, thus, may be shown to depend exclusively on the geometry (relative positions) of the points to which the curve is fitted. These procedures necessarily exhibit desirable invariance properties under large families of transformations such as the family of all affine transformations because the underlying geometry of the fit points behaves predictably under the families of transformations. Although the procedures are defined geometrically in order to make proofs of invariance straightforward, nevertheless, the algebraic representations of the fitted curves may be drived analytically. Conversely, whenever one can show that a curve-fitting procedure has an underlying geometric definition, then one may take advantage of geometric invariance properties.

GEOMETRY AND TRANSFORMATIONS

Geometry, which refers to the intrinsic relations among points, lines, curves, and areas, may be examined from the point of view of transformations of the plane. Geometric properties may be described and even defined in terms of those transformations. A classic illustration of such a property is the fundamental notion of **congruence** in classical Euclidean geometry. Two geometric figures are said to be **congruent** if one can be moved onto the other; or, perhaps more accurately, if the whole plane can be moved with a **rigid motion** or isometry so that one figure is brought into perfect alignment with the other. Euclidean geometry is concerned with those properties which are preserved by rigid motions of the plane (i.e., rotations, translations, reflections, and combinations of the three). Congruence involves nothing more than belonging to the same equivalence class of figures under all rigid motions. The statements of the Euclidean theorems, axioms, and propositions are given in terms of properties and descriptors which are left unaffected by rigid motions: those properties include **being parallel, being perpendicular, bisecting angles or line segments,** and **lying on a straight line** or **lying at a fixed distance from some other object.** For example, if line **A** is perpendicular to line B, and if a rigid motion is applied to the space, transforming line **A** into line **A'** and line B into line **B'**, respectively; then line **A'** will be perpendicular to line **B'**.

Euclidean geometry in the plane is simply the study of those properties which do not change when rigid motions are applied. Classical spherical geometry deals with invariant properties under rigid motions of the sphere. Hyperbolic geometry, another non-Euclidean geometry, does not involve rigid motions in the usual sense. Nevertheless, it does involve the study of properties which remain invariant under a family of transformations; and by a kind of duality, one could define a kind of albeit unnatural "rigidness" in terms of those transformations.

The family of rigid motions is not the only family of transformations one might wish to apply to the plane. Other families give rise to other geometries. One might add the family of contractions and dilations (uniform shrinking and expanding maps) to create a geometry of similar figures (same shape but different sizes). In this expanded geometry, two similar figures are equivalent. In this geometry of similar figures, absolute distances are lost; only relative distances have meaning. Different families of transformations preserve different properties or relations of lines, points, curves, and areas. By recognizing the properties preserved by a specific family of transformations, one may use those properties to define **geometrically stable entities**, entities which themselves do not vary under the transformation family. A curve-fitting procedure can be such an entity.

CURVE-FITTING PROCEDURES

A curve-fitting procedure will be understood to mean an **exact** curve-fitting procedure throughout. An exact curve-fitting procedure has as its input an ordered finite sequence of **n** points $(p_1, p_2, p_3,...,p_n)$ and returns a continuous one-dimensional curve which passes through the **n** points $(p_1, p_2, p_3,...,p_n)$ in order. Exact curve-fitting procedures are used in cartography, for example, to reconstruct smooth representations of irregular linear features from shape files containing only coordinate values for critical or extreme points. Considerable storage can be saved if only a few points on a river feature are kept on the file and the river is drawn from those points using a curve-fitting algorithm.

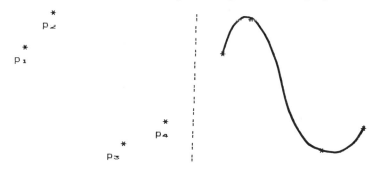

Figure 1. Four Fit Points and Cubic Polynomial Fit.

The above figure illustrates the result of fitting a cubic polynomial curve: $Y = AX^3 + BX^2 + CX + D$, to the four fit points $p_1, p_2, p_3,$ and p_4.

The cubic fit is not a very good curve for a number of reasons. In the first place, a cubic polynomial will not fit every possible sequence of four points in the plane. Because the value of **Y** is expressed as a function of **X**, there can be only one **Y** value for each **X** value. The curve can only move from right to left or from left to right, but not both ways; there can be no doubling back. Nevertheless, if the points $p_1, p_2, p_3,$ and p_4 have their X coordinates strictly increasing or strictly decreasing when the points are expressed in their sequential order, then (and only then!) can a cubic curve be fitted to the four points.

PROCEDURES WHICH COMMUTE WITH TRANSFORMATIONS

Given a sequence of points, a curve-fitting rule or procedure, and a transformation of the plane, one may perform two different composite operations:

First, one may apply the curve-fitting rule to the original points to produce a curve in the plane; then one may transform the plane to create a transformed image of the curve, which will be a new curve.

Alternatively, one may transform the original points by looking at their images under the transformation of the plane; then apply the curve-fitting rule to the transformed points.

If the resulting curves of the two composite procedures are the same, then the procedure is said to **commute** with the transformation.

In some cases the procedure will not only fail to commute with the transformation, it may even fail to be applicable to the transformed sequence of points. A simple example of such a complete lack of commutativity is the cubic polynomial fit rule or procedure with a 90° rotation of the plane for the transformation applied to the four points given in the example on the previous page:

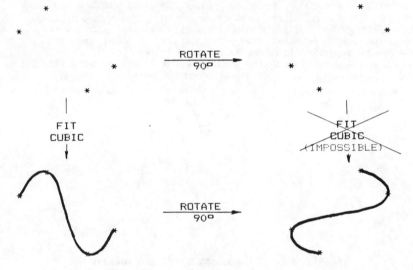

Figure 2. **Transformation and Procedure do not Commute.**

The cubic polynomial fit procedure will commute with some transformations of the plane. Reflections in the X-axis or in the Y-axis will commute with fitting a cubic to an acceptable set of four points. Indeed, reflections in any vertical or horizontal line will commute with the cubic fit procedure. Uniform (i.e. linear) stretching or shrinking along either axis is a transformation which will commute with the cubic polynomial fit procedure. Rotations in general will not commute, although a rotation of 180° will. The collection of all transformations which commute with a curve-fitting procedure for every suitable sequence of fit points will form a **group** of transformations. The larger the group of transformations, the better the curve-fitting procedure. A good curve-fitting procedure has for its group of transformations a group which contains all of the rigid motions. A good curve-fitting procedure also works for an arbitrary sequence of points in the plane. If this is the case, then congruent sequences of fit points will result in congruent fitted curves, no matter where in the plane the fit points are positioned (as a congruency class).

A curve-fitting procedure which can be applied to any sequence of points in the plane and which commutes with all rigid motions will be called **Euclidean** or **geometric**. Euclidean curve-fitting procedures are coordinate-free in the sense that they do not depend on the positioning of the coordinate axes. Although a Euclidean procedure may be defined in terms of X's and Y's, a change of origin or axes will not alter the final product, a fitted curve through the given points. This is due to the fact that coordinate changes correspond to rigid motions of the plane as well.

The notion of rating a curve-fitting procedure in terms of the transformations that it commutes with can be specialized even further. Instead of focusing on the quantity of transformations which commute, one may seek procedures which commute with a few distinguished transformations. It may be particularly desirable to commute with a special family of transformations (for example, transformations which define an entire class of map projections). Curve-fitting procedures which commute with all such map projections will produce consistent results on every map image belonging to the projection class.

DEFINING CURVE-FITTING PROCEDURES

So far, "good" and "better" curve-fitting procedures have been described in terms of external associated groups of transformations. No constructive direct approaches to curve-fitting procedures have been given. The only example given so far, the cubic polynomial fit, has fallen far short of being a Euclidean procedure. In this section the relation between Euclidean procedures and Euclidean geometry and its invariants provides the key to designing good procedures.

Large groups of transformations of the plane.

Two very important groups of transformations of the plane are:

(1) The **affine** group; or **affine** transformations; and

(2) The group of **rigid motions** or isometries.

The affine group properly contains the group of rigid motions. Every rigid motion is an affine transformation.

Every affine transformation, **T**, has the form:

$$T((x,y)) = (a_1 x + b_1 y + c_1, a_2 x + b_2 y + c_2),$$

for some real constants a_1, b_1, c_1, a_2, b_2, and c_2,

such that $(a_1 b_2 - a_2 b_1)$ is not equal to zero.

Every rigid motion <u>further</u> satisfies:

(i) $a_1 a_2 + b_1 b_2 = 0;$
(ii) $a_1^2 + b_1^2 = 1;$
(iii) $a_2^2 + b_2^2 = 1.$

Every rigid motion is a translation (vertical and/or horizontal shifting), a rotation, a reflection, or a composite or combination of two or more of these.

Every affine transformation is either a rigid motion, a stretching or shrinking on each of the axes (with possibly unequal stretching or shrinking on each of the axes), or a composite or combination of the two.

Affine transformations preserve lines, collinearity, and parallelism. Affine transformations do **not** preserve (but rigid motions **do** preserve) perpendicularity, distances, angles and angle bisectors. This is clear in the example:

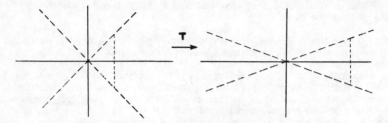

Figure 3. The Affine Transformation: $T(x,y) = (3x,y)$.

In order to illustrate the manner in which known invariant properties can be used to define curve-fitting procedures, the following elementary example is presented:

Given a sequence of points $\{ p_1, p_2, p_3, \ldots, p_n \}$, let the "curve" for that sequence be the polygonal line made up of straight line segments linking successive points p_i and p_{i+1} for $i=1, 2, \ldots, n-1$. The curve-fitting procedure given here could be described as, "link the successive points by straight-line segments."

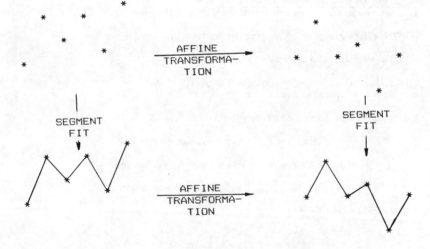

Figure 4. Commutative Diagram for Polygonal Line Fit.

This procedure commutes with all affine transformations **precisely because affine transformations send straight line segments into straight line segments** and a line segment is uniquely determined by its end points. A transformation which did not preserve straightness of lines would not commute with this fit procedure.

Admittedly the polygonal-line fit procedure is not a very interesting or elegant procedure; nevertheless, it illustrates one Euclidean (even affine) fit procedure and a simple means of proving that the procedure indeed commutes with all affine transformations. The next example is more complex and the curves produced are more attractive. The underlying approach is similar, however; and the resulting curve-fitting procedure is both Euclidean and affine.

A GEOMETRIC CURVE-FITTING CONSTRUCTION

This section describes a geometric construction for adding points to a curve one at a time. The procedure may be iterated in order to place points along a curve with any desired density. In particular, for raster plotters or raster display devices, the point generation procedure may terminate when a connected sequence of pixels or raster dots has been selected.

In order to demonstrate the commutativity of this procedure with all affine transformations, the identical construction steps are carried out on the image points of the original fit points; and the corresponding constructed points and line segments are verified to be carried over by the affine transformation at each step.

Assume for this first case that the sequence of points, $p_1, p_2, ..., p_n$, does not include **inflections** in the following sense:

For all $i = 2, 3, ..., n-2$, the points p_{i-1} and p_{i+2} lie on the same side of the line determined by p_i and p_{i+1}.

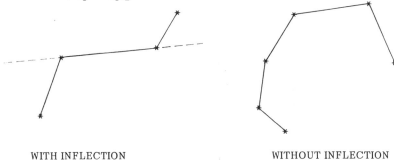

WITH INFLECTION   WITHOUT INFLECTION

Figure 5. Point Sequences With and Without Inflections.

For this first example it is also assumed that the section of the curve that is being built is an interior section linking the points p_i and p_{i+1}, where i is neither 1 nor n-1. (The construction required at inflections and at end segments is different.)

Observe that if a point sequence is without inflections in the above sense, then the image sequence under any affine transformation is also without inflections.

Let $p_1, p_2, ..., p_n$ be a sequence of n points without inflections, and let $p_1', p_2', ..., p_n'$ be the corresponding sequence of image points under some affine transformation. Link all successive pairs of points in each sequence with a straight line segment; and then link all alternating pairs of points with additional straight line segments (shown dashed in figure 6). Next construct line l_i through each point p_i parallel to the straddling secant $p_{i-1} p_{i+1}$. Do the same for each p_i'. These lines will serve as tangent directions at p_i and p_i' respectively for the curves to be defined. Notice that the corresponding constructions in the original space and in the affine image space are preserved by the affine map; that is, the parallel lines, for example, constructed in the affine image space are the affine images of the parallel lines constructed in the original space. Clearly the point of intersection of a pair of constructed lines in the original space is mapped by the affine transformation to the intersection of the image lines.

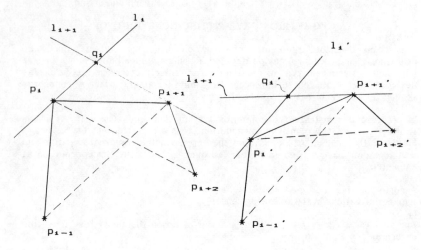

Figure 6. First Stage of Geometric Construction.

Let q_i (resp. q_i') be the intersection of l_i and l_{i+1} (resp. l_i' and l_{i+1}'). Consider the triangles $p_i q_i p_{i+1}$ and $p_i' q_i' p_{i+1}'$. The second triangle is the image of the first under the affine transformation. The next drawing illustrates the construction of a smooth curve through p_i and p_{i+1} which is tangent to $p_i q_i$ at p_i and is tangent to $p_{i+1} q_i$ at p_{i+1}.

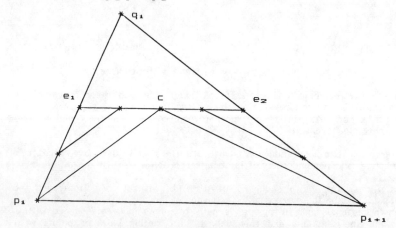

Figure 7. The Triangle for Enclosing the Curve.

Once a triangle has been constructed (in any fashion), the remaining procedure is a lopping or whittling procedure described entirely in terms of geometric characteristics of the triangle, characteristics which are preserved under affine transformations. A locus of points is described which is readily verified to be a smooth curve.

The curve described will be the upper boundary of the figure obtained by successive reduction of the triangular region. First remove a small similar triangle containing q_i by constructing a line parallel to $p_i p_{i+1}$ halfway between q_i and $p_i p_{i+1}$ and discarding the upper triangle and keeping the trapezoid below it. (See figure 7) Let c be the center of the upper base of the trapezoid, e_1 and e_2 the end points of the upper base as shown in figure 7 above. Continue by "lopping" the ends at e_1 and e_2 from the triangles $p_i e_i c$ and $p_{i+1} e_2 c$ as was done at q, again producing two new trapezoids. Construct triangles at the upper corners of each trapezoid, continue lopping until edges become rounded more and more.

With each successive lopping, add the center point of the new upper base of the trapezoid formed to the locus of the curve, since that center point will never be lopped off in any subsequent stage of the locus definition.

Whether one "lops triangles" or "adds" successive center points of trapezoid bases, the result is the same smooth curve. The curve is the closure of constructed points.

It is useful to observe that the triangles, trapezoids, and midpoints of the upper bases described in the construction above are all transformed to corresponding elements by affine transformations. Therefore, the curve fitting procedure described for the piece of the curve between p_i and p_{i+1} is both Euclidean and affine. It is straightforward to verify that the curve will be tangent to $p_i q_i$ at p_i and tangent to $p_{i+1} q_i$ at p_{i+1}. By the earlier specification of the tangent directions for all pieces of the curve, it is clear that the tangent directions fit together at all interior points without inflections.

In order to finish the curve-fitting procedure description, one needs to describe how to handle inflections and end points. Procedures to handle inflections and end points which commute with all affine transformations can be developed, but the ones that the authors have been experimenting with are too elaborate to be presented here. One possible approach to constructing an affine procedure involves building triangles (using affine-invariant triangle constructions) about the fit points so that a triangle lopping procedure may be applied to each triangle.

Figure 8. Possible Outcome of A Triangle Building Procedure.

After triangles have been constructed about fit points, another means of building affine-invariant curves involves finding piecewise parametrized cubic polynomial coordinate functions:

$$X_i(t) = A_{i0} + A_{i1}t + A_{i2}t^2 + A_{i3}t^3, \text{ and}$$

$$Y_i(t) = B_{i0} + B_{i1}t + B_{i2}t^2 + B_{i3}t^3;$$

where the parameter t above is chosen to be total cumulative straight line distance between successive points in the fit point sequence, the end points of the function pieces are the fit points, and the tangent directions match the slopes of the triangle legs. By specifying the parameter, the end points and the tangents at those end points, the cubic equations are completely defined on each interval between points p_i and p_{i+1}; and the equations mesh at the interval end points. (See White, Fagan, and Saalfeld, "On Fairing a Curve Through a Sequence of Points," for details.)

More generally, the building blocks may be higher order piecewise parametrized polynomial coordinate (PPPC) functions:

$$X_i(t) = A_{i0} + A_{i1}t + A_{i2}t2 + \ldots + A_{in}t^n,$$
$$Y_i(t) = B_{i0} + B_{i1}t + B_{i2}t2 + \ldots + B_{in}t^n,$$

defined for **t** in some interval, $[\ t_i, t_{i+1}\ ]$.

The collection of PPPC functions is especially useful for building affine-invariant curve-fitting procedures because of the following result:

Lemma. **If $(X(t), Y(t))$ is a curve given by parametrized polynomial coordinate functions of degree less than n in the parameter t on the interval $[\ t_i, t_{i+1}\ ]$, and if T is an affine transformation; then $T((X(t), Y(t)))$ is also a curve given by parametrized polynomial coordinate functions of degree less than n in the same parameter t on the same interval $[\ t_i, t_{i+1}\ ]$.**

The proof follows from the explicit representation of an arbitrary affine transformation seen earlier:

$$T((X,Y)) = (a_1 X + b_1 Y + c_1,\ a_2 X + b_2 Y + c_2),$$

The lemma guarantees that affine transformations will commute with the family of building blocks (PPPC functions) as a whole, moving one building block of the family into another building block of the same family. The trick of a good curve fitting procedure is to match building blocks to sequences of points in such a way that the affine transformations move the point sequences in the same way they move the associated building blocks. Differentiability and, therefore, the existence of a tangent direction for curve points is always preserved by affine transformations because the composition of a linear map with a differentiable map is also differentiable.

BIBLIOGRAPHY

Brown, K.Q., 1979, "Voronoi Diagrams from Convex Hulls," Information Processing Letters, 9(5), pp. 223-228.

Lee, D.T., and B.J. Schachter, 1980, "Two Algorithms for Constructing a Delaunay Triangulation," International Journal of Computer and Information Sciences, 9(3), pp. 219-242.

Pavlidis, T., 1982, <u>Algorithms for Graphics and Image Processing</u>, Rockville, MD, Computer Science Press.

Saalfeld, A., 1984, "Building Penplotter Operational Characteristics into Smoother Curve-Drawing Algorithms," Conference Proceedings: Computer Graphics 84, V.2, Fairfax, VA, National Computer Graphics Association, pp. 598-606.

White, M., and A. Saalfeld and J. Fagan, 1984, "On Fairing a Curve Through a Sequence of Points," internal Census Bureau document, submitted for publication in IEEE Computer Graphics and Applications.

A DATA MODEL FOR SPATIAL DATA PROCESSING

Robin G. Fegeas
National Mapping Division
U.S. Geological Survey
Reston, Virginia 22092

ABSTRACT

In the past, a wide variety of methods have been developed to represent spatial data digitally. Some of these methods can be categorized as access-oriented implementations of a specific data model. The spatial data of concern are limited to those either defining a surface or occurring upon a surface at a given point in time. All objects in the data model, including surfaces, may be called features. Data describing or defining features are classified as one of three types: (1) locational, (2) attribute, or (3) topological. These three descriptive data types in turn define various access paths for specific implementations (sometimes called data structures) of the model. Various existing data structures used to represent spatial data are analyzed with respect to access orientation and specific methods of representing the three descriptive data types. The completeness of the model is illustrated by outlining methods for generating all of the three access orientations from known structures. Finally, a single implementation is presented for accommodating all access orientations of the model. It is hoped this implementation will serve as a basis for a planned U.S. Geological Survey-designed integrated spatial data processor.

Publication authorized by the Director, U.S. Geological Survey, on January 2, 1985.

MEMORIAL UNIVERSITY OF NEWFOUNDLAND
St. John's, Newfoundland, Canada A1C 5S7

Department of Geography

Telex: 016-4101
Telephone: (709) 737-8~~753-120~~
Home :(709) 579 4

Contouring with Small Computers:
A Review.

David Forrest
Memorial University of Newfoundland

Gridding and contouring are of major concern to surveyors, geologists and cartographers. There has been a considerable amount of debate in the last few years about the most appropriate techniques for computer solutions to these problems, and several analitical tests have been carried out on mainframe programs.

The recent increase in the availability of micro-computers has lead to the development of several gridding and contouring programs for these small computers, but to date little has been published about the utility of these programs, so to this end some of the routines available for the IBM PC are reviewed and compared.

The tests are carried out with two distinctly different topographic surfaces, data being derived from photogrammetrically plotted contours. Data sets containing 50, 100 and 200 points, selected both randomly over the surface and selectively at critical points (ridges, valleys, etc.). Contour maps produced by the different programs will be compared with the original map and with each other.

Mathematical descriptions of the solutions employed are discussed (when known) as well as other aspects of the programs, including ease of use, flexibility, devices supported, related software and documentation.

TRIANGULATIONS FOR RUBBER-SHEETING

Daniel W. Gillman
Statistical Research Division
Bureau of the Census
Washington, DC 20233

ABSTRACT

This paper focuses on the application of triangulation and rubber-sheeting techniques to the problem of merging two digitized map files. The Census Bureau is currently developing a map merging procedure called conflation. Reproducibility, quality control, and a desire for mathematical consistency in conflation lead to a need for well-defined procedures. The Delaunay triangulation is well-defined and in some sense the 'best' triangulation on a finite set of points. It leads naturally into an efficient rubber-sheeting algorithm. The discussion starts with triangulations and rubber-sheeting in general and well-defined triangulations. This leads to the Delaunay triangulation, an algorithm for that triangulation and a specific rubber-sheeting technique. Finally, some problems that require further research are mentioned in an appendix.

INTRODUCTION

The Statistical Research Division of the Census Bureau is currently applying rubber-sheeting techniques in developing a system for merging two digitized map files of the same region. Our work began with an evaluation of algorithms for triangulating plane regions. Our needs led us to choose the Delaunay triangulation and to design and implement new algorithms for its maintenance and manipulation. This particular triangulation permits specialized operations that are not possible with other triangulations. This paper details the algorithms we have developed, describes the underlying theory used, and outlines some ongoing research.

The first section describes rubber-sheeting and triangulations in general and how we employ these techniques. Rubber-sheeting, which in our case is known technically as Piecewise Linear Homeomorphism (PLH), is used for transforming the coordinates of digitized map files. The PLH is a mathematical transformation that preserves topology and is linear on subsets of the map. We use triangulations to divide the map into the subsets on which the linear transformations are constructed.

The second section explains the need for a well-defined triangulation and discusses the consequences, i.e., uniqueness, and some of the specialized routines that are possible with such a process. There are two principal benefits arising from a well-defined triangulation. The first is the ability to exploit the underlying mathematical theory used in defining the triangulation. The other is the security of knowing the output will be unique. The specialized routines arise as consequences of the mathematical theory. These routines include local operations that are globally consistent and correct.

The third section defines the Delaunay triangulation and describes the underlying mathematical theory. This triangulation is well-defined and is in some sense the 'best' triangulation on a set of vertices. We present in this section several equivalent characterizations of the Delaunay triangulation, focusing on local properties.

The fourth section contains a detailed description of the triangle determination routines that have been developed for implementing the theory. This section also contains a description of the data structures we used.

The final section describes the rubber sheeting transformations which operate on the triangles of the Delaunay triangulation. These routines use the fact that a triangle is an elementary simplex and efficiently generate the convex coefficients for any point inside.

An appendix addresses two areas for continuing research. These areas are 1) whether a triangulation can be extended in a topologically consistent manner to another set of vertices, and 2) how one handles the case of four or more co-circular vertices.

SECTION 1. TRIANGULATION AND RUBBER-SHEETING PRELIMINARIES.

There are many instances in cartography where dividing a region into smaller subregions is beneficial. One useful technique for subdividing a region is triangulation. At the Census Bureau, triangulations are used to define rubber-sheet transformations which are then employed in the process of merging digitized map files.

First, it is necessary to define the term triangulation. Consider $N \geq 3$ points in the plane which are not all colinear. Let R be the region bounded by the smallest simple convex polygon* containing the N points. A triangulation is a maximal subdivision of R into triangles where the N points are the vertices. Every point in R is contained in one and only one triangle, except if the point lies on a triangle edge then it may be in more than one. (See figure 1) So, a triangulation may be viewed as a jigsaw puzzle where each piece is a triangle, and, using only the N points as vertices, no triangle may be divided into smaller triangles. The picture that is made by piecing the puzzle together is that of the region R. Note that for five or more points in the plane there is more than one way to triangulate these points. Also, every triangulation on a given set of points has the same number of triangles. (See Lee and Schachter, 1980)

A rubber-sheet transformation is a mathematically defined function from one region to another both of which have been divided into subregions. Each subregion in the first has a unique counterpart in the second. Also, every point and line of the boundaries and the subregions themselves maintain their relative positions with one another from the first region to the second, i.e., topology is preserved. The transformation is specified in pieces—a specific subtransformation for each subregion. Thus, the name "rubber-sheet" should be clear. Each subregion is transformed differently, much like taking a piece of rubber and stretching it in sections to make it fit over some object (see White and Griffin, 1985).

The specific rubber-sheet transformation that we use is called piece-wise linear homeomorphism, or PLH. The 'pieces', subregions, that the PLH is defined over are the triangles from a triangulation. It is quite easy then, to define a linear transformation from a triangle on one map to its counterpart on the other. Each linear transformation is a homeomorphism, i.e. it preserves topology. Combining this with the fact that topology is preserved between the triangulations, the resulting composite function is piece-wise defined, linear on each piece, and a homeomorphism over the entire map. Thus, we get a PLH.

---

* A simple convex polygon is a polygon where the edges meet only at the vertices and all exterior angles at the vertices are $\geq 180°$.

The procedure being developed for merging two digitized map files is called conflation. In part, it is an iterative process, each loop consisting of the following steps: 1) selection of pairs of vertices, one from each map; 2) triangulating the vertices of one (rubber-sheet) map; 3) transferring the triangulation to the vertices of the second (stable base) map; and 4) generating and performing a PLH transformation on the nodes of the rubber-sheet map. The result of a loop is that the features of the rubber-sheet map are in closer alignment with those of the stable base map. In addition, the node pairs which were selected as vertices are made to coincide.

SECTION 2. WELL-DEFINED TRIANGULATION.

There are many ways to triangulate a set of points in the plane and some ways depend on the ordering of the points. As a result of there being options, some problems arise. Use of a well-defined triangulation and procedure will eliminate the problems and also gives rise to other benefits.

A well-defined triangulation is a set of properties or rules used for building triangulations that has a mathematically precise formulation and does not depend on the order in which points are processed. These properties allow us to determine which vertices are connected by edges. The effect is the output from a well-defined triangulation procedure is unique. For example, take the triangulation defined by the rule that the total length of all the edges is minimum. This is well-defined, it doesn't matter what order points are processed. The minimum length of edges is always the same for the entire set of points. Note the two uses of the word 'triangulation'. Earlier we meant the set of triangles created on the points. Now we also use it to refer to the properties or rules used to build the triangles. The context will make the meaning clear.

The definition above is one that is global, i.e. we ask whether or not the resulting triangulation is unique. We can also look at the problem from another point of view. A locally well-defined triangulation has the property that given any three vertices it can be determined whether they form a triangle without processing the entire triangulation. Since the local condition clearly implies the global one, it is convenient to work with locally well-defined triangulations. An example of such a triangulation, the Delaunay triangulation, will be described in the next section.

The problem associated with the use of ambiguously defined triangulations is that the output may not be unique. The output depends on the order the data are processed. This has ramifications in a process such as conflation. Maps are merged using rubber-sheet transformations which are first defined by triangulations. A change in a triangulation will result in a change in the rubber-sheet transformation. This might result in different pairs of features being identified. Thus, quality control and reproducibility are very difficult if not impossible to manage.

We can see the problem in another way by looking at a simple example. Suppose we start with a square divided into two triangles. For each additional vertex, find the triangle containing the point and create three new triangles formed by the new vertex and each of the three pairs of vertices from the old containing triangle. (See figure 2) It is easy to see that the resulting triangulation (after just two new points are added) depends upon the order in which data are processed. (See figure 3)

The local property allows us to create some specialized routines that otherwise would not be possible. There are many such routines that could be developed. Here we will briefly describe two of them: DELETE and LOCATE.

DELETE is used to update a triangulation after removal of a vertex. Clearly, such a routine can be built for any triangulation procedure, however, the output may not be unique. Here, the local characterization is used to build the new triangles inside the simple polygon created by removal of the vertex. Therefore, the new triangles will be defined in the same way as the rest of the triangulation. Moreover, the resulting triangulation will be the same as that created from scratch given the modified set of vertices.

LOCATE builds the triangle surrounding a given point without building the entire triangulation. An iterative procedure produces triples of vertices which are tested. The local defintion determines whether the three vertices form a triangle. If so, it is easy to test whether the point lies in the triangle. If it does not, the procedure picks another triple in the appropriate direction. The triangle that is constructed at the end is the same as the one found to contain the point from the entire triangulation. This has to be the case, due to the local property.

SECTION 3. THE DELAUNAY TRIANGULATION.

The Delaunay triangulation is a globally and locally well-defined triangulation. It has properties which make it particularly useful for conflation. We give five equivalent definitions and properties of the Delaunay triangulation. For a more detailed discussion of the definitions, see Lee and Schachter, 1980.

Before we define the Delaunay triangulation, we need to restrict the set of vertices slightly. Assume that no four vertices are co-circular. The reason for this will become clear shortly.

We also need to define the Thiessen polygons on the set of vertices. Imagine that around each vertex a circle grows out like a wave, all at the same rate, starting at the same time. As a wave grows out, it freezes at all points of contact with other waves. The effects are that there is a unique simple convex polygon around each vertex which is not on the boundary of a triangulation and a unique unbounded polygonal region for each boundary vertex. These polygons are the Thiessen polygons. (See figure 4)

The following equivalent statements each characterize the Delaunay triangulation. The proof of their equivalence is in Lee and Schachter.

(local) 1) Three vertices form a triangle if and only if the circumcircle of these vertices contains no other vertex.

(local) 2) Two vertices are connected by an edge of the triangulation if and only if there exists a circle containing the vertices and no other vertex.

(global) 3) Given any triangulation on the set of vertices, change it in the following way. Consider each pair of triangles that shares an edge. Determine whether the circumcircle of one of the triangles contains the vertex not shared by the other. If so, then consider the pair of triangles as a quadrilateral. Replace the diagonal (the common edge) with the other, forming two different triangles. Continue until no more swaps can be made. (The updating criterion is called the local optimization procedure, LOP, in Lee and Schachter, 1980).

(global) 4) The triangulation formed by drawing an edge between any two vertices whose Thiessen polygons share an edge. (See figure 4)

(global) 5) Maximize the minimum measure of angles of all the triangles.

SECTION 4. OUTLINED ALGORITHM OF DELAUNAY TRIANGULATION.

PRELIMINARIES

A) Lines: Each line is expressed as $\{ (x,y) \mid Ay + Bx + C = 0 \}$ where the line passes through the points (x_1, y_1), (x_2, y_2) and $A = x_2 - x_1$, $B = y_1 - y_2$, $C = y_2 x_1 - x_2 y_1$.

B) Arrays:

    1) VERTEX = stores pointers to each vertex in counter-clockwise order for each triangle

    2) NABORS = stores pointers to each neighbor in counter-clockwise order for each triangle

    3) EDGES = stores A,B,C for each line of each triangle

    4) CIRCLE = stores x,y coordinates of center and square of radius of circumcircle for each triangle

Note: For all the arrays, the i-th record points to the i-th triangle.

ALGORITHM

I) Initialize = start with 4 corner points each 10% beyond N&E, S&E, S&W, and N&W boundaries of map respectively. Take 1st vertex from map and create 4 triangles. Initialize VERTEX, NABORS, EDGES, and CIRCLE arrays with data from these triangles.

II) For each vertex (NEW)
    A) Find triangle containing NEW (1st is found if point is on boundary). For each triangle constructed so far

        1) Perform LOP* on triangle and the point NEW
        2) Pass - test each line to see if NEW lies in positive half-plane
          a) All positive, found triangle, go to II-B
          b) First negative, neighbor of this line is next triangle, go to II-A-1
        3) Fail - go to II-A

    B) For each new triangle (3 new ones if from II-A-2-a, 2 new ones if from II-C-2-b)
Update VERTEX, NABORS, EDGES, CIRCLE

    C) For each adjacent pair of triangles, only one of which has NEW as a vertex

        1) Perform LOP on triangle of pair without NEW with the point NEW
        2) Pass
          a) Switch diagonals of quadrilateral formed by triangle pair
          b) go to II-B
        3) Fail - go to II-C

---

* See Lee and Schachter, 1980, for details about the subroutine LOP.

SECTION 5. THE PLH RUBBER-SHEETING TRANSFORMATION.

The rubber-sheeting transformation, a PLH, is particularly easy to compute. The method and underlying theory will be presented as a list of facts below. For details, see Saalfeld, 1985. Keep in mind that a separate linear transformation is defined for each triangle.

1) A triangle is an elementary simplex, i.e., any point p inside a triangle can be expressed uniquely as $p = \alpha_1 V_1 + \alpha_2 V_2 + \alpha_3 V_3$ where $\alpha_1 + \alpha_2 + \alpha_3 = 1$, $\alpha_i \geq 0$, and V_i are the vertices of the triangle. The α_i are called the convex coefficients of p.

2) A line in the plane passing through two points (x_1, y_1) and (x_2, y_2) can be expressed as $L = \{ (x,y) \mid Ay + Bx + C = 0 \}$ where $A = x_2 - x_1$, $B = y_1 - y_2$, $C = y_2 x_1 - x_2 y_1$.

3) Each line divides the plane into two half-planes:

$L^+ = \{ (x,y) \mid Ay + Bx + C \geq 0 \}$ and

$L^- = \{ (x,y) \mid Ay + Bx + C \leq 0 \}$.

4) If the vertices of a triangle are ordered counter clockwise, then the interior of the triangle is $L_1 \cap L_2 \cap L_3$ where L_i is the line on which the i-th edge lies.

5) Let L_i be the line of the edge opposite vertex $V_i = (x_i, y_i)$. If the formula for L_i is $Ay + Bx + C$, then the area of the triangle $A_T = \frac{1}{2} \mid A y_i + B x_i + C \mid$.

6) For any point p in a triangle T, the i-th convex coefficient for p is $\alpha_i = A_{pi}/A_T$ where A_{pi} is the area of the triangle formed using the i-th edge and p.

7) If the vertices are in counter clockwise order, the absolute value sign in 5 is not necessary. Moreover, for any point p and any triangle T, A_{pi} is non-negative for all i if and only if T contains p.

8) Let p be a point in a triangle, T. Let T' be the triangle corresponding to T in the other map. Let α_1, α_2, α_3 be the convex coefficients of p in T. Let V_1', V_2', V_3' be the vertices of T'. Then the image of p under the PLH is $p' = \alpha_1 V_1' + \alpha_2 V_2' + \alpha_3 V_3'$.

CONCLUSION

Of the different kinds of triangulations on a finite set of points, well-defined triangulations are best suited for applications such as conflation. These triangulations enable the user to build specialized subroutines that operate locally but are globally consistent. The well-defined Delaunay triangulation is easily described and implemented. Moreover, the implementation requires ideas that can be used to define a rubber-sheet transformation. The Delaunay triangulation and its associated rubber-sheet transformation have been implemented for use in the conflation process at the Census Bureau.

APPENDIX. - CONTINUING RESEARCH.

There are a number of issues that require further research. Some of these problems are data structure problems and others are mathematical problems related to the algorithms. Two of the mathematical problems will be discussed here.

The easier of the two is the problem of what should be done about the case of four or more cocircular points. Recall, the definition of the Delaunay triangulation specifies that no four points are cocircular. If four points are cocircular, then there is a choice as to which pair of triangles to form. Unfortunately, the choice is arbitrary; the theory will not allow a resolution of the problem. Allowing software to make an arbitrary choice is unsatisfactory since we are motivated by the desire to have a well-defined triangulation. So, a well-defined method for handling this case is needed to make the theory complete.

The harder question is the extensionality problem. During conflation the triangulation created for one map is extended to the second via the one to one correspondence that exists between the two sets of vertices. If the correspondence lifts to a homeomorphism between the triangulations, then there is no problem. However, this is by no means guaranteed. It is possible, and it happens in practice, that the vertices shift with respect to each other enough to cause triangles to reverse orientation or flip on top of each other. (See figure 5) Topology does not remain consistent in either case. We have implemented an algorithm that recognizes when topology is not preserved, but the crucial question before us is how to fix the triangulation so as to maintain topological consistency and remain faithful to the definition of the triangulation.

BIBLIOGRAPHY

Corbett, J., 1979, "Topological Principles in Cartography," Technical Paper 48, Census Bureau.

Lefschetz, S., 1975, Applications of Algebraic Topology New York: Springer-Verlag.

Lee, D. and B. Schachter, 1980, "Two Algorithms for Constructing a Delaunay Triangulation", International Journal of Computer and Information Sciences, Vol. 9, No. 3, pp. 219-241.

Saalfeld, A., 1985, "A Fast Rubber-Sheeting Transformation Using Simplicial Coordinates," to appear in the October issue of The American Cartographer.

Singer, I. and J. Thorpe, 1967, Lecture Notes on Elementary Topology and Geometry, New York: Springer-Verlag.

White, M. and P. Griffin, 1985, "Piecewise Linear Rubber-Sheet Map Transformations," to appear in the October issue of The American Cartographer.

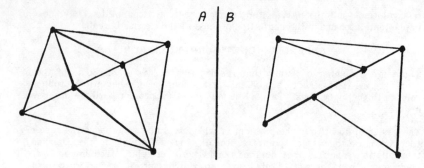

Figure 1. A) Maximal Set of Triangles; B) Non-maximal Set

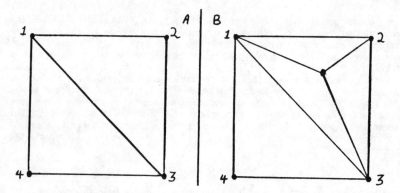

Figure 2. A Simple Ambiguous Triangulation Procedure, A) Before, and B) After Adding a Vertex.

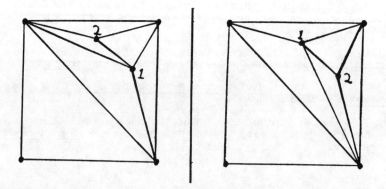

Figure 3. An Example of Order Dependence Using the Procedure of Figure 2.

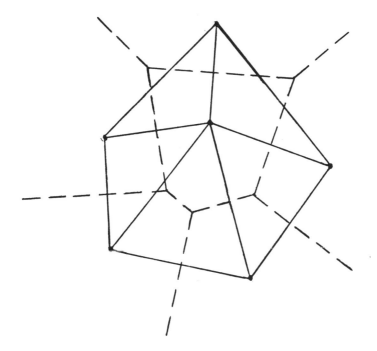

Figure 4. Thiessen Polygons on 6 Points (dashed lines) and the Resulting Delaunay Triangulation (solid lines).

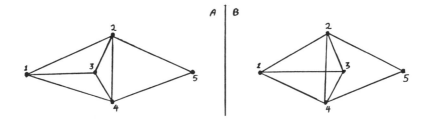

Figure 5. Topological Inconsistency in Extending Triangulation From One Set of Points (A) to a Second (B).

LASER DISC TECHNOLOGY APPLICATIONS OF
NAVIGATIONAL CHART COMPILATION AND
IMAGE DISPLAYS

Cdr. David J. Goehler
Chief, Requirements and Technology Staff
National Ocean Service
Rockville, Maryland 20852

BIOGRAPHICAL SKETCH

Cdr. Goehler is responsible for coordinating production specifications for Federal Aviation Administration aeronautical charting requirements and supervising aeronautical charting research projects. He received his B.S. degree in Industrial Engineering from Purdue University in 1967 and M.B.A. degree from the University of Santa Clara in 1972. Cdr. Goehler is a member of the National Oceanic and Atmospheric Administration (NOAA) commissioned officer corps and a member of the American Congress on Surveying and Mapping.

ABSTRACT

The National Ocean Service's (NOS), Office of Charting and Geodetic Services is exploring the applications of laser disc technology for its charting and geodetic programs. Two separate but related projects are currently underway.

The first is a cooperative government agency project to develop a prototype system to enhance communications and operations for an emergency response system. The system will use analog videodisc, microcomputer, and telecommunication technologies to improve information exchange. The NOS will use this opportunity to evaluate the potential of analog discs in computer assisted chart compilation, document storage and data retrieval applications.

The second project will investigate the potential of the digital disc to store and display aeronautical chart navigational data in an aircraft environment. The six month feasibility study will survey digital cartographic requirements of advanced avionic systems; analyze future NOS production and electronic distribution alternatives; conduct a technology conference with potential users and developers of new navigational systems; and formulate concepts for a low cost technology demonstration of digital cartography.

INTRODUCTION

During the last quarter century, the cost, the energy consumption and size of computers of comparable power have decreased by a factor of 10,000. Computational speed has increased by a factor of 200 (Toong and Gupta 1982). Until recently, advances in mass memory have not kept pace with these technological improvements in computer equipment. The 12 inch diameter analog laser disc can store up to 108,000 images; the same size digital laser disc is capable of storing 2 billion bytes of digital data. Manufactured by laser optical equipment, laser discs represent low cost, easily portable, high density storage media. When integrated with today's inexpensive microcomputer processors, laser discs may revolutionize the production and distribution of cartographic data and speed the wide spread use of electronic chart systems.

THE ANALOG DISC

Analog discs store information as television pictures. The discs are produced from a master videotape which directs a low power laser beam to expose a pattern of pits on the photosensitive surface of a glass master disc. Each image is recorded on one of 54,000 separate tracks per side. After additional processing, plastic production discs can be mass produced at nominal costs. The disks can be read by commercially available off-the-shelf laser disc players. As the disc spins at 1800 rpm, a lower powered laser detects light reflection from the pit pattern and displays the recorded image on a video monitor. The playback laser can randomly access any one image in less than 3 seconds. When interfaced with a microcomputer, the analog disc may represent an

> ....alternative to or interim measure for, the digital data base. The vast storage capacity of these discs would allow the equivalent of hundreds of map sheets to be placed on a single side disc..... This image can be displayed on a video monitor, digitized by the host minicomputer and used as a base product in the creation of specialized products..... In addition to the use of the videodisc map as a cartographic base, all of the standard map analysis can be performed with computer assistance (Loomer 1984).

In December 1983, the Aeronautical Charting Division (ACD) received a $6,000 grant from the NOS Science and Technology Council to study potential cartographic applications of videodisc technology. By February of 1984, ACD learned of a proposed project by the Army Corps of Engineers (COE), Water Resources Support Center. In cooperation with the United States Geological Survey and the Federal Emergency Management Agency, the COE was designing a videodisc prototype system to enhance communications and planning activities for an emergency response for the southwestern United States. The system would use videodisc, microcomputer and telecommunication technologies to improve information exchange. Since limited funding prevented full utilization of the videodisc's storage capacity, the COE was seeking additional support from other interested agencies. NOS used this opportunity to dedicate the original $6,000 grant plus an additional $14,000 from the NOS Aeronautical and Nautical Charting Divisions to investigate the application of videodisc technology in the compilation and revision of navigational charts.

As currently proposed, the two sided disc will be available in March 1985 and contain approximately 108,000 images of maps, charts, diagrams, satellite and aerial photographs, and related narrative data. The NOS portion will consist of nautical and aeronautical charts, aerial photographs, geodetic network diagrams, chart indexes and narrative navigational data. In addition to the disc, the prototype workstation will include a laser disc player, color monitor, graphics interface board, light pen and associated software to permit individual frame access, graphics overlay and telecommunications capabilities. The NOS has procured a 16-bit microcomputer with a 10 megabyte hard disk, color monitor and modem which will control the laser disc player in the merging of analog disc imagery with digital data overlays and permit communications between remoted workstations.

Using the NOS 1:250,000 scale Houston Terminal Area Chart as a test case, all nonregulatory, low rate of change data will be recorded on the videodisc. Specific regulatory, high rate of change data available in digital form will be downloaded from a remote main frame computer facility to the microcomputer. The ability to accurately overlay digital data from the microcomputer on the analog disc imagery will be investigated. Should this technique prove feasible, the compiler could update and position digital data changes interactively using the analog disc image as a reference background. Once all digital updates have been made, the digital data could be read to a laser plotter device to produce appropriate size negatives for printing. These negatives could be later registered to the corresponding base negatives in the photo-mechanical production of printing plates.

The large image storage capacity of the analog videodisc and the data base management and graphics potential of the microcomputer may provide an alternative to the labor intensive digitizing process and its associated mass storage requirements. The use of analog disc technology in automated cartography may give new meaning to an old adage. Since a single video image requires the same space as 22,000 bytes or characters, it may still be true that "one picture is worth a thousand words", even 16-character words.

THE DIGITAL DISC

Digital discs record and store data as binary values similar to magnetic disc systems. Digital discs are produced by low power lasers that burn holes or form bubbles ranging in size from 0.4 to 5 microns on the recording layer of the disc. The holes or bubble formations represent binary zeros or ones. This high density packing of binary coding permits the two billion byte storage capacity of each 12 inch diameter disc. Digital discs range in size from the 14 inch/4 billion byte to as small as the 2 inch/40 million byte version. The equivalent of a 5-1/4 inch floppy disc size capable of storing 550 million bytes may soon become a standard size for microcomputer applications. Random access times are in the range of 100 to 150 milliseconds.

The NOS with support of the Federal Aviation Administration (FAA) is exploring the potential of digital discs to facilitate the production and distribution of cartographic data as well as a new medium to display NOS aeronautical chart information and graphics in a cockpit environment. The application of map data to electric cockpits is already being evaluated in numerous government and private sector research and development efforts. All these programs, however, will require a readily available and reliably accurate navigation data base. The current perception of potential users and hardware developers is that electronic map displays will not be practical until a suitable data base is available - possibly not for 5-10 years (Elson 1984). Consequently, independent, specialized data bases supporting unique hardware requirements are beginning to appear in the marketplace. The eventual affect of various nonstandard electronic chart symbologies on public safety issues are as yet undefined.

Both the Defense Mapping Agency and the NOS have been building digital data bases principally to support automated paper chart production. With the current and projected flexibility of these digital data files and the commercial availability of portable mass storage devices, such as digital discs, automated chart production will be expanded to include the development of digital cartographic products. A demonstration of digital cartography may be necessary to merge the needs of potential users with the resources of hardware developers and NOS/FAA data base compilers to ensure the successful development of electronic charts.

This project is being contracted through the United States Air Force Wright Aeronautical Laboratories. Wright Patterson's mission in the design of advanced cockpit systems and its outstanding test facilities makes it an ideal site for follow on efforts should this initial six month feasibility study prove promising. Begun in January 1985, Phase I consists of a concept definition which includes:

- A survey and analysis of industrial and government programs planning to use digital cartography. This effort will include a review of the data needs of advanced avionics equipment and a measure of the positional data by kind, volume and accuracy level required. The insight provided may reduce the lead time and development costs of future NOS digital products.

- A technology assessment of alternative strategies for electronic publishing and distribution of future NOS cartographic products.

- A technology conference with potential users and hardware developers requiring digital cartographic data input. The conference will be a combination of briefings and workshops involving industry and government developers of avionic products and requirements. The goal of the conference will be to create a forum for the interchange of ideas; a mutual awareness of available resources; and an understanding of research and development in related technologies.

- A preliminary analysis and definition of a low cost concept for evaluating digital aeronautical data in a general aviation aircraft. The capability of the digital disc will be evaluated to store and display digital map data using low cost avionics. The necessary NOS production and distribution posture to support this new medium will also be defined.

CONCLUSION

Through carefully monitoring and evaluating evolving technologies such as analog and digital laser discs, the NOS is striving to keep pace with future avionic system digital data requirements, while continually improving its present paper chart production process.

ACKNOWLEDGEMENTS

The author wishes to thank Mr. Malcolm Murphy for his valuable suggestions and Mrs. Joan R. Goehler for typing and proofing numerous drafts of this manuscript.

REFERENCES

Ciampa, J. 1984, Optical Storage: <u>Photomethods - The Journal of Imaging Technology, Imaging Technology in Research and Development</u>, Vol. 27, No. 7, pp. 14-20

Elson, B.A. 1984, Electronic Displays Gain in Cockpit Use: <u>Aviation Week and Space Technology</u>, Vol. 120, pp. 225-239

Loomer, S.A. 1984, The Videodisc: Applications and Implications for Cartography, <u>Technical Papers 44th Annual Meeting ACSM</u>, pp. 331-340

<u>Technology Study and Development Plan for Next Generation Cartography</u>, 1984, No. 84-9380, The University of Maryland Research Foundation

Toong, H.D. and Gupta, A. 1982, Personal Computers: <u>Scientific American</u>, Vol. 247, pp. 86-107

Zuckerman, C. 1984, Advancing Office Operations Through Optical Memory Technology: <u>Government Executive</u>, Vol. 16, No. 10, pp. 48-52

CARTOGRAPHIC USE OF DITHERED PATTERNS
ON 8-COLOR COMPUTER MONITORS

Ann M. Goulette
Michigan State University*
Department of Geography
East Lansing, MI 48824

ABSTRACT

As microcomputers are increasingly used in cartography, their limitations are becoming more apparent. The limited numbers of colors available on inexpensive computer monitors can be artificially increased through the use of dithered patterns, a computer graphics technique. This paper addresses which types of dithered patterns are suitable for cartographic purposes and investigates some simple aspects of the perception of these patterns.

INTRODUCTION

Recent decreases in microcomputer prices have brought these devices within the reach of most organizations; however, most inexpensive color monitors can display only 8 or 16 colors. This limited palette restricts the cartographer in making complex or aesthetically-pleasing maps on these monitors. Dithering, a common technique of computer graphics, can be used to increase the range of perceptibly different colors on common color raster displays.

Dithering, sometimes referred to as halftoning, has typically been used to display three-dimensional graphics on computer monitors (Ryan, 1983) and for solid modeling. By varying pixel color or intensity gradually across the image, the impression of depth and shadow is achieved. The individual pixels merge visually to create the illusion of a new color or change in value. The dithering technique is not limited to the display of three-dimensional objects, however.

Truckenbrod (1981) recognized the potential of incremental color mixing on color monitors. By varying the proportions of two colors across a display plane, she effectively varied the color elements of hue, value and chroma in both graded and discrete steps. Her examples are particularly illustrative of what can be achieved through the dithering technique, but the patterns are a bit too busy for effective use in mapping.

Cartographic convention and research limit which patterns are suitable for mapping purposes. Jenks and Knos (1961), for example, found that map users prefer dot patterns rather than line or irregular patterns in a graded series.

---
* The author is currently affiliated with the U.S. Bureau of the Census, Geography Division, Washington, DC 20233.

In addition, users prefer uniform texture among patterns of a series and find finer textures most pleasing.

As little research had focused on dithered patterns, the preliminary investigation centered on creating patterns suitable for mapping. After appropriate patterns had been chosen, research questions then focused on the perception of dithered patterns. In particular, the ability of subjects to distinguish dithered patterns from one another was investigated. Later research will address the use of these patterns in a mapping context.

CHOICE OF DITHERED PATTERNS FOR MAPPING

On expensive color graphics monitors (i. e., those with 256 or more colors available), the possible combinations to create dithered patterns are almost infinite. On more inexpensive monitors, the number of possible patterns is still large, but is constrained by the number of available colors and limited ability to vary pixel intensity. By requiring the dithered patterns to be suitable for mapping, the number of possible patterns is limited even further.

The diterered patterns chosen for the cartographic application addressed in this paper were created by varying the proportions of two hues of pixels to fill polygons. Only patterns that resembled screened tones (as used in printing) were used. These patterns were created by turning on a regular lattice (usually a triangular grid) of pixels on a background of a contrasting color. Polygons were filled with the patterns using a program based on an algorithm by Pavlidis (1982).

In choosing the patterns, the intent was to create a value scale of patterns using blue and white pixels. For example, the patterns would range from 0% blue to 100% blue in discrete steps. The patterns were not difficult to create on the computer screen, but problems arose with the appearance of some patterns on the monitor.

While it was easy to create patterns at both ends of the value spectrum, the middle-range value patterns often had disturbing optical effects. For example, the 50% blue pattern, which consisted of a checkerboard of single blue and white pixels, appeared to have an "op-art" effect. The arrangement of pixels on some other middle-range value percentages created the appearance of either vertical or horizontal stripes. Apparently, when single pixels are rapidly varied in close proximity, disturbing optical effects are possible. As a result, the middle-range value percentage patterns were adapted to eliminate the unintended effects. Instead of varying single pixels of the foreground color, some patterns varied two adjacent foreground pixels to create the dithered patterns.

The dithered patterns chosen for the test represented a value range from 0% blue to 100% blue in discrete, but uneven steps. Twenty-three patterns were selected for testing. These patterns and their value percentages are shown in Figure 1.

METHODOLOGY

Each pattern was paired with itself and the five most adjacent value percentage patterns. The resulting 124 pairs of patterns were photographed as slides from the computer screen (a Texas Instruments Professional Computer with a color monitor). The slides were divided into two groups for ease of testing.

Subjects were undergraduate and graduate students and faculty of the Geography Department of Michigan State University. Subjects were tested in small groups (usually 3-4 subjects at a time) in a small room darkened for viewing the slides. Thirty-four subjects were tested on one group of slides; thirty-three were tested on the other group. Each slide was presented to the subjects for 10 seconds.

While viewing the slides, subjects were instructed to compare the two patterns to determine whether the patterns seemed to contain the same percentage of blue. If the patterns were judged to contain different percentages of blue, subjects were to indicate which pattern appeared bluer.

RESULTS

The results were tested using the normal approximation to the binomial distribution. Six of the original twenty-three patterns were excluded after the test as being too coarsely textured to be used on maps (Patterns 2, 3, 5, 19, 21, and 22). Of the remaining eighteen patterns, twelve were determined to be perceptibly different from one another ($p \leq .05$). These were patterns 1, 4, 7, 8, 9, 10, 11, 12, 13, 15, 20, and 23.

DISCUSSION AND SUMMARY

Dithered patterns suitable for mapping purposes can be created on inexpensive color monitors. The preliminary research on these patterns, however, shows that care must be taken in their selection for use in cartographic applications. Patterns to be viewed or photographed from a monitor should be free of disturbing optical effects, yet be as finely textured as possible. Cartographic conventions and the findings of psychophysical research such as that of Jenks and Knos (1961) and others should be followed when designing displays with the patterns.

In the present study, subjects were able to distinguish a surprising number of patterns. Although given only 10 seconds to view each slide, subjects seemed to have sufficient time to make very fine discriminations between the pattern pairs. The author does not believe the twelve

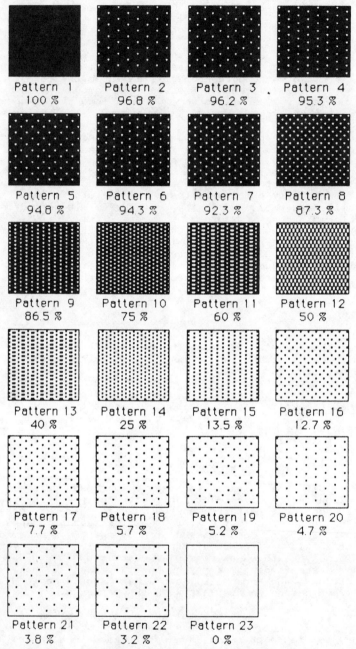

Figure 1. Arrangement of pixels and percentage of blue pixels in the 23 patterns chosen for study.

patterns found to be perceptually different in this study could be used simultaneously (as they would be in mapping) and still be easily distinguished.

Subjects were able to better differentiate patterns at the dark end of the value scale. White pixels appeared very bright on the computer screen and were easy to distinguish against the predominantly blue background. When the proportions of blue to white are reversed, however, the brighter white pixels may overwhelm the darker blue pixels.

Dithered patterns are potentially useful in many mapping applications. They readily lend themselves to situations where areal fill patterns of different colors or values are needed (e. g., choropleth or shaded isoline maps). In addition, they are potentially useful in continuous-tone applications. These patterns, however, remain to be investigated in a mapping context. The author will use the twelve perceptually different patterns found in this study to create choropleth maps as part of her future research.

ACKNOWLEDGMENTS

The author wishes to acknowledge the contributions of Dr. Judy Olson to this research and the encouragement of the Mapping Operations Branch staff at the U.S. Bureau of the Census.

REFERENCES

Jenks, G. F., & Knos, D. S. 1961. The Use of Shaded Patterns in a Graded Series: Annals of the Association of American Geographers, Vol. 51, pp. 316-334.

Pavlidis, T. 1982. Algorithms for Graphics and Image Processing, Computer Science Press, Rockville, MD.

Ryan, D. 1983. Computer Programming for Graphical Displays, Brooks/Cole Engineering Division, Monterey, CA.

Truckenbrod, J. R. 1981. Effective Use of Color in Computer Graphics: Computer Graphics, Vol. 15, No. 3, pp. 83-90.

TERRAIN ANALYST WORK STATION

(ABSTRACT)

LASLO GRECZY
ELIZABETH PORTER
RICHARD MARTH
US ARMY ENGINEER TOPOGRAPHIC LABORATORIES
FORT BELVOIR, VIRGINIA 22060-5546
Phone: (202) 355-2877

TERRAIN ANALYST WORK STATION (TAWS)

In todays modern Army, with its emphasis on mobility and rapid response, battlefield commanders need timely and accurate terrain and environmental information to assist them in making crucial tactical decisions. Current manual terrain analysis procedures are too cumbersome and inflexible to satisfy future requirements for rapid generation, update and dissemination of terrain information and terrain products. The Army is turning to digital terrain data bases and automated terrain analysis systems to address this problem.

Researchers at the U.S. Army Engineer Topographic Laboratories have initiated an effort to develop a prototype Terrain Analyst Work Station (TAWS) which will be used to demonstrate computer-assisted terrain data extraction, analysis and manipulation techniques in the field. The TAWS development effort capitalizes on recent advances in micro-computer technology, analytical photogrammetry, computer-assisted photo interpretation and geo-based information processing systems to provide an integrated terrain data extraction, analysis and display capability. The TAWS prototype incorporates off-the-shelf hardware and existing terrain analysis software to provide a multi-purpose work station which can be used for a variety of terrain information generation tasks.

The basic TAWS hardware consists of a state-of-the-art 32-bit micro-computer and peripherals, hard and soft copy data input devices and hard and soft copy output and display devices. The data input devices consist of alphanumeric terminals, a light table digitization and mensuration system, a graphics digitizing table and an analytical stereoplotter with dual-channel graphics super-position and profiling firmware. The output and display devices consist of a high quality color graphics plotter, a line printer, and color graphics display devices.

The operating system being used on TAWS is UNIX*. Applications software is being written in Fortran 77, Pascal and C, and will include geographic information system software, digital terrain elevation data manipulation software, and environmental effects software.

The primary function of TAWS will be to perform data extraction digitization and mensuration. However, the work station will also incorporate data manipulation, analysis and product generation capabilities. The system will provide Army terrain analysts with the tools needed to: (1) Create topologically valid digital terrain data bases; (2) Edit, update, revise and intensify existing data bases; (3) Merge data from various sources; (4) Manipulate, analyze and display digital terrain data; and (5) Generate and disseminate quick-reaction tactical decision aids and special purpose terrain products.

The TAWS prototype will be operational in the laboratory by mid-1985. After initial laboratory tests, the TAWS will be taken into the field for a series of demonstrations. The results of the laboratory tests and field demonstrations will be used to help define requirements and validate fielding concepts for data extraction, update, revision and analysis capabilities needed in a fieldable Digital Topographic Support System. The TAWS will also be used to support development and demonstration of new techniques and software under the Corps of Engineers AirLand Battlefield Environment program.

*UNIX is a trademark of Bell Laboratories.

STATISTICAL EVALUATION OF ACCURACY FOR DIGITAL CARTOGRAPHIC DATA BASES

Arnold Greenland & Robert M. Socher
IIT Research Institute
5100 Forbes Boulevard
Lanham, Maryland 20706

MAJ Michael R. Thompson
U.S. Army Engineer Topographic Laboratories
Fort Belvoir, Virginia 22060

BIOGRAPHICAL SKETCHES

Dr. Greenland is a Senior Analyst for the IIT Research Institute. He earned a B.S. in Mathematics from Case Western Reserve University in 1969, an M.A. in Mathematics from the University of Rochester in 1973, and a Ph.D. in Probability Theory from the University of Rochester in 1975. Dr. Greenland has applied statistical methods to a wide range of engineering, economic, and environmental problems. His current interest is in the application of statistical methods to digital geographic data bases.

Mr. Socher is a Senior Programmer/Analyst for the IIT Research Institute. He earned a B.A. in Mathematics from St. John's University, Minnesota in 1967. During his career, Mr. Socher has developed a broad base of knowledge in automated data processing for cartographic/terrain data. For the past seven years, he has served as project manager guiding the design and development of interactive/batch graphics software on the Digital Terrain Analysis Station (DTAS) for the U.S. Army Engineer Topographic Laboratories.

MAJ Thompson received a B.S. degree from the United States Military Academy in 1973 and was commissioned as a military officer in the U.S. Army. He has served in infantry assignments in both Europe and the United States and is a graduate of the Infantry Officers Advanced Course. In 1981, MAJ Thompson earned an M.S. in Photogrammetry and Geodesy from Purdue University and is currently assigned to the U.S. Army Engineer Topographic Laboratories (ETL) as an R&D Coordinator.

ABSTRACT

The major statistical techniques used by cartographers to measure accuracy in digital cartographic data bases -- namely, linear and circular probable errors -- were derived for application to analogue cartographic products. Digital products because of their use, application, and construction do not necessarily fit in the accuracy model of paper products. This paper documents the application of a relatively new measure, the Kappa statistic, for analyzing the accuracy of digital geographic feature data bases. The specific application was the comparison of two digital cartographic data bases provided by the Defense Mapping Agency (DMA).

BACKGROUND

In order to evaluate the Army terrain analysis requirements for digital terrain data, the adequacy of each of two DMA prototype data sets was assessed in terms of data content and completeness, absolute and relative accuracy (vertical and horizontal), resolution of the elevation and feature data, and the format or structure in which the data are recorded (including the coordinate systems and reference datums). This evaluation was aided by the use of a commercial, interactive graphics system, the Digital Terrain Analysis Station (DTAS), which is designed to automate tactical terrain analysis.

The first prototype data set which was provided by the DMA Hydrographic/Topographic Center (HTC) is limited to those natural and man-made features which are of tactical military significance. The data set consists of six feature topics which include Surface Configuration, Vegetation, Surface Materials (soils), Surface Drainage, Transportation, and Obstacles. The prototype is in a 12.5 meter gridded format and utilizes the Universal Transverse Mercator (UTM) coordinate system. Each grid point consists of 199 bits where the first 16 bits contain the elevation and the other 183 bits contain the associated codes for the features.

The second prototype data set which was produced by DMA Aerospace Center (AC) is the High Resolution Prototype Data Base enhanced for tactical terrain analysis applications with the addition of three new micro-descriptors (Surface Drainage, Transportation, and Vegetation). The data set is in a vector format and utilizes the World Geodetic System (WGS) coordinate system. Each feature record consists of two or three subrecords:

- A primary feature description
- Zero to six optional micro-feature descriptions
- A "delta set list" of relative coordinate pairs for the location information.

Elevation information is obtained from a DMA standard Digital Terrain Elevation Data (DTED) magnetic tape file. The data is in a 1201 x 1201 matrix with a grid spacing of 3 seconds of arc which equates to approximately 60 meters by 90 meters. This data also uses WGS coordinates. Both prototype data sets consist of two areas in the state of Washington, Fort Lewis and Yakima Firing Center.

To evaluate the ability of the two DMA prototype data sets to support terrain analysis requirements, software routines were developed to read and reformat the DMA data into the DTAS data base. With the DMA prototype data sets reformatted for the DTAS, terrain analysis models (products), developed as part of a software development program were executed using both DMA prototype data sets as input. These products were then compared to manually-prepared products produced for the evaluation. Operational suitability, in terms of the usefulness and the acceptability of the prototype data element features and the automated terrain analysis products, was determined by visual analyses conducted by the Terrain Analysis Center at ETL, with support from military terrain analysts.

This paper addresses the aspect of the data base evaluation during which objective measures were formulated to compare and to quantify the differences between digital and manual data features and products. The same techniques were used to evaluate features and products at degraded resolutions to determine the minimum acceptable data resolution necessary which could satisfy Army requirements. The end result is a viable objective method that can be used to statistically capture the subjective analysis performed in visual evaluation.

A thorough, in-depth statistical analysis of the elevation data was also performed and is documented in Herrmann, et al. (1984). The remainder of this paper, however, will concentrate on the statistical analysis of the feature data.

THE STATISTICAL EVALUATION OF FEATURE DATA

The selection of the statistical measure used for evaluation of feature data was driven by two considerations. The first relates to a common methodology, Circular Probable Error (CPE), applied to feature data. In this case, one obtains a sample of locations (monuments) between which one measures the distances in the two feature map representations to be compared. These distances (errors) are used to estimate the CPE statistics. The problem with this approach is that it ignores the fundamental character of the feature data. Feature data is categorical or nominal and not interval as is required for the CPE approach. This means that the variables do not take on numerical values measured on a continuum, but rather they are simply categories or names of the type of features. A more consistent approach is to use the methods of categorical analysis to obtain accuracy metrics. The second consideration flows from the way individuals normally evaluate a product visually. The most common reaction when confronted with a feature map produced by the DTAS was to overlay the map produced directly from the digital data onto an existing hard copy map, and see how it looked. We felt it was useful to find a statistical approach which was the realization of that visual comparison process. These two considerations led to the choice of a statistical technique noted in some cartographic contexts (see Chrisman (1982), Congalton and Mead (1981)) called Kappa. The details of this method will be discussed below.

A Measure of Agreement

The ideas discussed above were the impetus behind the formulation of the feature data metrics for the data base evaluation. The approach is best illustrated by Figure 1. The figure contains two realizations of a feature classification of a particular region. The solid polygon represents the region designated as category 1 by the first product and the interior of the dashed line represents the region designated as Category 1 by the second product. As one can readily discern, the two classifications do not agree completely. The disagreement can be described in the matrix shown in TABLE 1. Let p_{11} be the fraction of the total area displayed in which the first product or source shows category 1 and so did the second; let p_{12} be the fraction of the total area displayed in which the first source shows category 1 but the second shows category 2; etc. Let a subscript of "+" indicate summation over that index in the matrix. Of course $p_{++} = 1$, since that is the total area of interest.

One obvious measure of agreement is the sum down the diagonal, $p_0 = p_{11} + p_{22}$. This measure ignores the magnitudes of the "marginal" probabilities (the fractions shown as row and column sums in TABLE 1). Cohen (1960) suggested a measure based on the table just described which removes the effect of chance from the measure.

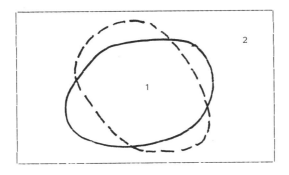

Figure 1

TABLE 1

		Classified by Second Source		
		1	2	
Classified by First Source	1	P_{11}	P_{12}	P_{1+}
	2	P_{21}	P_{22}	P_{2+}
		P_{+1}	P_{+2}	$P_{++} = 1$

The expected probabilities for each cell in the matrix are computed using the marginals as follows: $P_{1+} \times P_{+1}$ is the expected probability for cell (1, 1), etc. Thus the "expected" fraction of agreement is: $P_e = P_{1+} P_{+1} + P_{2+} P_{+2}$. The Kappa statistic is defined as: $k = (P_o - P_e)/(1 - P_e)$ and is interpreted as the proportion of agreement over and above chance agreement. A detailed explanation of this measure can be found in either Bishop et al. (1975) or Fleiss (1981).

A final methodological comment is required when the variables used for classification are something other than nominal, such as ordinal, in nature. A disagreement in this context may be more or less critical. An example is that classifying a point as type 1 in one method and type 2 in another may be much less problematic than if the classification types were, respectively, 1 and 7. To remedy this situation Cohen (1968) also introduced the weighted Kappa. A complete explanation of this concept is beyond the scope of this paper, though the reader is referred to Fleiss (1981) for an excellent explanation. The interpretation is essentially the same, i.e., the fraction of weighted agreement over and above chance. The "weightings" are often displayed in a matrix such as the one shown in

TABLE 2. The values there show for example that an exact agreement between "control" and "product" maps is given full weight as demonstrated by 1's down the main diagonal. A disagreement of only one category (such as 2 for control and 3 for product) is given a weight of 1/2. Of course, the weight matrix for the unweighted Kappa is an identity matrix.

TABLE 2
KAPPA WEIGHTS

Control	PRODUCT						
	1	2	3	4	5	6	7
1	1	.5	0	0	0	0	0
2	.5	1	.5	0	0	0	0
3	0	.5	1	.5	0	0	0
4	0	0	.5	1	.5	0	0
5	0	0	0	.5	1	.5	0
6	0	0	0	0	.5	1	.5
7	0	0	0	0	0	.5	1

The Operational Issues

The first step in the evaluation of a set of feature polygons or product polygons is to obtain a control product for comparison. To that end ETL undertook the task of hand digitizing the feature and product overlays which were to be evaluated. The included sets of feature and product polygons were:

- Soil

- Slope

- Vegetation

- Cross Country Movement (CCM).

CCM is an analytical model that predicts off-road speed potential based on vehicle characteristics and the terrain (soil, slope, and vegetation).

The second step in the process was to register the matching feature or product, from either the HTC or AC source, to the digitized control set of polygons. The computational task required to compute Kappa or weighted Kappa is to obtain the cross-classification matrix such as the one shown in TABLE 1. The matrix values p_{ij} are estimates for the areas of that type of agreement or disagreement in the maps. One method is to use polygon processing capabilities to identify the various regions and then to calculate the enclosed areas. A second method is to use an underlying grid structure where each grid point represents a region of area surrounding it. For this application, the grid approach was chosen.

The grid structure allowed for the estimation of fractional agreement by comparing the files position by position in the grid. For details of the mechanics of the grid approach and an example of an agreement matrix, see Herrmann et al. (1984).

Any human errors in polygon labeling in the hand digitized, HTC, or AC data sets that were detected during the creation phase or visual analysis phase of this evaluation were corrected. Also, as the statistical analysis phase was executed, the agreement matrices were examined for any abnormally large numbers off the main diagonal which could be an indication of mislabeled polygons. The input data was closely scrutinized and any mislabeled polygons corrected. Then the statistical analysis was performed again.

One aspect of this approach that has not yet been considered in the literature is the evolution of "rules of thumb" to help interpret the Kappa results. In order to help remedy this situation, this study included two separate attempts to begin this evaluation process. First we generated a "benchmark" value as follows. The soil polygons for Yakima were digitized two separate times. The value of Kappa derived from those two sets of polygons was used as a reference point from which to assess the other values of Kappa that occurred. As shown in TABLE 3, the benchmark was 0.967, a very high value. It has the interpretation that approximately 97% of the grid points (an estimate of area) were in agreement over and above what would be expected to agree by chance alone. The second thing done to evolve the rules of thumb for interpeting Kappa is described below in the section on granularity.

Results

Following the methods described above Kappa and weighted Kappa coefficients were computed and are shown in TABLES 3 and 4. The former table contains coefficients for features. They are shown grouped by feature type (soil, slope, and vegetation) crossed with region (Yakima and Fort Lewis). Within that classification, four values of HTC data granularity (12.5m, 25m 50m and 125m) and the lone AC format are shown. The latter table contains the four products at different HTC granularities and the one AC product. The major cells shown in the table contain weighted Kappa statistics broken down by three different weighting schemes. The first one, unweighted, is identical to the method used for the feature polygon evaluation in that only those points along the main diagonal are considered to agree. The second weighting scheme takes into account those grid points where the two sources differ by one speed category, and weights these points 1/2 as great as a perfect match (see TABLE 2). In the final weighting scheme, proposed by Cicchetti and Allison, a linear approach is taken. Weights are assigned with respect to relative positioning in the matrix as applied through the formula: $W_{ij} = 1 - [1 - (i-j)/(h-1)]$, where h = the number of categories. Therefore, in a seven by seven matrix the two minor diagonals closest to the main diagonal are assigned weights of 5/6, the next two diagonals are assigned weights of 4/6, etc. These weightings are crossed with location (Yakima and Fort Lewis) in TABLE 4.

TABLE 3
FEATURE KAPPA STATISTICS*

		Spacing	Yakima	Ft. Lewis
Soil	HTC	12.5	.961	.946
		25	.960	.942
		50	.954	.941
		125	.924	.920
	AC	N/A	.944	.910
Slope	HTC	12.5	.900	.954
		25	.900	.948
		50	.895	.939
		125	.858	.891
	AC	N/A	.921	.953
Vegetation	HTC	12.5	.960	.938
		25	.956	.937
		50	.957	.922
		125	.937	.855
	AC	N/A	.964	.928

*Benchmark = .967

TABLE 4
CCM WEIGHTED KAPPA STATISTICS

		Spacing	Yakima	Ft. Lewis
Unweighted	HTC	12.5	.925	.709
		25	.919	.702
		50	.916	.699
		125	.883	.661
	AC	N/A	.911	.774
"1/2" Adjacent Diagonal Weighting	HTC	12.5	.928	.790
		25	.921	.782
		50	.918	.778
		125	.888	.739
	AC	N/A	.913	.825
Cichetti & Allison Weighting	HTC	12.5	.931	.801
		25	.925	.793
		50	.922	.789
		125	.893	.747
	AC	N/A	.918	.829

The two tabulations just described were then used to address the following issues (shown in order of importance):

- the extent to which the two prototype data bases support feature and product generation

- the appropriateness of the DMA data bases in producing CCM

- the granularity of data required to produce acceptable features and products.

The following discussion of these issues is presented in the reverse order, for reasons of logical development. We will then discuss the extent to which the Kappa statistic fulfilled those goals.

Granularity. It would be desirable to have statistical measures which bring with them accepted "rules of thumb" for interpreting their magnitude. Unfortunately, because the Kappa statistic has not been used extensively in this area, it was necessary to evolve acceptable levels of Kappa by close reference to the visual analysis. The visual analysis by ETL technical personnel was accomplished completely independently of the statistical analysis. The hope was that when the two analyses were compared they would be mutually supportive. Indeed, this was the case. The visual analysis also allowed one to evolve a working level of Kappa below which one is likely to see unacceptable results. Refer first to TABLE 3. The most striking thing about the table is the uniformly high percentages of agreement (Kappa coefficients) shown. Only a few values on the page are below .900. The visual analysis of features found essentially that the computer generated feature plot for Yakima slope (with k = .858) and for Fort Lewis vegetation (k = .855) were the only unacceptable features. Those were in fact the two lowest Kappa coefficients. The next lowest value was .891 for Fort Lewis slope which was acceptable. The logical conclusion is that somewhere between .850 and .900 the level of agreement becomes unacceptable.

The specifics of the granularity study are as follows. It would be quite justifiable to assert that spacing as high as 50 meters between raw data values can be tolerated in producing acceptable feature overlays. However, spacing of 125 meters between data points is not universally acceptable, though when the number of polygons is small the results (as one would expect) were reasonable.

There were also some general things to be said about the differences between Yakima and Fort Lewis which were detected by the percentage agreement figures. First of all, in Yakima where the terrain is much more rugged than Fort Lewis, the slope coefficients were uniformly worse at each granularity by approximately 5 percentage points. However, in Fort Lewis, where vegetation is complex, the coefficients were uniformly worse than Yakima where vegetation is sparse. The difference between Yakima and Fort Lewis widened as the data was degraded.

CCM. The results for the CCM analysis are shown in TABLE 4. The first point to make is that the Fort Lewis results are uniformly not acceptable. There are, however, several pieces of information which allow one to explain the discrepancy for Fort Lewis. The problem is categorical representation of stem spacing and stem diameter in the DMA data bases. The DMA data format has the categorical values for stem spacing broken down into groups that are no smaller than .5 meters wide and stem diameter broken into groups that are no smaller than 2 centimeters wide. Consequently, the algorithm could only use a mid-range estimate for the stem spacing and diameter instead of an exact value which was available to the analysts creating the manual products. The results were quite dramatic as the Fort Lewis CCM (from the HTC source) got no better than 80% agreement (even when one gives partial credit by non-exact agreements as is done in the weighting schemes).

The contention that the error is due to stem spacing and diameter is reinforced by the other results. For the Yakima CCM, one would expect that the vegetation values would not hamper the results since Yakima contains no significant vegetation; and indeed, the Yakima CCM Kappa values are above 90% in most cases. In fact, the resulting values fall roughly between the Kappa values for soil and slope in Yakima which is what one would expect.

A second piece of evidence is the CCM results for Fort Lewis. It just happened that the DMA data in the AC format included one additional category of stem spacing than does the HTC data; therefore, one could expect just slightly better results for the AC data. Indeed, for each weighting scheme used, the AC CCM showed the best level of agreement among all done in that group.

There are two points to emphasize. First that the CCM algorithm is quite sensitive for some variables and therefore would be much improved if the coding scheme allowed a more accurate description of these variables. Second, it is likely that when the data is provided that the results will be quite good since the Yakima CCM products (which do not rely on vegetation) were, with the exception of the 125 meter spacing, well above 90% agreement and thus acceptable products.

AC and HTC Comparison. Both data sources are quite good in representing the features with neither data format showing a uniformly superior performance. For example, for the soil features, HTC results were slightly better than those for the AC source. However, for slope the AC source was about the same or better than the HTC. For vegetation, it was mixed with AC better for Yakima and HTC better for Fort Lewis. Neither pattern of the differences nor their magnitudes were significant enough to suggest that one data base was "better" than the other for features or feature dependent products (like CCM); and both were acceptable. The fact that the Fort Lewis CCM was superior for the AC source is worth noting; but neither product was really in the acceptable range. Indeed, the upgrade of stem spacing and diameter to numeric quantities would be helpful for both data bases.

Conclusions

The use of the Kappa and weighted Kappa statistics was a very useful and credible methodology for analyzing the relative strengths and weaknesses of the two data bases; and as such provided the Army with a good case for the conclusions reached in the prior paragraphs. The major area of work will be in educating users in map products about the meaning of the statistics and "rules of thumb" for interpreting the strength of the results.

FURTHER RESEARCH

There are several areas of further research that can be mentioned here. The Kappa statistic needs to be applied in more situations so that rules of thumb about the magnitude can be evolved. In addition, other modifications to the Kappa which are more spatially motivated need to be developed. The current form of the statistic does not take the size and extent of the boundary into consideration. Therefore, as pointed out, the regions with many polygons are, because the opportunity is there, more likely to have lower levels of agreement.

ACKNOWLEDGEMENTS

We gratefully acknowledge the support of the U.S. Army Engineer Topographic Laboratories for this work as well as programming contributions of Mary Armstrong and James Span of the IIT Research Institute. Finally, we thank Jacqueline Rota and Chris Magagna for their accurate typing and continuing patience during preparation of this paper.

REFERENCES

Bishop, Y.M.M., Feinberg, S.E. and Holland, P.W. 1975, Discrete Multivariate Analysis, MIT Press, Cambridge, MA.

Chrisman, N.R., 1982, Beyond Accuracy Assessment: Correction of Misclassification: Proceedings ISPRS Commission IV, p. 123-132.

Chrisman, N.R., 1984, Alternatives for Specifying Quality Standards for Digital Cartographic Data: Issues in Digital Cartographic Data Standards Report # 4, ed. H. Moellering, National Committee for Digital Cartographic Data Standards, ACSM, Columbus, Ohio.

Cohen, J., 1960, A coefficient for nominal scales: Educ. Psychol. Meas., 20, p. 37-46.

Cohen, J., 1968, Weighted Kappa: nominal scale agreement with provision for scaled disagreement or partial credit: Psychological Bulletin, 70, p. 213-220.

Congalton, R.G. and Mead, R.A. 1981, A Quantitative Method to Test for Similarity Between Photo Interpreters: Technical papers 47th Annaul Meeting ASP, p. 263-266.

Fleiss, J. L., 1981, Statistical Methods for Rates and Proportions, Second Edition, John Wiley & Sons, New York.

Greenwalt, C. R. and Shultz, M.E. 1962, Principles of Error Theory and Cartographic Applications ACIC Technical Report No. 96, DMA-AC, St. Louis, MO.

Hermann, R. et. al., 1984 Army Digital Topographic Data Requirements, ETL-GSL-2. U.S. Army Engineer Topographic Laboratories, Fort Belvoir, VA.

POST-CENSUS CARTOGRAPHY IN FRANCE :
PRODUCTION INCREASE
DESPITE BUDGETARY RESTRICTIONS
Jean-Philippe Grelot
Institut Géographique National
2, Avenue Pasteur
94160 Saint-Mandé - France

SUMMARY

The knowledge of population phenomena is determinant for defining a development policy. The whole aspect is insufficient : the geographic distribution must be apprehended. With computerized census processing, automatic cartography is the privileged way of diffusing the information needed by geographers and planners. For all that this tool should be known and financial arrangements should be made as soon as the census budgetary conception.

INTRODUCTION

Though a government organization, the Institut Géographique National-France provides a wide range of commercial services both in the French and international markets. Among these activities are to be found the thematic cartography generally speaking, and the statistical cartography in particular. For this purpose, the IGN has equipped itself in 1979 with an automatic cartography system, which has been gradually enriched by six years of experience and production, both in the hardware and software fields (Pasquier 1980, Pasquier 1982, Grelot 1984).

Among the suppliers of statistical data, the French Institut National de la Statistique et des Etudes Economiques (INSEE) holds a particular place in France. It coordinates the whole national statistical system, and carries out itself the heavy census operations, alone or with the assistance of other ministerial departments (Ministry of Agriculture and Home Office). After gathering and processing phases, INSEE publishes the census results.

Contrarily to other statistical organizations, INSEE does not possess any cartography unit. If it uses more or less summary softwares to produce very simple maps or study maps, its appeals most often to external organizations for the publication of detailed maps for the general public.

PRODUCED MAPS

Exploitation of past censuses

In 1978, IGN and INSEE made together an experiment by producing an atlas of some fifty maps over a region covering about one tenth of the French territory (Rhône-Alpes 1978). It was the first large production of statistical maps through data-processing.

This led to carry on this joint work over the whole France. For this purpose, IGN undertook the constitution of geographic reference files including the 36,500 "communes" (French smallest territorial division): a first file gives the boundaries of these communes, digitized from a schematization specially conceived at the scale of 1:500 000 for bearing considerable reductions ; two other files give the coordinates of particular points located inside each commune, the centroid or visual centre, without rigorous mathematic definition, and the chief town (county town), position of the townhall.

Then two books were published, each including some twenty maps along with remarks. Data came from the 1975 general population census and from previous censuses (1936, 1954, 1962 and 1968) ; they were mapped either at the commune level (36,500 geographic units), or at the immediately higher administrative level, the canton one, with a file obtained by aggregating communes into cantons (3,500 geographic units). The first book deals with demography, and shows especially the distribution of the population and its evolution, as well as its structure per sex and age (IGN-INSEE 1980). The second book deals with activity, showing in particular some socio-professional distributions and housing ones, whether main or secondary dwellings (IGN-INSEE 1981). On the map itself appears a short comment of a few lines, inviting the reader to reflect on a particular aspect or make comparisons between several maps. More important comments give a geographic interpretation of each map or group of maps. An administrative map and a physical map go with statistical maps. The used scale is 1:2 500 000, which corresponds to the A2 standard size (42 cm x 59.4 cm).

These two books were published respectively in 1980 and 1981. Despite their unquestionable interest and their acknowledged aesthetic quality (Muller 1984), they have not been best-sellers, partly because of the data remoteness (5 or 6 years), the more considerable as the census was to take place in March 1982.

1982 census

For this census, arrangements were made to associate cartography with the publication of the first figure tables. Thus was carried out, five months after census, a first series of 17 maps, from the first estimates made public simultaneously (INSEE 1982). As the accuracy of the results improved, other maps were drawn up at finer geographic levels, with provisional results then with final results. As a whole, more than 40 maps will have been drawn in some thirty months. Other maps were also produced over more reduced geographic areas.

STATISTICIANS AND MAPS

Statisticians are not always sensitive to the charm of maps. They mainly reproach this means of expression with degrading information, with its possible perversion and the production time and cost of maps.

Degradation of information

Degradation of information is undeniable. Statistical cartography, mainly when geographic units are of reduced size, compels one to regroup raw data into classes, represented by various colours or symbols of various sizes and shapes. With this end in view figured data must be represented, then a choice must be made among the possible representations, a choice which can be made but after an interpretation necessarily impressed with subjectivity. But this finally represents but the passing from data to information, which after all is the characteristic of the communication of the knowledge holder towards a public. If accuracy is lost, on the other hand the geographic expression of a phenomenon is won, which no figure table can show with the efficiency of a map. If the number is the archetype of statistical data, the map remains the archetype of the representation of localized data, of geographic data.

Perversion of information

Perversion of information is a more serious risk. It may come from a wrong choice of thresholds, which leads to transmit a wrong impression of the statistical distribution, to an illusion. This is feared by statisticians who often ignore that cartographers fear in the same way the wrong use of visual variables and graphic language rules : there is no more risk of illusion, but of incommunicability. One only has to glance through documents advertising very popular commercial softwares to be convinced of the justice of this remark (Carter and Meehan 1984). Statisticians and cartographers must necessarily work jointly to avoid such dangers.

Production time

Map production time is an important commercial obstacle. Cartography brings a knowledge the larger and more interesting as the represented geographic level is small. By the force of circumstances, this requires a long time in the availability of data on the one hand, and in the choice of the most pertinent statistical variable and the most performing graphic expression on the other hand. Then the quick time limitation of data can be felt. However one can note that generally speaking demographic phenomena evolve slowly, at least for a number of them. Then the mapped geographic level must be adapted to the data accuracy, processing accuracy and data estimated variation between the census date and that of map production. This notion of processing accuracy was used in France for 1982 census : the "first estimates" were mapped at the regional and departmental level, then the "provisional results" at the departmental and cantonal level, then the "final results" at the cantonal and communal level.

Cartography cost

Obviously cartography cost is not to be neglected. Statisticians are sensitive to it, particularly when they are compelled to appeal to external organizations for this task. However it would be righter to compare this cost with that of works of publication of census results, and with the whole census budget. But in time of budgetary restrictions, it is quite obvious that to begin with what is not deemed essential and comes the latest in the work schedule will be sacrificed, and maps quite often fall into this category.

Nevertheless we verify that statisticians appeal more to graphic and cartographic expression. Evidence of it is given by the insertion of maps in the first official publication of 1982 census ; another evidence is a book drawn up by INSEE and entitled "Données Sociales 1984" (1984 Social Data), in which texts explain images and images do not illustrate texts (Bonin 1984). The constitution of a public interest group "Reclus" intended, among other things, for publishing an atlas of France within the coming years, must also provoke a leading effect and spread the use of statistical maps.

HARDWARE AND SOFTWARE DEVELOPMENTS

Directed since the beginning towards publication for the general public, the system used by IGN in thematic cartography has required from the outset important developments in statistical cartography particularly, for softwares available on the market at this time did not correspond to aims looked for.

Softwares

Softwares cover two different aspects : on the one hand a statistical analysis helps to choose the representation and threshold ; on the other hand symbolization softwares offer a wide range of representations.

Thus for mapping an absolute magnitude, one disposes of representations in pinpoint signs :
- circles, squares, or pie charts, the total surface of which is proportional to the statistical variable ; in case of superposition, the smallest symbol, surrounded by a white ring, deletes automatically the biggest one ;
- cluster of points in a number fixed by the data value, so that the points might be distributed in an aleatory manner inside the corresponding area ; curiously, this representation does not obtain the favour of our usual interlocutors.

A density or proportion will be interpreted by a zonal representation, by varying colour and intensity ; if need be, hatchings are used when drawn surfaces are large enough.

Relations from commune to commune, like exchange flows or attractions, are represented by lines the width of which is modulated by the statistical data.

Being a country of high administrative tradition, France is not divided into 54 million Frenchmen, for actualizing Georges Bernanos'famour remark, but into a multitude of administrative wardings. Softwares of commune aggregation and symbolization of hierarchized or tangle boundaries have been then conceived to produce these maps of administrative references.

Hardwares

With these software developments are associated extensions of material configuration. The Sémio system (Pasquier 1980), provides screened films, insolated by a laser plotter, ready for printing ; the quality of these products is excellent. But the issue of films is followed by the constitution of a proof,obtained by Cromalin process, necessary for judging both the aesthetic rendering and the pertinence of the statistical choice, which results in a relatively high cost. Consequently complementary equipment has been acquired, for which we are developing proper softwares. We have acquired on the one hand an electrostatic colour plotter (Versatec ECP 42), which produces in one or several copies proofs of quality indeed lower than the Sémio system, but nevertheless quite satisfactory, and on the other hand a graphic colour console of high resolution (Lexi-data 3 700) which brings some interactivity at the expense of a display by sampling when the number of geographic units is important. These two materials are connected with a Vax 11/750 computer from Digital Equipment Corporation.

Further developments will consist in acceding, through telematics, to data bases established by national statistical organizations.

FUTURE CONDITIONS

The hope of seeing budgetary restrictions disappear soon is thin. The increase of statistical cartography publication must be looked for in other ways : improvement of productivity, dynamic cartography, budgetary discipline, training.

Productivity improvement

Map production profits by data-processing productivity gains, design and drawing assisted by computer. Equipment allowing intermediate visualizations offer a great interest, though it does not show really what the printed map will be. The following stage will be the direct production of printing plates from digital files, without passing through films and proofs.

Dynamic cartography

By dynamic cartography I mean two different notions.

The first one is rapidity of production : maps must be published at the same time as figured data tables. This means that cartographic representation must be adapted to data accuracy, particularly for the two stages already evoked which provide estimates then provisional results. Then one must choose properly the basic geographic level, and may be give up the notion of exactness linked to a map by tending to unsharp representations.

The second notion covered by the expression "dynamic cartography" is the selection, among phenomena to be mapped, of those which bring an help to anticipation. The verification map, like the one which shows the distribution of the population or its density, must be established, but not necessarily updated with censuses : it will be more interesting to show the various components of the population evolution, the structure per age on which depends the life or extinction of a region or an economic system, to represent birthrate, urbanization, evolution of jobs, agriculture... These maps will be used for planning, much more than verification maps which reveal the past more than they give tendencies for the future.

Budgetary discipline

A census budget generally provides for three stages which are data gathering, data processing and publication of results. It is quite obvious that any side-slipping of the cost of the two stages results in reducing the financial envelope available for publications.

It appears clearly that the cartographic edition can almost never be added a posteriori to budgetized publications. Setting up supplementary estimates takes time and jeopardizes the simultaneous publication of figures and maps, an essential factor already emphasized. Then maps must be part of publications scheduled in the budget, and technical arrangements must be examined in time. This is still extremely unusual and requires an effort of persuasion and training.

Training

The map is a precious tool for communicating the statistical information. It illustrates it for it is first of all an image and its aesthetic quality attracts the reader ; it is in full harmony with our time in which are developing means of visual expression at the expense of textual expression. It synthesizes, but much better than any figure table with linear and sequential reading, it links data geographically and reveals at first sight the force lines of the represented phenomenon. At last, the map teaches for it brings information in an attractive form which facilitates its storage.

To provide for the future of census cartography, a training effort is to be made, supported by the preceding remarks, to statisticians, planners, educators and general public. Reading a map must be as usual, as easy as reading a text or a figure table ; then understanding geographic phenomena and assimilating knowledge will be easier.

CONCLUSION

The next general census of the French population will take place at the end of the present decade. This leaves time to undertake, through sensitizing and training actions, the preparation of a budgetary item devoted to post-census cartography. Works of research in dynamic cartography are to be undertaken by then, specially by taking advantage of the publication, scheduled for spring 1985 i.e. 3 years after census, of results related to the socio-professional structure of the population. The required financing will be granted but if the statistical cartography proves that it is a planning tool, a tool helping to decide.

REFERENCES

Bonin S. 1984, Une double expérience : la graphique à la base d'un recueil d'articles socio-économiques ; la graphique utilisée pour exploiter un questionnaire : 12° Conférence Internationale de l'ACI, vol 1, pp. 211-220.

Carter J.R. et Meehan G.B., 1984, SAS/Graph Mapping Capabilities and Cartographic Education : Austra Carto One, Technical Papers, pp. 259-276, Perth.

Grelot J.P. 1984, Cartographie statistique assistée par ordinateur : Conférence Micad 84, vol 3, pp. 1051-1057.

IGN-INSEE, 1980, La Population Française, IGN-INSEE, Paris.

IGN-INSEE, 1981, Activité et Habitat, IGN-INSEE, Paris.

INSEE, 1982, Recensement Général de la Population - 1982 - Premières estimations, INSEE, Paris.

Muller J.C., 1984, La Population Française 1980 et Activité et Habitat 1981 : The American Cartographer, vol 11 n°1, April 1984, pp. 91-92.

Pasquier B. 1980, Une chaîne informatique pour la production intensive de cartes thématiques : 10° Conférence Internationale de l'ACI, in Bulletin d'information de l'IGN n°80/3, Paris.

Pasquier B. 1982, Statistique et cartographie thématique : PCM 79°année n°2, pp. 63-65, Paris.

Rhône-Alpes, 1978, Rhône-Alpes en 50 cartes, INSEE - Direction régionale de Lyon, Lyon.

A Fuzzy And Heuristic Approach To Segment Intersection Detection And Reporting

by

Jose Armando Guevara Ph.D.
GIS Software Engineer - Assistant Professor of Computer Science

and

Dave Bishop
Senior Software Analyst

Environmental Systems Research Institute
380 New York St.
Redlands, California 92373
Tel. (714) 793-2853

Abstract

One of the fundamental Geographic Information System Procedures is that of Polygon Overlay. Within the theme of Polygon Overlay there are many interesting geometric and topological problems not yet efficiently solved. Of particular interest and key to the whole overlay process is that of segment intersection: the problem of dealing with the pairwise intersections among a set of N segments. In practical terms, the intersection process calls for detecting, reporting and doing segment coordinate modification(s) when the intersection is found. Theoretical algorithms for segment intersection have been developed, yet most of them do not address implementation problems. Another issue that is related to numerical error, and also not discussed in the literature, is that of segment representation. The implementation problem is intimately connected to the numerical error problem. Both have an impact in the efficiency of the segment intersection algorithms. The importance of efficient implementation is becoming more and more apparent as data volume in geo-processing grows increasingly ambitious: a single digital coverage may contain hundreds of thousands of segments.

In this paper the authors, based on notions of computational geometry [band sweeping], introduce an approach to segment intersection which yields an efficient algorithm for the detection and reporting of N segments in the plane. The segment intersection per se is based on a set of heuristic rules and the concept of fuzzy intersection. Unlike ´exact´ intersection, where two segments intersect only if there is a mathematical point that defines the intersection, the fuzzy intersection defines an intersection point or points when the two segments are within a ´fuzzy distance´ away from each other. Four problems are addressed and resolved in this paper: 1) fuzzy creep, 2) subtolerance segments, 3) segment intersection scanning order, and 4) colinearity. Concurrently with the fuzzy intersection concept, a data structure is presented that efficiently models the band-sweep idea. Since the system presented here has been implemented as part of the ARC/INFO software, statistical profiles are given of its computational behavior.

FUNCTIONAL COMPONENTS OF A SPATIAL DATA PROCESSOR

Stephen C. Guptill
National Mapping Division
U.S. Geological Survey
Reston, Virginia 22092

ABSTRACT

A study to determine the feasibility of developing a spatial data processor (a standalone, office environment, image processing and geographic information system) capable of manipulating imagery and digital cartographic data has been completed. A key phase of this study involved determining the set of operational functions required for spatial data processing. A set of 16 functional components was identified. This set provides useful criteria for not only evaluating existing systems but also determining the capabilities of new systems. The extent to which each functional component is used is dependent on the user application. A review of several existing systems determined that no one system provides all functional capabilities working with either raster or vector data. Although the majority of current image processing and geographic information system applications can be handled by using raster data, the functions of a spatial data processor require both raster and vector data types. The most expeditious way to build a spatial data processor would be to utilize existing raster-processing software functions; identify functions that require vector capabilities and develop and incorporate software modules to perform those tasks; and integrate the raster and vector functions.

BACKGROUND AND OBJECTIVES

Digital cartographic data bases and digital imagery holdings are increasing rapidly in size and becoming more widespread in use. Image processing and geographic information system (GIS) technology is used to merge, integrate, and analyze data from these data bases, and the demand for such systems is increasing. Although a number of systems exist, they are implemented on a wide variety of hardware configurations and do not have an integrated approach to performing both image processing and GIS functions with both raster and vector data. Two additional desirable criteria, that the software be in the public domain and that the hardware be relatively low-cost and able to operate in an office environment, are met by few systems. The lack of a system that meets these general requirements has led towards multiple efforts to develop standalone systems

---

Publication authorized by the Director, U.S. Geological Survey, on January 4, 1985. Any use of trade names and trademarks in this publication is for identification purposes only and does not constitute endorsement by the U.S. Geological Survey.

to do either GIS or image processing functions. In response to this situation the National Mapping Division (NMD), U.S. Geological Survey (USGS), conducted a study of these topics.

The study determined the feasibility of developing a spatial data processor (SDP)(a standalone, office environment, integrated, image processing and geographic information system) capable of manipulating imagery and digital cartographic data. The portions of the study concerning assumptions, functional requirements, hardware characteristics, and findings (after studying a number of existing systems) are presented here.

ASSUMPTIONS

In conducting the SDP feasibility study, certain initial assumptions about the system were made. These assumptions concerned issues relating to users, functions, data, hardware, operating environment, and system architecture.

Users
The users of an SDP are assumed to represent a wide range of interests and expertise. They could range from scientists in a research environment to technicians in a production center. Although an SDP design could be optimized for specific applications, a single design cannot optimally perform tasks in research, production, demonstration, and application with USGS and other data sources. These four tasks conflict in optimization requirements. Production operations must be focused to attain maximum throughput and allow adequate data security. Research operations must be diverse and varied to allow the interplay of ideas and operations to stimulate new developments. A system to use USGS digital cartographic data must be flexible and accessible. Where design conflicts were discovered, preference was given to performing research operations rather than production operations.

Functions
The varied users of an SDP are assumed to require practically the entire range of GIS and image processing functions. These functions have been limited, however, to those available today, proven possible by a demonstrated implementation. The first implementation of an SDP is not expected to extend the state-of-the-art functionally in a significant way, except possibly to integrate functions previously available only in separate existing systems. However, it is assumed that the sum of all functions required by all users is possible in one system. This one system may actually be a family of systems, with a given function modularized in perhaps more than one implementation, each implementation optimized for a given set of user/performance specifications.

Data
The SDP will handle any data considered image, geographic, or cartographic and will readily accept USGS digital cartographic and image products. The SDP is expected to accommodate all or most existing models of these spatial data

to be able to utilize and integrate data from the widest possible range of sources. A full range of non-spatial-attribute data handling facilities is also assumed.

The internal data representations employed by the SDP might, by necessity, imply innovative design approaches. The integration of multi-user requirements and various data models will place an enormous demand on flexibility and interchangability of data. The design of the internal data models and data management approaches is a key to creating an SDP for a wide range of user environments.

Hardware
The hardware necessary to perform all functions for all users at satisfactory performance levels is assumed to be available at a cost of $50,000 to $100,000. A basic system configuration can be outlined, with various hardware options added or interchanged to optimize the configuration for a given class of users. Once again, modularity is important. It may be that the only hardware component which remains constant through all SDP configurations is the central processing unit.

Operating Environment
The set of SDP software is expected to operate as an application within an existing vendor-supplied operating system and using standard data processing support facilities. The operating system or utilities will not be modified except perhaps with respect to device drivers. One operating system will be chosen within which all SDP configurations will run.

System Architecture
An SDP is assumed to be a standalone system having the capability of performing all functions locally. With the increasing availability of various local-area-networking and longer-range distributed processing options, the system design incorporates the ability to participate in a distributed processing environment. For some requirements, such as those associated with production tasks, distributed processing may be preferred.

Modularity is a possible key to solving the problem of multiple user scenarios. Modularity is assumed to be required for both hardware and software components. Conflicting requirements may be resolved by the same or similar function implemented in different, optimized, hardware and (or) software configurations.

FUNCTIONAL REQUIREMENTS

A functional analysis forms the basis of the SDP feasibility study. Once the objectives and assumptions are specified, the functional components of an SDP need to be determined. The functional components list (and companion set of data models needed by the functional components) is a generic description of the capabilities that an SDP should have.

The investigation into the operational functions required for spatial data processing began by detailing the components of five major GIS processes: data input, data editing, data storage and retrieval, data manipulation, and map, image, and report generation. Each component has counterparts that work on vector or raster data types.

The resulting lists of components lacked a unifying theme and varied in level of detail. Therefore, another approach was taken in which the underlying spatial data models were defined, and then the processes that operate on the data models were developed.

Data Models
Two schemas are suggested for defining spatial data model types--one for vector data, the other for raster or grid data. With the vector schema, two factors are considered in examining the data model: the model's suitability for handling point data, line data, area data, and mixed cases of these data types; and the degree of topological structuring (the ability to address points of intersection among lines, common edges between areas, etc.) The vector schema may also employ alternate representations of two-dimensional (cartesian) space, such as Generalized Balanced Ternary (Gibson and Lucas, 1982) addresses, or Peano keys (Marvin White, 1984, written commun.).

Two factors are also considered with the raster schema: the cell shape (rectangular, triangular, hexagonal, or some other n-sided shape); and whether the data are explicitly coded cell by cell or whether some run-length encoding is employed.

With either schema, three additional factors need to be examined: the model's accommodation of features; the model's accommodation of attribute information; and the model's accommodation of temporal changes.

There are three ways each schema may accommodate features: data oriented to explicit features (data may be directly accessed by individual feature); explicit feature tags (individual features may be extracted by exhaustive search of the set of vector topological elements, raster pixels, or grid cells); and feature encoding included within attribute classes (in which case individual features are not explicitly encoded).

Among the factors to consider in handling attribute information are: the number of codes or amount of attribute data which may be associated with a given feature, topological element, or raster pixel/grid cell; and the types of attribute information handled--either numeric (with measurement level specified as nominal, ordinal, interval, or ratio) or alphanumeric (text).

The manner in which temporal data are handled may be specified by the types of other data that carry temporal information: entire map or overlay set; spatial or other subsets of a map/overlay; individual features or topological elements; and individual attributes.

Functional Components
The earlier lists of spatial data processes and sub-
processes were regrouped into generic functional components
which are applicable to either a raster or vector data
model. This list of functional components is given in
Table 1. These functional components and seven system-
level evaluation criteria (functionality, operating
environment, performance, long-term support, modularity,
software transportability, and expandability) were used to
evaluate various existing spatial data processing systems.

Hardware Considerations
A general set of hardware requirements was developed
during the study. These requirements include: 32-bit CPU
with virtual memory; color raster and vector displays (512
x 512 resolution for imagery, 1024 x 1024 resolution for
line graphics); and bus design to support mass storage
devices (Winchester disk, tape drives), local area
networks, and various graphic and alphanumeric input and
output.

FINDINGS

The systems examined during this study covered the gamut
of functions desired in an SDP. No single system provides
all functions; however, some generalizations may be drawn
from an examination of several existing systems. It is
evident that raster-based systems are more standardized in
functionality and more consistent in data base design and
structure. For these reasons the raster processing components of a SDP can probably be extracted directly from
existing systems. A system that ranks high in most of
these functional components is the USGS Mini Image
Processing System (MIPS) (Chavez, 1984).

The vector-based systems are extremely diverse both in
functionality and data base structure. The systems have
various methods for encoding and storing the attribute,
coordinate, and topological information. It will be difficult to draw from these existing systems to support an SDP.
Certainly, no existing vector system can supply all the
requirements of the SDP vector schema in a modular and
easily transportable form.

The development of the SDP functional components, the
existing system studies, and the list of hardware considerations allowed a consensus to be reached in a number of
areas. These findings are summarized as follows:

- Functionality and software base are more important
  than the performance specifications of any given
  hardware/operating system configuration.

- Acceptable hardware configurations are available for
  $50,000 to $75,000.

- Final choice of hardware and operating environment
  will be driven by the existing software modules that
  are chosen to form the basic components from which a
  system will be constructed.

Table 1.--Functional components of an SDP

<u>Data Capture</u>: assembling analog source data in digital form; for example, line digitizing, raster scanning.

<u>Structuring</u>: processing data from intial digital form into a resident model; for example, skeletonizing scanned line data, deriving topology, polygon chaining.

<u>Editing</u>: inserting, deleting, and changing attribute and geometric elements to correct and (or) update model; for example, node snapping, sensor noise removal.

<u>Representation/Structure Conversion</u>: moving between representations and the structures associated with them; for example, raster-to/from-vector, polygon-to/from-grid, digital elevation model-to/from-contour, polygon-to/from-arc-node.

<u>Geometric Correction</u>: fixing model to ground or image space in some referencing system; for example, adjustment of map or image to control points.

<u>Projection Conversion</u>: transforming coordinates between alternative referencing systems; for example, geographic-to/from-UTM.

<u>Spatial Definition</u>: paneling and clipping to achieve the spatial limits for data in a model; for example, limiting data to within a county boundary.

<u>Generalization</u>: reducing detail in the model; for example, resampling to larger spacing, reduction of points in a line.

<u>Enhancement</u>: modification of detail in the model; for example, edge definition.

<u>Classification</u>: analysis and interpolation of the model to form classes; for example, classification of spectral response data, choropleth mapping.

<u>Statistical Generation</u>: deriving descriptive statistics and (or) measurements from model; for example, histograms.

<u>Retrieval</u>: selective extraction of data from the model by attribute and (or) spatial searches or neighborhood analysis; for example, categories within a circle of given radius from a point.

<u>Overlaying</u>: relating two models in a Boolean and (or) arithmetic manner; for example, creating composite maps, image ratios.

<u>Display</u>: generating a graphic image from the model; for example, color CRT displays, symbolized line maps.

<u>Analytical Technique Support</u>: using analytical manipulations and computations on data model; for example, Markov chaining, network analysis, location/allocation.

<u>Data Management</u>: managing access and archiving of data models; for example, storage, retrieval, update, security protection, data base sub-schemas, transaction records.

- Digital Equipment Corporation's VAX computers and VMS operating system probably offer the best environment in which to implement existing software.

- Both raster and vector data models must be supported.

- No one existing software system provides all functional components for both data types.

- Automated cartography applications require a full set of functional components that use a vector data model.

- Existing public-domain vector software systems offer very little software from which to build a system.

- Raster data model functions are supported by a number of image processing software systems.

- Vast majority of image processing/GIS applications can be met by functional components that use a raster data model.

- By using a raster data model to satisfy most of the functional capabilities, an operational SDP could be created in a timely fashion and with limited resources by taking the "best" components from existing raster systems and adding necessary vector data handling functions.

RELATIONSHIP TO OTHER SPATIAL ANALYSIS RESEARCH ACTIVITIES

The development of an SDP serves a dual role in USGS's spatial analysis research activities. First, it would provide a system with some basic functionality in image processing, GIS, and automated cartography. These capabilities can be used to conduct specific applications projects. Secondly, it would provide a software base and hardware configuration conducive for future enhancements to the system.

SDP represents an important component in the effective design of a geographic information system. It would provide the general-purpose geoprocessing capabilities about which user specific applications may be integrated. The future development and enhancement of an SDP will draw from developments in software, hardware, firmware, and artificial intelligence.

Research is directed along several fronts. Under study are existing systems for graphic, geographic, and image processing to define a fundamental set of capabilities that respond to a range of user queries. Offering these basic capabilities as spatial operators, much like arithmetic and relational operators in programming languages, provides a sparse syntax for expressing desired types of manipulations. Additionally, by defining generic forms of spatial representations, spatial data stored in a particular structure may be reduced to its generic form, rather than the more extensive process of restructuring. An enhanced SDP with capabilities developed about generic representations and their combinations will result in less restructuring.

Developments in logical structures and expert systems provide a likely shell for an enhanced spatial data processor. With the hypergraph-based data structure (Bouille, 1978), for example, the concept of topological structuring, which had advanced digitial cartography in earlier years, is extended to other features of the map model. Rule-based systems may utilize these relationships, together with patterns of previous queries, to direct processing in the most effective manner.

In short, SDP development does not replace more long-term research activities. It provides a near-term capability as well as a hardware and software base for implementing enhancements resulting from ongoing studies.

CONCLUSIONS

A study to determine the feasibility of developing a spatial data processor (SDP) (a standalone, office environment, image processing and geographic information) system capable of manipulating imagery and digital cartographic data has been conducted. In the first phase of this study a set of the functional components required for spatial data processing, as well as the characteristics of the underlying data models, were determined. The functional components and seven other system evaluation criteria (functionality, operating environment, performance, long term support, modularity, software transportability, and expandability) were used to evaluate various existing spatial data processing systems. The general hardware characteristics of the system also were developed. After these studies, it has been concluded that an SDP using Digital Equipment's VAX hardware, the VMS operating system, utilizing the raster processing capabilities of the MIPS software, and adding necessary vector data handling functions offers the best opportunity, considering the resources and time available, for creating an SDP.

REFERENCES

Bouillé, François, 1978, Structuring Cartographic Data and Spatial Process with the Hypergraph-Based Data Structure, in Dutton, Geoffry, ed., First Advanced Symposium on Topological Data Structures for Geographic Information Systems: Cambridge, Mass., Harvard University

Chavez, Pat, 1984, U.S. Geological Survey Mini-Image Processing System: USGS Open-File Report 84-880, Reston, Va.

Gibson, L., and Lucas, D., 1982, Vectorization of Raster Images using Hierarchial Methods: Computer Graphics and Image Processing, v. 20, p. 82-83

ABSTRACT

METHODS FOR COMPARING DATA ENTRY SYSTEMS

by
Don Hansen, FWS
Mike Dwyer, BLM, CSO
Marilyn Mogg, BLM, DSC

A need to standardize many automated geo-processing capabilities has prompted the U.S. Fish and Wildlife Service (FWS) and the Bureau of Land Management (BLM) to evaluate the systems currently in use within the two agencies. The Automated Digitizing Systems (ADS) and the Analytical Mapping System (AMS) are geographic data entry systems developed by BLM and FWS respectively for later processing by the Map Overlay and Statistical System (MOSS). ADS and AMS are similar in many respects in that they produce comparable digital map products. The methods used to generate those product are, however, generally very different. These methods were analyzed in terms of their inherent efficiencies and deficiencies. The performance of each system was evaluated on the basis of several benchmark tests which addressed user time requirements, computer loading factors and data accuracy. The results of all analyses, in addition to responses to a user survey, were incorporated into recommendations for improvement of each respective system. The implications associated with using ADS and AMS in a mapping program were evaluated in terms of established cartographic standards.

A task was initiated in 1984 through a joint agreement by the U.S. Fish and Wildlife Service (FWS) and Bureau of Land Management (BLM) to study and compare the capabilities and characteristics of geographic data entry systems used within the Department of the Interior (DOI). The primary objective of the study was to identify those components of data entry systems that contribute most significantly to their ability to compile high quality geographic data bases. Information collected during this comparison would be very useful in efforts to standardize geoprocessing techniques throughout DOI. The purpose of this paper is to report the methods that were used to evaluate the Automated Digitizing System (ADS) and the Analytical Mapping System (AMS).

The collection and analysis of large volumes of spatial data is a major effort of most land and resource management concerns. Traditionally, the inventory and planning tool has been pertinent data plotted on a physical map. This, however, can be cumbersome to work with, especially when complexity magnifies with additional themes and layers of data. Geographic Information Systems (GIS) provide a facility for the manipulation of large volumes of digital spatial data without the limitations encountered with a physical map. Both FWS and BLM are involved in extensive database construction and cartographic modeling projects to assist managers with their land use decisions. Digital databases can be accessed easily to obtain inventory data and manipulated in some logical method to generate useful information. Most spatial data is not, however, immediately available in digital form (an exception to this is digitally recorded, remotely sensed data). Spatial data must typically be manually transcribed onto a physical map incorporating geographic control. Lines are to describe the locations of features such as roads, or the boundaries between areal features, such as vegetation communities. The problem then becomes the transformation of a physical map to a digital map compatible with a GIS. This transformation is achieved by manually digitizing map lines with the assistance of specialized computer hardware and software packages.

A map digitizing system consists of several components each contributing an important function to the whole process. A certain hardware configuration is necessary to run the program. Programming techniques and user documentation affect the maintainability and usability of a system. Data storage file structures impact computer resources, as well as maintainability of established databases, and interfaces with a variety of GIS's permits transportability of databases. The data entry process itself, can be further categorized into 6 distinct stages of sub-processing encountered by the operator in the following order of occurence:

Project Set-up
Map Registration
Data Capture
Editing
Polygon Formation/Verification
Database Management

These ten components/processes were analyzed with the goal of addressing four topics of primary importance as identified by users of geographic data entry systems:

- Accuracy of the geographic database
- Efficiency of the data collection process
- Impacts on the computer environment
- Interface between the program and it's users

Accuracy of the Geographic Database: The results of a cartographic inventory or modeling effort are only as accurate as the data that were used. Land use decisions involving legal boundaries necessitate strict controls over data throughout its processing. Three types of analyses were performed to identify sources and causes of error introduced to the database and to compare the significance of program generated error with that introduced during the original map drafting process.

First, the sub-processes that potentially impact accuracy of data were analyzed to determine what programming procedures are being used. For instance, one of the major sources of error can be attributed to map registration which establishes the link between physical coordinates of the digitizing tablet and geographic coordinates of a map projection. The link is a set of mathematical expressions derived using sampling methods. Polynomial regressions or geometric relationships are used to best approximate or "rubber-sheet" a spherical coordinate system using input from a flat map. The algorithms used to do this are characterized by inherent deficiencies that can be analyzed mathematically.

The data capture stage of digitizing introduces the primary source of human error to a database. Digital map quality is dependent on how accurately the operator traces map lines with the cursor. However, other factors such as digitizing mode (point, increment or stream coordinate measurement), weeding of points to reduce data volume, and "adjustments" to coordinate locations made automatically by the program, can also contribute significantly to the accuracy of the final product.

Polygam formation/verification and editing are other sources of error, as again, adjustments are made to the data by the program and/or operator. Giving the operator a large amount of editing flexibility could also allow further degradation of the data originally collected during data capture.

After reviewing the methods used in the computer programs to retain accuracy, a test was conducted to measure the amount of error being produced during the registration process. This test consisted of measuring the deviation of a digitized point from the true geographic location taken from a map. Maps representing 3 map projections, 6 representative scales and 4 geographic regions were selected. The maps were generated on stable medium with a grid of tic marks located at known latitude/longitude coordinates using a GIS map plotting function.

A set of digitizing tablet coordinates (in inches) were entered into the registration programs of each respective data entry system, thereby eliminating human biases from the test. After the maps were registered, coordinates (in inches) of the internal tics were likewise entered into each system using options that return the transformed, geopgraphic location (in latitude/longitude) of the points. The returned latitude/longitude of each sample point was compared to the latitude/longitude that was expected and the deviation (in seconds of latitude and longitude) was computed. The sampling design permitted statistical analyses of the effects of various combinations of projection, scale and geographic region on accuracy of the database. An example is the strong correlation that was found between map scale and accuracy.

Based on the results of this test an analysis of the "cartographic implications" of error being introduced to geographic databases was performed. The intent was to assess the significance of documented errors from the test in terms of impacts on cartographic products developed from these data bases. Arc second deviations were converted to map inches at various scales. The difference in performance of the two systems was analyzed in terms of significance levels, and a regression was computed to describe the relationship between scale and accuracy.

Efficiency of the Data Collection Process:

A second set of performance tests were used to measure the amount of operator effort required to complete the digitizing process. Factors that influence efficiency include volume and frequency of entering information at the data entry terminal, the logical sequence of digitizing lines and entering polygon attributes, and digitizing mode.

Processes that are implemented automatically by the program require less operator involvement than operations carried out manually. Besides the initial capture of lines, editing capabilities and procedures can contribute significantly to the efficiency of the data collection process.

A test was initiated to compare data entry systems in this respect. The procedure consisted of digitizing maps of varying levels of complexity and measuring the time necessary to

complete each step of digitizing. Three levels of map
complexity were tested. Low complexity was defined as
containing a moderate number of straight lines requiring a
minimum number of X, Y coordinates. A public land survey (PLSS)
map was chosen to represent low complexity. High complexity was
defined as containing a high density of extremely irregular
lines bounding many polygons. A 1:24,000 scale map of
vegetation cover was chosen to represent high complexity. A
third map delineating the locations of large range allotment
boundaries was chosen as moderate complexity for its few number
of moderately irregular lines.

Experienced data entry operators were selected to digitize the
maps. Strict controls were implemented to ensure uniformity of
operating conditions throughout the testing. Time was accrued
each time the operator logged onto the computer and stopped
accruing when logged off. Each time the operator moved to a
different stage of processing, time spent on the processor was
recorded. Functions such as polygon formation, which did not
require operator involvement, were not measured for this test.
(Statistics were collected for use in analyzing impacts on
computing environment. See below.)

Results of this test permitted analysis of each digitizing
process separately. Insight was gained into the potential for
enhancements to specific procedures that required input of
unnecessary or redundant data.

Impacts on the Computing Environment:

In addition to the clock time that was being recorded during the
digitizing test, measurements were also being recorded on
variables that indicate the demand on the computer's processing
resources. Some data entry tasks are inherently "CPU
intensive," demanding dedication of the computer's processor to
complete the function. Polygon formation is one such task.
Volume of calculations as well as the amount of data that must
be transferred between terminal, memory and disc storage
severely affect computer response being felt by other users.
Programming techniques can be used to maximize the efficiency
with which data is handled. Variables measured to evaluate the
programs in this respect included: "CPU seconds" (central
processor time dedicated to a specific task) and number of I/O
calls (requests the processor to input or out put data to
devices). As in the time test, CPU statisitics were measured for
each processing stage. The resulting analysis provided the
basis for critiquing programming techniques used in the two data
entry systems.

Interface between Program and Its Users:

This topic addressed the so-called "user friendliness" of a
computerized process. It is generally preferred to obtain

computer programs that require a minimum amount of training to become proficient. This is especially important with large and complex geographic data entry systems. Factors that contribute toward usability include: user documentation outlining speicific procedures to be followed; "online documentation" available for quick reference; clarity of online menus and instructions; and functional capability. Powerful editing functions, for example, not only minimize time requirements but may also lower the frustration associated with highly intricate work. Program documentation is essential to a system analyist attempting to maintain large volumes of program source code necessary to operate one of these systems. All these factors affect the user's overall ease in using the system. Therefore, the analysis included frequent consultation of users of the two systems. Two surveys were written and distributed to operators and programmer personnel. Responses to the survey provided insight not normally available or utilized by developers of systems such as these. The conclusion quickly becomes evident that the end user should play an important role in the development and enhancement of geographic data entry systems.

Conclusions:

This study not only provided FWS and BLM personnel objective evaluations of two data entry systems, but also relied on a systematic framework for the evaluation of other similar systems. Additionally, the knowledge obtained by comparing two established systems generated many ideas that can be used in future development and enhancement efforts.

Donald B. Hansen
Remote Sensing Specialist/Systems Analyst
Technicolor Government Services, FWS Operations
P.O. Box 9076
Fort Collins, Colorado 80525 Tel. 303/226-6183

Marilyn A. Mogg
Cartographer
Bureau of Land Management
Division of Surveying and Mapping Systems
Denver Federal Center, Bldg 50 D-417B
Lakewood, Colorado 80215 Tel. 303/236-0112

Michael F. Dwyer
Cartographer
Bureau of Land Management
Colorado State Office
1037 20th Street
Denver, Colorado 80202 Tel. 303/844-3066

AN OPERATIONAL GIS FOR FLATHEAD NATIONAL FOREST

Judy A. Hart, David B. Wherry
Washington State University Computing Service Center
Pullman, WA 99164-1220
and
Stan Bain
Flathead National Forest
P.O. Box 147, Kalispell, MT 59901

ABSTRACT

Flathead National Forest is utilizing Geographic Information System techniques to establish an updateable, forest-wide, geographic database. Hungry Horse Ranger District, discussed in this paper, is the first of five ranger districts to be completed. GIS data layers for the Hungry Horse District include: digital terrain data (elevation, slope and aspect); Landsat data classified by spectral class and then stratified by elevation and aspect to depict vegetation associations; timber compartments; timber harvest history; land types; land ownership; administrative areas such as wilderness areas, recreation areas, wild and scenic areas, and experimental forest boundaries; mean annual precipitation; drainage basins; streams, rivers, lakes and islands; roads and trails; as well as data planes depicting distances to water and distances to roads. Techniques used to compile the database are discussed; various data planes are depicted through illustrations; and, applications of the GIS database are described.

INTRODUCTION

Flathead National Forest (FNF) encompasses approximately 2.6 million acres in northwest Montana (Figure 1) with Forest headquarters in Kalispell. Bordered on the north by Canada and on the northeast by Glacier National Park, FNF is confronted with the problem of managing an area with diverse concerns. To balance the needs of timber, wildlife, and recreation in an economically and ecologically sound manner, it is increasingly important that map and tabular information be available to resource managers. FNF has adopted the use of a Geographic Information System (GIS) as one means of supplying information in a timely and economical fashion.

Primary data compilation for FNF's GIS database started in 1982 in the Hungry Horse Ranger District (HHRD). The completed HHRD database has served as a demonstration of GIS capabilities to other ranger districts and regional Forest Service offices. In 1984, Forest managers scheduled the establishment of 17 to 25 primary data planes for each of FNF's four remaining ranger districts over the next three years. Management problems differ from one ranger district to the next. As such, GIS data planes will be compiled to meet the needs of each individual ranger district.

Among Forest Service GIS efforts, this project is unique in two respects: the Forest Service has not acquired any additional computer hardware or software to reach the present operational stage; and,

Forest Service personnel perform all project work rather than having the project contracted to an outside service agency. Personnel at FNF headquarters perform Landsat classification, map digitization, data management, project implementation coordination and applications through data plane manipulation. As the database for each ranger district reaches the operational stage, ranger district personnel receive training in data plane combination and query techniques enabling data manipulation capabilities at the ranger district office. During the planning and startup stages, the overall project design and guidance with implementation techniques required collaboration with data processing and image analysis consultants at Washington State University Computing Service Center (WSUCSC). The need for consulting from WSUCSC staff diminishes steadily as Forest Service personnel gain skill and familiarity with the systems.

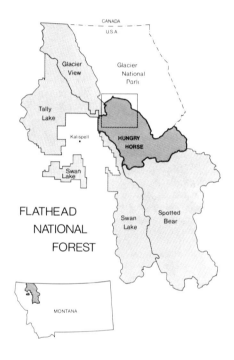

Figure 1: Flathead National Forest. The boxed area at the north portion of the Hungry Horse Ranger District delineates the area shown in the remainder of the figures.

METHODS

Processing and manipulation of the database is accomplished through the facilities in the Digital Image Analysis Laboratory (DIAL) at the Computing Service Center in Pullman, Washington. DIAL systems include Video Image Communication and Retrieval/Image Based Information System (VICAR/IBIS), a batch oriented system, which runs on an Amdahl 470/V8 mainframe; and, an International Imaging Systems-System 511, an interactive system, which is resident on a dedicated PDP 11/34 mini-computer. The FNF's Data General NV8000 is used as a Remote Job

245

Entry/Remote Job Output station via teleprocessing to WSUCSC's mainframe environment.

VICAR/IBIS, the primary processing system used in this project, utilizes raster-based GIS techniques. Raster format data can be conceptualized as a two-dimensional matrix of data elements called pixels. Each pixel stores one data value and the location of the pixel in the data matrix denotes its geographic location. The GIS data planes, a collection of spatially corresponding raster files, are manipulated mathematically using digital image processing algorithms (Jaffray, Hansen & Hart, 1984).

Data Acquired in Digital Form

Defense Mapping Agency (DMA) digital terrain data and Landsat Multispectral Scanner (MSS) data were acquired in computer compatible format on magnetic tapes. Both DMA and Landsat datasets are standard products distributed by U.S. Geological Survey and National Oceanographic and Atmospheric Administration respectively. For compatibility with VICAR/IBIS, the format of each dataset was restructured with specialized VICAR/IBIS "logging" programs (Hart & Wherry, 1984).

Digital Terrain Data Processing. Four 1 x 1 degree DMA digital elevation blocks were geometrically corrected to a Universal Transverse Mercator (UTM) map projection and mosaicked into a 2 x 2 degree, 50 by 50 meter pixel image encompassing the HHRD. Best results were achieved in subsequent processing when DMA 16 bit pixel accuracy was preserved during this process.

Images depicting slope magnitude and slope direction (aspect) were modelled from the DMA elevation mosaic. Slope and aspect models were compressed to 8 bit pixel accuracy for storage efficiency and data generalization purposes. Data were represented as 40 vertical feet per pixel value; 1 degree of slope per pixel value; and 5 degrees of declination per pixel value on the final elevation, slope, and aspect GIS planes.

Landsat 2 MSS Data Processing. Landsat acquisition 22354-17420 (03 July 84, EDIPS format) encompasses all of the HHRD and Glacier National Park. This scene exhibited no cloud cover or data anomolies. Landsat classification and subsequent stratification procedures were performed by FNF personnel in cooperation with Glacier National Park researchers who are developing a GIS database in parallel with FNF (Haraden, 1984, and Key, et al., 1984).

The multispectral classification techniques employed in this project most nearly resemble the multi-cluster blocks technique described by Hoffer (1979). Numerous training areas of known cover type were processed through a clustering algorithm and the consequent spectral statistics were evaluated and compared. From several hundred original spectral classes, 99 were chosen for the final classification statistics set. The resulting classification of the entire Landsat scene did not adequately discriminate between land cover types. The spectral classes contained confusion between water and shadow, inadequate distinction between agricultural classes and natural grasslands, and confusion between certain timber categories and shrubs.

The 99 spectral classes were combined with elevation and aspect data forming 3,600 unique spectral-elevation-aspect categories. Contingency tables were constructed to correlate spectral-elevation-aspect categories with ground data and each spectral-elevation-aspect category was assigned to one of 189 land cover classes. This stratification process alleviated the confusion problems inherent in the multispectral classification and improved the quality of the classified Landsat product.

The southern two-thirds (the United States portion) of the classified, stratified Landsat scene was geometrically corrected to a UTM map projection at a 50 by 50 meter pixel resolution. A set of 38 control points (approximately one control point per 7-1/2 minute quad area) were selected in the DIAL facility by FNF and Glacier National Park personnel. Figure 2 shows a black and white rendition of the 189 land cover categories in the geometrically corrected, stratified, Landsat classification.

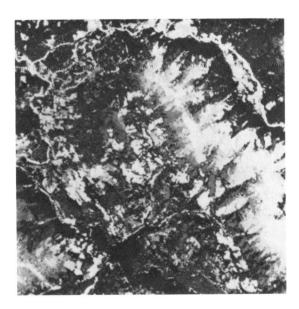

Figure 2: Landsat Classification. Stratified by elevation and aspect, registered to UTM map projection; 189 classes.

Data Digitized from Maps

Maps are digitized at the FNF headquarters offices in Kalispell using a Numonics model 2401 coordinate digitizer which is on-line to the Data General computer. Line segments are recorded in tablet inches, converted to latitude/longitude and subsequently to UTM coordinates, then edges are matched between separate quad sheets. The data files are mailed to WSUCSC on computer tape for incorporation into the GIS database. All but two data planes originated from 1:24,000 USGS quads.

Timber Harvest History. Records of timber harvest in the HHRD date back to 1943. The timber harvest data plane identifies each area by year and type of activity (thinning, selective cut, clearcut, etc.). For purposes of manipulation, summation, or display, data can be grouped as needed by any permutation of year and type of activity. Examples of year/type combinations are: all clearcuts of 1982; all areas which were thinned between 1970 and 1980; or, all types of logging operations for 1954.

Landtypes. Landtypes, outlined in Table 1, identify the opportunities and limitations associated with the physical characteristics of FNF. The landtype classification system is based on geologic processes (as reflected by physiography), soil types, and the factors which determine the behavior of ecosystems (i.e., climate, vegetation, relief, parent materials, and time) (Proposed Forest Plan, 1983). The Landtype data was derived from 1:63,360 scale maps.

```
Table 1: Landtype Classification System

    Group  1 -- Flood Plains
    Group  2 -- Wetlands
    Group  3 -- Avalanche Fans, Talus
    Group  4 -- High Alpine Basins
    Group  5 -- Mass Failure Lands
    Group  6 -- Alluvial Fans, Outwash Plains, and Reworked Tills
    Group  7 -- Silty Glacial Tills
    Group  8 -- Sandy Glacial Tills
    Group  9 -- Clayey Glacial Tills
    Group 10 -- Steep Glacial Tills
    Group 11 -- Residual Soils, 20 to 40% slopes
    Group 12 -- Residual Soils, 40 to 60 % slopes
    Group 13 -- Alpine Ridges and Rocklands
    Group 14 -- Glacial Trough Walls
    Group 15 -- Fluvial Breaklands
    Group 16 -- Structural Breaklands
```

Land Ownership. The Land Ownership data plane delineates private land holdings within the FNF boundaries. In the HHRD there are 16,600 acres of private land.

Administrative Areas. The Administrative Areas data plane delineates areas such as the Coram Experimental Forest (8,020 acres), Wild and Scenic River System (13,838 acres), and wilderness and primitive areas. The HHRD shares administrative responsibilities with other ranger districts for the Great Bear Wilderness (286,700 acres) and the Jewel Basin Hiking Area (15,368 acres).

Precipitation. Mean annual precipitation data derived from snow surveys and Soil Conservation Service historical records were drafted on a 1:63,360 scale map. The precipitation isolines were then digitized and entered into a GIS data plane. Precipitation in the FNF ranges from 20 inches to 120 inches per year.

Timber Compartments. Timber Compartments (Figure 3) are management units of timber stands used for timber inventory purposes. Boundaries of these units were determined by ridges and streams. The Timber Compartments data layer correlates this database to the Forest Service Timber Stand Database at Fort Collins, Colorado.

Drainage Basins. The Drainage Basin data plane shows the watershed boundaries in the HHRD (Figure 4). Information from the GIS database is often summarized by watershed area using a polygon overlay process.

Roads and Trails. Figure 5 illustrates the road network in the HHRD. Line segments denoting segments of road are classified by seven different schemes as shown in Table 2. Trails occupy a separate data plane as do each of the seven road classification schemes.

Figure 3: Timber Compartments. Management units for timber inventory.

Figure 4: Drainage Basins. Watershed boundaries.

Figure 5:
Road System.
All existing roads.

Table 2: Road Classification Schemes

Road Function Class
1 Other
2 Arterial
3 Collector
4 Local

Road Service Level
1 paved, double lane
2 gravel, double lane
3 gravel, single lane
4 dirt

Road Management Reason
1 public safety
2 resource protection
3 soil and watershed
4 management direction
5 wildlife
6 administrative

Road Management Jurisdiction
1 County
2 State
3 Forest Service
4 State forest
5 private
6 Class D permit

Road Management Device
1 Barrier
2 Controlled gate
3 Service barrier
4 Gate
5 Signed
6 Controlled by sign
7 Controlled by barrier

Road Management Date
1 closed year long
2 closed 4/1 - 7/1
3 closed 10/15 - 11/30
4 closed 12/1 - 5/15
5 closed 10/15 - 5/15
6 closed 10/15 - 7/1
7 closed 9/1 - 7/1

Road Maintenance Level
1 custodial care only
2 limited passage
3 low usage/care
4 moderate usage/care
5 maximum usage/care

Hydrology. Streams are recorded as first through fifth order, perennial or ephemeral. Line segments denoting the South and Middle Forks of the Flathead River are given different values; and, islands are differentiated from lakes. Figure 6 illustrates the hydrology data plane.

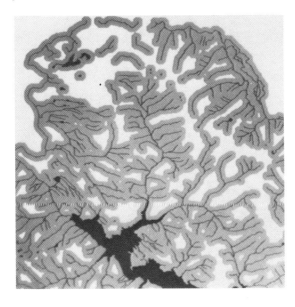

Figure 6: Hydrology. Streams, rivers, lakes and islands; 600 meter distance to water is shaded.

Distance Function Data Planes. Graduated distance corridors were modelled from the hydrology and roads data planes. Pixel values in the distance corridor image indicate how far any pixel is from a road or hydrologic feature. The distance modelling is accomplished with iterative applications of a convolution filter to the data plane of interest. Corridors identify distances between 0 and 1,200 meters.

APPLICATIONS OF THE GIS

In one application utilizing the HHRD database, a data plane depicting distances of zero to 600 meters from water was digitally overlaid with another data plane delineating timber harvest areas. The resulting image and acreage summary quantifies areas which present erosion hazards in that particular watershed.

A road viewshed plan presently being constructed will be used by the FNF landscape architect to design scenic and recreational qualities along a heavily traveled reach of highway. In this application, a distance corridor around arterial roads is superimposed with information from the data planes delineating hydrology, timber harvest, and generalized Landsat classification. Information from these images will be used to plan management of timber harvest methods along the road to enhance the recreational and visual experience and provide a pleasing spatial arrangement to the traveler.

Generation of a map showing areas critical for elk calving habitat is a good example of the utility of the FNF GIS. Modelled after Langley

(1983), Table 3 outlines the selection criteria used in manipulating the HHRD data layers. An acreage summary of the three suitability categories revealed 253 acres of Optimum, 1,798 acres of Acceptable, and 94 acres of Marginal elk calving areas in the HHRD. The data plane depicting elk calving areas will be used in conjunction with the distance to road data plane to evaluate road closure in areas of critical habitat.

Table 3: Elk Calving Criteria

Optimum
    elevation: 4,000 - 4,800 feet
    aspect: flat or south facing
    proximity to water: less than 300 meters
    vegetation: non-forest; less than 50 acres in size

Acceptable
    elevation: 3,500 - 5,000 feet
    aspect: southwest or southeast facing
    proximity to water: 300 - 750 meters
    vegetation: non-forest; greater than 50 acres in size

Marginal
    elevation: above 5,000 feet
    aspect: west or east facing
    proximity to water: 750 - 1,200 meters
    vegetation: timber

CONCLUSIONS

The demand for information from the FNF GIS database grows steadily with each successful application of manipulation and query capabilities. The implementation of this GIS has passed the all important test of user acceptance and is becoming an integral part of strategies for accessing resource management alternatives.

The resolution and accuracy of the GIS database meets or exceeds the quality of FNF databases constructed by traditional methods. In addition, the computer assisted GIS approach affords FNF personnel a quicker and more economical means of updating the database. Manual cartographic updates are replaced by data file editing and rerunning a series of computer jobs.

FNF personnel estimate that the GIS will save a significant amount of money and time and provide the ability to explore a larger variety of management alternatives. The GIS will allow planners and managers to analyze alternatives on critical lands where vegetative manipulation and other project proposals can achieve the desired objectives of land management economies and efficiencies.

REFERENCES

Haraden, R.C. 1984, "Development of Non-Renewable Resources" in Proceedings of Conference on the Management of Biosphere Reserves, Great Smoky Mountains National Park, Gatlinburg, TN

Hart, J.A. and D.B. Wherry ed., 1984, VICAR/IBIS User Reference Manual, Volumes 1-4, Washington State University Computing Service Center, Pullman, WA 99164-1220, 876 pages

Hoffer, R.M. 1979, "Computer-aided Analysis of Remote Sensor Data -- Magic, Mystery, or Myth?" in Proceedings of Remote Sensing for National Resources, University of Idaho, Moscow, ID, pp. 156-179

Jaffray, M.J., R.S. Hansen, and J.A. Hart 1984, "A Geographic Information System Application for Spokane County" in Proceedings 22nd Annual Conference of the Urban and Regional Information Systems Association, Seattle, WA, pp. 154-163

Key, C.H., S. Bain, D.R. Killerude, and D.B. Wherry 1984, "Regional Databases through Interagency Cooperation" poster presentation at Conference on the Management of Biosphere Reserves, sponsored by UNESCO MAB Secretariat and National Park Service, Great Smoky Mountains National Park, Gatlinburg, TN.

Langley, P.G. 1983, Multiresource Inventory Methods Pilot Test, Final Report, U.S. Department of Agriculture, Forest Service, Nationwide Forestry Applications Program, Houston, TX, Report No. NFAP 309

Proposed Forest Plan 1983, U.S. Department of Agriculture, Forest Service, Flathead National Forest, Kalispell, MT, p. II-28

BIOGRAPHICAL SKETCH

Judy A. Hart is an Image Processing Analyst, Graphics and Image Analysis Group, Washington State University, Computing Service Center, Pullman, Washington.

David B. Wherry is the Project Manager of the Graphics and Image Analysis Group at Washington State University, Computing Service Center, Pullman, Washington.

Stan Bain is Supervisory Cartographer Technician, Flathead National Forest, USDA Forest Service, Kalispell, Montana.

THE MAP ANALYSIS PACKAGE AS THE CORE
OF AN INTEGRATED WATER RESOURCE INFORMATION SYSTEM

Christopher Heivly
Social and Behavioral Sciences Lab
Gambrell Hall
University of South Carolina
Columbia, South Carolina 29205
(803) 777-7840

Alfred H. Vang
Richard E. Rouse
William C. Campbell
South Carolina Water Resources Commission

For the past several years, various public and private entities within South Carolina have been cooperating to develop a comprehensive automated information system. This information system is needed to assist in the management and inventory of the states water resources. By integrating hydrologic, demographic and economic variables the supply and demand of the states water resources can be efficiently monitored and projections for the future made. This paper will describe the process of integrating point, line and area data from a variety of sources to a uniform digital data base. The core of this process is the Map Analysis Package (MAP). This software, developed at Yale University, performs numerous analytical tasks and will be used to overlay a series of digital files for analytical purposes.

The South Carolina Water Resources Commission through the University of South Carolina Social and Behavioral Sciences Lab has recently integrated several geographic files for the building of a digital data base. At this time, this base consists of three files. The first file, a polygon based file, is composed of 694 water basins for the state of South Carolina, and parts of North Carolina and Georgia. This file was converted to a grid cell file using the USGS GIRAS polygon to grid conversion. The second file composed of average stream flow levels for over 250 sites was created with the Surface II mapping software and consists of an identical grid cell base of 1000 meters. The Surface II program was used to interpolate these average flows over the entire state. The third file is the state boundary and has also been referenced to these two files.

Other files which are being created are, a river reach file, the census county division file and the industrial site file. The river reach file provided by the U.S. Environmental Protection Agency (EPA), consists of digitized stream segments between 5 and 20 miles long and there are over 3,000 segments for South Carolina. The state CCD (Census County Division) file contains the coordinate boundaries for 294 CCD's within the state and will be used for demographic analysis. The USGS place name file is being used to extract latitude/longitude coordinates for industrial sites from the South Carolina Industrial Directory and all coordinates will then be converted to UTM coordinates and directly assigned row-column locations within a MAP data file.

Current development efforts include the use of new hardware and software at the SBS Lab. An IBM-XT370 will be used to run the MAP interactively. This process will greatly enhance the analysis aspect of the information system by being able to quickly manipulate various files available in the MAP package. The SBS Lab hardware configuration of a high resolution color frame buffer system (Vectrix) and comparable color monitor (Aydin), interfaced with the IBM-XT370 will be used to display these digital files. This color system will allow for the display of over 512 colors at one time at a resolution of 480 x 640.

"THE UTILITY OF A GIS IN EVALUATING THE ACCURACY OF CLASSIFIED
LANDSAT LAND COVER TYPE MAPS ON THE KENAI PENINSULA"

James Henderson
U.S. Fish & Wildlife Service
Anchorage, Alaska 99504

ABSTRACT

In 1981, the U.S. Fish & Wildlife Service (USFWS) developed maps of the wetlands on the Kenai Peninsula as part of the National Wetland Inventory (NWI) Program, which was mandated by Congress. The NWI maps depict various types of wetlands according to the classification scheme developed by Cowardin, Carter, et al in 1979. The wetland types were interpreted from recent 1:60,000 scale high altitude color infrared photography and mapped at a scale of 1:63,360 to register with the USGS topographic base maps.
During the same period, another division of USFWS collected baseline environmental data to support the development of a Comprehensive Conservation Plan for the Kenai National Wildlife Refuge. The plan was developed in response to a Congressional mandate as part of the Alaska National Interest Lands Conservation Act (ANILCA). An essential part of the environmental database was land cover type. This layer of data was developed from the analysis and classification of digital LANDSAT images and digital terrain data. The classification process involved a "modified clustering technique" which employed a "maximum likelihood classifier" to generate a land cover data set of 17 unique types. Some of these types are related to wetland classes, as defined in the National Wetland Inventory.
A study area consisting of one 1:63,360 USGS quadrangle on the Kenai National Wildlife Refuge was selected for the evaluation. The NWI data was digitized and converted to a raster format at the same resolution (50M X 50M) as the land cover data, so as to exactly register the two data sets. Both raster files were loaded into the same grid cell database to facilitate manipulation and analysis. The Geographic Information System (GIS) was used to (1) selectively retrieve particular land cover types and wetland classes, (2) composite the selected data sets through spatial analysis techniques, and (3) determine the location and number of occurrences of various combinations of land cover type and wetland class. An analysis of the composition of each combination revealed occurrences which were defined as "logical mis-matches", or "errors" in classification of land cover type. The basis for this evaluation was the acceptance of the wetland class map as representative of the "real world".
The results of the evaluation demonstrated that (1) the GIS was a logical tool for making the comparison between the two data sets, (2) the GIS was efficient in determining the nature and number of classification "errors", and (3) there was a significant agreement between most land cover types and wetland classes in the study area, but there were classification "errors" to varying degrees in all the land cover types compared.

INTRODUCTION

Purpose & Scope

The purpose of this paper is to present the results of a small project designed to study a technique for quickly assessing the accuracy of land cover type maps derived from LANDSAT data. The project had as its major objective, to investigate a methodology for quickly evaluating the accuracy of LANDSAT derived landcover type maps using the analytical capabilities of a Geographic Information System (GIS) in an area for which a digital land cover type map existed. It was not intended to conduct a thorough and detailed assessment of the accuracy of the landcover type maps.

The scope of the project involved a small area on the Kenai Peninsula of Alaska which covered about 75% of one USGS 1:63,360 scale quadrangle map. The project was very limited in available time so the study was confined to a subset of the land cover types, those normally associated with wetland environments.

Background

As a background to this project, the U.S. Fish & Wildlife Service (USFWS) developed maps of the wetlands on part of the Kenai Peninsula as early as 1981 as part of the National Wetland Inventory Program (NWI). The Kenai Peninsula was chosen partly because of its importance for both wildlife and human use values and partly because of the need to provide environmental baseline data to support the comprehensive planning effort for the Kenai National Wildlife Refuge. The development of a Comprehensive Conservation Plan for the Kenai refuge was mandated by Congress in 1980 as part of the Alaska National Interest Lands Conservation Act (ANILCA). An essential element of data that was missing on the Kenai refuge was a comprehensive map of landcover types. It was a typical situation over much of the State of Alaska, especially on National Wildlife Refuges. Due to the vast expanses of land involved, the lack of extensive high altitude photography, the inaccessibility of the refuge for intensive field surveys, and the short time frame for the comprehensive planning effort, it was decided to utilize LANDSAT multispectral scanner data and automated classification techniques to produce the landcover type maps.

The USFWS has supported the development and utilization of LANDSAT derived landcover type maps on all the refuges currently in the comprehensive planning process throughout Alaska. This extensive land cover mapping effort has been a cooperative one with the USGS. The digital landcover maps have been produced by the staff of the EROS Field Office in Anchorage. EROS has produced landcover type maps for National Wildlife Refuge lands covering over 20 million acres in the last three years. Landcover type maps covering another 30 million acres of refuge lands are currently in production or planned in the near future. There is a strong commitment to the use of these digital land cover maps by the USFWS, not only in the comprehensive planning process but also in refuge management operations as well.

So far there have only been limited efforts to assess the accuracy of the digital landcover maps produced for the refuge lands. Plans have been made in the past to incorporate detailed accuracy assessment steps into the process of producing landcover type maps, but as of today no such steps have been completed. There exists a real need to find a reasonably quick and simple way to assess the accuracy of the land cover type maps while there is still the mechanism for utilizing the results of such an assessment.

The data management and analysis for the investigation utilized the facilities of the USFWS regional computer center in Alaska, which is part of the Office of Information Resources Management (IRM). The facilities include a Data General MV8000 minicomputer, high speed drum plotter, digitizing tables, and several interactive graphics terminals. The regional computer center currently supports two major GIS software packages, "GRID"/"PIOS" from Environmental Systems Research Institute and "MOSS/"AMS"/"COS" from Autometric, Inc. Portions of both major systems were used in the project to conduct the study of the Kenai Peninsula land cover type maps.

Location of Study Area

The Kenai Peninsula is located in Southcentral Alaska, bounded by the Gulf of Alaska to the east and south, Cook Inlet to the west, and the Chugach Mountains to the north. It extends from about 30 miles to 100 miles south of Anchorage, and covers an area of approximately 6 million acres. The Kenai National Wildlife Refuge occupies roughly the western half of the Peninsula. The project study area lies within the refuge boundary and covers about 3/4 of the Kenai C-3 USGS 1:63,360 scale quadrangle map. It is predominately flat to slightly rolling country covered with lowland conifer or mixed forest and deciduous shrubs and extensive areas of lakes and wetland habitats.

LANDCOVER MAPPING AND WETLANDS INVENTORY PROGRAMS

Landcover Mapping

A major data element required in the refuge planning process is landcover type. It forms the basis of most wildlife habitat suitability and natural resource development capability models. Maps of land cover type are generated from the analysis and classification of digital LANDSAT images, in conjunction with digital terrain data from Digital Elevation Models (DEM). The process of generating the final landcover type map is a complex process involving several tasks organized into three major phases. (Figure 1)

FIGURE 1

PHASES	TASKS
(I) Pre-Processing	(A) Select & partition LANDSAT scenes of Refuge
	(B) Correct radiometric distortion
	(C) Correct geometric distortion and register scenes with 50 Meter UTM Grid
	(D) Mosiac selected scenes or sub-scenes
	(E) Generate strata mask of refuge boundary
(II) Image Classification	(F) Select training blocks from high altitude photography
	(G) Extract raw MSS spectral data for each training block
	(H) Cluster the raw MSS spectral data in each training block and generate training statistics
	(I) Conduct field surveys; verify and evaluate training blocks
	(J) Classify and evaluate spectral data in

	training blocks; revise statistics

(III) Post Classification
- (K) Classify spectral data for entire image using revised training statistics; generate "initial" landcover type map
- (L) Determine additional landcover types needed for refuge planning data analysis
- (M) Identify and acquire ancillary data for stratification
- (N) Geometrically register ancillary data and generate additional strata masks
- (O) Stratify "initial" classified landcover image using appropriate strata masks; generate "final" classification
- (P) Produce final landcover type map

    In summary, LANDSAT digital images (scenes) selected for use in generating landcover type maps often originate from different years and/or different seasons, due to the general lack of "cloud-free" scenes for an entire refuge in any one year. Once a set of scenes is acquired for a refuge, they are geometrically registered to the USGS 1:250,000 or 1:63,360 scale topographic base maps covering the refuge. All geographic coordinates are referenced to the appropriate UTM zone.
    Following registration of the scenes, "training blocks" representing areas of typical landcover types are selected on available 1:60,000 high altitude color infrared photography. A clustering operation is performed on the spectral data from the training blocks involving the use of an algorithm called "ISOCLASS" to form groups or spectral classes, where the optimum number of spectral classes is determined by minimizing the transformed scatter ratio. Each spectral class is assigned a landcover type based upon a comparison of its location in the training block image to the initial landcover types interpreted on the high altitude photography. The objective of the clustering operation is to define, to the extent possible, unique spectral classes, each of which represents no more than one landcover type. However, several spectral classes may represent the same landcover type.
    Field studies in the training blocks are designed to verify the landcover types interpreted from the high altitude photography, and training statistics are developed for each training block. Using the training statistics, the spectral data in the training blocks are classified and the results evaluated with the field data and high altitude photography. Following the evaluation, appropriate revisions are made to the training statistics, and the revised statistics are used to classify the entire image. An "initial" landcover type map is the result.
    The "initial" landcover type map inherently does not differentiate between certain landcover types adequately. For example, certain spectral classes are often inseparable due to a degree of "confusion", such as in areas of ice/snow, clouds, barren ground, and lichen. In addition, a few landcover types are not particularly suited to mapping by classification of spectral data, such as airfields, townsites, roads, or commercial land use. Additional ancillary data about infrastructure, physiographic provinces, hydrography, etc. are geometrically registered to the classified image and combined to stratify the initial classification. The results of the stratification yield additional landcover types and produce a "final" landcover type map for the refuge. A list of the final landcover types for the study

area is located in Appendix A.

Wetlands Inventory

In 1974, the U.S. Fish & Wildlife Service was directed to conduct a new inventory of the nation's wetlands. The inventory was designed to provide basic data on the characteristics and extent of wetlands and deep-water habitats, which would facilitate the management of these areas on a sound, multiple-use basis. The Fish & Wildlife Service elected to design a new classification scheme in order to provide uniformity in concepts and terminology in mapping ecologically similiar wetland habitats throughout the country.

There are seven major steps in producing wetlands inventory maps:
(1) Preliminary field investigations
(2) Interpretation of photographs
(3) Review of existing wetlands information
(4) Quality control of interpreted photographs
(5) Production of draft maps
(6) Interagency review of draft maps
(7) Production and distribution of final maps

The inventory and mapping process begins with field investigations in which sample plots are located in several areas representing each of the major wetland habitats. Color infra-red photography at a scale of 1:60,000 is obtained from the Alaska High Altitude Photography Program for the second step. With this imagery, the photointerpreter is capable of detailed wetlands mapping to a minimum size unit of 3 acres. The photo- interpretation is accomplished with the use of a large stereoscope. The photo-interpreter identifies, maps, and classifies each wetland by analyzing vegetation, landform, slope, and drainage patterns, in conjunction with other available data, such as soil surveys, topographic maps, and the field investigations. The boundaries of each wetland are drawn on a mylar overlay to the photograph. All adjacent boundaries on other photographs are "edge-matched" to assure the accuracy of mapping between photographs.

Once all the photographs covering a complete quadrangle have been interpreted, the boundaries of the wetlands on each photo mylar overlay are transferred to a mylar overlay of the USGS 1:63,360 quadrangle map through the use of a Stereo-Zoom Transfer Scope. During this process the individual photographs are registered to the quad map prior to the cartographic transfer. It usually requires 6 - 8 photos to cover a typical 1:63,360 quad map. The result is a 1:63,360 scale mylar overlay showing the location, shape, and classification of the wetlands. It is reviewed to assure that it meets national mapping standards, and verified a second time with the field data. Following the review of the draft product, corrections are made and a final "map" generated. Copies of the final map are sent to Corps of Engineers and the appropriate resource management agencies.

Recently, the National Wetland Inventory maps for selected areas have been digitized. Several interested agencies are funding this conversion to a digital format in order to allow them to incorporate this data into existing digital databases. Once in a digital database, the wetlands data can be integrated with other environmental data for specific project objectives.

A list of the major wetland types occurring within the study area can be found in Appendix B.

METHODOLOGY

Data Preparation

At the beginning of the project, the landcover type data and the wetlands inventory data were in two radically different formats. The landcover type data came directly from the raw LANDSAT image and in a "raster" form, whereas the wetlands data was digitized as lines in a "vector" form. In order to compare the two data sets it was necessary to have them in the same form. The most efficient way to accomplish this task was to convert the wetlands data from its vector format to a raster format, since the reverse involves a much more complex and less reliable procedure. It was decided that the GRID system would be used as the GIS for the study, so the wetlands data was converted to a "Single Variable Grid" format (SVG). There are several different variations of raster formats, and the landcover type data was in one known as "Interagency Transfer Tape" format (ITT). To make the two raster data sets compatible, the landcover type data was converted from the ITT version to the SVG, a relatively simple process.

Not only did the two data sets have to be the same format, they also had to be of the same "resolution". In other words, the pixel size (size of the grid cell) had to be the same dimensions in each case. Since the pixel size of the landcover type data was already 50 meters by 50 meters, the size of the pixels for the conversion of the wetlands data was set to 50 meters square also. In terms of the area represented on the ground by that level of resolution, each pixel covered approximately .6 acres. Both data sets were previously registered to the same UTM coordinates and were therefore registered to each other automatically. This enabled a point on the ground to be referenced in both data sets by the same pixel location (row and column number). At this time the two data sets were loaded into the same grid cell database in order to facilitate the manipulation and analysis of the data simultaneously. The data was now prepared for the next phase of the study.

Data Selection and Comparison

Once the two data sets were in the GRID system database, individual landcover types and wetland classes were selectively retrieved for the study area, using the extraction function of the GIS system (ie. all gramminod marsh landcover type and all persistent emergent marshes wetland class). As stated earlier, the study was confined to landcover types that were correlated with wetland environments. After the selected data were extracted from the database, they were combined through boolean logic to form categories representing various possible combinations. For example, a simple extraction might be that of gramminoid marsh landcover and persistent emergent marsh wetland. The possible combinations would be (1) gramminoid marsh + persistent emergent marsh, (2) gramminoid marsh alone, (3) persistent emergent marsh alone, and (4) neither. As the number of landcover types and/or wetland classes increases, the number of possible combinations increases considerably, though usually not all of the possible combinations will occur, and not all the combinations that occur will be logical ones. The extraction and combination procedure (compositing) is an analysis technique to determine the degree to which the two selected data sets spatially correspond. In other words, for each landcover type in the study area, the compositing procedure determined which wetland classes occurred in the same location and the amount of area they covered. As an example, there might be 300 acres of landcover type A in the study area. Within the area covered by type A there might be 40 acres of wetland class 1, 10 acres of wetland class 2, and 250 acres of wetland class 3, or three combinations that occur.

The compositing procedure was accomplished in the GRID system by programming a "model", which extracted the appropriate landcover types and wetland classes and combined them into new data categories whose values reflected the particular combination that occurred. The modelling program then calculated the number of pixels (grid cells) for each combination that occurred. The final result of the compositing procedure consisted of two items. First, a tabular report showing the combinations of landcover type and wetland class with the number of acres that occurred. Second, a Single Variable Grid cell file (SVG) in which the value stored for each pixel was a unique number representing each particular combination. The SVG was used later as the input file to a raster plotting routine, which generated a plot file for graphic output on a high speed drum plotter.

Accuracy Assessment

In assessing the accuracy of the landcover type map for the study area a fundemental assumption was made concerning the criteria for assessment. It was assumed that the wetlands inventory data was the most accurate representation of the actual landcover on the ground, with respect to the wetland environment. In other words, the wetland data was the standard by which the landcover type map was to be measured or compared. It was felt that this assumption was valid, considering the greater level of detail in the classification and mapping of the wetland classes, as well as the higher incidence of field verification and accuracy assessment performed on the wetlands data.

The accuracy assessment procedure for the landcover type map consisted of identifying combinations of landcover types and wetland classes according to their degree of logical correspondence or "match". Combinations were then grouped into one of three assessment categories based on their degree of match.

Category 1 : High probability of error in classification
(logically mis-matched)

Category 2 : Potential error in classification (possible match under certain conditions or assumptions)

Category 3 : High probability of correct classification (logical match under most conditions and assumptions)

In the first category were combinations such as a landcover type of "soil/rock/sediment" and a wetland class of "persistent emergent marsh". The second category included combinations where the landcover type could conceivably contain small, isolated occurrences of the particular wetland class or was so broadly defined as to allow for the occurrence of several, more detailed wetland classes. An example of the second category would be a combination of "dwarf shrub - lichen tundra" landcover type and "saturated shrub bog" wetland class. The last category contained all the combinations in which it was very likely the two data sets were in close agreement, such as "string bog - wetlands" landcover type and "saturated, emergent, bog-type marsh" wetland class.

Once the combinations were identified and grouped into one of the three categories, the total area of each combination and its assessment category were summarized by the landcover type. By comparing the totals in each assessment category, a relative measure of accuracy was determined for each landcover type. A ranking among landcover types was made based upon comparing their relative accuracies. The results of both comparisons were summarized in tabular display. Limited time prevented any further work with the results, such as more sophisticated statistical analysis between landcover types.

RESULTS AND CONCLUSIONS

To reiterate, the purpose of this study was to investigate a methodology for quickly evaluating the accuracy of LANDSAT derivved landcover type maps, and not to conduct a thorough and detailed assessment of the classification accuracy of the Kenai Peninsula landcover type maps. In view of the limited scope of the study, the results of testing the methodology on the Kenai Peninsula landcover type maps are very preliminary. A summary of the initial results of the comparison between the landcover types and wetland classes is included in Appendix C.

The comparison of the landcover type map and the wetland inventory for the Kenai C-3 quadrangle yielded three conclusions with respect to the task of assessing the classification accuracy.

(1) The use of the GIS provided the capability to determine where the two data sets were logically "mis-matched", the nature of the possible classification errors, and the extent of the conditions.

(2) The GIS performed the assessment functions relatively quickly and in an efficient manner, particularly in view of the fact that the study area covered over 180 square miles, at a resolution of about 1/2 acre.

(3) The preliminary results of the testing of the methodology showed a significant agreement between many of the landcover types and the wetland classes. In addition, there were classification errors to varying degrees for all landcover types. It should be noted that this conclusion is preliminary and is not supported by a rigorous statistical analysis of the results yet.

Although the project seemed to demonstrate the utility of a GIS to facilitate an accuracy assessment of LANDSAT derived landcover type maps, there remains a great deal of work yet to be accomplished. In particular, a need exists to develop specific techniques for assessing accuracy, specific criteria for measuring accuracy, and strategies for utilizing the results. A GIS is a valuable tool in such a task.

APPENDIX A

KENAI C-3 LANDCOVER TYPES

Code	Description
1	Conifer Forest
2	Conifer Woodland
3	Mixed Deciduous / Conifer Forest
4	Deciduous Scrub - Sub-Alpine
5	Deciduous Scrub - Lowland & Montane
6	Dwarf Shrub - Low Shrub Peatland
7	String Bogs & Wetlands
8	Dwarf Shrub - Tundra
9	Dwarf Shrub - Lichen Tundra
10	Lichen Tundra
11	Gramminoid & Disturbed Areas
12	Snow & Ice
13	Water - High Sediment
14	Water - Moderate Sediment
15	Water - Clear
16	Soil/Rock/Sediment

APPENDIX B

KENAI C-3 WETLAND INVENTORY

Code	Description
PSS4/1B	Saturated, Open Canopy Black Spruce Bog
PSS1/EM5B	Saturated Shrub Bog
PSS4B	Saturated Black Spruce Bog
PEM5B	Saturated, Emergent, Bog-type Marshes
PEM5C	Seasonally Flooded, Persistent Emergents
PSS1/EM5F	String Bogs and Reticulate Bogs
POWH	Permanently Flooded, Small Open Ponds
PEM5F	Semi-Permanently Flooded, Emergent Marshes
PSS1B	Saturated Shrub Bog (70% Canopy)
PSS1C	Seasonally Flooded, Dense Shrub
PFO4/EM5B	Saturated, Black Spruce Bog - Emergent Layer
PEM5H	Permanently Flooded, Emergent Marshes
PAB4H	Permanently Flooded, Floating Aquatics
PSS/EM5B	Saturated Shrub Bog (30% canopy)
PSS1/4B	Saturated Deciduous Shrub Mixed Black Spruce
PEM5/OH	Permanently Flooded Open Water, Emergent Marsh
PSS4/EM5B	Saturated Black Spruce Bog W/ Emergent Layer
PSS4/EM5C	Seasonally Flooded Areas W/ Black Spruce
PFO4B	Saturated Black Spruce Bogs
PSS1/EM5C	Seasonally Flooded Areas W/ Deciduous Shrub
L1OWH	Permanently Flooded, Open Water Areas of Lakes
L2AB4H	Permanently Flooded, Shallow Lakes - Aquatics

APPENDIX C

SUMMARY OF COMPARISON RESULTS

Landcover Type	5	6	7	8	11	14	15	16
Total Acres	13955	15674	3414	636	2231	131	4333	103

| Wetland Class | \multicolumn{8}{c}{Acres by Landcover Type} |

Wetland Class	5	6	7	8	11	14	15	16
PSS4B	* 33	? 890	+ 60	? 1	* 27		* 9	
PSS1/EM5B	? 3284	? 3394	+1069	? 6	* 219		* 138	* 1
PSS4B	* 10	? 23	? 14				* 3	
PEM5B	* 97	? 164	+ 42	? 4	? 5		* 22	
PEM5C	* 232	? 269	+ 9	? 2	? 3			
PSS1/EM5F	* 198	? 290	+ 722		* 17		? 36	
POWH	? 258	? 126	? 139	? 4	? 6		+ 223	
PEM5F	* 132	? 158	+ 247	? 1	* 16		? 19	
PSS1B		? 3	+ 1		* 1			
PSS/EM5B					? 1			
PEM5/OH		* 7	+ 17					
PSS4/EM5B	* 77	? 48						
PSS4/EM5C	* 214	? 49						
L1OWH	* 124	* 109	* 372	* 3	* 35	+ 114	+ 3039	* 16
L2AB4H	* 13		* 22		* 6	+ 6	+ 147	* 6

Category 1 :   *   (High Probability of Error in Classification)

Category 2 :   ?   (Potential Error in Classification)

Category 3 :   +   (High Probability of Correct Classification)

ARMY REQUIREMENTS FOR DIGITAL TOPOGRAPHIC DATA

R.A. Herrmann
US Army Engineer Topographic Laboratories
Fort Belvoir, VA 22060-5546

ABSTRACT

The Engineer Topographic Laboratories (ETL) recently completed a two year study which assessed the digital topographic data (DTD) needs of the US Army. The study identified the Army's DTD requirements for tactical terrain analysis, the Army analysis community, and for known and anticipated Army systems/programs. The overall objective was to define Army DTD requirements for the Defense Mapping Agency (DMA) in order to allow them to consolidate their work efforts and to plan their future production requirements. Both subjective and analytical evaluations were performed using two DMA prototype data sets, with investigations focused on the data sets abilities to support terrain analysis, the analysis community, and existing and emerging Army systems/programs covering tactical, combat modeling, simulation, training, testing, and developmental applications. In addition, the existing or anticipated DTD requirements of Army systems or programs were documented, summarized and evaluated. These DTD requirements were defined and evaluated in terms of data content, accuracy, resolution, and format, and a data base specification for DTD, encompassing the Army requirements for terrain analysis, and all other system/programs, was prepared to provide a total Army requirement for DTD.

INTRODUCTION

Increasing demand exists throughout the Army for comprehensive DTD to support known and anticipated systems and programs. Previously, Army needs for DTD remained undefined due to a lack of stated/validated requirements in developing Army systems. However, in the past several years, knowledge and use of DTD expanded within various Army systems and programs to the point that users realized the capabilities and benefits of using DTD to provide timely, accurate, and complete information in support of decision making, planning and combat operations. Increasing demand for DTD occurred within the Army on a system by system basis which resulted in individual users and developers beginning to define specific DTD requirements. However, these DTD requirements were being developed independent of one another, were similar enough in nature to warrant concern over duplication of efforts, and were not supportable by DMA. Developing separate data bases to support individual Army systems or programs is not only redundant and costly, but also effects an inefficient utilization of DTD production resources. This inefficiency endangers the full development of the rapidly expanding DTD technology by creating an increased workload for DMA. The best and most cost effective solution, therefore, is to satisfy the DTD requirements of many systems/programs with a single, unified data base. To address this challenge, the Department of the Army (DA) requested that DMA develop a prototype digital terrain data set which the Army could then evaluate with its overall DTD needs. The intention was to determine if a single DMA product could support the combined needs of a number of Army agencies and, if so, define these needs together. In response to the DA request, DMA agreed to produce two digital terrain data prototypes and officially requested the Army develop a plan to evaluate these data sets. DA also requested that the plan include the identification of all known and anticipated Army systems/programs which are, or will be, requiring DTD.

Through a series of command taskings, ETL was appointed the lead agency in developing and in conducting the Army's overall DTD evaluation plan. The plan had three objectives. The first objective was to investigate and to assess the adequacy of the two DMA prototype data sets in supporting tactical terrain analysis. The second objective was to identify and to articulate the known and the future requirements for DTD in existing and emerging tactical systems, combat models, simulators, training devices, and operational displays. Both of these objectives focused on the data content, accuracy, resolution, and format necessary to successfully meet the Army's needs, and constituted the Army's requirements analysis process. The third and final objective was to consolidate the large variety of known and anticipated Army DTD requirements into a single, unified statement formulated as a specification for a digital topographic data base which could support as many Army users as possible.

The Army's evaluation plan, as developed with the three objectives mentioned above and as approved, was a two-phased DTD requirements assessment. Phase I mainly addressed the Army's terrain analysis requirements for DTD and was conducted primarily at ETL. Phase II concerned both known and anticipated near- and long-term requirements for fielded tactical systems and also for the Army's simulation, modeling, and training activities. The requirements of the Army's analysis community were addressed in both phases of the evaluation plan. The DTD requirements evaluation within the Army analysis community was the responsibilty of the Army's Training and Doctrine Command's System Analysis Activity (TRASANA) and covered the data requirements for existing and developing combat models, simulators, and training devices. The Army analysis community comprises the Operations Research and Systems Analysis (ORSA) cells within the Army's various proponent schools, major Army commands' systems analysis agencies, and several Army research elements. The analysis community's requirements were evaluated independently from ETL's evaluation efforts; however, ETL closely coordinated with TRASANA and was ultimately responsible for integrating the analysis community's part of the Army's evaluation into the overall report. The combined needs from both phases yielded a total statement of Army DTD needs, which, coupled with a cross-comparison with the two DMA prototype data sets, allowed for the preparation of the digital topographic data base specification which describes the Army's requirements. Each phase of the Army's plan is discussed below.

PHASE I

To evaluate the Army's terrain analysis requirements for DTD, the adequacy of each of the DMA prototype data sets was assessed in terms of data content completeness, absolute and relative accuracy (vertical and horizontal), resolution of the elevation and feature data, and the format or structure in which the data are recorded (including the coordinate systems and reference datums). That is, were the prototypes of sufficient accuracy and resolution and did they have sufficient data content to be adequately utilized for tactical terrain analysis and thus support Engineer Terrain Teams in their duties? These terrain analysis requirements for DTD result from the development of automated terrain analysis capabilities which are being designed to help the Army meet the urgent demand for timely, comprehensive, and accurate information about the military aspects of terrain to support decision making, mission planning, and combat operations. Such capabilities are an integral part of the Army's current doctrine. Many of these automated terrain analysis capabilities have already been demonstrated in ETL's Geographic Sciences Laboratory where the Digital Topographic Support System (DTSS) is under advanced development. The DTSS is a battlefield system which will provide Engineer Terrain Teams the capability to produce complex terrain products in a quick, automated mode as opposed to the time-consuming, manual process of today.

The DTSS and other future battlefield systems will require accurate digital topographic data to operate. Recognizing that these data will be provided by DMA, Phase I of the Army's evaluation focused on determining how well the two DMA prototype data sets could support digital terrain analysis. This evaluation, conducted at ETL, was aided by the use of an in-house, commercial interactive graphics system called the Digital Terrain Analysis Station (DTAS). The DTAS is a research and development program designed to automate tactical terrain analysis from digital sources through software development.

To evaluate the two DMA prototype data sets' ability to support terrain analysis requirements, software routines were developed to read and reformat the DMA data into the DTAS data base. With the DMA prototype data sets reformatted for the DTAS, terrain analysis models or products (e.g. cross country mobility, concealment, river crossing, terrain profile, masked area, etc.) developed as part of the referenced R&D software development program, were executed using both DMA prototype data sets as input. These products were then compared to manually prepared products produced for the evaluation. These manual products were hand-compiled from DMA-supplied data feature overlays using DMA accepted terrain analysis, synthesis, and modeling techniques.

Operational suitability, in terms of the usefulness and the acceptability of the prototype data element features and the automated terrain analysis products, was determined by visual analyses conducted by the Terrain Analysis Center at ETL, with support from military terrain analysts. Statistical evaluations of both the DMA prototype data element features and the automated terrain analysis products from each data set were performed to objectively compare and to quantify the differences between the digital and the manual data features and products. The accuracy of the elevation data from both prototypes was also evaluated statistically, using a second-order ground truth survey as control. The visual and statistical evaluations also included data degradation analyses to determine the minimum acceptable resolution which could satisfy Army requirements. To provide further validity to the overall Phase I evaluation, ground truth exercises were conducted by ETL for field verification of the elevation data, the data element features, and the synthesized terrain analysis products.

The evaluation of the suitabilty of the DMA prototype data sets for combat modeling, simulation and training systems within the Army analysis commmunity was the responsibilty of TRASANA, and included input on mobility from the Army Corps of Engineers Waterways Experiment Station. TRASANA's efforts were primarily accomplished using the DMA prototype data sets in their current modeling and simulation programs. The results were reviewed and the suitability of the prototype data sets to support these modeling and simulation programs was determined by analytic comparisons of the resultant products. Finally, a review of the terrain analysis and Army analysis community requirements with respect to the data content that should be present within a digital data set was conducted and the results compared to the data content from the two DMA prototype data sets. Thus the Army's specification for DTD requirements includes the data content necessary for both terrain analysis and the analysis community.

PHASE I RESULTS

For the most part, the Army's tactical terrain analysis, requirements for DTD, as well as those requirements for the analysis community, can be satisfied with digital products (both elevation and feature data) derived from and comparable to the content, accuracy, and resolution of 1:50,000 scale equivalent sources and products. The DMA specifications for their prototype data sets could support these requirements; however, potential deficiences do exist in data, accuracy, resolution, and format as well as in data feature content. As the availability of acceptable DTD increases, Army terrain analysts, working with maps, charts and other sources, will still have to take changes into account. Even after automation, troops in the field will still need to update and revise terrain data to reflect current conditions. The soldiers who man the topographic units of the future must also be equipped to create new terrain data bases should they be called upon to support combat operations in areas for which DMA data are not available.

PHASE II

Phase II of the Army's evaluation of DTD requirements, for known and anticipated tactical systems and other systems (simulation, modeling, training), was a coordinated effort among the Army's major commands, research facilities/laboratories, proponent schools, analysis groups, and project/system managers. Requirements for the Army analysis community were also addressed. Known Army requirements for DTD are those which have been stated in specific system and program requirement documents or have been validated for support from DMA. Anticipated Army requirements apply to those Army activities which foresee the use of DTD as either an integral or supporting part of their system or programs under development. Generally, specific requirements for DTD relating to these activities have neither been officially stated nor validated by DA. These future requirements encompass emerging tactical systems, combat models, simulators, training devices, aviation tactical mission training devices, cockpit displays, operational displays, and terrain analysis.

Phase II of the Army's evaluation plan was directed towards identifying and articulating known and anticipated DTD requirements of Army systems and program offices, in both the near-term (two year) and the long-term timeframes and was accomplished concurrently with Phase I of the Army's evaluation effort. As with Phase I, contractor support was used to assist in this evaluation effort. Liaison was established with all identified potential Army DTD users through a literature search and by ETL telephone and letter correspondence. A questionnaire was then developed to assist participating Army activities in identifying DTD requirements for their systems. The questionnaire was mailed to each activity and on-site interviews were scheduled. During these on-site interviews, an understanding of the system's or programs's specific requirements for DTD was obtained from those developers who could clearly articulate their needs. In many cases, the interviews served to educate the developers about the use of DTD and about ways in which they can determine their requirements for these data in the future. Based upon the completed questionnaires and the on-site interviews, individual summaries for each identified Army system/program were prepared. These summaries identify each system or program by name and present the key findings on their individual DTD needs. Each summary also provides a description of the system/program, it's status and applicable military functions, and lists available documentation. Also included in the summaries are details on both terrain elevation and feature data requirements as well as information relating to the on-site visits. A system summary verification process followed to ensure that the summaries accurately reflected the viewpoint of the respondents. Once verification and validation of all system/program data was accomplished, the DTD requirements were organized into a workable format for final evaluation by dividing these data into three categories of systems/programs: 1) tactical, 2) simulation, training, test, and development, and 3) analysis community. DTD requirements matrices were then constructed for each category, and the four main evaluation criteria for both terrain elevation and feature data (data content, accuracy, resolution, and format) were evaluated systematically. In addition, an analysis was performed of system/program functions requiring DTD to determine the types of applications that will utilize the data. Geographic coverage and data requirements were summarized across all systems/programs to obtain an overall, general picture of data volumes and production requirements.

PHASE II RESULTS

Army systems/programs were identified that either require or anticipate a requirement for Digital Topographic Data. In general, these systems and programs requirements for content, accuracy, and resolution exceed the specifications for standard products currently produced by DMA. The majority of Army organizations stated DTD requirements based on current or developing systems and programs. Many potential Army users had neither used terrain elevation nor feature data in digital form. Therefore, requirements were often stated in reference to analog map products. In the event DTD had been utilized previously, terrain elevation data from DMA or another source was most often the type used.

There existed a great variation in the degree of specification with which Army users were able to articulate their requirements. Often, the degree of specification closely corresponded to the level of development of the system or program. Efforts at a more advanced level of development had a better idea of their supporting DTD requirements than efforts just beginning development. Nevertheless, very few systems/programs have conducted error budget analyses or other studies to derive data base requirements. Clearly, monitoring all Army systems and programs with stated requirements will be important as the developmental process continues.

Examination of accuracy and resolution parameters for DTD did not yield a clear-cut specification for a digital topographic data base. In general, specifications for 1:50,000 scale equivalent topographic maps and terrain analysis studies are sufficient to support most Army users. However, there are some very strict accuracy and fine resolution requirements stated by several Army users, and these requirements may need to be accomplished by the producers of a digital topographic data base. Among these stringent requirements, tactical systems and programs involve worldwide data coverage while simulation, training, test and development efforts typically require data for test sites or other small geographic areas.

A comparative analysis between the resultant DTD requirements of the Phase II evaluation and the two DMA prototype data sets indicates that generally, requirements for the majority of Army DTD users could be satisfied by the current DMA prototype specifications once deficiencies present in the data feature content and also the data accuracy, resolution and format are corrected.

OVERALL RESULTS AND RECOMMENDATIONS

As previously stated, one objective of the Army's DTD evaluation plan was the consolidation of the results of both the Phase I and Phase II evaluations into a single Army statement or specification of DTD requirements. This Army specification is based upon the analysis and evaluation of DTD requirements in support of tactical terrain analysis and the Army analysis community. The specification is also based upon an analysis of system/program DTD requirements applicable to all identified Army systems/programs users.

Through the requirements analysis, conducted as part of the Phase I and the Phase II evaluations, a set of recommendations were derived in the form of a prescribed specification for a digital topographic data base to support Army users. The strictest requirements of the known and anticipated Army systems/programs identified, particularly for data accuracy and data resolution, were carefully analyzed using the following criteria:

- What are the applications for which DTD will be used?
- Have DTD been used in prior efforts?
- What is the status of the system or program?
- What type of system or program is it? (tactical, simulation, training, test and development, etc.)
- What are the data coverage requirements of the system or program?

Based on these criteria, a determination was made on whether the strictest requirements were capable of "driving" an eventual data base specification toward very accurate or high resolution levels. Each of the systems/programs containing stringent requirements were analyzed in this fashion. The final product of this analysis was a recommended specification for terrain elevation data (and their component parts) as well as for feature data and their component parts.

The merging of results from the two-phase evaluation indicated that the majority of Army user community requirements, specifically most tactical applications (including terrain analysis) and these applications of the Army analysis community, can be satisfied with digital products (elevation and feature data) roughly equivalent to the 1:50,000 scale specifications for standard topographic maps/terrain analysis products. That is, digital data derived from and comparable to the content, accuracy and resolution of 1:50,000 scale sources. The DMA prototype data sets, with modifications, could potentially support these requirements.

Requirements for simulation, training, test and development applications are generally much more stringent, as are a few of the tactical and analysis community programs. These applications require accuracy, and resolution for terrain, elevation, and feature data at significantly higher levels than 1:50,000 scale sources. The requirements can be met with 1:12,500 scale, equivalent specifications; however these specifications exceed those of both DMA prototype data sets.

For the reasons stated above, the Army's specifications for DTD requirements are stated at two levels. This two level description, which applies only to the accuracy and resolution of terrain elevation and feature data requirements, is explained in the Army Specification and is summarized briefly below:

Level I is comprised of digital data set relatively equivalent to a class B, 1:50,000-scale topographic map/terrain analysis study. This data set will support most of the Army's DTD requirements and therefore, the majority of the identified Army systems/programs.

Level II is a data set comparable to a class B, 1:12,500-scale topographic map/terrain analysis study. All Army DTD requirements, with a few exceptions, are met by the Level II data, and for the most part, requirements for Level II data within the Army are for very small geographic areas.

The Army's recommended digital topographic data base specifications are stated at relatively high levels. Items such as file characteristics and feature identification codes are not detailed in the manner of typical DMA digital data base specifications. However, general recommendations on format are discussed. The specifications are designed to provide guidelines for the production of digital products to support various Army weapons systems and other Army applications. As such, the specification will serve to assist DMA in its development of DTD products to serve the Army and other military users.

CONCLUSIONS

There is a growing requirement throughout the Army for digital topographic data, and the need for DTD will certainly increase as greater use of digital processing is introduced into Army systems. This increased need will be both in terms of the number of systems using the data and the degree of data content, accuracy, and resolution required. Formats must be standardized to the largest extent possible and intermediate transformations not accommodated by DMA must be resolved within the Army. The consideration of these needs is not only critical, but DMA's capability to support them must be clearly defined. In many cases, the support capability of DMA has been assumed by the proponent developer without any consideration for the considerable amount of time and resources required to produce DTD products. Such a situation can dangerously lead to fielded systems that cannot be supported. Therefore, system developers must be provided with DTD specifications that meet their requirements and can be realistically supported by DMA. A continued effort will be required to insure that emerging systems requiring DTD can be supported, and DMA must be kept advised of new requirements as they emerge if the end result is to be a true force multiplier in support of the field Army.

Distribution of the report is limited to US Government Department of Defense Agencies only.

ACKNOWLEDGEMENTS

In addition to ETL, the following government agencies contributed to this report: US Army Engineer School (USAES); TRADOC Systems Analysis Activity (TRASANA); US Army Engineer Waterways Experiment Station (WES); and the 359th Engineer Terrain Team (USAR). In addition, contractual services were provided by the IIT Research Institute (IITRI) and PAR Technology Corporation.

DISCLAIMER

This paper is not to be construed as an official Department of the Army position.

Constructing Shaded Maps with the DIME Topological
Structure: An Alternative to the Polygon Approach

Michael E. Hodgson
Geography Department
University of South Carolina

ABSTRACT

Creating a choroplethic map with a vector plotter is routinely performed by filling the regions on the map with a regular set of lines of a specified angle and line spacing. The traditional approach to creating a shaded region map has been to shade each region individually. While considerable effort has gone into designing efficient algorithms to shade regions in a polygon format, little, if any, research has examined the possibility of creating choroplethic maps from other cartographic data structures. This paper presents an algorithm for shading a map of regions described by the Dual Independent Mapping and Encoding (DIME) topological data structure.

INTRODUCTION

Most modern polygon data bases begin simply as boundaries between regions in the digitizing phase; usually in the form of arcs, DIME segments, or DIME chains. These elemental entities are later linked in a "polygonization" step to form closed polygons, primarily for the purpose of mapping. However, polygonal data bases are inefficient in that they require more storage space (almost twice as much) than arc or "chain" data sets. If editing of the data base is necessary, as in the case of areas undergoing constant change (e.g., street patterns, land use/cover files, etc.), the updates are made using the original elemental boundary file. The boundaries must be linked into polygons after each update for choropleth mapping.

This traditional process of maintaining up-to-date digital cartographic files requires storing both the elemental boundaries and the polygonal files, as well as continuing the intermediate step of polygonizing the boundaries. Since the databases typically begin in the elemental boundary format, it seems unnecessary to polygonize the data merely for constructing shaded maps. Nonetheless, few, if any, efforts have been made for creating choroplethic maps directly from the elemental boundaries. This paper presents a method for constructing such a map using DIME boundaries.

POLYGON AND DIME BOUNDARY CARTOGRAPHIC DATA STRUCTURES

The creation of choroplethic maps with vector plotters are routinely performed by constructing the regular lines for each region individually. Regions are termed **polygons** in this approach, where the polygon is defined as an ordered sequence of coordinates, either clockwise or counterclockwise around its boundary. Boundary **edges** are implicit in the order of the points. If any islands (holes) occur inside the polygon, they too must be similarly described with the containing polygon as a ordered sequence of coordinates (Figure 1a). Generally, the islands are defined by a closure opposite to the containing polygon. As defined, a polygon retains no information about its contiguous neighbors.

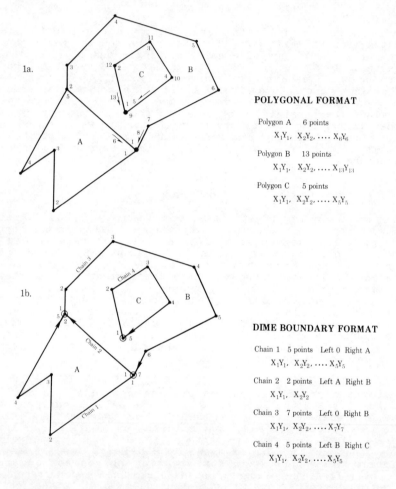

Figure 1. Polygon and DIME boundary representations of an identical area.

In the Dual Independent Mapping and Encoding (DIME) system of boundary representation, the topological relationships are maintained. The basic element of the DIME structure is a directed line segment defined by two points, a beginning and ending point. The **DIME segment** represents a boundary or portion of a boundary between two regions. Codes for the regions on the left and right sides of the segment are included in its description. A **DIME "chain"** is similar to a DIME segment in that it also is the boundary between two regions, but it may be composed of more than two sequential points in defining the boundary (Puecker and Chrisman, 1975) (Figure 1b). Like the elemental DIME segment, the DIME chain has a beginning and ending point and includes codes for the regions on the left and right sides. Regardless of the boundary complexity, a single DIME chain can adequately describe it. For the purpose of this paper, both the DIME segments and DIME chains will be referred to as **DIME boundaries.**

TYPICAL POLYGONAL SHADING ALGORITHM

Several different procedures have been developed for improving the efficiency of the single-polygon shading algorithm (Brassel and Feagas, 1979; Lee, 1981; Cromley, 1984). The procedures for shading the polygon are essentially the same. The typical polygon shading algorithm requires the complete set of coordinates for the containment polygon and any islands along with the parameters for the shading line construction (e.g., line spacing, pen number, angle). The shading lines for a polygon are constructed either independently or as an entire set. For the immediate discussion, the shading lines will be restricted to horizontal lines.

In the "independent approach," the intersections for a single shading line and all boundary edges of the polygon are computed and stored. They are then sorted by their X-coordinate, and implicitly linked and plotted in a "move-to-odd intersection number, draw-to-even" manner. As the intersection coordinates for any given shading line are sorted from left (lowest X-coordinate) to right (largest X-coordinate), moving the pen to the first intersection begins the shading line. The pen is down while moving to the next intersection point to the right, and then back up to the following intersection, and so on. The same procedure for computing intersections, sorting, and plotting is performed for all other shading lines.

In the "entire set approach," the intersections of all the shading lines with every boundary edge of the polygon are computed and stored. They are then sorted by their Y and then X-coordinates, implicitly linked, and plotted. As the shading lines are horizontal, sorting the intersections by their Y-coordinates groups intersections together on their common shading line. Advantages of the "entire set method" is that each edge of the polygon need only be examined once. However, this approach requires more memory and sorting.

In either approach to polygon shading, shading lines at a certain rotation angle may be used if the coordinate points of the polygon are first rotated so that the shading line is parallel with the X-axis (Figure 2). (It is much easier to calculate the intersection of a horizontal line with any

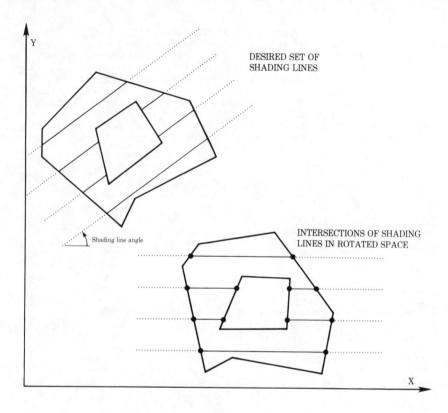

Figure 2. Polygon rotation for calculating shading line intersections.

other line than the intersection with two arbitrary lines.) After computing the intersections, the intersection points are sorted by their X-coordinate and are rotated back into the original space before plotting.

Variants to these basic concepts have been presented by Brassel and Fegeas who decompose the polygon into convex subpolygons in the form of trapezoids during the shading procedure (Brassel and Feagas, 1979). Lee and Cromley each provide additional improvements to the trapezoidal extraction step (Lee, 1981; Cromley, 1984). Convex polygons have the property that a shading line may only intersect the polygonal boundary at two points. This decomposition procedure thus eliminates sorting the intersection points along a shading line by their X-coordinates. It also allows the shading line intersections to be computed in unrotated space. This method uses an incremental displacement for both the X and Y directions.

DIME SHADING ALGORITHM

One alternative to shading regions, other than from polygons, is to use DIME boundaries, as suggested here. The algorithm to create a map of shaded regions from DIME boundaries is graphically presented in Figure 3. Basically, the method involves two general steps: 1) computing the entire set of intersections points for all of the boundaries and storing them, and then 2) sorting the entire group of intersection points with an angle-Y-X sort, linking the intersection points into shading lines and plotting.

Similar to most traditional choroplethic mapping programs a cross-reference table of shading parameters (i.e., angle, spacing) for each region code has been constructed. This table matches data values to be displayed to a set of shading parameters. As the DIME boundary contains region codes for the left and right sides, the shading parameters for a side are found by looking into the cross-reference table.

Arbitrarily beginning with the left side, the boundary is first rotated counterclockwise by the angle of the referenced shading line so that the shading line set is horizontal to the X-axis (Figure 2). The boundary is then decomposed into segments and each segment is then oriented so that it points downward (Figure 4). Each segment now has a beginning Y-coordinate which is greater than or equal to the ending point. Beginning with the first boundary segment, the slope of the segment is calculated. Next, the possible range of shading lines are found by the minimum and maximum Y-coordinates of the points of line segment are found. For each of the shading lines from the maximum Y-coordinate downward, the intersection points are computed.

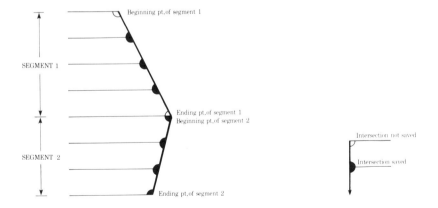

Figure 4. Example of oriented segment and inclusion rule for intersections.

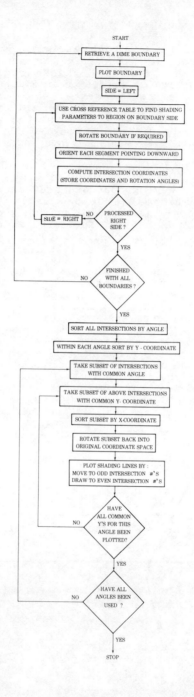

Figure 3. DIME boundary shading algorithm.

By orienting the boundary segments downard before computing intersections, the unique cases where a shading line intersects a beginning or ending point of a segment are handled by the following rule. Intersections created by a shading line with the beginning point of the segments are disregarded. The directional nature of the oriented segments insures the top point of the next segment in the boundary will pick up the intersection of the shading line at this point, and thus, only one intersection will be found (Figure 4). Similarly, horizontal segments are not processed for the same reasons. This inclusion rule also takes care of the cases where a shading line intersects a local peak or pit or a junction of three or more DIME boundaries. As they are computed, the intersection points are then stored along with the appropriate shading angle depicting the shading line set for that boundary. All segments in the boundary are processed in the same manner. The right side of the boundary is then processed.

After all intersection points for the entire map have been computed, they are sorted and plotted. First, they are sorted by their corresponding angle. For each angle, the intersections are sorted by their Y-coordinate and then X-coordinate withing common Y-coordinates. Processing each group of intersections with an identical angle and Y-coordinate, the points are rotated back into the original coordinate space, linked together into shading lines and plotted in. Like the "entire set" approach to shading single polygons, the intersection points are implicitly grouped as shading line segments because of the sorting.

If multiple pens are used (as in the case of pen colors), the pen number for each point would also have to be saved. The pen intersection points should then be sorted by pen number first, then angle, Y-coordinate, and X-coordinate.

An example map using the elementary DIME street segments from the U.S. Bureau of the Census is shown in Figure 5. Note the street segments terminating at a point instead of an intersection. The nature of the DIME shading algorithm allows terminal segments to exist in the regions.

SUMMARY

This algorithm has been presented not as a "better" or more efficient algorithm for shading single polygons, but as an alternative to polygonizing the elemental boundaries simply for the purpose of creating a shaded map. As the DIME boundary shading algorithm is radically different from polygon shading algorithms, constraints common to the traditional shading algorithms are avoided. Typical shading algorithms have limits on the number of islands, the order and closure direction of polygons and islands, and inability to handle terminal segments such as in DIME street files. Region boundaries are only processed once, thus sliver lines are eliminated.

Furthermore, the algorithm is directly applicable to 'coloring' polygons in raster graphics. In such an application, the shading lines would be horizontal and analagous to a scan line on the display screen. Memory requirements for storing intersection coordinates would be significantly less as there is a finite number of scan lines on a display screen.

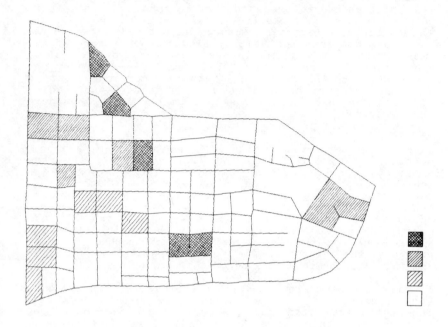

Figure 5. Example map produced using DIME shading algorithm and DIME street segments.

REFERENCES

Brassel, Kurt E., and Robin Fegeas, 1979. "An Algorithm for Shading of Regions on Vector Display Devices," Computer Graphics, vol. 13, pp. 126-133.

Cromley, Robert G., 1984. "The Peak-Pit-Pass Polygon Line-Shading Procedure," The American Cartographer, vol. 11, pp. 70-79.

Lee, D.T. 1981. "Shading Regions on Vector Display Devices," Computer Graphics, vol. 15, pp. 37-41.

Puecker, Thomas K., and Nicholas Chrisman, 1975. "Cartographic Data Structures," The American Cartographer, vol. 2, pp. 55-69.

ANALYTIC AND CARTOGRAPHIC DATA STORAGE: A TWO-TIERED
APPROACH TO SPATIAL DECISION SUPPORT SYSTEMS

Lewis D. Hopkins
University of Illinois at Urbana-Champaign
1003 West Nevada
Urbana, Illinois 61801

Marc P. Armstrong
University of Iowa
Iowa City, Iowa 52240

ABSTRACT

Data can be analyzed from a data structure that is organized around the relationships that are required to support decisions. This approach is more efficient than analyzing from cartographic data structures. Especially for streams data, the sensitivity to fractal dimension makes direct encoding of analytical relationships more efficient than computation from digital geographic coordinates. Using an analog cartographic tier and a digital analytic tier is one good way to implement such a database.

INTRODUCTION

The purpose of a geographic information system (GIS) is to support data manipulation, analysis, synthesis, and display for decision making. The purpose of display is, in turn, to present data in such a way that the powers of human perception can be brought to bear on interpretation of data. The success of a GIS should, therefore, be evaluated not as a cartographic reproduction device with a fixed level of spatial and attribute resolution, but as a particular kind of decision support system. Bonczek et al. (1981:12) describe a decision support system as "an information-processing system embedded within a decision making system." A database should be designed so that data can be manipulated efficiently. Display of data in cartographic form need not be planimetric, if its purpose is to support interpretation. For these reasons it may be more efficient to organize data to be used for decision support into a separate tier in which they are stored and retrieved from an analytical structure, rather than directly from a cartographic structure.

This analytic tier must be supported in some degree by a cartographic tier, but the cartographic tier may be different in form from a traditional GIS. In particular, it need not be computerized, which may result in significant savings in initial costs compared to traditional approaches. An analytically structured system cannot do everything a cartographic system can, but it can support a prescribed set of tasks more efficiently than a cartographic system because the data can be collected and stored for efficient retrieval for particular analytical relationships.

The two-tiered approach is described here by using the Illinois Streams Information System (ISIS), an information system developed for streams data, as an example. The ISIS system is intended to support decisions on stream classification for management and regulation and to support decisions on permit applications for specific projects. The intended tasks are first identified in Section 2. The use of cartographic data structures to support these tasks is described in Section 3. A two-tiered approach is described in

Section 4. A preliminary evaluation of the two-tiered approach in Section 5 suggests that it can provide substantial savings in storage, retrieval, and initial investment costs.

The gradual development of a computerized cartographic tier, after development of the analytic tier, may be an effective strategy for GIS development. This apparently backward strategy may help to alleviate a problem identified by Dueker (1979): Many land resource information systems in the past were not effective because of mismatches between spatial/attribute resolution and the kinds of decisions the systems were intended to support. Developing the analytic tier first assures a satisfactory match because the database design starts from the decisions, not from the data.

STREAMS INFORMATION SYSTEM TASKS

A data structure designed to support streams management and planning should provide for efficient implementation of the following generic tasks: 1) retrieval of thematic information about a place, including its relationship to other places, and 2) selection of places meeting a specified set of characteristics. In the streams case the places may be points on or in relation to streams, reaches of streams, entire streams, or watersheds. The relationships frequently used, which should therefore be supported efficiently in the data structure, are:

1. upstream of
2. downstream of
3. tributary of
4. distributary of
5. flow distance between
6. right bank, instream, left bank

These relationships are simpler to support than more general geographic data relationships because streams form a nearly hierarchical network. Streams data are, therefore, an appropriate case to illustrate the potential advantages of a two-tiered system. Armstrong (1984) has addressed analytical data structures for other types of geographic data.

These generic capabilities can support a wide variety of streams planning and management tasks, as illustrated by the following examples. If a toxic spill occurs, water intakes within a given downstream distance can be identified. All streams with given percentages of bankside landcover can be identified. Flow at a point can be predicted as a function of arbolate sum (total length of channel) above the point. All characteristics surrounding the proposed location for a new bridge can be listed. The nearest water quality sampling station can be found. The above relationships are sufficient for analysis of stream charateristics given sufficient topological resolution and precision of measurement of flow distance. Hydraulic engineering studies, or detailed fish habitat studies require data with greater spatial resolution than stream classification based on aggregate characteristics, but the same data structure would support each task.

A clear understanding of the relationship between resolution and valid interpretation, which is different from resolution and apparent quality of display, is necessary to compare a cartographic tier based on hard copy manuscripts and field notes, as used in ISIS, to a digitally encoded cartographic tier. The difficulties of determining relationships that depend on the scale of the original source, which determines fractal dimension, are especially critical in streams data. The primary relationships of flow length and right vs. left bank are much more sensitive to fractal dimension than is the area of a polygon or error in whether points are in or out of a polygon. Consider a square, as illustrated in Figure 1. The area will remain almost constant, regardless of the fractal dimension of the lines forming its edges. The location of most points as in or out of a polygon is also relatively insensitive to fractal dimension, especially for large polygons. The length of an edge, on the other hand, increases systematically with an increase in fractal dimension. Also, a point immediately adjacent to the edge, such as on its bank, is likely to switch from one side of the edge to the other as fractal dimension changes.

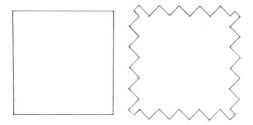

Figure 1: Fractal dimension and accuracy in areas and lines

In the next two sections we describe how the streams analysis tasks can be supported, first by a cartographic data structure then, by an analytically data structure.

CARTOGRAPHIC DATA STRUCTURES TO SUPPORT STREAMS DECISIONS

A widely accepted approach to organizing cartographic data is to use a chain-node data structure. Chains are ordered sets of identifiers for coordinate pairs, and nodes are defined as the meeting of two or more chains. Pointers are used to identify the nodes at which a chain ends, the chains emanating from a node, and the polygons bordered (left-right) by a chain (see Peucker and Chrisman, 1975). Streams in such a system are described as chains of coordinates representing reaches, which connect nodes located at confluences, stream ends, or some other noteworthy break in the stream network. In streams data, chains have among their attributes the direction of flow.

Consider the implementation of Tasks 1 and 2 — Upstream of and downstream of. To find a data item, such as a sampling station, upstream of point A, which is identified by geographic coordinates:
1. Determine which chain (reach) point A is on.
2. Check each successive pair of coordinates in the chain to see if any sampling station coordinates fall within the line between the two points.
3. If the upstream end of the chain is reached, continue search on upstream chains.

Note that at any given resolution, the determination of whether a point is on a line will be ambiguous. Points near confluences or very sinuous reaches will be especially problematic. For consistent analysis, the users must know and apply a set of rules for resolving these ambiguities. For these reasons most geographic information systems store the relationship "point on reach" directly, rather than computing it from geographic coordinates. This relationship is not only stored directly for more efficient retrieval, but is also frequently encoded directly based on determination by a human that the point is on the line by definition.

Next, consider the implementation of Tasks 3 and 4 — Tributary of and distributary of. Given a chain, the following steps are required to identify a tributary:

1. Determine flow direction of the chain from attributes.
2. Identify chains linked to upstream end using pointers.
3. Determine which chain is the mainstem from attributes.
4. All other chains are tributaries.

To identify all tributaries of a given stream the process is repeated. An analogous process can be used for identifying distributaries, by looking at the downstream end of the given chain.

Task 5 — Distance between two points, A and B, requires several distinct operations. For each point:

1. Determine which chain (reach) the point is on.
2. Compute the distance from the point to its nearest neighbor in the chain.
3. Sum this distance and the distances between all other succeeding points in the chain.

Then,

4. Determine which chains are between points A and B.
5. Sum distances for all points in the chains.
6. Sum chain lengths for these chains and distances within the two chains containing points A and B.

It is readily apparent that is is more efficient to compute each length once and store it as an attribute of the chain. Distance computed from the strings of coordinates is inherently at some given resolution (fractal dimension). Distance along a stream is an increasing function of fractal dimension at which it is measured; there is no true distance (See for example, Gardiner, 1982; Klein, 1982).

Task 6 — Determination of right bank, left bank, or in the stream, is very sensitive to resolution. Experience with trying to correctly position points that are near streams, suggests that it is generally necessary to specify the relative position directly. Very small errors in location can lead to incorrect results. If the point location is done first on a manuscript that also contains the stream centerline, then the placement of the point can be purposely moved to avoid ambiguity. Doing so, however, means that the coordinates of that point indicate only relative position and cannot be relied on for determining spatial relationships with items from other manuscripts. Although the actual positioning error is very small, the topological error with respect to streams is obvious. It is difficult to be confident in a dataset that places your town on the wrong side of the river.

Description of the cartographic approach has already suggested some aspects of the analytic approach. The basic idea is to encode directly and store directly the relationships that are frequently used or ambiguous in determination. There is little surprising in analytic data structures, but, importantly, they can be created without the expense of first creating a computerized cartographic GIS from which to compute these relationships. If stream relationships are only to be computed once (or even infrequently for updating), then it is more efficient and accurate to do so external to the computerized GIS environment. This conclusion is made even stronger if one recognizes the resolution required to accurately determine whether points are on lines and the lengths of lines at a fractal dimension sufficient for most data analysis needs.

In the analytic approach the six generic tasks are inherent in the data source. For flow distance between two points it is necessary to do one subtraction operation. If the flow continues to or from a tributary an addition operation will also be required. Relationships of right bank, left bank, instream are not only used to structure the data efficiently for retrieval, but are also the form in which data are encoded. The accuracy of field notes and fully interpreted cartographic manuscripts is thus maintained.

IMPLEMENTATION OF A TWO-TIERED APPROACH

Both tiers, the cartographic and the analytic, must be present in some degree. The cartographic tier might consist only of field notes describing relationships; it might be detailed map manuscripts; or it might be a computerized cartographic GIS. The questions of implementation are:

1. What is stored in machine readable form?
2. What data structure(s) are used?
3. What spatial and attribute resolution is supported?
4. What tasks - manipulation, analysis, and display - are supported from what stored data and at what initial and operating cost?

The crux to understanding the two-tiered approach is that decisions are supported not by the digital (cartographic) representation of the map itself, but by interpretations of it. Careful specification of the generic tasks to be supported makes it possible to store the interpretations themselves. For tasks in which the means of interpretation are not well defined, so that a human must make the interpretation, we can still store only the minimum data required for a sufficient thematic map display. Data for thematic display may be much less costly to store and manipulate than planimetrically accurate data of equivalent resolution. If very detailed planimetric data are needed for interpretation by humans, it is now feasible to store the map or photograph itself on video disk at low cost.

The analytic relationships must be computed from field data or some accurate cartographic source. In ISIS, distances are determined from a set of USGS 1:24,000 quadrangle maps. The relationships tributary of and direction of flow are determined from these maps and related sources using many definitions and precedures to resolve ambiguities (see ISIS, 1983, 1984). Most other data relationships are taken directly from field reports or special purpose maps. Cartographic information added to the USGS base is recorded on mylar overlays to allow additional interpretations, and updating of information.

In ISIS, relationships determined from this analog, not digital, cartographic base, are encoded, and stored, using the data structure shown in Figure 2. ISIS currently operates on SIR/DBMS (SIR, 1983). No geographic coordinates are stored, only hydrologic relationships and positions. The stream numbers are hierarchical so that tributary relationships are inherent in the sorted records. RMI is River Mile Index, which identifies position along the stream centerline. ISIS uses streams labeling in which an entire named stream is indexed as a single entity, rather than labeling each reach (chain between two confluences) as an entity. The positions of points on a stream are then given in reference to its mouth and each tributary is referenced to such a point. All data about a named stream can, therefore, be retrieved directly without using pointers to link together reaches. If it is necessary to know the length of a river, it is stored as an attribute of an entity; it need not be computed by linking reaches and summing the lengths. There is no clear advantage derived from using a reach system, because the stream entity can be divided into either arbitrary or meaningful reaches by designating river mile break points. RMI INFLOW is the

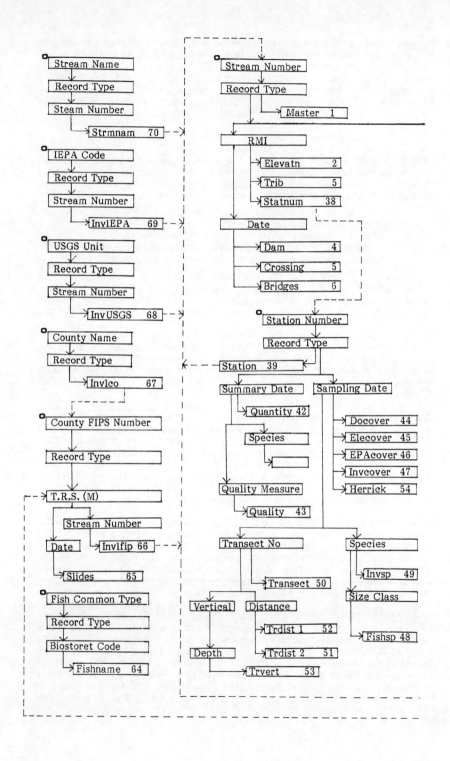

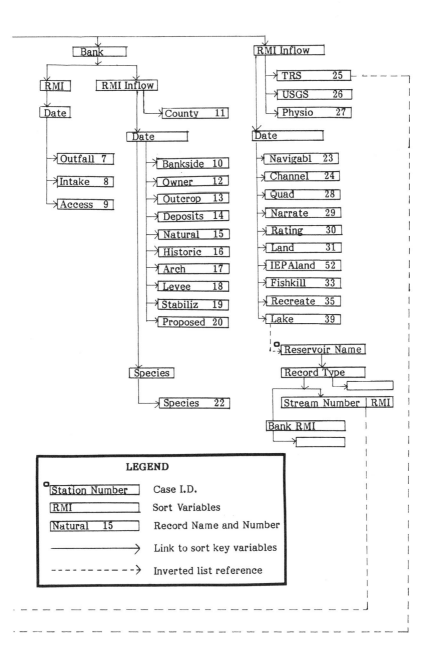

upstream end of a section of stream with a given attribute. Data records sorted by RMI INFLOW identify attributes of sections of stream between two RMIs.

In an analytic data structure, multiple levels of positional resolution can be maintained. The position of a bridge can be specified to the nearest 0.1 mile. Points within a station sampling for fish can, however, be located <u>relative to the bridge</u>, and therefore to each other, to the nearest one foot using field data.

This approach allows detailed transect data to be stored. It is difficult to handle such multiprecision relationships within most available GIS's. Multiple precision analytic structures for streams, on the other hand, can be handled straightforwardly in most data base management systems.

Topological relationships and detailed relative postions (e.g., bank position of discharger) were encoded directly from field data, rather than computed from geographic coordinates, even though digital cartographic data were available. This digital cartographic tier could not be used because the digital data were at 1:250,000. Empirical comparisons of potential errors showed that 1:250,000 USGS stream lines were on the average .85 of the length of 1:24,000 stream lines. Also, some streams are not included on the 1:250,000 data. Because of severe resolution requirements, the streams case in particular is not effectively supported by computing topology and stream distance from a cartographic GIS.

COMPARATIVE EVALUATION

If all generic tasks are supported by the analytic tier, there can be no advantage in supporting decision tasks directly from a cartographic system. Indeed, the chain encoding approach can be viewed as an analytic tier for a particular set of tasks, which emphasizes the point that efficiency is gained by storing directly <u>only</u> the <u>needed</u> relationships at a necessary accuracy. If more elemental relationships are stored than are needed, storage requirements will increase and retrieval efficiency will decrease.

A comparison of tasks shows that for streams, the cartographic approach is inefficient both for storage and retrieval. Finding the characteristics of a place, finding the places with given characteristics, and determining the six relationships listed in Section 2 are more efficient in the analytic structure for both retrieval and storage. From the above comparison we can conclude that the analytic tier should be stored in the computerized database and used to support decision-making. Whether the cartographic tier should also be stored depends on its value in computing analytic relationships, the relative costs and accuracy of an analog vs. computerized cartographic tier, the justification of the initial cost of the cartographic tier for related decision support systems, and the value of the cartographic tier for computerized display.

The value of a cartographic tier in computing relationships is highly dependent on resolution. An evaluation for ISIS of a computerized cartographic tier at 1:250,000 indicated that none of the relationships being determined and recorded at 1:24,000 could be computed satisfactorily from the 1:250,000 data. Most of the relationships could be computed from a computerized cartographic tier at 1:24,000, but, at a cost on the order of two to five times what it cost to use a hard copy cartographic tier. A computerized cartographic tier might be justified if it supported a sufficient number of different analytic data structures designed to support efficiently different sets of decision tasks. In other words, if the full capabilities of the cartographic tier were used sufficiently frequently from the elemental level to create many analytic databases, the computerization might be justified.

Computerized display for human interpretation may be supportable directly from an analytic tier. A tree of hydrologic relationships could be constructed with straight lines. Such a display would be quite sufficient to scan for flow relationships. This display could be enhanced through fractalization to improve comprehension (Armstrong and Hopkins, 1983). In general, much lower resolution will be adequate for thematic map display than for computing relationships. Relationships stored in an analytic tier can be correctly displayed, regardless of the resolution or fractal dimension of forms used for thematic display. It may, therefore, be more efficient to create a map display system that is dependent on the analytic tier, than to use a cartographic tier for map display. Data capture, storage and manipulation will be much less costly because the resolution required will be much lower. For many uses of displayed data, video display of a high quality map manuscripts and aerial photographs will be more useful for interpretation.

CONCLUSIONS

Two tiers, an analytic tier and a cartographic tier should be distinguished. The decision to computerize each tier is distinct and must be made using different criteria. If the relationship between the two tiers is carefully structured, it will be possible to modify the two tiers independently. As additional decision support systems based on analytic tiers are created, eventually it may be worthwhile to computerize the underlying cartographic tier. Careful specification of a topological accuracy standard is required to assure valid computation of relationships from digital data. Relationships determined from field data and encoded as relationships can not necessarily be computed from the cartographic tier, even if it is at the same resolution as the hard copy cartographic tier. Data recorded on the original hard copy cartographic tier may have been based on topologic relationships determined in the field. Thorough documentation of data collection and interpretation procedures must be provided to assure valid use of data.

The two-tiered approach will encourage much more rapid adoption of computerized spatial databases to support decision making because it greatly reduces the initial cost of computerization, but still does not preclude evolution toward the full range of GIS capabilities. Very little effort is wasted because the analytic tier, implemented initially on the basis of an existing, hard copy cartographic tier and field data, will continue to be the most efficient storage of data for direct support for decision making. Even if a computerized, cartographic tier is created later, the design and implementation of the analytic tier is therefore justified.

ADKNOWLEDGMENTS

The work reported in this paper is being performed in part under contract with the Illinois Department of Conservation and funded in part by the U.S. Department of the Interior.

REFERENCES

Armstrong, M.P. 1984, An Extended Network Data Structure for Spatial Analysis, paper presented at Annual Meeting, Association of American Geographers, Washington, D.C.

Armstrong, M.P., and L.D. Hopkins. 1983, Fractal Enhancement for Thematic Display of Topologically Stored Data: Proceedings, Sixth International Symposium on Automated Cartography, Volume II, Ottawa, pp. 309-318.

Bonczek, R.H., C.W. Holsapple and A.B. Whinston. 1981, Foundations of Decision Support Systems, Academic Press, NY.

Dueker, K.J. 1979, Land Resource Information Systems: A Review of Fifteen Years Experience: Geo-Processing, Vol. 1, pp. 105-128.

Gardiner, V. 1982, Stream Networks and Digital Cartography: Cartographica, Vol. 19, No. 2, pp. 38-44.

ISIS, 1983. Data Manual, Illinois Streams Information System, Department of Landscape Architecture, University of Illinois, Urbana, Illinois.

ISIS, 1984. Stream Number and River Mile Index Manual, Illinois Streams Information System, Department of Landscape Architecture, University of Illinois, Urbana, Illinois.

Klein, P.M. 1982, Cartographic Databases and River Networks: Cartographica, Vol. 19, No. 2, pp. 45-52.

Peucker, T. and N. Chrisman. 1975, Cartographic Data Structures: The American Cartographer, Vol. 2, No. 1, pp. 55-64.

SIR, 1983. SIR/DBMS Users Manual, SIR, Inc., Evanston, Illinois.

VECTORIZING SURFACE MODELING ALGORITHMS

David L. Humphrey
Steven Zoraster
ZYCOR, Inc.
2101 South IH-35
Austin, Texas 78741
(512) 441-5697

BIOGRAPHICAL SKETCHES

David L. Humphrey is a Project Manager in ZYCOR's Research Division. Mr. Humphrey received a Master's degree in Mechanical Engineering from the University of Texas in 1978. He has previously worked for Lawrence Livermore Laboratory. At ZYCOR, Mr. Humphrey is responsible for development of gridding and grid processing algorithms and is in charge of developing cartographic applications for array processors.

Steven Zoraster is Manager of ZYCOR's Research Division. Mr. Zoraster received a Master's degree in applied Mathematics from UCLA in 1975. He has previously worked in the field of radar and signal processing. At ZYCOR, Mr. Zoraster has been responsible for research in cartographic automation, software optimization, and advanced algorithm development.

ABSTRACT

Recent reductions in the cost of computer hardware makes the use of "supercomputers" and array processors to execute cartographic algorithms more practical. However, considerable reorganization in algorithm implementation may be required to make efficient use of these architectures. The purpose of this article is to acquaint the reader with the general concepts of vector computers and some of the design issues which arise in adapting cartographic algorithms for execution on these machines. An example involving modelling of a faulted surface is presented.

INTRODUCTION

Though the speed of computer hardware has doubled approximately every three years since the mid-1950's, this trend has become difficult to maintain for standard sequential machine architectures. Recent advances in computing speeds have been made using alternative architectures such as array processors (APs) and "supercomputers" which employ multiple computing elements in parallel operation. These machines will be collectively termed vector processors in this paper.

Until recently, vector machines were so expensive that their use was justified only in special situations for which they were designed such as the solution of partial differential equations encountered in continuum mechanics calculations. However, the ever decreasing price/performance ratios for these machines, particularly for APs, have made it more feasible to use them in selected cartographic applications. Investigations currently in progress at ZYCOR are aimed at assessing the suitability of the following algorithms for execution on vector hardware:

o   gridding from point data,

o   gridding from contour data,

o   grid refinement

o   grid smoothing, and

o   grid contouring.

The purpose of this paper is to acquaint the reader with the general concepts of vector computers and some of the design issues which arise in adapting cartographic algorithms for execution on these machines. An example involving modelling of a faulted surface is presented.

CONCURRENCY IN SOFTWARE

The successful application of a parallel machine to a real-world problem depends strongly on the amount of parallelism inherent in the algorithm being executed. Hardware designers have analyzed typical programs for various applications to determine baseline requirements for their machines, i.e., software has driven the hardware design. In this section some of the characteristics of these constructs are examined. Some seemingly sequential processes may be reorganized to enhance their parallel nature while others are inherently sequential and do not use parallel hardware facilities efficiently.

Strongly Parallel Processes
Parallel processes are often executed on sequential machines in a DO loop. Consider the process of multiplying each element of an array A by the corresponding element of an array B to produce an array C. In FORTRAN this might be accomplished by

```
          DO I=1,N
              C(I) = A(I)*B(I)
          ENDDO
```

where N is the number of elements to be produced. This coding is equivalent to

```
          C(1) = A(1)*B(1)
          C(2) = A(2)*B(2)
                  .
                  .
                  .
          C(N) = A(N)*B(N)
```

A cursory inspection of these operations shows that N independent pairs of operands are used to produce N results. A conventional computer will produce C(1), C(2), ..., C(N) sequentially. Since the operands are independent, that is, they are all defined before execution of the loop begins, the calculations could all be done simultaneously on a computer with N multiplier units. It is these kinds of constructs that APs are designed to execute efficiently.

Reorganization To Enhance Parallel Content
Some kernels may be coded in such a manner as to mask their parallel nature. One such example is

```
set J
DO I=1,N
    D = A(I)*B(I,J)
    C(I) = 1.0 + D**2
ENDDO
```

Note that the parallelism of this kernel is destroyed by the calculation of the temporary scalar D on each iteration of the loop. While this is probably a good implementation for a sequential machine (subscript calculation and primary storage are minimized) an alternative implementation which exploits the parallelism is:

```
set J
DO I=1,N
    C(I) = A(I)*B(I,J)
ENDDO
DO I=1,N
    C(I) = C(I)*C(I)
ENDDO
DO I=1,N
    C(I) = C(I) + 1.0
ENDDO
```

This triple loop configuration requires more coding than the original version but is more desirable in a concurrent processing environment. On some computers the length of the machine code necessary for this implementation is shorter than for the former since each loop generates instructions which invoke hardware rather software implementations of loops.

Constructs Which Reduce Parallelism
Some kernels are inherently sequential in nature and usually cannot use concurrent processing hardware efficiently. Sources of interference in a kernel include:

1. A subroutine call
2. An I/O statement
3. Nonlinear indexing of array subscripts
4. Branches to other sections of code
5. Recurrence relations (use of the results of an iteration on subsequent iterations)

Implementations with nonparallel kernels may often be restructured to move the interference source outside the parallel areas. This may involve additional coding and storage but the result may be much higher execution speeds.

AN EXAMPLE FROM FAULTED SURFACE MODELLING

Processing of fault data in filtering of subsurface grid models presents an example of the need for a new programming perspective in using vector processors efficiently.

Biharmonic Grid Filtering

A well accepted goal in gridding random point data is a surface with minimum total curvature consistent with the input data. The minimum curvature criterion leads naturally to the use of a biharmonic smoothing filter (Briggs, 1974). The biharmonic operator is implemented using a 13 point stencil as shown in Figure 1. Normalized weights are applied to each of 12 nodes surrounding the current target node to obtain a new "filtered" value for the target node. The operator may be applied in a number of ways. One method which provides rapid convergence on sequential machines is successive over-relaxation.

Faults in Subsurface Modeling

Gridding of random point data is a standard cartographic problem. In the energy exploration industry the input data and the resulting models usually represent subsurface geological formations. In this domain, geological faults mark discontinuities in the subsurface. They are drawn by geologists or geophysicists on basemaps and hand digitized to provide input to gridding and filtering algorithms. Information should not propagate across a fault during mathematical operations such as grid filtering. In situations where a standard biharmonic operator would extend across a fault a special operator must be developed. Figure 2 shows an example of this situation.

While it is not hard to write software to perform grid filtering in the presence of faults, it is difficult to make such code run efficiently. Since everyday problems can involve grids with more than 500 rows and columns and thousands of fault segments, efficiency is important.

As can be seen in Figure 2, a fault which impacts the filter operator at a grid node will cross one of four search lines radiating from that node. Therefore, in performing biharmonic smoothing in the presence of faults it would seem that the "natural" order of operations is:

Loop over grid nodes

    Loop over search directions to gather
    surrounding node values

        Loop over fault segments to detect fault crossings
        of search lines

            Apply appropriate biharmonic operator.

While this approach is acceptable on sequential hardware, different methods are required for best performance on array processors or supercomputers.

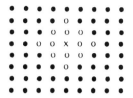

Figure 1(a). Biharmonic Operation Stencil. Point to be filtered is at x. Values at points marked by o are used in filtering x

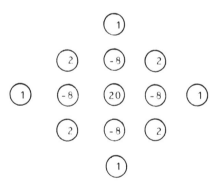

Figure 1(b). Weights of Points in the Biharmonic Operator Stencil

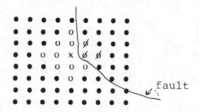

Figure 2. Biharmonic Operation Stencil. Point to be filtered is at x. Values at points marked by o are used in filtering x. Values at points marked by ∅ are used in standard operation but are blocked by fault in this case.

The Solution for Vector Hardware
A difficulty with the algorithm described in the previous section is that it requires looping over many fault segments, no more than one of which is likely to intersect a search line at a critical point. This requires use of conditional logic, something which should be avoided in vector software.

The problem description along with the programming guidelines provided in Section 2 provides hints on how to develop better vector code. First, the fault data tends to occur in short segments. If all segments were short enough it would be possible to guarantee that each individual segment crossed no more than one search line. Second, search directions for nodes on the same grid row or column or the same 45 degree profile overlap. Therefore, initial intersections finding can be organized on the basis of shared search lines rather than on individual grid nodes. Finally, loop reorganization is often an effective "trick" for optimizing performance on vector hardware. In this case, the best loop reorganization involves processing the fault data first rather than almost last. Combining these ideas leads to the following approach which is more suitable for vector processing. Three software modules are used. The first is:

  Normalize the fault data to the grid coordinates

    Loop over fault segments to detect and mark fault crossings of search lines

      Sort and store fault crossing for each search direction

The second module is:

  Loop over grid nodes

    Loop over search directions with fault crossings to develop special biharmonic operators for each grid node position.

The third module is:

Loop over grid nodes

    Loop over search directions to gather surrounding node values

    Apply appropriate biharmonic operator.

For each node, values of zero are assigned to biharmonic filter weights which correspond to data points across faults. This assignment is completed in the second module so that the third module requires no direct information about the fault structure beyond what is encoded in the filter operator for each node. Many filter operators are developed, but a simple data structure allows them to be shared by different grid nodes.

In the first module fault data is processed to guarantee that each segment is shorter than the separation between adjacent grid rows and columns. Longer segments are subdivided. Fault data is also normalized to the minimum grid X coordinate and intercolumn spacing so that the coordinates of each column in the grid and the corresponding vertical search profile are numbered 1 to N. Combining short segments and this normalization mean it is possible to generate significant information by performing the following operation for each fault segment $((x_i,y_i),(x_{i+1},y_{i+1}))$.

K_i = ABS (INT(x_{i+1})-INT(x_i)) * MAX (INT(x_{i+1}),INT(x_i)).

Here ABS, INT, and MAX are respectively the absolute value function, the integer truncation function, and the maximum value function. A little thought should convince the reader that K_i is 0 if the segment does not cross a vertical search profile (because INT(x_i) = INT(x_{i+1})) and is the number of the vertical search profile crossed otherwise.

The ABS, INT, and MAX functions are implemented as vector operators in nearly all supercomputer and array processor subroutine libraries. After the calculation of the K_i, it is possible to continue with calculation of search line intersection points by simple, vectorizable, algebraic operations. These intersections can be sorted and stored for latter retrieval. Equivalent operations can be performed on horizontal and 45 degree offset search profiles.

OBSERVATIONS AND CONCLUSIONS

Though the investigations are still in progress, several observations may be recorded at this time.

To take full advantage of these specialized hardware architectures, major organizational changes are usually required to the software. These changes are usually associated with processing information in large groups or with avoiding conditional processing.

Algorithms organized for vector computers usually require

significantly more memory for the storage of intermediate results such as in the preceding example. As such, only machines which support a large memory address space are should be considered for cartographic applications such as grid filtering and contouring. Hardware-software configurations which include sustained data transfers to a disk drive or host computer will probably leave the computational units idle a large fraction of the time.

Implementation of an algorithm for vector processing requires that the implementation be tailored for the hardware rather than the programmer. As such, such programs are usually harder to comprehend which imposes an extra burden on the maintenance programmer. However, once the programmer has shifted his thought processes to vector mode and the algorithm has been reformulated, transportation of the code to other vector machines should be straightforward.

A final observation is that the 64-bit accuracy offered by some vector computers is not necessary for most cartographic computations such as grid initialization, filtering, and contouring.

REFERENCES

Briggs, I.C., "Machine Contouring Using Minimum Curvature", Geophysics, Vol 39., No. 1, Pages 39-48, February 1974.

AUTOMATED DERIVATION OF HYDROLOGIC BASIN CHARACTERISTICS FROM DIGITAL ELEVATION MODEL DATA

Susan K. Jenson[*]
Technicolor Government Services, Inc.
Sioux Falls, South Dakota 57198

BIOGRAPHICAL SKETCH

Susan Jenson is a Senior Applications Scientist at the U.S. Geological Survey's EROS Data Center in Sioux Falls, South Dakota. She received her B.S. in mathematics from the University of South Dakota in 1973.

ABSTRACT

Digital elevation model (DEM) data in a raster format can be used to automatically derive the drainage characteristics of an area. A procedure has been designed that is capable of operating on matrices of elevation data having no algorithmically imposed size limit, while performing within the resolution and accuracy tolerances of the DEM data.

Each cell is processed as the center of a 3- by 3-cell spatial window in the raster elevation data. If a cell is a local minimum in comparison with two of its non-adjacent neighbors, it is labeled as a drainage cell. The linkages of the drainage cells within user-specified distance and elevation thresholds are established in a separate process. The products of these processing steps are digital masks of the drainage cells and the watershed basins, both in raster format.

A drainage cell mask derived using this procedure is useful in computing slope values for a raster data base. Slope has traditionally been calculated for each cell by fitting a plane through the eight nearest cells. However, if the terrain represented by these cells is V-shaped, such as a gully, a plane does not fit well; in fact, the desired slope value is the slope along the bottom of the gully, regardless of the steepness of the gully sides. The automated drainage process will label such a cell as a drainage cell, and its slope can then be computed from the elevation values of neighboring drainage cells.

INTRODUCTION

The U.S. Geological Survey (USGS) has developed an operational procedure to digitally delineate watershed

---

Publication authorized by the Director, U.S. Geological Survey on November 15, 1984.

[*]Work performed under U.S. Geological Survey contract no. 14-08-0001-20129.

boundaries from raster elevation data and gaging station locations. The project had the following conditions:

1. The procedure must be able to process matrices of elevation data having no algorithmically imposed size limit.

2. The procedure must be adaptable to the accuracy and noise level of the data.

3. The procedure must be able to recognize noncontributing areas of a watershed. For example, a quarry or sinkhole and the area that drains to it must be labeled differently than the surrounding area.

4. The computer implementation must be transportable and memory-efficient.

ALGORITHMS AND IMPLEMENTATION

The algorithms described here have three major steps: identifying drainage cells, grouping drainage cells, and linking groups of drainage cells. Two raster products, a mask of drainage cells for an area and a mask of the watershed basins for an area, are produced. The two products are closely related in that the drainage mask is necessary for delineation of basins and the basin-making process establishes the drainage cell linkages. The computer program makes approximately 10 passes through the data, reading lines of elevation data and reading and writing lines of drainage labels.

The data source for figures 2 through 7 is USGS DEM data collected with the Gestalt Photo Mapper II system and within a vertical accuracy of 7 meters weighted root mean square error (Elassal and Caruso, 1983).

In order to define basins based on gaging stations, the elevation value at each gaging station location is made negative. For instance, if a gaging station location has an elevation of 1,100 meters, that elevation value is changed to -1,100 meters. This has the effect of introducing a hole, or pit, in the surface. Since the algorithms function within elevation tolerances, they recognize these pits and their watersheds as being disconnected from the surrounding watersheds by the same logic that isolates natural non-contributing areas.

Identifying Drainage Cells

The first step is to label each cell in the raster elevation data by its drainage characteristics. The premise for this step is that if an area's elevation profile is V-shaped, it will channel water and should be part of a drainage network. Each cell is isolated as the center of a 3- by 3-cell neighborhood and each neighborhood's elevations are examined in cross-section. As shown in figure 1, there are 12 possible cross sections which intersect at least a portion of the center cell. Cross sections such

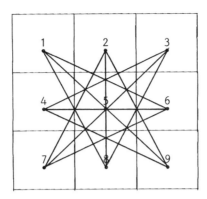

Symmetric Cross Sections	Asymmetric Cross Sections
1-9	1-6
2-8	1-8
3-7	2-7
4-6	2-9
	3-4
	3-8
	4-9
	6-7

Figure 1.--The 12 possible cross sections intersecting at least a portion of the center cell of a 3-cell by 3-cell neighborhood.

as 1-3 which do not intersect cell five, are not considered because it is cell five's characteristics that are being examined. Cross sections are further categorized as symmetric or asymmetric as shown in figure 1. The symmetric cross sections are more valuable than the asymmetric because they are more strongly controlled by cell five's elevation and hence are more indicative of cell five's characteristics. If both end points of a cross section are higher in elevation than cell five, then the area is V-shaped and cell five is a local minimum. Figure 2 shows a small matrix of elevation data with its symmetric and asymmetric local minima. This matrix is 120 by 120 cells in size with each cell representing 50 by 50 meters on the ground. The 50-meter cell size was produced by bilinearly resampling the original 30-meter USGS DEM cells in order to match the cell size of an existing spatial data base. The symmetric minima visually correlate to paths of drainage and the asymmetric minima thicken the paths. At this point, the mask of drainage cells has been established. This intermediate product may be retained and used on its own merit or processed further to define watersheds.

The drainage cell identification process is implemented as a 3- by 3-cell moving window. The process requires only one pass through the data and minimal processing time.

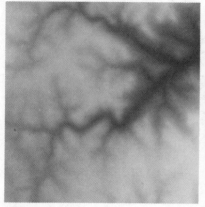 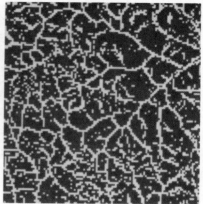

Elevation Data   Symmetric Minima in White

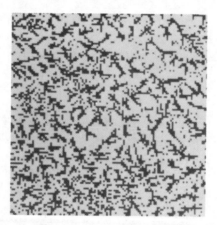

Symmetric and Asymmetric Minima in White

Figure 2.--A 120- by 120-cell matrix of elevation data with its symmetric and asymmetric local minima.

Grouping Drainage Cells

The second step is to group the drainage cells by drainage channel. The premise here is that if two drainage cells are adjacent, and the cell that is lower in elevation belongs to channel X, then the other cell also belongs to channel X. To begin the grouping process the drainage cell with the lowest elevation is found and given channel label one. This cell's eight adjacent neighbors are then tested. If a neighbor cell is a drainage cell with no channel label and is not lower in elevation than the original center cell, it is given a label of one. The neighbor cells of all cells so marked are tested, as are neighbors of neighbors, and so on, until no more cells qualify to receive the label one. The process is then repeated, with the unlabeled drainage cell with the lowest elevation as the new starting point, until all drainage cells have channel labels. Figure 3 shows elevation data and the drainage cells with their channel labels. The circled cells are defined to be roots for the channels. A root is the lowest elevation cell for its channel and is a starting point in one of the iterations of the channel labeling process.

It is necessary to limit the cells grouped in this step to drainage cells only. If all cells were considered in the process, a channel beginning on one side of a hill could grow through a saddle area and erroneously continue to grow upwards on the other side of the hill.

Elevation Data

102	101	101	100	100	102
101	101	99	98	99	103
100	101	97	97	100	101
94	95	101	98	99	100
100	102	103	97	98	99
102	102	100	99	99	99

Channel Labels

0	0	0	1	1	0
0	0	1	1	1	0
0	0	1	1	0	0
①	1	0	1	0	0
0	0	0	②	2	0
0	0	2	2	0	0

Figure 3.--A 6-cell by 6-cell matrix of elevation data is shown with its corresponding channel labels. The zeroes in the matrix of channel labels correspond to cells which are not minima. The two circled cells are channel roots. The computer implementation of this step is based on a last-in/first-out (LIFO) stack. The stack is reinitialized for each channel label. When a cell is labeled, its location is pushed up on the stack. When the stack is completely processed, a new root is sought.

If the elevation values rose smoothly from the root of a
drainage up to the tops of all the branches, the channels
labeled in the second step would be spatially extensive
and only one label would be needed to cover a watershed.
In practice, however, the elevation values undulate along
a drainage, especially in areas of low gradient. Each
undulation causes the labeling process to begin another
channel label, as illustrated in figure 3. In figure 3,
drainage line two is actually a tributary of drainage line
one, but the root cell of line two is artifically low at
the joining location. This grouping step uses approximately 4,000 labels to process a 512- by 512-cell data
set. Finally, all non-drainage cells are given the label
of the channel to which they drain, shown in figure 4.
This is accomplished by iteratively labeling cells uphill
from the labeled cells until all cells are labeled. The
labels now correspond to watersheds for channels which are
artificially small due to noise and inaccuracies in the
elevation data. A third processing step is now required
to establish the linkages among these mini-basins.

Elevation Data

102	101	101	100	100	102
101	101	99	98	99	103
100	101	97	97	100	101
94	95	101	98	99	100
100	102	103	97	98	99
102	102	100	99	99	99

Extended
Channel Labels

1	1	1	1	1	1
1	1	1	1	1	1
1	1	1	1	1	2
①	1	1	1	2	2
1	1	2	②	2	2
1	1	2	2	2	2

Figure 4.--The 6- by 6-cell matrix of elevation values
from figure 3 is shown with its corresponding extended
channel labels. The circled cells are channel roots.

Linking Groups of Drainage Cells

The basic premise for basin linking is that if the root
cell of basin X is sufficiently close in distance and in
elevation to a cell with a different label, for example Y,

and there is a path of sufficiently low elevations connecting the cells, then basin X is actually a continuation of basin Y. In the linking step, these distance and elevation tolerances are empirically determined for a given topographic data set and basin generalization requirement.

To determine basin linkages, each root's spatial neighborhood is examined for cells with a different label, with a path of acceptable elevations, and within the distance and elevation tolerances. In figure 4, channel two will link to channel one with an elevation tolerance of zero, a path tolerance of one, and a distance tolerance of two cells. The linkage could also be made with an elevation tolerance of one, a path tolerance of zero, and a distance tolerance of one cell. The user typically begins with small tolerances and gradually increases them in each iteration until the tolerances approach the accuracy of the data. When channel two links to channel one, all cells with a label of two are relabeled one. Channel two's root is then a member of channel one and has lost its distinction as a root.

When the process is completed, the entire matrix has been subdivided into watersheds. Some will have gaging stations for their roots. Others will have the lowest elevation in a non-contributing area for their roots, or they will be draining off the edge of the data set so the root will be on the data set edge. Still others may be erroneously segmented into sub-watersheds because errors in the DEM data exceeded the ability of the algorithms to establish linkages. In these latter cases, the user must relabel the watersheds using a mapping or renumbering program.

Figure 5 shows the three watersheds that were found for the data in figure 2, with the symmetric local minima superimposed in black.

APPLICATION TO SLOPE CALCULATION

A drainage cell mask derived using this procedure is useful in computing slope values in a raster data base. Slope has traditionally been calculated for each cell by fitting a plane through the eight nearest cells. However, when a cell is part of a drainage path, the desired slope value is the gradient along the drainage path regardless of the steepness of the side slopes. Since drainage cells can be identified in the automated process, the slope calculation can be modified to recognize drainage cells and use only other neighboring drainage cells to calculate their slope. Figure 6 shows a 12-cell by 8-cell matrix of elevation data with slopes calculated traditionally, as well as by special treatment of symmetric local minima drainage cells. Revised percent slopes are shown for symmetric local minima only. The dashes in the table of revised percent slopes represent the cells which are not symmetric local minima. On average, the slope values for drainage cells dropped by 7 percent. Individually, however, there was a great deal of fluctuation. For

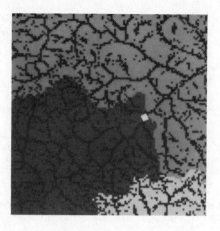

Figure 5.--The three watersheds found for the data in figure 2 are shown here as three shades of gray. The symmetric local minima are superimposed in black. The white cursor (☐) shows a gaging station location.

example, note the behavior of the drainage cell at line 8, column 2. This cell's 3-cell by 3-cell neighborhood is examined in figure 7. By the traditional method, the slope was 17 percent. Using only the elevations of other neighboring drainage cells, the slope was 24 percent. Since each cell represents a 50-meter square area, a 17-percent slope is equivalent to a drop per cell of 8.5 meters; a 24-percent slope is equivalent to a drop of 12 meters. The example in figure 7 shows an area which drains to the right. When a plane is fit through the elevation values of drainage cells only (figure 7b), the plane slopes to the right with an average drop of 12 meters. When a plane is fit to all elevation values (figure 7a), the steepness of the plane is mitigated by the values at line 7, columns 2 and 3, and line 9, columns 2 and 3; and the average drop is lessened to 8.5 meters.

Elevation Data, in Meters

COLUMN

	1	2	3	4	5	6	7	8
1	1720	1728	1719	1727	1735	1736	1725	1717
2	1725	1735	1725	1726	1732	1732	1725	1721
3	1730	1735	1729	1723	1726	1724	1720	1718
4	1739	1737	1727	1722	1724	1722	1717	1711
L 5	1739	1733	1726	1722	1724	1721	1715	1706
I 6	1735	1732	1722	1712	1716	1713	1712	1698
N 7	1729	1727	1717	1716	1707	1703	1696	1687
E 8	1729	1719	1712	1708	1694	1684	1686	1693
9	1739	1729	1718	1700	1691	1692	1706	1723
10	1744	1742	1729	1702	1689	1702	1725	1743
11	1739	1738	1732	1714	1691	1694	1719	1737
12	1732	1730	1726	1720	1699	1692	1711	1732

Percent Slopes by Traditional Method

COLUMN

	1	2	3	4	5	6	7	8
1	23	1	10	15	8	10	18	19
2	18	8	10	7	10	12	12	10
3	14	7	12	4	7	11	12	9
4	10	10	13	3	2	7	12	11
L 5	11	13	15	9	8	10	16	16
I 6	11	15	17	11	15	19	24	27
N 7	11	17	16	16	22	28	24	13
E 8	15	17	16	22	23	9	14	34
9	26	27	30	28	11	22	46	48
10	16	16	36	38	11	30	40	25
11	12	13	24	39	18	27	43	28
12	17	19	20	27	24	16	41	36

Revised Percent Slopes for Symmetric Local Minima
(Dashes indicate cells which are not minima)

COLUMN

	1	2	3	4	5	6	7	8
1	-	-	10	-	-	-	-	-
2	7	-	8	8	-	-	-	-
3	7	-	-	3	-	-	9	-
4	-	-	-	-	-	-	14	14
L 5	-	-	-	7	-	-	-	14
I 6	-	-	-	15	-	-	-	-
N 7	10	-	-	-	23	-	-	14
E 8	-	24	15	-	19	27	6	19
9	20	-	-	18	9	11	-	-
10	-	-	-	-	4	-	-	-
11	-	-	-	-	5	7	-	-
12	-	-	-	-	-	1	-	-

Figure 6.--Comparison of slopes calculated traditionally versus along drainages.

Elevation Values Used to Compute Slope

Traditional Slope

	1	2	3
7	1729	1727	1717
8	1729	1719	1712
9	1739	1729	1718

Figure 7(a)

Drainage-Based Slope

	1	2	3
7	1729	--	--
8	--	1719	1712
9	1739	--	--

Figure 7(b)

Percent Slope

Traditional Slope

	1	2	3
7	11	17	16
8	15	17	16
9	26	27	30

Figure 7(c)

Drainage-Based Slope

	1	2	3
7	10	--	--
8	--	24	15
9	20	--	--

Figure 7(d)

Figure 7.--An example of the difference between traditional and drainage-based slope calculation based on the data in figure 6.

CONCLUSION

Digital elevation data have potential for use in creating hydrologic categories in digital spatial data bases. It is now possible to automatically derive spatially referenced drainage cell and watershed masks, and these basic derivative products enable improved calculation of slopes.

REFERENCES

Elassal, A. A. and Caruso, V. M., 1983, "Digital elevation models." U.S. Geological Survey Circular 895-B.

Capabilities for Source Assessment

John A. Kearney and James R. Muller
Synectics Corporation
111 East Chestnut Street
Rome, NY 13440

Abstract

The Source Assessment System (SAS) is a newly developed photo interpretation system consisting of three computer controlled high resolution video cameras, a dual screen high resolution image processing system, and a minicomputer driven workstation. The high resolution cameras are integrated into two map stations and one film station. Software control of the cameras permit an analyst to remotely control magnification, position, focus, and rotation. A custom engineered video signal digitizer enables rapid conversion of analog video data to a 1024x1024x8 bit digital image file. Multicolor lithographic maps are digitized via use of color filters. The image display subsystem consists of a video switch, color and monchrome monitors, and a multiple memory image processor coupled with a geometric transformation unit for performing spatial transformations. The image processor offers the analyst control of image roam/zoom, image enhancements, image/graphic superimposition, and soft copy feature tracing capabilities. System output includes hard copy graphics recorded using a large format flatbed plotter, as well as textual reports. All hardware and software resources of the SAS are controlled by an analyst who utilizes a graphical interface at a minicomputer driven workstation.

The combined resources of the Source Assessment System provide photoanalysts and cartographers a set of capabilities to permit overlay and inspection of diverse image and map materials. The application of state-of-the-art video and digital technology combined with automated procedures permit visual change detection using diverse inputs. Vertical, oblique and panoramic imagery, maps of varying projections, scales, and formats are placed into a common frame of reference and manipulated such that the presence of new features can be rapidly detected, identified, and recorded.

Source assessment functionality can be partitioned into the following:

1. **Data input** capabilities, to allow input of vertical, oblique or panoramic imagery, maps of various projections, scales and formats in addition to textual information.

2. **Digital and video display** capabilities for high resolution images include digital/video mixing, superimposition, split screen and side by side displays.

3. **Digital image manipulation,** for video to digital data conversion, use of a hardware geometric transformation unit for image warping, color and monochrome viewing, roaming and zooming and image enhancements.

4. <u>Human interaction</u> is aided by utilization of a graphic interface for specification of commands, expandable help files, extensive error checking, and screen formatted keyboard data entry.

5. <u>Input station capabilities</u> include user control of magnification, focus, roaming, rotation, color filters, as well as coordinated movement between input stations.

6. <u>Graphical and reporting output</u> provide hard copy plotting and assessment evaluation report.

This paper will further detail these functional capabilities and describe the implementation of capabilities into operational procedures and system hardware and software components.

CONSIDERATIONS IN THE DESIGN AND MAINTENANCE
OF A DIGITAL GEOGRAPHIC LIBRARY
Hugh Keegan and Peter Aronson
Environmental Systems Research Institute
380 New York Street
Redlands, California 92373
(714) 793-2853

ABSTRACT

A digital geographic library is a structure for organizing large digital geographic data bases for efficient storage and retrieval. Its successful implementation requires considerable planning and possibly significant file manipulation. The issues associated with the effective design and organization of a digital geographic library are described.

THE NEED FOR DIGITAL GEOGRAPHIC LIBRARIES

Increasing numbers of government, private and academic organizations are assembling large digital geographic data bases. In these efforts, primary attention has been given to data collection, and efficient data automation and manipulation. Secondary attention has been given, when considered at all, to the compatibility and long-term management of digital geographic products. As a consequence, data is either functionally lost from the system because it cannot be retrieved or it is used in a manner which is inconsistent with its attribute and intended use.

Frequently insufficient attention is given to a systematic organization of geographic data based on issues such as thematic content, spatial coverage, precision, resolution, age, scale, classification systems and end use. These factors contribute to inadvertent misuse of geographic data and result in a digital geographic data base that is considerably less useful and reliable than it could be.

In addition, the volume of data assembled to support most organization's requirements far exceeds its on-line data storage capacity. Data transfer to magnetic tape is typically performed only when storage requirements compel it, and frequently in a haphazard manner.

Documentation of file contents are generally maintained mentally by the tape's creator. The association between tape documentation, file names, and file contents become increasingly obscure over time. The result is that data becomes unretrievable.

DIGITAL GEOGRAPHIC LIBRARY GOALS

A digital geographic library (map library) rectifies these problems through the imposition of a consistent format for organizing data by location and attribute (for a complete discussion refer to Aronson and Morehouse, 1983; Morehouse, 1985). It provides the user with search, retrieval, and perusal capabilities of both on-line and archived data sets based on both thematic and spatial criteria. The map library gives the user access to map data as well as information about the maps themselves. With a map library, a user is able to browse through digital map library catalogs and locate map data by subject as well as by location. He can determine the map's scale at automation and the tolerances used in its processing. He can determine the maps' age and resolution. He is also able to determine whether a map is in circulation or has been archived to tape. He can leave instructions for the map librarian to retrieve archived material for use. The user is able to make copies of digital maps from unrestricted areas of the map library for his own use. He is not, however, allowed to recopy these maps back into the library after use. Instead, he can nominate maps for addition to the library.

The actions of the user are restricted so that the integrity of the map library is always assured. Decisions about what maps to include in the map library are the responsibility of the library administrator(s).

The map library should enhance the user's access while maintaining data base integrity.

A MAP LIBRARY DATA MODEL

A number of data models can be applied to the organization of a digital geographic library; however, for the purpose of this discussion, the model used in the ARC/INFO MAP LIBRARY (Aronson and Morehouse, 1983), will be used.

In the MAP LIBRARY model, geographic data bases are organized simultaneously in two dimensions -- by subject or content into layers and by location into tiles (see Figure 1).

Tiles. The geographic area represented by the map library is divided into a set of non-overlapping tiles. Although tiles are generally rectangular (e.g., 30 deg. squares), they may be any shape (e.g., counties or forest administration units). Tiles are the digital analogue for the map sheet of a conventional map series. All geographic information in the map library is partitioned by this tile framework.

Layers. A layer is a coverage type within a library. All data in the same layer have the same coverage features (e.g., points, lines, polygons) and feature attributes. Examples of layers are land sections, roads, soil types, and wells. It is useful to think of a layer as a coverage which spans several tiles.

Map Sections. Once a layer has been subdivided into tiles, it consists of a set of individual units called map sections.

In terms of a map library, a geographic data base may be organized as a set of tiles and layers. The tiles are defined in a special INDEX coverage, where each polygon in the INDEX coverage represents a single tile in the library. Layers are defined by the feature classes present (points, lines or polygons) and the thematic items associated with each feature class.

The specification of the tile structure and layer content for a map library which optimally fulfills the requirements of a given user community requires thorough documentation and planning.

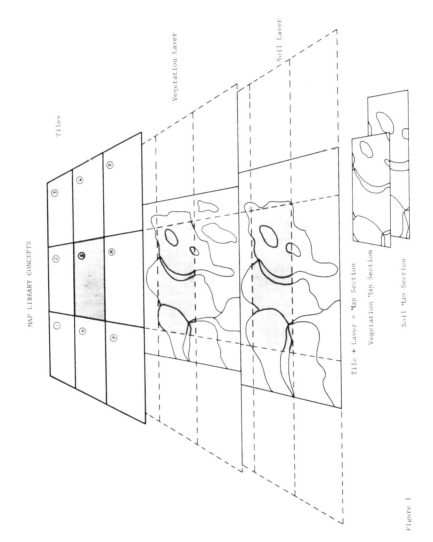

Figure 1

THE MAP LIBRARY DESIGN PROCESS

The principle consideration in designing and implementing a map library is to respond to the typical user inquiry as efficiently as possible. This is essentially an exercise in optimization and compromise. The degree to which an efficient response can be fulfilled is a function of the total amount of disk storage which can be devoted to a library and a sensitive balancing of tile size and layer organization. Both affect retrieval time and subsequent processing time. Additional considerations not to be overlooked throughout the design process are: 1) the cost in both manpower and computer time of implementing the library; and 2) the cost of maintaining the library once created.

Before proceeding, however, it should be pointed out that there are situations when a single map library may be inadequate to fulfill the requirements of a broad user community or prove inefficient for specific user groups. Among these are:

<u>The data base is composed of data differing in resolution and precision.</u>
Data captured at different scales may not be suitable for use together. Although digital maps are essentially without scale, they still have precision and resoultion. If a user has data sets at widely varying degrees of precision, it may be necessary to store them separately. The precision of a map produced by any given process is that of the least precise input into the process, therefore separation into libraries may be warranted.

<u>The data base is composed of incompatible data types.</u>
Separate libraries might be created from a data base including both natural history and detailed taxation data.

<u>The data may be extremely non-contiguous.</u>
When areas of interest are widely separated such as the producing oil fields of the world, it may be desirable to treat them as separate libraries rather than have numerous tiles with no data.

<u>Portions of the data are proprietary or sensitive.</u>
Even though a particular group of data might be well accommodated by an existing library's tile and layer organization, it may be the basis for a separate library if it is confidential in nature.

<u>A user group has highly specialized requirements.</u>
A single group within a larger user community may have highly focused data requirements. Ease of access and processing efficiency may indicate that a separate library is warranted.

With these considerations in mind, the steps in the design of a map library are outlined in Table 1 and discussed in the following sections. It is assumed that extensive data bases are in place.

<u>Step I   Assess User Needs</u>

The fundamental step in designing a library is performing a user needs assessment. This is true whether a broad user group is envisioned as the library consumer or a single department. The user needs assesment provides the conceptual basis for defining tile size and layer content in addition to verify that the construction of a library is warranted or

Table 1
Library Design Sequence

Steps

I   Assess User Needs:

     Quantify volume and type of user requests (simple reference or will data be processed?)
     Determine areal extent of typical user inquiry
     Document attribute data being processed
       - identify logical relationships between data elements
       - identify frequency of specific data item utilization
       - identify missing data elements

II Evaluate Cartographic and Thematic Data
       - scale of data when automated
       - mapping resolution
       - processing precision
       - content of existion files
       - spatial organization of coverages
       - original intent or use of the data
       - classification systems used

III Select Tile Structure and Size

IV Select Layer Content

V   Prepare Pilot Study
       - select study area
       - set up library
       - evaluate library construction time
       - evaluate retrieval time
       - evaluate tile size (too small, too large, just right?)
       - evaluate layer contents (determine number of reselects and appends
         required in operation)
       - quantify manpower and computer resources expended to implement the pilot study and perform at least a rough cost/benefit analysis.

VI Re-evaluate Tile System and Layer Content Based On Pilot Study

VII Implement the Real Thing

feasible. The volume and nature of user interaction with the data is
established during this step in the design process.

A.   Establish volume of user interaction with the data base. How many
different coverages are currently being manipulated and processed? A
recording period of at least several months should be used to avoid
anomalies in the sampling process.

One should also make an estimate of whether this interaction with the
data base would increase if a map library were installed and by how
much. This evaluation of use should be scrutinized carefully to
determine if the construction of a library is justified.

B.   The aerial extent of the user's typical inquiry should be
identified. Care should be taken to assure that the current coverage
organization does not influence the definition of the area. For
example, the fact that data are currently stored in coverages based on
state boundaries should not prevent the user from indicating county
level definition, if processing at the county level comprises the bulk
of user manipulations.

C.   Attribute data being manipulated by the user should be listed both
by variable and feature type (point, line or polygon). Logical
relationships between the variables being used should be identified and
recorded. For example, if the user is selecting marshlands from one
file and bogs from another these should be grouped together.

The frequency of use of different variables should also be noted,
particularly when the same groups of data are being used repeatedly. It
is also important to identify variables for which there are only partial
thematic coverage. The way the data are being manipulated has direct
impact on the way map library layers should be subsequently defined and
organized.

Step II Evaluate Cartographic and Thematic Data

   At the same time the use of the data is being evaluated the
qualitative nature of the data should also be documented.

A.   The scale of the data when automated is extremely important to
record. It has ramifications on both the resolution of the data as well
as its precision.

B.   Equally important to record is the current overall precision of the
processed digital geographic file. Tolerances used in processing a file
may result in a file whose precision is considerably less than when it
was first automated. Ideally, tolerances used in processing the
original data file should be recorded.

C.   Extreme care should also be taken to document compatibility between
systems used to classify thematic data. This is particularly true for
spatially and temporally separated digital maps. It also applies to
documenting spatial fidelity between features recorded at different
times for the same location (for a complete discussion see Chrisman,
1982). Otherwise, the information content of digital maps having the
same resolution, precision, and scale can be severely corrupted when
assembled as composites or when identical processing operations are
performed in spatially separate areas.

D. The boundaries of all coverages currently being used should be explicitly plotted as a composite. This permits the identification of irregularly shaped coverages and coverages which overlap. It also permits islands of non-contiguous coverages to be located. This graphic documentation of existing spatial data base can be thought of as a crude INDEX cover for the later definition of tile shape and size.

Step III   Selection of Tile Structure and Size

A good tile definition is one which corresponds to the typical usage of spatial data. If 90% of user requests are 7-1/2´ quad sheet based, then that should prove to be a good tile size. However, if operations frequently result in multiple tiles having to be extracted for processing, a larger tile size should be considered. The goal is to select a tile which is large enough to accommodate the user requirements, but not so large as to cause unnecessary overhead in file transfer and processing.

The selection of a tile structure should be based upon the user needs assessment. It should, however, also be tempered by a realistic evaluation of the potential costs involved in restructuring the original data base, since coverages will need to be either aggregated together or subdivided.

Step IV Selection of Layer Content

Like the tile structure the creation of data layers should reflect typical usage. A layer is a map type or an association of related map types. Examples of a single map type include: streams, deciduous forests, lease areas, and zoning. Associations of types might include: an aggregate of deciduous forests, mixed forest, coniferous forest, land use and waterbodies. A good layer has a logical single nature. If one is conducting a highway transportation analysis, it would not make sense to have streams in the same layer as roads, as one would constantly have to reselect the desired data. Likewise one would want to make sure that roads, highways and interstate highways are all in the same layer rather than extracting them from separate layers and appending them. It should be remembered, morever, that the data in a layer must be sympathetic in feature type; lakes and streams are both water, but one is a polygon layer and the other is a line layer.

Another consideration in constructing layers, particularly when several items may be merged into a single layer, is the volatility of any of the data items. Items which change frequently, such as land ownership or zoning in transition urban areas, should probably be segregated into a single layer to reduce the need to update several layers because of the presence of a rapidly changing elements.

Step V   Prepare a Pilot Study

Prior to embarking on a full scale data base conversion effort, a pilot study should be conducted on a limited study area. This allows the library designer to quantify cost, and response time, and to empirically test tile size and structure and the layer organization.

A. The cost of the data base conversion effort should be carefully tracked throughout the pilot study. While some of the energy expended in the pilot data base conversion will be expended in learning conversion procedures, the balance of work will be useful for projecting final data base conversion cost. This may cause a redesign of the library.

B. The appropriateness of the tiling structure can also be evaluated by carefully tracking the number of occurrences where multiple tiles must be extracted to fulfill user requirements. Likewise, if there are never situations in which more than a single tile must be extracted then the tiles may be too large.

C. The layer organization should also be evaluated by determining the frequency with which unwanted variables must be dropped out or appended to include missing items. The user should also perform revisions of the pilot layers to document the ease or difficulty with which changes can be made.

D. A thorough pilot study might include several different tile structures and layer organizations, which should be compared and contrasted with regard to labor in conversion, processing requirements, retreival time, flexibility of usage, and ease of data revision.

Step VI   Redesign

It is almost certain that the final map library which is implemented will differ from the pilot study's original specifications. The primary question which should be asked is whether the pilot study library meets the user's typical needs.

Based on an evaluation of the pilot study results, design decision should be revised and incorporated into the final design. The cost of total data conversion may force the reduction of the scope of the library by either increasing tile size and reducing the number of layers or by reducing the aerial extent of the map library's coverage.

Because the end design is both both difficult and expensive to revise once implemented, conducting one or more revisions of the pilot study may be prudent.

SUMMARY

A digital geographic library is a useful organization tool for managing and accessing large geographic data bases. The speed of access to specific thematic data or cartographic data can be enhanced by careful design. The design of a digital geographic library is much simpler if the data are homogeneous in type, scale, precision, resolution, and classification. The design becomes more complex give permutations among these. This may require large dissimilar collections of digital maps to be organized in separate libraries.

REFERENCES

Aronson, P., Morehouse, S., 1983, "The ARC/INFO MAP LIBRARY: A DESIGN FOR A DIGITAL GEOGRAPHIC DATA BASE"., Presented at AUTO-CARTO VI.

Chrisman, R., 1981, <u>Methods of Spatial Analysis Based on Maps of Categorical Coverages,</u> Unpublished Ph.D. Dissertation, University of Bristol.

Morehouse, S., 1985, "ARC/INFO: A Geo-Relational Model For Spatial Information"., To be presented at AUTO-CARTO VII.

ACKNOWLEDGMENTS

The authors wish to acknowledge the kind assistance of Susanne Rohardt and Scott Morehouse for their comments and suggestions.

MIS-APPLICATION OF AUTOMATED MAPPING, AN ASSESSMENT

Poh-Chin Lai
Department of Geography
University of Maryland
College Park, MD 20742

BIOGRAPHICAL SKETCH

Poh-Chin Lai received her Ph.D. from the University of Waterloo in Ontario, Canada. She is presently an Assistant Professor in the Department of Geography, University of Maryland. Her research interests are computer-assisted cartography, geographic information system, map transformation, tactual mapping, and Southeast Asia. She is a member of the AAG, ACSM, ACA, and a Fellow of ICAS.

ABSTRACT

The on-going trend of software boom is becoming more pronounced with the increased affordability of personal computers. More packaged programs than in the past have been used by individuals with insufficient backgrounds in the subject matters, so much so that the computers are merely toys which generate "pretty" pictures but not acceptable by professionals in the discipline. The author attempts to address the practical problems in relation to thematic mapping software for micro-computers and to enumerate the desirable and undesirable features for such use.

INTRODUCTION

The word "mis-application" denotes wrong or unjust use of a resource. This trend of mis-application is an epidemic disease in the field of automation, more so now than ever before, particularly with the introduction of an increasing number of user friendly micro-computers which encourage users of "average" intelligence to "guess their way through." I recently came across an interesting definition on computer in Peter Lyman's article <Lyman, 1984: 20>, quoting Alan kay, Chief scientist at Atari, to say the following:

> The computer is not a tool — that is a weak characterization of the thing. The tools on the computers are the programs that make it into various kinds of levers and fulcra. The computer itself is a medium like paper — zillions of degrees of freedom, used in many ways that the inventors of it can't and don't need to understand, making a fundamental change in the way people think about the world.

What strikes me as interesting is the analogy between the "computer" and the "paper" medium which suggests both the democratization of knowledge via technological advancement and the underlying danger of misuses from an immense degree of flexibility.

COMPUTER-ASSISTED MAPPING

The use of computers and related technology to create visual images instead of numbers and words is not new to some of us. Up until the recent offering of slick, powerful micro-computers and a wide selection of mapping (as in cartographic and as opposed to CAD drafting) software at affordable prices, computer-assisted mapping has been rather cumbersome to operate - one having to invest much time and labor in setting up input and program files by adhering to rigid rules and structures. It was also painstakingly true that the making of one map via automation is hardly worthwhile and mass production is essential to take advantage of the computer's speed and greater dexterity. The above viewpoint is rapidly changing with the availability of low-cost and highly detailed geographic coordinate base files for mapping by means of micro-computers.

Two broad classes of mapping programs for micro-computers can be identified: single-purpose versus multi-purpose application programs. Examples of the former include DIDS (Domestic Information Display System), STATMAP, and MULTIMAP I; all of which create color-coded or choropleth maps. The latter includes Golden Software's graphics system for three-dimensional perspective drawing and contour mapping, and DOCU-MAP which permits a selection of display formats: land cover, contour, choropleth, proportional circle symbol, prism, and three-dimensional perspective.

CARTOGRAPHIC RESPONSIBILITIES

Single-purpose application programs share the following characteristics that they are menu-driven to facilitate usage and that they are designed for one specific use. Menu instructions no doubt reduce user frustrations and are extremely helpful in getting the beginning users over their fear of using computers. But it has two major flaws. In order to provide menu instructions, the program must be restrictive to a number of choices thereby restraining users of exceptional calibre to be creative beyond the permissible options. Also, the experienced users will probably find it tedious to proceed through menu after menu. A more serious implication of the ease of use of these mapping programs is the possibility of increased experimentation without an adequate understanding of the mapping process involved.

Programs such as DIDS, STATMAP, and MULTIMAP I provide the following interactive features:

1. data manipulation to transform continuous variables into categorical variable to create map classes;
2. changing the number of classes;
3. changing the class interval breaks or ranges;
4. enlarging or "zooming" in on portions of the overall map;
5. changing the colors for different data classifications;
6. including or suppressing the boundaries of the geographic areas;
7. customizing the title, legend, and peripheral information.

Any decision making involved in this mapping process requires a prior knowledge of the cartographic language, in particular that of map design and cartographic objectives. Keates <1982: 111>, in refering to map making activity, points out that "Although the initiation of the map, the formulation of its general concept or plan, the specification of content, and the collection of necessary data are all essential states in the preparation of the map, there still remains the construction of the map in its final form. This must include the design and specification of the cartographic symbols, and its technical production." From the perspective of communicative effectiveness, choropleth maps of fewer classes tend to communicate much better to the general audience. From the accuracy point of view, no class choropleth mapping is unquestionably the ultimate solution. The decision on the number of classes (or no classes at all) and the division of these classes will be one of the first questions that comes to mind in the event of choropleth mapping. To users with some cartographic background, the decision is of primary importance because the eventual result varies with different methods of classification and numbers of classes. To others, any arbitrary value will suffice the purpose as long as a map appears as the end product. The difference here is a definite lack of professional responsibility in the latter case.

In constrast, multi-purpose mapping programs confine their users to a lesser degree. What is needed here is the skills to give "commands" to the computer and to give them in an orderly manner. If we assume that every mapping system has its equivalents for the commands MOVE, DRAW, COLOR, ROTATE, SCALE, etc. to achieve three-dimensional drawing, transformations, curve drawing and surface representation, it suffices to illustrate that the sequence MOVE→DRAW→COLOR produces a different result from COLOR→MOVE→DRAW (Figure 1).

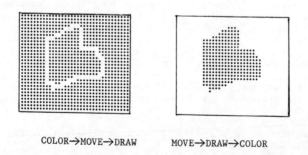

COLOR→MOVE→DRAW   MOVE→DRAW→COLOR

Figure 1: Results of Different Mapping Instructional Sequence

Here, a greater degree of choices expands the mapping application and provides the users freedom to make their own judgement. But it also demands an adequate mapping "literacy" to decide what one wants and then make it happen. The task is not always easy. For one thing, there is a wealth of mapping techniques to graphically present data in ways that make it easily understood and the mapping programs offer a rich store of such choices (such as land cover, contour, choropleth, proportional circle symbol, prism, three-dimensional perspective, etc.). Because of a certain level of flexibility inherent in many of these programs for user-supplied values, it has become necessary for the users to have some familiarity with map displays of varied forms in order to provide the appropriate user specifications. The selection of the most suitable mapping methods for presenting data needs to be carefully matched not only to the data but to the audience for which it is intended. Accepting the fact that certain types of map portrayal may be more appropriate and effective than others, users with insufficient backgrounds have a greater tendency to generate "absurd" maps of no significant value.

CONCLUSION

A micro-computer working environment encourages the "hands on" model of learning in which the user is active in working with the tool (e.g., mapping programs). Most micro-computer software are menu-driven to relieve the users from the agony of mechanical and operational difficulties, thus enabling them to focus more on the activity involved. What is lacking here is the teacher who "consults". Unless the computers are made "smart" enough to detect mistakes or until such time has come, a basic understanding in the constructions and functional uses of specific map displays is imperative to alleviate examples of poor or badly designed maps.

There is no magic formula to mapping. After all, a cartographic product is a work of art, within which its internal order and harmony reflect its creator's own effort. Despite the general lack of specific rules for design specifications, the users need to be conscious of the inherent constraints in mapping for maps convey locational facts via selected mathematical and cartographic manipulations. The ultimate use of a map is for communication and it makes no sense to create unmeaningful, incomprehensible, or illegible maps.

ACKNOWLEDGEMENTS

The author wishes to acknowledge Dr. Derek Thompson and Allen Eney for their assistance in material provision.

REFERENCES

Ganesa Group International, Inc., n.d., "STATMAP".

Golden Software, n.d., "Golden Software's Graphics System".

Keates, J.S., 1982, Understanding Maps, New York: John Wiley and Sons.

Lai, P.C., 1984, "Geocartographic Education Responsibilities in Automated Mapping", Technical Papers of the 44th Annual ACSM Meeting, Pp. 467-474.

Lyman, P., 1984, The Book and the Computer in an Age of "Computer Literacy", ACLS Newsletter, Vol. XXXV, Nos. 1 & 2, Pp. 17-28.

Marshall, G., 1983, Programming with Graphics, Englewood Cliffs, N.J.: Prentice-Hall, Inc.

Morgan-Fairfield, n.d., "MICRO-MAP / DOCU-MAP".

Muehrcke, P.C., 1978, Map Use - Reading, Analysis, and Interpretation, Madison, Wisconsin: JP Publications.

Planning Data Systems, n.d., "Overview of MULTIMAP 1 - A Comprehensive Thematic Mapping Package for Microcomputers".

Sammamish Data Systems, Inc., 1984, "DIDS News and Notes".

MICROCOMPUTER MAPPING SYSTEMS FOR LOCAL GOVERNMENTS

Alven H.S. Lam
Lincoln Institute of Land Policy
1000 Massachusetts Avenue
Cambridge, MA 02138

ABSTRACT

Computer mapping provides various assistance in data analysis and decision-support functions for local public or private sectors who need tools to research spatial data. A generic mapping system not only can present and manipulate different types of areal data, but also can assist intelligent decision-support algorithms.

Local governments have a wide variety of operational and managerial functions which are related to land information. Computer mapping technique, such as thematic mapping, entity overlay, surface interpolation, data conversion, and contour mapping, can be implemented in a microcomputer environment to provide flexible, inexpensive, and intelligent ways to support land information management and policy making.

INTRODUCTION

One of the primary functions of local governments is to monitor area development by establishing appropriate land policy. The goals, mostly from an economic stand point, are to maximize the utilizations of natural resources and to allocate proper resources to proper locations. Conceptually, there are three groups of variables involved in the decision making process: political, social/economic, and physical environments. The policies are established based on the numerous iterations of inter- or intra-confrontations and reconciliations among these three variables.

By understanding the procedures and factors involved in policy making, the information used by the policy makers and the tools to determine alternatives then can be specified. Political environment is always uncertain, but aggregated information is always needed to represent the majority interests. Social/economic environment, which describes the structure and the characteristics of different human factors, has balancing functions for different interests and behavior; physical environment represents the resources with fixed natural locational characteristics and changable artificially added-on values.

Computerization of operational and managerial functions of these environments within a local government has been widely accepted after the "User Friendly" microcomputers were marketed. New technology has upgraded the applications from the level of word processing to information management, and even higher level of policy analysis and decision making.

Computer mapping has been used as a decision-support tool because it presents information in an aggregated and easily understandable form. Direct impact from graphic output prompted further input of in-depth rethinking and reconsideration to the cycle of decision making process. Two dimensional and three dimensional images are more acceptable to human intelligence than linear data when dealing with two dimensional or three dimensional world.

Projects experimenting computer mapping capabilities were done on IBM PC XT and Radio Shack TRS 80 Model 16 microcomputers. Operating systems were MS DOS 2.0 and TRS DOS 4.2.6 from IBM and Radio Shack respectively. BASIC was the language. The target area was the Enterprise Zone in the City of Cleveland, Ohio, covering 41 census tracts. Geographic data base had 1100 points, 500 line segments, and 70 polygons. Attribute data base was downloaded from 1980 census data. Net storage space consumed by cartographic data was 33K bytes.

Based on the essential functions of local government agencies, three categories of mapping applications - mapping for land use planners, assessors, and utility managers, will be discussed.

MAPPING FOR LAND USE PLANNERS

Because of the complexity of the physical environment and a great variety of land-related information land use planners have to deal with, different kinds of maps are frequently used, such as topological, thematic, soil, 2-dimensional surface, 3-dimensional contour, etc. Inside this complicated world, planners not only have to remember the physical characteristics, but also to understand the interrelationships among physical elements, vertically and horizontally. How to store and retrieve different types of map efficiently and to use those maps together to construct the base of decision making are the major issues here.

Vector Maps
Topological maps, street maps, administrative boundary maps, planning area boundaries, zoning maps, railroad and water, land use maps, etc. are presented in vector form because precise boundary descriptions are required in planning operations.

Digitizer was used to store the source maps into vector form. Scanning and downloading remote sensed data were not considered because of the limitations of facilities and the computing power of microcomputers. The map output was on a 6-color pen plotter in vector format(See Figure 1). The system also overlays attribute information, or social/economic figures on top of geographic boundaries and become thematic maps(See Figure 2). These maps can be "manually" overlayed to obtain integrated information of the land. Planners then have the ability to integrate more or digest more knowledge to make decision.

Figure 1: Vector Map (Boundary)

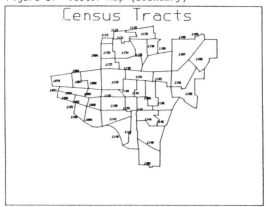

Figure 2: Vector Map (Thematic)

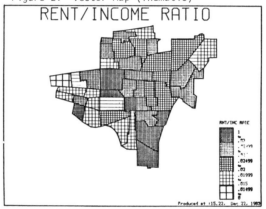

Figure 3a: Grid Map (Thematic)   Figure 3b: Grid Map (Boundary)

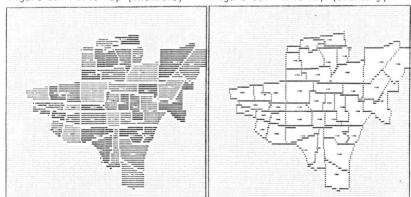

Grid Maps
Grid maps are used when area boundary precision is not essential. Some examples are regional planning and environmental impact studies. The advantages of a grid map are efficiency of processing, ease of vertical overlays, and fast in other spatial search applications. Two basic techniques are used:

Vector-to-Grid Conversion: A vector-based map can be displayed as a grid format by a conversion process. The conversion consists of two steps: boundary conversion and infilling conversion. Grid map then can also present either the boundaies or thematic infilling. The method used to present different combinations of grid maps (see Figure 3a and 3b) is the logical manipulation of Set theories.

Set Functions: The logic functions in Set theory are AND, OR, NOT, XOR, etc. If a polygon on a grid plane can be recognized as a "Set", then by using logic functions to manipulate the Set with other Sets will result in some polygon overlays. For example, Set A and Set B are shown on Figure 4a and 4b respectively. Figure 4c, 4d, and 4e are the OR, AND, XOR functions implemented to Set A and Set B. Same applications of this Set manipulation are used to present the grid maps on Figure 3a and 3b. This technique create a quick solution for polygon overlay problems. For microcomputers, we need solutions to overcome complicated geo-processing problems.

Surface Processing
Trend Surface is generated by using the altitude of known control points to interpolate other unknown points. Planner can comprehend the prospective view of the whole area by just knowing the characteristics of limited key locations. Two dimensional surface is usually presented in grid form (Figure 5a). Three dimensional surface is usually in vector form (Figure 5b). This processing is time consuming based on the factors of surface resolution and weighting and scaling algorithms.

Contour Maps are valuable for land planners because slope, orientation, and terrian roughness are some key factors affecting the suitability of development in many cases. The maps can be presented in two dimensional or three dimensional forms. Two dimensional maps need other forms of intelligence to configurate the slope and the true impacts of physical terrain. Three dimensional maps, although have better visibility, carry disadvantages of slow processing and difficulties of integrating with other spatial information(Figure 6).

MAPPING FOR ASSESSORS

In the United States, there were 13,439 assessing units or jurisdictions operating under local governments in 1979, and 77.7% of local tax revenue, or 29.5% of total local revenue was generated from property taxes. The importance of property taxes, as well as the importance of modernization of assessement operations, has been emphasized in local

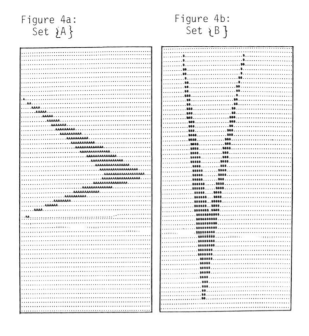

Figure 4a: Set {A}

Figure 4b: Set {B}

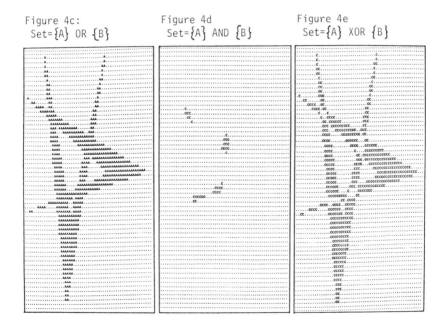

Figure 4c: Set={A} OR {B}

Figure 4d: Set={A} AND {B}

Figure 4e: Set={A} XOR {B}

Figure 5a:
2-D Trend Surface

Figure 5b:
3-D Trend Surface

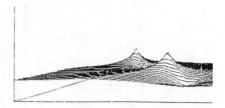

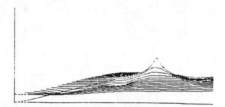

Figure 6: 3-D Contour Map

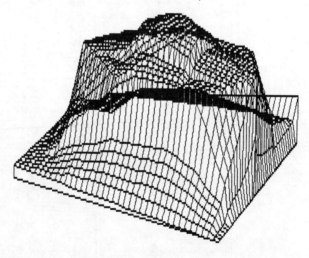

governments by using new technology to improve the
efficiency and effectiveness of assesment activities.

Assessor's maps, traditionally, are used to show locations,
dimensions, and configurations of individual parcels.
Recently, assessors are expecting wider definitions and
functions of their maps to become "a data base that informs
about the land, its geographic and non-geographic
attributes, and about relationships among these attributes"
(Costello, M. and Garvin, L., 1982). Maintaining a good
performance of a mapping system, or a "multi-purpose
cadastral" system, can assist assessors in monitoring their
assessment activities, such as tracing the transaction,
collecting deliquencies, mass appraisal and reappraisal of
properties. Computer mapping can easily presents
interesting areas such as properties with tax deliquency, or
properties with assessed value higher or lower than the
selling price.

Parcel Mapping
The components of an assessors' map, usually used for legal
descriptions, include parcel boundaries, dimensions,
bearings, sub-division names and boundaries, easement and
right-of-way boundaries, location and name of street, parcel
identifiers, street adress. Mapping these components
requires accurate input and output devices in order to
support legal statements.

There are two ways to input/store parcel maps: digitizing
the source maps and typing geographic references from the
land survey into the computers. The second method although
prevents the error caused by the physical limitations of a
digitizer, human errors will still occur and need additional
editing.

Accuracy requirements for parcel mapping are +0.35'',
+0.17'', and +0.08'' in rural area, rural and urban ring
area, and urban area respectively (National Research
Council, 1982). Once the source maps are accurately
surveyed and recorded at these standards, computerized
mapping should be able to store and display these maps with
the same accuracy standards. If a microcomputer can not
handle enough significant digits to process these numbers or
an output device has less graphic resolution to meet the
requirements, other written documentations or descriptions
should be attached with the computer maps for legal
purposes.

MAPPING FOR UTILITY MANAGERS

Availability of sufficient utilities is an important factor
for evaluating area development priority. Sophisticated
utility management systems provide operational and
managerial functions of design, maintenance and
construction, as well as decision functions of resource
allocation. Whenever an area or a particular location needs
new facilities or infrastructures, or making decision to
allocate resources, utility managers must take into
consideration other physical characteristics of the

surrounding area, such as abutting parcels, soil types, street right-of-way, easement boundaries, and other utility systems.

Utility mapping assists managers to comprehend the physical relationships of all different utility systems and infrastructures underground or on or above the surface. The accuracy required, however, is higher than other mapping systems in terms of displaying, searching, or aggregating spatial data. Important information such as length, depth, width, capacity is crucial when engineering design and maintenance are underway.

EVALUATION OF A MICROCOMPUTER MAPPING SYSTEM

Hardware performance, software flexibility, and satisfaction of user's needs are fundamental evaluation standards. Local public agencies computerize not only spatial information, but other non-spatial data as well to improve their operation, managerment, and decision making. Hardware and software are tools to achieve these goals. The data, or the knowledge accumulated from data manipulation, however, is the most valuable treasure to the agencies. Therefore, the capabilities of data transferring from machine to machine, or system to system, is another key factor in the evaluation. Generally, factors for evaluating and implementing a microcomputer mapping system are:

Cost of the Hardware and the Software:
Total hardware, including computer, storage device, printer, digitizer, and plotter, should be under $10,000. Software cost should be at the range from $600 to $3000. Software may have two parts: Attribute data base management($300 - $800) and cartographic data base management($300 - $2,000).

Portability and Compatibility:
Revolutions of revising microcomputers' hardware and software occur frequently. The system should be easily updated and upgraded without losing data, delaying users' project schedule or paying great expenses on new hardware, software and personnel training.

Abilities of Downloading Data from Large Computers:
There are many geograpgic and cartographic data files maintained on mainframe or mini computers by US Bureau of Census, USGS, etc. Intracting those files, or part of the files, to a microcomputer mapping system is essential because the redundancy of data gathering, editing, and error checking is expensive and not necessary.

Abilities of Interfacing with Other Systems:
Cartographic data and attribute data are usually maintained independently. Many attribute data base management systems are marketed with great sophistication. A mapping system can become more "knowledgeable" if interfacing with a good attribute data base management system.

Accuracy of the Mapping Data:
Mapping for different applications needs different level of accuracy. A planner's map needs moderate levels of accuracy because the maps are presenting spatial relationships or a comprehensive view of the spatial characteristics of physical elements. An assessor's map has legal responsibilities, so it needs high accuracy corresponding to field survey results and other legal statements. Utility mapping, which affects engineering operation, construction, maintenance, and capital investment, needs the highest accuracy in the mapping systems.

Storage Space:
Storage space is critical when mapping a complex area. The vast volume of cartographic data may consume the secondary storage very fast. The experiment project used 450 bytes to a polygon averagely. A 15M byte hard disk can store approximately 34,000 similar polygons.

Ease of Learnning and Implementing:
Ease of learnning and implementing in terms of map input, editing, labeling, scaling, windowing, and overlaying is important for non-technical personnel to operate the system.

CONCLUSIONS

Microcomputer mapping systems have rapidly expanded their horizon and become more sophisticated as new technologies are introduced everyday. Technically, computing speed, computer graphic algorithms, data retrieval algorithms, and adequate graphic output devices are key elements that can be improved significantly. It is important for users to keep up with the trend. Microcomputer is a tool for self-training and self-education. Realizing the purposes of using computers and computer maps, practically, is the first and foremost consideration when implementing or choosing a system. If local officials are using computer mapping for planning, operation management, and policy stratification, it is worthwhile implementing or using a microcomputer based mapping system in their daily operations.

REFERENCES

Costello, M. and Garvin, L., 1982, A Map-based Information System to Support Reappraisal: 48th International Conference on Assessment Administration, The International Association of Assessing Officers, Chicago, IL.

Goldburg, M. and Chinloy, P., 1984, Urban Land Economics, pp. 140-142 & 198-200, John Wiley & Sons, Inc., New York.

Lam, H.S.A., 1984, Geographic Information Systems (GIS) and Spatial Analyses in Microcomputers: The Changing Role of Computers in Public Agencies, Urban and Regional Information Systems Association, Washington D.C.

Monmonier, M.S., 1982, Computer-Assisted Cartography Principles and Prospects, pp. 45-99, Prentice-Hall, Inc., Englewood Cliffs, NJ.

National Research Council (U.S.), Panel on a Multipurpose Cadastre, 1983, Procedures and Standards for a Multipurpose Cadastre, pp. 66-84, National Academy Press, Washington, D.C.

Peucker, T.K., 1972, Computer Cartography, Association of American Geographer, Washington D.C.

U.S. Department of Commerce, 1980, State and Local Ratio Studies, Property Tax, Assessment, and Transfer Taxes: State and Local Government Special Studies No.99, pp. 1-3, U.S. Dept. of Commerce, Bureau of Census, Washington D.C.

PATTERN RECOGNITION PROCEDURES FOR AUTOMATIC
DIGITIZING OF CADASTRAL MAPS

Werner Lichtner
University of Hanover
FRG

ABSTRACT

The author reports about the necessity of pattern recognition
software for automatic digitization of cadastral maps by
scanners in the FRG. This software has to support the feature-
tagging after the raster-vector-conversion. He describes own
software developements for this purpose briefly and two examples of successful algorithms.

INTRODUCTION

The new edition of cadastral maps and the establishment of a
so called "Automated Cadastral Map" are two of the main tasks
of german cadastral authorities. There do exist different
methods to get digital data of the features represented in the
cadastral maps. Some federal states take the measurement elements to calculate point coordinates, some states digitize
the maps only and transform the table coordinates with the
aid of fix-points into the official coordinate system. In the
case of old cadastral maps based on a bad geodectic network
the authorities digitize the cartographic features in the map
and mark them for the storage in a data bank as graphic data.
After updating the geodectic network and the measurement elements the marked coordinates will be replaced by calculated
coordinates. So for both tasks it is necessary to digitize a
lot of cadastarl maps.

BRIEF EVALUATION OF ALTERNATIVE METHODS
OF DIGITIZING CADASTRAL MAPS

The evaluation of the potential of certain digitizer systems
for cartographic purposes depends on the data form, in which
the geometry of a map has to be represented digitally. For
cadastral purposes vector-data are preferred. With vector-data
you can get a data structure by tagging the cartographic
features with special codes and the data are stored in a feature
selected way. These facts facilitate the storage, updating and
selection of features in a data bank and support the application
of feature related user software.
In the Federal Republik of Germany(FRG) for the digitization of
cadastral maps table digitizer are used generally. This method
is economic for that special purpose today/LICHTNER 1984/. In
relation to semi-automatic and automatic digitizer systems the
amount of investment is low and acceptable for many small
engineering offices. So you can divide and distribute the work
of a digitization program to several offices.
Semi-automatic systems like line-followers offer real advantages
only in the case of isoline representations. Scanner systems

produce raster data. Therefore a raster-vector-conversion is
necessary.For that purpose theredo exist several software systems
which generally produce in a batch-process a line- and node-file
first. Then in a second step the feature-tagging is realized in
a very time-consuming procedure at an interactive graphic working
station by an operator. This method is not economic for cadastral
maps today, but author believes that new software developements
will offer more capabilities with this technique in the future.
So his Institute of Cartography at Hanover University is engaged
in the developement of software to support the application of
raster scanners and digital image processing systems for cartographic purposes. In the discussed case of digitizing cadastral
maps software is desired to support the feature tagging after the
raster-vector-conversion with the aid of pattern recognition procedures.

PRACTICAL APPLICATIONS OF PATTERN RECOGNITION PROCEDURES

General. Fig.1 demonstrates the typical graphic of a german
cadstral map. Each corner of a real estate is marked by a circle
feature and the building areas are structured by a system of
parallel lines. The parcels are coded with parcel numbers.

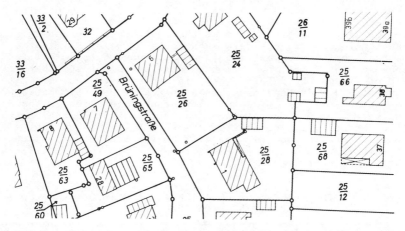

Figure 1: Typical graphic of a german cadastral map.

A handicap for pattern recognition procedures is the fact that
the graphic of old cadastral maps is not of the best quality
and you have to accept gaps in lines and damaged house-corner
pattern.
In the first step the image(cadastral map) is scanned in order
to get an image matrix. With the aid of a threshold operation
the grey value matrix of the image is transformed into a binary
matrix. Before starting the raster-vector-conversion it is advantageous to preprocess the binary matrix in order to reduce
noise. For that purpose we use non-linear operations(mask operations)/NIEMANN 1974, PAVLIDIS 1977, PRATT 1983/ to delete
noise in a binary pattern. We delete noise in a size of one
pixel only.
For the raster-vector-conversion our own software developement
is used. The skeletonizing is realized by the topologic method

combined with the information from a distance matrix. The result
is a line-and node-file extracted from a binary skeleton matrix.
At this point existing pattern recognition software generally
starts with the preprocessing, the investigation of topologic
and geometric characteristics of the vectors and the detection
of certain cartographic features by statistical classification
methods. If we scanned cadastral maps of a good graphic quality
this procedure was very successful in our tests. But the reason
of the new edition of cadastral maps is their bad graphic quality.
That means the pattern recognition procedures have to work with
damaged and destroyed topologic line structures. That is the main
reason of unsatisfactory results in the case of old maps with
a bad graphic quality.
So we modified the detection procedure for certain features,
which have been very sensible of a damaged topologic line structure. A typical example is the circle feature, which appears in
mentioned maps very often. A safe detection of this feature is
very helpful for the structuring of the real estate borderlines.
It is not the intension of the author to describe all pattern
recognition procedures of his software system. but two examples
may demonstrate two typical problems and tasks.

Example: Detection of circle features.
In the case of damaged circle features(gaps in the circle line)
the pattern classification of vectors is very sensible. So we
tried to get better results by the application of a correlation
technique in the binary matrix before skeletonizing. With the aid
of map design description you can define an ideal pattern $\underline{s}$ of
a certain feature. $\underline{f}$ is an existing pattern in the binary matrix.
For getting characteristics we calculate the correlation between
$\underline{f}$ and the ideal pattern $\underline{s}$. If the ideal pattern $\underline{s}$ is in $\underline{f}$ we will
get a top of correlation at pixel (i,j).

$$h_{i,j} = \sum_{k=0}^{k_o-1} \sum_{l=0}^{l_o-1} f_{i+k,j+l} \, s_{k,l} \qquad (1)$$

If the value h at pixel (i,j) is more than a certain threshold
we can classify the pattern.
With this simple correlation technique we detected about 91% of
all circle features(incl.the damaged ones) in our tested old
cadastral maps. With the classification technique for vectors
we got 60-65% only. The correlation technique is not so sensible
to gaps in circle lines and that is very useful in our special
case.
The reason of 9% non-classification was not a damaged circle
pattern, but the influence of the neighbouring cartographic
features. So we looked for a modification to reduce this influence. The developed method is very practical and successful.
In a first step we calculate a difference mask $\underline{d}$:

$$\underline{d} = \underline{s} - \underline{f} \qquad (2)$$

The mask $\underline{d}$ is the multiplied with a special weight mask $\underline{q}$ and
the weighted values of all pixels are summed up (see equ.3).
The comparision if $t_{i,j}$ with a threshold decide in the classification process.

$$t_{i,j} = \sum_{m=0}^{m_0} \sum_{n=0}^{n_0} d_{i+m} \, g_{j+n} \qquad (3)$$

You need a little bit more computer time for the described modification, but with a good definition of g for a special feature this algorithm is very safe. In tests with four cadastral maps and with about 35% damaged circle features we got a 100% correct classification for this symbol. It was possible to differentiate the circle pattern from number characters like "2","3","6","8" and "9" very safe. So we got a successful support for the structuring and further classifications later on.

Example: Detection of house corners.
The classification of pattern in the binary matrix is very useful only in certain cases as mentioned before. For other features (i.e.buildings) we prefer the pattern classification with vectors.
The graphic representation of a building has two typical characteristics. There don't exist circle symbols at the corner-points of the building outline and the area of a building feature has a texture of parallel lines (see Fig.1). If the outline of a building feature is classified you need to detect the house-corners on this outline. For this task we developed the following procedure.
The extracted outline of the building is a skeleton of the representation of a polygon line. For the detection of the polygon-points(building corners) it is important and helpful to find first one corner point of the building outline. This point can be found easily by determing the line-point with the longest distance to a free selected starting point(any point of the outline). This so found line point is always a corner point of the building outline. You can repeat this procedure-if you want- and can find a second corner by using this procedure again. A repetition is impossible if the starting point was already a corner point. In this case you will find your first corner point again.
After the detection of a first corner point you can continue with a subroutine for the determination of straight lines in the building outline. Starting with the first corner point and the next point on the extracted outline the parameters of a straight line equation are computed. If the distance between the line and a new point is smaller than a threshold, this point is used to compute an updated line; otherwise a new line is introduced.

CONCLUSION

The mentioned software system is in the developement and there do exist a test version only for optimazing the used algorithms and procedures. First tests with four old cadastral maps have been very successful. The next task is the modification of the test version into a software package for users at cadastral authorities.

REFERENCES

LICHTNER,W.   1984   Investigations and Experiences on Automatic
                      Digitization of Cadastral Maps.
                      Invited Paper,EURO-CARTO III,Graz 1984

NIEMANN,H,   1974   Methoden der Mustererkennung.
                      Akad.Verlagsgesell., Frankfurt a.M. 1974

PAVLIDIS,T   1977   Structural Pattern Recognition.
                      Springer Verlag, Berlin-Heidelberg-NY, 1977

PRATT,W.K.   1983   Digital Image Processing.
                      Wiley-Interscience Publ., New York 1983

Summary of proposed paper for AUTO-CARTO 7,
Washington, D.C. March, 1985

Contact: Dr. Joseph C. Loon
Dept. of Geodetic Science and Surveying
The Ohio State University
1958 Neil Ave.,
Columbus, Ohio 43210-1247

Tel. Work: 614-422-6753
Home: 614-235-4741

Title: " Evaluation procedures for testing contour-to-grid interpolation methods using synthetic surfaces"
Authors: Joseph C. Loon
Petros G. Patias
The Ohio State University
Betty A. Mandel
Engineer Topographic Laboratories

This study deals with ways for the evaluation of contour-to-grid interpolation algorithms. The basic scenario is as follows : One starts off with terrain contours and then digitizes these contours. From the digitized contours, a Digital Elevation Model (DEM) is to be constructed using some particular interpolation method. One now needs to know how accurate this derived DEM is. This can be done if true elevations are known at points where interpolation takes place. A useful tool for an investigation of this type is a synthetic surface. This is defined as a surface of the form $Z=f(X,Y)$. That is, we have an exact equation whereby we can calculate an elevation (Z) for each X, Y coordinate. This value can then be compared with the interpolated value and the true error will be known.

A method is described for deriving a synthetic surface from a portion of a topographic map. Although the resulting surface map may not look like the original topographic map, it contains much of the original characteristics. Six synthetic surfaces, of differing degrees of complexity, were derived. A set of contours for each surface was generated in raster format. A DEM was derived from each set of contours using three interpolation methods. These methods were: cubic interpolation along the steepest slope; linear interpolation in the direction of steepest slope; and linear interpolation parallel to the X and Y axes. The interpolated grid values of the elevations were compared with the true values (calculated from the synthetic surface equation) and the following measures were then obtained for each case : mean error ; maximum absolute error ; and standard deviation of the error. In addition, residual maps were constructed for each case. To obtain a better understanding of the practical significance of these errors, for each case a map was plotted showing the original (true) contours and the contours as plotted from the derived DEM.

An analysis of the results obtained is given.

We have also found indications that, although the synthetic surface does not look like a "real" topographic surface, the errors obtained with synthetic surfaces are larger than one would get if a "topographic synthetic surface" was used.

CONFLATION: AUTOMATED MAP COMPILATION
—A VIDEO GAME APPROACH—

Maureen P. Lynch and Alan J. Saalfeld
Statistical Research Division
Bureau of the Census
Washington, DC 20233

ABSTRACT

An interactive system for semi-automatic map matching/merging (conflation) is being developed at the Census Bureau. By means of that system an operator matches together features of two maps on the terminal screen. Several features of video games facilitate the operator's interaction with the screen image. Those features include movement simulation (rubber-sheeting), cross-hair targeting and pointing, color coding to identify status, flashing highlights, system prompts, system constraints to permit only "legitimate" moves, and a screen-clearing (partial image removal) strategy. Other game-like capabilities include preview and review of operations.

Full automation of map merging appears ever more feasible as testing continues with the prototype conflation system. Plans are underway to conflate United States Geological Survey maps and Census Bureau maps of all jointly mapped regions in the entire United States.

THE SHORT HISTORY OF MAP CONFLATION

The word "conflation" (from the Latin **con flare** meaning "blow together") was coined for use with automated cartography in 1979 by the wife of James Corbett, the father of the Census Bureau's DIME files. "Conflation," many cartographers were surprised to discover, is a legitimate word traditionally used to describe the merging of two manuscripts into a third combined version. Most dictionary references to conflation mention the Bible, an all-time favorite text for conflating. Conflation of maps refers to a combining of two digital map files to produce a third map file which is "better" than each of the component source maps.

Much of the initial theorizing about the proper way to conflate maps centered on the subjective definition of the "best" features of each individual map and how to extract them and recombine them into a unique improved map. Many arguments arose about what constituted a match of map features; and counter-examples were presented to every conflation approach proposed. In the purely abstract and theoretical framework of philosophizing about conflation, too many alternatives precluded the possibility of a global theory; and, hence, for a long time there was no theory at all, only assorted ideas for handling specific cases. Map matching theory, or map conflation, really began developing in the course of a concrete experiment undertaken at the Census Bureau in 1984. Only when real digital files were compared could the experimenters begin to measure what would work and what would not work in terms of merging the two files into one.

The Census Bureau's interest in implementing some type of conflation system arose when the United States Geological Survey (USGS) agreed to provide laser-scanned line files from their 1:100,000 scale cartographic data base. The Census Bureau currently has a set of hand digitized maps covering approximately 5 percent of the land area of the United States and 60 percent of the country's population. The Census Bureau's Statistical Research Division began implementing some newly developed rubber-sheeting techniques on a high-resolution color graphics terminal to begin an experiment attempting to conflate the maps from the USGS and the Census Bureau.

As the experiment got underway, the research team decided to limit the scope of the file matching and merging task. Their goal was not going to be total conflation. Several factors entered into that decision, probably the greatest of which was that conflation in its broadest sense did not seem possible, whereas several subtasks undertaken did produce tangible and useful results. Moreover, the subtasks moved the Census Bureau closer to its goal of facilitating the GBF/DIME file updating/correction procedures. The final product of the current conflation effort will not be a unique third "better" map; however, it will be two "improved" map files which are linked in many ways, and which could be used to create one or more "better" maps at some point or to transfer feature-based information from one map to the other.

SCOPE AND OBJECTIVES OF THE JOINT USGS/CENSUS PROJECT

Although total conflation (one new improved map) was not the objective, information transfer between files remained an important goal. The Geological Survey had made their digital files available to the Census Bureau with the understanding that the Bureau would tag the features in exchange for the use of the files. Tagging involved two levels of information transfer: tagging features by type, such as "class I roads", and tagging features by street name and house number ranges. Tagging by type is being accomplished by a highly interactive manual screen-based graphics system. Adding names and address ranges is more complex because of the quantity and quality of matches required. It requires a preliminary step of matching enormous numbers of linear features in a **consistent** and **comprehensive** manner to prevent duplicates and missed coverage. That preliminary matching step will be one of the products of the present limited conflation package being developed and used at the Census Bureau.

In addition to providing the USGS with information, the Census Bureau also expects to benefit from using the USGS files. The Census Bureau wants to compare the two digital representations of the same area to find discrepancies that would permit the Bureau to focus on possible recent housing development or on map drawing errors of one file or the other. In addition, the Census Bureau wants to improve appearance of its maps. The rendering standards used to create the USGS files give USGS maps an appearance far superior to the Census Bureau's manually digitized maps. The Census Bureau wants to borrow that appearance if possible, to straighten the lines, smooth the curves, and thereby gain publication quality maps in place of the "stick maps" that are currently produced from DIME or DIME-like files.

The appearance enhancement of Census maps will be accomplished by rubber-sheeting Census maps to align them with the USGS maps. Rubber-sheeting involves moving some of the Census map's points to align exactly with corresponding points on the USGS map. The remaining Census map points are then moved in a proportional fashion so that the relative positions of all of the Census map's points are not changed. The Census map is shifted and stretched to fit the USGS map. More points are then aligned and the process of rubber-sheeting continues until all features that can be identified are identified.

During the rubber-sheeting of the Census map, the USGS map does not move. In some sense the USGS map behaves as the underlying "truth" to which we fit our map. The USGS map does not move because to distort it would destroy the straightness of lines and smoothness of the curves, which we have already recognized as important to preserve. One conceivably could move both maps some average distance to bring them together at points that can be matched; however, the matching of features would not be improved and both maps would lose their original shapes.

Another reason to leave the USGS coordinates alone and change only the coordinates of the Census map points is the Census Bureau's promise to **add** information to the USGS digital file and to change only errors of an agreed upon type and level on the USGS files.

The Census Bureau's more limited objectives for the conflation system may be summarized as follows:

1. Improve coordinates of the Census Bureau's Geographic Base File by exactly adopting the USGS coordinates wherever possible, which may include the addition of points that define shape.

2. Establish match flags and pointers between Census map files and USGS map files for all line and point features that can be paired on the two files.

3. Establish non-match flags for those line and points features which appear on one map and not the other.

4. Create maps which highlight matches and non-matches in a manner which facilitates secondary verification of controversial features.

The Census Bureau will create a new set of coordinates for all of the point features on its maps. The Census maps eventually may be drawn with either the new or the old coordinates. Additional information will be available at the end of the conflation process; no old information will be destroyed or overwritten.

Even with the new coordinates the Census maps will maintain their topological integrity. The whole point of triangulating and rubber-sheeting is not to alter the relative positions of map features. If a Census map requires correction of topological relations, that adjustment must occur elsewhere.

Rubber-sheeting of the Census maps is accomplished by computationally fast piecewise-linear local transformations defined on a triangulation of the map space. Coordinates of 0-cells (intersections) are recomputed and 1-cells (streets) are redrawn. Rubber-sheeting is applied iteratively to increase the number of map points aligned at each step. Each rubber-sheeting step leaves Census map features in the same relative positions but more closely aligned to the USGS map.

EVOLUTION OF THE COMPUTERIZED ALIGNMENT SYSTEM

The iterative process of map alignment was initially designed to be highly manual because of the expected difficulty in deciding what to match. An operator was expected to apply complex decision rules to a large variety of situations. The initial design called for real-time operator interaction with a color graphics screen displaying the features of each of the maps. That design produced a video game type of situation. The similarity to video games was further exploited by adding system features which rewarded or reinforced correct choices by the operator via the screen display. The operator could be challenged to move quickly and accurately and could see visible results of moves made.

Recent experience with real maps has pointed to the feasibility of a more fully automated approach to matching, and such an approach is underway. The situations and decisions facing an operator are not as complex in general as was first imagined. The initial fears that a fully automated rubber-sheeting system would completely distort the Census map have been allayed. A stepwise iterative alignment procedure now matches a few points based on very strict criteria on the first pass, then relaxes match criteria with each successive iteration to match additional points. A first-pass match of a few points gives, in general, a very good initial alignment of the maps, and after an initial alignment of maps is accomplished, future distortions may be avoided entirely by simply not allowing large point movements.

The title of this monograph reflects the initial idea of endowing the sytem with powerful manual tools for the operator. Experience suggests that manual intervention is not needed to the degree originally believed. The initial prototype system was almost entirely manual which tested the operator's skill and speed at making matches of point features of the two maps. Almost immediately it was realized that the speed of a purely manual operation would never accommodate even a small fraction of the large quantity of maps that must be processed. With sufficient machine guidance, however, the operator could be led to a significant number of error-free matches as experienced in our test cases. In later prototype systems, instead of leading the operator to error-free matches, the matches were made automatically, in batch, without operator intervention. To complete the analogy, the second prototype system was like the video game "Space Invaders" with an automatic pilot and an automatic "aim-and-fire" added to the operator's weapons. The operator merely presses the "start" button and sits back. Such an automatic system may take the fun out of the game by fighting all of the battles alone; nevertheless, it plays consistently and never loses!

BASIC CHARACTERISTICS OF THE SYSTEM

The following properties of the map alignment/matching system were in the initial design. The system is:

(1) **Interactive.** A menu of commands allows the operator to make successive changes to the Census GBF map to improve its alignment with the USGS map and to select matching or non-matching features on the two maps.

(2) **Iterative.** By applying a fixed procedure repeatedly, the operator improves the number of match and non-match classifications at each stage.

(3) **Semi-automatic.** The machine applies topological and geometric tests after each iteration by the operator and then recommends, based on the results on those tests, additional matches or identifications. The operator then accepts or rejects all or some of the proposed automatic changes or makes additional manual changes.

(4) **Screen based.** "Screen based" means that the map image(s) on the screen are the principal (at times the only) data source used by the operator. The manipulation and selection of those images by the operator using highlighting, panning, zooming, and cursor positioning are the only interaction required. The computer programs take care of updating the data base accordingly. The programs also provide the tools for operating most efficiently and successfully with the screen image, including highlighting and prompting.

(5) **Overlay oriented.** This means that the matches depend on the ability of the operator and the system to position a feature of the

Census map directly above a feature of the USGS map; and, after such positioning is accomplished, the features remain coincident from that point on in the iterative procedure.

(6) **Full color coding.** Color is used to identify each of the map sources the node and segment matches and unmatchables, triangle areas, and feature types. By utilizing color, each of the distinct map components is made clearly distinguishable.

(7) **Simplistic design.** The overlay of maps are shown as simply as possible without any clutter for the easy identification of matching nodes. More details are available on request by the operator, including the highlighting of 0-cells or 1-cells, the display of previously matched or unmatched segments, and the display of statistical information that is available from the files.

(8) **Menu/Window driven.** Part of the display screen is reserved for the continuous display of available operator commands. (A tablet and mouse may be used instead to free more screen area.) A series of menus may be used to give the operator options for selecting a particular conflation process.

(9) **Subtractive reduction.** A procedure which tags and can (temporarily) remove from display items which have been identified as matches (or as non-matchable) facilitates the matching process. Such a procedure terminates allows the operator to "clear the screen" in order to focus attention on unresolved sections of the maps.

The following even more specialized capabilities of the system were added to improve efficiency and accuracy:

(10) **Subset selection and highlighting.** The operator must be able to find and see clearly the features on each map. Flashing displays and highly contrasting colors are employed to view such subsets as matched points or segments in a small neighborhood of a point, block boundaries, non-street features, nearest matched pair, and other subsets.

(11) **Separate image maintenance of each map with identifying links.** Each map must maintain its separate identity even as its features are transformed onto those of the other map. Image decomposition may be necessary to accomplish the identifications. Unmatching (correcting mistaken matches) is possible with this approach.

(12) **Nearest point subroutines to facilitate identifications.** Many of the search procedures used in later stages of matching are based on proximity of features. Effective local searching in specified subsets speeds nearest neighbor searches.

(13) **Geometric/topological editing to prompt likely matches.** A series of tests for similarity of map features assist the operator to make match/non-match selections. Test results are used to prompt the operator or to make a match decision that the operator need only accept or reject.

(14) **Topological edits to prevent unreasonable manual identification.** Unless specifically over-ridden, the system does not permit the operator to nail down two points which are grossly different in terms of their topological and geometric make-up.

(15) **Local screen rubber-sheeting and editing.** In conjunction with a zooming capability, the manual identification procedure should be able to operate locally. Changes to the image in the window (including accepting/rejecting the look-ahead automatic alignment) need not affect the rest of the image. Special software for windowing and subfile creation permits the operator to process the map locally.

(16) **Variable (adjustable) tolerance settings for automatic look-ahead alignment.** Points may be close in terms of distance or some topological measure; segments may be close in terms of direction. Because there are no a priori rules for determining how close is close enough and there are no summary results available on how close is typical or reasonable, variable tolerance settings permit different acceptance levels of matches as the process is iterated.

ILLUSTRATIONS OF THE CONFLATION PROCESS

The technical details of the process are presented in several other papers included in the **AUTOCARTO 7 Proceedings**; and the mathematical theory behind the process is described in detail in those papers and in the references given in this paper's bibliography. This overview of the system will conclude with several illustrations of the stages of alignment. Colors are used on the screen display to differentiate line types and node types. The illustrations contained herein differentiate lines by type: solids, dots, and dashes; and differentiate points by different point symbols. All of the maps shown in the illustrations below are of Fort Myers, Florida.

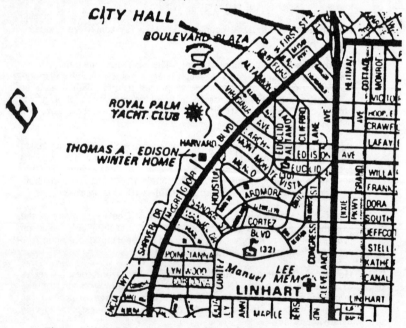

Figure 1. Road Map of Area South of Caloosahatchee Bridge.

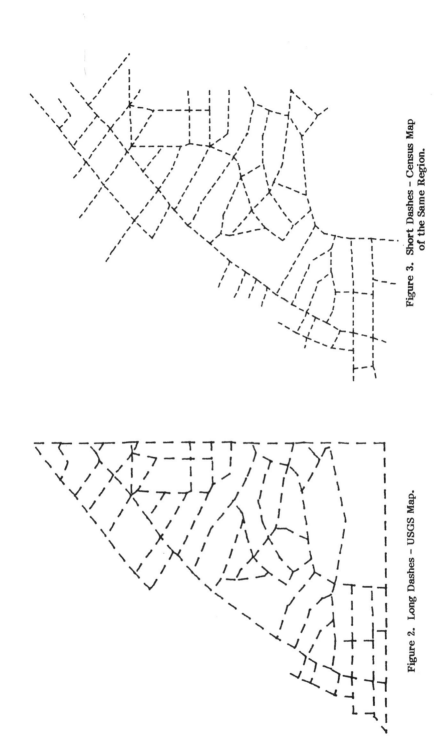

Figure 3. Short Dashes - Census Map of the Same Region.

Figure 2. Long Dashes - USGS Map.

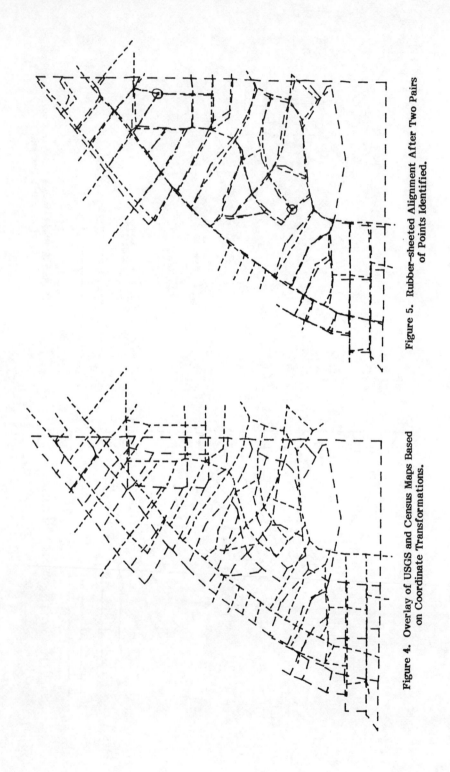

Figure 5. Rubber-sheeted Alignment After Two Pairs of Points Identified.

Figure 4. Overlay of USGS and Census Maps Based on Coordinate Transformations.

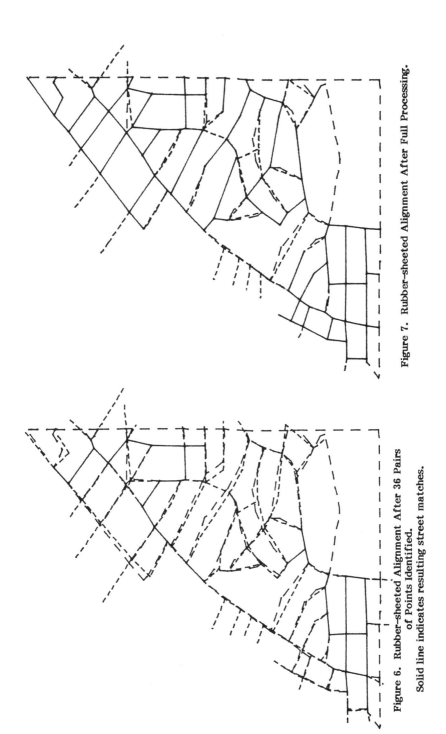

Figure 7. Rubber-sheeted Alignment After Full Processing.

Figure 6. Rubber-sheeted Alignment After 36 Pairs of Points Identified. Solid line indicates resulting street matches.

ONGOING RESEARCH AND DEVELOPMENT IN CONFLATION

The principal focus of research in conflation at this time at the Bureau of the Census is in the area of map file storage and manipulation. The large data sets involved with digitized map files must be organized to interact with local rubber-sheeting routines of the prototype system. Data structures will be made compatible with windowing features so that appropriate data subsets are automatically selected when working windows are chosen by the operators. In principle, the entire map, whatever its size, will be processed as a whole, although local routines will be applied to chosen subsets while the remainder of the map is simply "fixed by the identity transformation."

The working prototype programs utilize arrays to store point and line information; and the size limits of those storage methods are currently 400 to 500 points and 400 to 500 line segments per map. It has been observed that a screen image of approximately 250 line segments and 150 intersection points per map approaches the limits of managability both in terms of the limits of screen resolution and operator visual ability to discriminate features without a great deal of magnification.

Because the expected work unit will be a managable screen image of approximately 400 total line and point features per map, the existing rubber-sheeting routines in the prototype package will not require rewriting for processing larger maps. Additional refinements of the triangulation procedures will be necessary to interact with local editing and local windowing, and those refinements are part of the current research plan.

BIBLIOGRAPHY

Gillman, D., 1985, "Triangulations for Rubber Sheeting," **AUTOCARTO 7 Proceedings.**

Griffin, P., and M. White, 1985, "Piecewise Linear Rubber-sheet Map Transformations," to appear in October issue of The American Cartographer.

Keates, J.S., 1973, Cartographic Design and Production, New York, NY, Halsted Press.

Murch, G.M., 1984, "Physiological Principles for the Effective Use of Color," IEEE Computer Graphics and Applications, 4(11), pp. 49-55.

Rosen, B., and A. Saalfeld, "Matching Criteria for Automatic Alignment," **AUTOCARTO 7 Proceedings.**

Saalfeld, A., 1985, "A Fast Rubber-Sheeting Transformation Using Simplicial Coordinates," to appear in the October issue of The American Cartographer.

U.S. Geological Survey/Bureau of the Census, 1983, "Memo of Understanding for the Development of a National 1:100,000 Scale Digital Cartographic Data Base," Washington, DC.

White, M., 1981, "The Theory of Geographical Data Conflation," internal Census Bureau document.

DEVELOPMENT OF GEOGRAPHIC INFORMATION SYSTEMS
FOR POORLY MAPPED AREAS OF THE WORLD

Barry D. MacRae, Research Engineer
Charles L. Wilson, Manager - Earth Resources Data Center
Robert H. Dye, Senior Research Engineer

Environmental Research Institute of Michigan
P.O. Box 8618
Ann Arbor, Michigan 48107

(313) 994-1200

Mr. MacRae has been working for more than twelve years in the processing of multispectral data for use in earth resources analysis and applications studies. Formerly with Bendix Aerospace Systems Division where he managed and participated in the software development program for the Bendix M-DAS system. Conducted training programs in use of image analysis systems in Japan, Egypt, India, Argentina, and Brazil. Past responsibilities have included implementation of a geographic information system for the Government of Peru to provide land use/resource management assessment and management capability for the entire country.

ABSTRACT

Recent advances in both hardware and software have opened up an exciting new area for the application of GIS technology - land assessment and resource management for areas of the world where current cartographic information does not exist or is inadequate.

Through the use of satellite doppler receivers and state of the art geometric correction software, Landsat satellite data can be used to generate planimetric base maps and digital data bases meeting national map accuracy standards at scales of 1:200,000 or smaller.

The consequence for GIS technology from this ability to generate planimetric maps utilizing Landsat data is the capability to provide land resource managers with a tool to assess and control future resource utilization in areas where no viable alternative exists. The Landsat derived map becomes the base map onto which the manually interpreted data is transferred. In addition, computer-assisted land cover classification of the Landsat data may be merged with the manually interpreted data in the GIS data base.

The new capability means that third world or underdeveloped countries may generate a set of maps conforming to national map standards, produce digital data bases corresponding to the map format, and produce land assessment and resource management decisions which may be referenced directly to the national map series. This ability to achieve GIS technology transfer to nations which are very much in need of the technology for a relatively low cost in a short time-frame is a most exciting aspect of the GIS future potential.

APPROACHES FOR QUADTREE-BASED GEOGRAPHIC INFORMATION
SYSTEMS AT CONTINENTAL OR GLOBAL SCALES

David M. Mark and Jean Paul Lauzon
State University of New York at Buffalo
Department of Geography
Buffalo, New York 14260

ABSTRACT

Several alternative strategies for defining quadtrees and related recursive tesselations over the sphere are examined as possible bases for a global, quadtree-based geographic information system (GIS). We propose that the UTM coordinate system provides the most appropriate basis for such a GIS. The general structure of such a GIS is described.

INTRODUCTION

Quadtrees are an effective way to store grid-cell based information, including geographic data (Gargantini, 1982; Lauzon, 1983; Abel, 1984; Mark and Lauzon, 1984; Samet and others, 1984; Lauzon and others, in press). Briefly, a quadtree first encloses the area of interest within a square, and subdivides the square into for quadrants. Each quadrant is then recursively subdivided into subquadrants until all subquadrants are uniform with respect to image value, or until some predetermined lower level of resolution is reached. With their variable resolution and natural subdivision into hierarchical patches, quadtrees are ideal for handling very large geographic areas. The papers mentioned above all describe _linear_ quadtrees. In such structures, relations among quadrants (nodes in the tree) are indicated not by pointers but rather by linear key numbers based on an ordered list of the node's ancestors.

Because quadtrees are based on square or rectangular grid cells, difficulties may be encountered when the concept is applied to large areas of the Earth's surface (continental scale or larger), since squares cannot tesselate a sphere. The purpose of this paper is to review previous work on this subject, and then to outline a proposed GIS system based on the Universal Transverse Mercator (UTM) coordinate system.

PREVIOUS WORK ON HIERARCHICAL PARTITIONING SYSTEMS
FOR LARGE AREAS

Quadrangle-based Systems
One class of solutions to the problem involves partitioning the globe into areas which may be termed _quadrangles_, cells which are 'square' in latitude-longitude terms. Such systems have a variety of analysis and display problems, since cells are not geometrically square, and furthermore change size and shape with latitude. However, the frequent

use of latitude-longitude quadrangles for non-computerized mapping makes them attractive, since printed maps are often used as a primary data source for GIS. Two such systems are reviewed briefly below.

Canada Geographic Information System and Related Structures. A good example of a quadrangle-based GIS which is hierarchical in the quadtree sense (but only down to a certain scale) is the Canada Geographic Information System (Tomlinson and others, 1976; Comeau, 1981). The Canada Geographic Information System was the first full geographic information system. Interestingly, it used what are now seen as quadtree concepts a decade *before* quadtrees were formally developed as a data structure.

A coordinate reference system suitable for the project had to meet the following criteria (Comeau, 1981, p.2):

1) the system had to accomodate storage of a vast amount of mapped data covering all parts of Canada;

2) the system had to allow for a multitude of mapping scales, without loss of accuracy or the storage of redundant points;

3) the area and shape of the input regions had to be preserved; and

4) the system had to facilitate map to map processing and allow for retrieval of entire regions within the country.

The resulting design was a frame referencing system which allowed for frames of variable size so that the variable scale requirement was met. The storage sequence for the frames on sequential magnetic tape was developed by G.M. Morton and is fully explained in Morton (1966). Each frame is assigned a location key which maximizes the likelihood that its geographic neighbors are also its neighbors on magnetic tape. If each frame is given x (column) and y (row) coordinates, numbered from zero, then the Morton key of the frame can be obtained simply by interleaving the bits of the coordinates (see Morton, 1966; also Lauzon, 1983; Lauzon and others, in press). Within each frame, the hierarchical system is abandoned; data are stored as points and lines, with lines being represented by direction codes (see Tomlinson and others, 1976, p.50-55). Similar 'tiling' systems for geographic information systems are described by Cook (1979) and Weber (1979).

TOPOG. TOPOG is an example of a worldwide *topographic* data base which uses the quadrangle approach. This system was developed by the U.S. National Telecommunications and Information Administration (Department of Commerce; see Spies and Paulson, 1981). While TOPOG is not a quadtree system, it has many similarities to one.

TOPOG is a hierarchical system. The globe is divided into 36 TOPOG *zones*, each $5°$ of latitude in extent. These zones are then subdivided into a total 3060 units termed TOPOG *regions*. TOPOG regions are $3°$ of longitude wide between

$50°$ south and $50°$ north latitude, but are wider (in longitude degrees) toward the poles. Each TOPOG region is then subdivided into 15 TOPOG districts, in a 3 (east-west) by 5 (north-south) pattern. Note that in the mid-latitudes and tropics, these districts are $1°$ by $1°$ quadrangles. Next, each district is subdivided into 64 TOPOG blocks, as an 8 by 8 array of quadrangles. For the 48 contiguous states, as well as other mid-latitude and tropical areas, TOPOG blocks are identical to 7 1/2 minute quadrangles. The lowest level of the TOPOG system defines an array of 151 by 151 TOPOG points within each block. For the area between $50°$ north and south latitudes, these points are 3 seconds apart in both latitude and longitude. Elevations are stored for these points as integer meters above an arbitrary local datum selected to minimize storage requirements and stored at a higher level in the data structure.

In their introduction, Spies and Paulson (1981, p. 3-7) review some alternative terrain storage methods, but did not consider quadtrees, which were not well-known in the GIS and computer mapping literature when their report was being written. Although TOPOG is hierarchical, the hierarchy is irregular, with only one step (districts to blocks) being a power of two (8 by 8). Unfortunately, TOPOG thus missed the possibility of incorporating many of the advances in quadtree handling developed over the last 5 or so years.

Summary of quadrangle-based systems. Both the TOPOG system and CGIS partition the globe into quadrangles. In CGIS, the hierarchical arrangement is one of binary partition, that is, at each step, a cell is split into four quadrants. However, CGIS and the related systems described above partition the region down to some fairly large cell size, and then represent detail in vector form. For a pure quadtree system, the fact that quadrangles are neither square nor of equal size would cause problems for display. In the TOPOG system, partitioning continues down to indivisible grid cells, but the hierarchy is neither binary nor even regular. The problems associated with non-square cells are present, and the potential advantages of quadtree partitions are not available.

Polyhedral Systems
A second class of solutions represents the globe at the most generalized level as a simple polyhedron with triangular faces (a tetrahedron, octahedron, or icosahedron), and then hierarchically decomposes each triangular face into sub-faces (again, a tree, often of out-degree four). A coordinate notation for such a system begining with a tetrahedron was published more than a decade ago by Wickman (1973). Dutton's (1983, 1984) GEM system (Geodesic Elevation Model) begins with an octahedron. At each step in the hierarchy, each triangle is divided into nine faces. Elevation data can be stored in internal nodes of the tree as well as in leaf nodes, leading to efficient generalization. Polyhedral systems are highly appropriate for modelling the Earth and other planetary bodies as solids; their advantages over other

methods for mapping on the Earth's surface are less obvious.

Systems Based on Map Projections

The third class of solutions maps the globe onto a plane or set of planes, using some map projection, and then defines a grid cell network in cartesian coordinates on the plane(s). Since no map projection can be both equal-area and conformal, square cells on the map would represent areas in the real world which vary in size, shape, or both. Smyth Associates (1983) developed their own projection, mapping the Earth onto six squares which form a cube; quadtree decompositions were then applied to the faces of the cube. Since equations for map projections can easily be implemented on computers, and since the resulting cartesian coordinate space can be treated as a Euclidean plane for the quadtree, we believe that map projections systems are the most useful basis for a global quadtree-based GIS.

THE PROPOSED GLOBAL QUADTREE SYSTEM

System Overview

This section outlines a continental or world-wide scale GIS based on quadtree concepts and the Universal Transverse Mercator (UTM) coordinate system. In the proposed system, three hierarchical levels are used. The highest divides the world into UTM zones and subzones. Each UTM subzone is then divided into square patches, which are numbered according to the Morton sequence described above. Finally, within each patch, a 256 by 256 array of cells is the basis for the quadtrees or other geographic data files.

For each patch, a variety of types of data may exist. We propose to recognize four fundamental classes of data: points, lines, regions (coverage data), and surfaces. The highest level of the GIS would consist of a data base management system (DBMS) which would contain a directory of patches, data types, and data sets actually available, with summary statistics relating to the file contents. In fact, the _patches_ can be treated as pixels, and summary statistics can be mapped at this highly generalized level.

UTM Coordinates

The use of UTM coordinates is recommended because: conversion of geographic (latitude-longitude) coordinates to UTM is well known and computer programs or formulae are readily available; UTM coordinates are in general use by the armed forces of the United States, Canada, the United Kingdom, and other countries; and the U.S. Geological Survey (USGS) distributes digital data either already in UTM coordinates or with coefficient for conversion to UTM contained in the file headers (see Elassal and Caruso, 1983; Allder and Elassal, 1984).

The UTM system divides the globe into 60 UTM zones, each 6 degrees wide (east-west), which extend from 84 degrees north latitude to 80 degrees south, and are numbered from 1 to 60 eastward from the 180th meridian (see Figure 1).

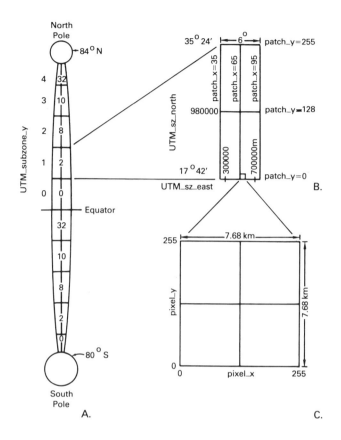

Figure 1: Relations among UTM zones (A), subzones (B), and patches (C), as well as pixels and coordinates from the proposed quadtree-based GIS.

Additional special UPS (Universal Polar Stereographic) zones are used to cover the polar regions. UTM coordinates are given in meters, and are unique within each zone. The central meridian of each zone is given an arbitrary east coordinate ('easting') of 500,000 meters; the equator is the zero point for north coordinates in the northern hemisphere, and is given an arbitrary value of 10 million meters for southern hemisphere 'northings'. It is important to note that the cartesian coordinate systems of adjacent zones are not alligned, except at the equator.

For the proposed system, a standard grid cell size (SGCS), in meters, must be established. Within each subzone, UTM_cell coordinates are obtained by dividing the UTM coordinates by the SGCS. For convenience of processing on a 16-bit processor, UTM_cell coordinates are represented by 16-bit unsigned integers. For the cell sizes anticipated, the north-south UTM_cell coordinate will often be greater than 65536 (2^{16}); we thus propose to divide the UTM zone into **subzones**. UTM subzone numbers are obtained by dividing the UTM coordinates by (65536 * SGCG), with truncation; then, the UTM_cell coordinates are (UTM / SGCG) modulo 65536. These UTM_cell coordinates are then split into two parts. The upper 8 bits define the **patch** x and y coordinates. These can be obtained by dividing the UTM_cell coordinates by 256, with truncation. Patch x and y coordinates are then combined into a **Morton number** for the patch (Tomlinson and others, 1976; Comeau, 1981; Lauzon, 1983; Lauzon and others, in press; see also above). Within each patch, individual cells or pixels are defined in **pixel coordinates**. Pixel coordinates are obtained from UTM_cell coordinates by taking the UTM_cell x or y modulo 256. Within-patch pixel coordinates may be combined to form a Morton number of the pixel. Figure 2 illustrates the relations among the patch and pixel coordinates and Morton numbers. Direct conversion of UTM_cell coordinates to patch_number and pixel_number can be readily accomplished in a computer language such C, in which individual bits can be addressed through the use of structures. Eventually, such procedure would probably be performed by specially designed chips.

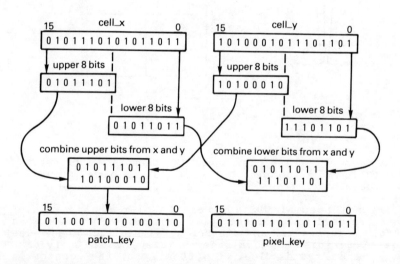

Figure 2: Relationships among the patch and pixel coordinates and Morton numbers, with emphasis on these at the bit level.

Given the UTM zone and UTM coordinates of a point, the following equations are used; for the proposed system for the United States, SGCS = 30 meters (see below):

UTM_subzone_x = floor (UTM_east / (65536*SGCS));

UTM_subzone_y = floor (UTM_north / (65536*SGCS));

UTM_subzone = morton (UTM_subzone_x,UTM_subzone_y);

UTM_sz_north = UTM_north modulo (65536*SGCS);

UTM_sz_east = UTM_east modulo (65536*SGCS);

UTM_cell_x = UTM_sz_east/SGCS;

UTM_cell_y = UTM_sz_north/SGCS;

patch_x = floor (UTM_cell_x / 256);

patch_y = floor (UTM_cell_y / 256);

patch_number = morton (patch_x,patch_y);

pixel_x = UTM_cell_x modulo 256;

pixel_y = UTM_cell_y modulo 256;

pixel_number = morton (pixel_x,pixel_y);

Any square patch in the world of size SGCS by SGCS is uniquely denoted by three or four numbers: the UTM_zone number (plus UTM_subzone if necessary, depending on the cell size), the patch_number, and the pixel_number.

Implementation for the United States

The U.S. Geological Survey is a major producer of digital cartographic data, an important element in a geographic information system. The USGS currently distributes two classes of cartographic products: Digital Line Graphs (DLGs; see Allder and Elassal, 1984)) and Digital Elevation Models (DEMs; see Elassal and Caruso, 1983). The former contain information on non-topographic line work from USGS 7 1/2 minute quadrangles; locations are presented in arbitrary cartesian coordinates, with coefficients contained in the file header which allow for conversion to Universal Transverse Mercator (UTM) or to State Plane coordinates. The DEMs are presented as regular square grids. Grid cells are 30 by 30 meters, and are allinged with the UTM coordinate system. It is therefore proposed that for the United States, an SGCS value of 30 meters be used.

Section Summary

In summary, three hierarchical levels are used. The highest divides the world into UTM zones and subzones. For the 30 meter cell implementation, each UTM subzone is divided into square patches of side-length 7680 meters (30 x 256), which are numbered according to the Morton sequence. Finally, within each patch, pixels are numbered in the Morton sequence.

DATA TYPES

For each patch, a variety of types of data may exist. It is proposed to recognize four fundamental classes of data, based mainly on their dimensionality. For each patch, several files may exist, each containing a different data type.

Type 0: Point Files

Point files would be held by Morton number. Point locations would be reported to the nearest pixel or grid cell (in this application, 30 meters), and converted to within-patch pixel coordinates. The coordinates would then be combined to form the Morton number (key) of the point. Each record in the key-ordered file would contain the key of the point and its attributes/values. It may be desirable to retain the pixel coordinates as data elements, although these can be obtained from the Morton number algorithmically. If points have a low spatial density, it may be advantageous to represent the points within a 2x2, 4x4, or higher-level node within the quadtree of the patches. This could be handled fairly readily within the proposed system.

Points would be held in key-order for ease of interfacing with other data types, and also because White (1983) has shown that this is an effective data structure for approaching closest point problems and other computational geometry questions.

Type 1: Line Files

It is well-known that quadtrees do not handle line data well. We suspect that it may be best to handle geographic line features as vectors or direction codes within patches, as has been implemented within CGIS and the systems described by Cook (1979) and Weber (1979). Note however that boundary lines would not be stored as line data; rather, the bounded regions are stored as coverage files (Type 2, below).

Weber (1979) in fact viewed his structure as a hybrid between vector and grid-based data structures which capitalizes on the advantages of each. Weber suggests working in one data structure, as long as it is 'economic', and transferring to the other structure once it is more 'economic'. The frame referencing is based on a gridded (in fact, quadtree-like) structure, while the details of a map are recorded in vector format.

Type 2: Coverage Files

Coverage data, such as land use, soils, or rock types, will be stored as linear quadtree files. In most if not all cases, two-dimensional run-encoding (2DRE) will be used as the file structure. This structure, developed by Lauzon (1983) and Lauzon and others (in press), run-encodes a linear (key-based) quadtree file. 2DRE files are an efficient way to store quadtrees, especially for GIS applications, such as overlay (Lauzon, 1983; Mark and

Lauzon, 1984; Lauzon and others, in press). Although 2DRE files are usually viewed as compacted linear quadtrees, they can be constructed simply by visiting all the pixels of an image in Morton order, storing the Morton key of the last pixel of a run of consecutive pixels of the same value (color). The Morton numbers thus form both a basis for subdivision into patches and a link between different data types.

Type 3: Surface Files
Quadtree representations are not efficient if neighboring cells seldom have identical values. Good examples would be a topographic surfaces (gridded digital elevation models) and LANDSAT or other MSS satellite imagery. For data such as these, it is more efficient to store a value for every grid cell. For ease of interfacing with other data types, the proposed system would use the Morton number (key) of each pixel as a virtual address for referencing the elevation within a contiguous binary file. The integration of the topographic component into a quadtree-based GIS is described in detail by Cebrian and others (this conference).

The System: Summary
For each patch (7680 by 7680 meters in this example), any number of files of one or more of the above file types might exist within the GIS. The upper level of the GIS would consist of a data base management system (DBMS), perhaps relational, which would contain a directory of patches, data types, data sets actually available, and summary statistics for each. Many GIS queries could be answered from data contained in this non-spatial data base. The overall strategy for the system is to examine the actual geographic data files only when necessary.

REFRENCES

Abel, D.J., 1984, A B^+-tree Structure for Large Quadtrees: Computer Vision, Graphics, and Image Processing, 27: 19-31.

Allder, W.R., and Elassal, A.A., 1984, Digital Line Graphs From 1:24,000-scale Maps: U.S. Geological Survey Circular 895-C, 79pp.

Cebrian, J.A., Mower, J.E., and Mark, D.M., 1985, Analysis and Display of Digital Elevation Models Within a Quadtree-based Geographic Information System: Proceedings, Auto-Carto VII (this conference).

Comeau, M. A., 1981, A Coordinate Reference System for Spatial Data Processing: Canada Land Data Systems Technical Bulletin No.3, Lands Directorate, Environment Canada, Ottawa.

Cook, B. G., 1979, The Structural and Algorithmic Basis of a Geographic Data Base: International Advanced Study Symposium on Topological Data Structures for Geographic Information Systems Vol. 4, Dedham, Mass., G. Dutton (ed.), Harvard University Laboratory for Computer Graphics and Spatial Analysis.

Elassal, A.A., and Caruso, V.M., 1983, Digital Elevation Models: U.S. Geological Survey Circular 895-B, 40pp.

Gargantini, I., 1982, An Effective Way to Represent Quadtrees: Communications of the ACM, December, 25, 12: 905-910.

Lauzon, J.P. 1983, Two-dimensional Run-encoding for Spatially Referenced Data: M.A. Project, Department of Geography, State University of New York at Buffalo.

Lauzon, J. P., Mark, D. M., Kikuchi, L., Guevara, J. A., Two-Dimensional Run-Encoding for Quadtree Representation: Computer Vision, Graphics, and Image Processing, in press.

Mark, D.M., and Lauzon, J.P., 1984, Linear Quadtrees for Geographic Information Systems: Proceedings, International Symposium on Spatial Data Handling, Zurich, Switzerland, August 1984, vol. 2, pp. 412-430.

Samet, H., Rosenfeld, A., Shaffer, C.A., and Webber, R.E., 1984, Use of Hierarchical Data Structures in Geographic Information Systems: Proceedings, International Symposium on Spatial Data Handling, Zurich, Switzerland, August 1984, vol. 2, pp. 392-411.

Smyth Associates Inc., 1983, EWorld Computer Atlas (TM): Software Product Description, Smyth Assocaites Inc., Seattle, Washington.

Spies, K.P., and Paulson, S.J., 1981, TOPOG: A Computerized Worldwide Terrain Elevation Data Base Generation and Retrieval System: U.S. Department of Commerce, National Telecommunications and Information Administration, NTIA-Report-81-61, 214pp.

Tomlinson, R. F., Calkins, H. W. and Marble, D. F., 1976, Computer Handling of Geographical Data: An Examination of Selected Geographic Information Systems: Paris: The UNESCO Press.

Weber, W., 1979, Three Types of Map Data Structures, Their ANDs and NOTs, And a Possible OR: in International Advanced Study Symposium on Topological Data Structures for Geographic Information Systems Vol. 4, Dedham, Mass., G. Dutton (ed.), Harvard University Laboratory for Computer Graphics and Spatial Analysis.

White, M., 1983, N-trees: Large Ordered Indexes for Multidimensional Space: Presented at the Lincoln Institute of Land Policy's Colloquium on Spatial Mathematical Algorithms for Microcomputer Land Data Systems. May 9-10, 1983.

APPROACHES FOR QUADTREE-BASED GEOGRAPHIC INFORMATION
SYSTEMS AT CONTINENTAL OR GLOBAL SCALES

David M. Mark and Jean Paul Lauzon
State University of New York at Buffalo
Department of Geography
Buffalo, New York 14260

BIOGRAPHICAL SKETCHES

David M. Mark received the degrees of B.A. (1970) and Ph.D. (1977) from Simon Fraser University, and M.A. (1974) from the University of British Columbia. Since 1981 he has been a member of the Department of Geography, State University of New York at Buffalo, where he is currently Associate Professor and Director of the Cartographic Laboratory. His current research interests include quadtrees, digital elevation models, artificial intelligence, and theoretical geomorphology.

Jean Paul Lauzon received his B.A. from the University of Western Ontario, and his M.A. from the State University of New York at Buffalo. He is currently a Ph.D. candidate in the Department of Geography, State University of New York at Buffalo, and also Senior Database Engineer with Wild Leitz Canada Ltd., Willowdale, Ontario. His research interests include quadtrees, geographic information systems, relational databases, and artificial intelligence.

Database Issues in Digital Mapping

David M. McKeown, Jr.
Department of Computer Science
Carnegie-Mellon University
Pittsburgh, Pa. 15213*

(412) 578-2626 Office
(412) 687-2532 Home

Extended Abstract

Although databases for geographic information systems (GIS) have been developed to manage digital map data, the integration of aerial imagery and collateral information is rarely performed. For the most part, the use of sophisticated intelligent spatial databases, in which a user can query interactively about map, terrain, or associated imagery, is unknown in the cartographic community. In standard GIS systems, the ability to formulate complex queries requiring dynamic computation of factual and geometric properties is severely limited, often reflecting its origin as collections of thematic map overlays. Spatial database research requires the integration of ideas and techniques from many areas within computer science such as computer graphics, image processing, artificial intelligence, and database methodology as well as from the traditional area of photogrammetry. The problem is complex along many dimensions.

First, digital cartography requires a massive amount of raw data: image, map, textual, and collateral data. Recently, various estimates of the amount of data associated with cartographic production[1] have been revealed by the three major users of remote sensing and aerial imagery in the United States: the U.S Geological Survey (USGS) estimates 10^{12} to 10^{14} bits; the National Aeronautics and Space Administration (NASA) estimates 10^{16} bits; and the Defense Mapping Agency (DMA) 10^{19} bits. In a further projection, DMA estimates that by 1995 they will require support for 1000 on-line users, each having a local database context of 10^{12} bits.

Second, the number and variety of spatial entities, their description, attributes, and the rich set of spatial relationships between entities can not be handled by precomputed tables of relationships. The extraction of man-made and natural features from source imagery using automated image analysis tools will require a map database as a source of knowledge to guide the extraction and interpret the processing results. Thus, in order to perform image analysis using knowledge-based techniques[2,3], we must have some *a priori* information that allows us to constrain computation and search to reasonable levels. This is a chicken-and-egg problem of grand proportions.

The development of intelligent spatial databases addresses two problems in digital mapping. First, from a database perspective, the explosive increase in the availability of imagery and image related information makes finding some small piece of relevant information increasingly difficult. On-line storage of tens of thousands of images does not help unless the user can quickly locate a feature or landmark of interest in many

---

*This research was partially sponsored by the Defense Advanced Research Projects Agency (DOD), ARPA Order No. 3597, and monitored by the Air Force Avionics Laboratory under Contract F33615-78-C-1551. The views and conclusions in this document are those of the author and should not be interpreted as representing the official policies, either expressed or implied, of the Defense Advanced Research Projects Agency or the U.S. Government

different images simultaneously.

Second, the same underlying problem exists from an automatic image interpretation standpoint; symbolic indexing and addressing of images for automated analysis requires many of the same techniques as in interactive analysis, except that the human image analyst provides the guidance. Facilities such as on-line image/map databases, signal and symbolic indexing of natural and man-made features, and spatial reasoning can be viewed in the short term as an semi-automated tool for increasing productivity of human photo interpreters and analysts, and in the long term, as the knowledge base for automated systems capable of detailed analysis including change detection and automatic update of map descriptions.

This paper gives a brief overview of current research at Carnegie-Mellon University in the area of spatial database systems for digital cartography and aerial photo interpretation. A brief overview of MAPS, the Map Assisted Photo Interpretation System, is presented[4]. MAPS is a large integrated database system containing high resolution aerial photographs, digitized maps, and other cartographic products, combined with detailed 3D descriptions of man-made and natural features in the Washington, D. C. area.

In this system, the user can formulate queries into the spatial database using high resolution imagery to specify an area of interest for spatial indexing, or to specify a generic type of feature. If an appropriate user interface is provided and explicite image-to-map correspondence is performed, the image can act as a map for a user unfamiliar with an area of interest. Further, the ability to represent and index spatial knowledge allows us to begin to develop image processing techiques to recover roads, buildings, and terrain features used to update or refine existing maps[5].

1. References

1. Tatalias, Kosmo D., "Mass Data Base/Knowledge Base System," Tech. report DMA Technology Base Symposium, Defense Mapping Agency, August 1983.

2. McKeown, D. M. and Lukes, G. E., "Digital Mapping and Image Understanding," *Archives of the XVth Congress on Photogrammetry and Remote Sensing*, International Society for Photogrammetry and Remote Sensing, Rio de Janeiro, Brazil, June 1984, pp. 690-697.

3. McKeown, D.M., Harvey, W.A. and McDermott, J., "Rule Based Interpretation of Aerial Imagery," *Proceedings of IEEE Workshop on Principles of Knowledge-Based Systems*, Dec 1984.

4. McKeown, D.M., "MAPS: The Organization of a Spatial Database System Using Imagery, Terrain, and Map Data," *Proceedings: DARPA Image Understanding Workshop*, June 1983, pp. 105-127, Also available as Technical Report CMU-CS-83-136

5. McKeown, D.M., Denlinger, J.L., "Map-Guided Feature Extraction from Aerial Imagery," *Proceedings of Second IEEE Computer Society Workshop on Computer Vision: Representation and Control*, May 1984, Also available as Technical Report CMU-CS-84-117

THE U.S. GEOLOGICAL SURVEY 1:100,000-SCALE DIGITAL CARTOGRAPHIC DATA BASE

by John D. McLaurin
Digital Cartography Program Manager
National Mapping Division
U.S. Geological Survey
516 National Center
Reston, Virginia 22092
(703) 860-6222

ABSTRACT

The U.S. Geological Survey has initiated a major new production program to produce a 1:100,000-scale digital cartographic data base for the conterminous United States. The initial objective of the effort is the completion of the transportation and hydrography data sets by 1987. This phase of the program is being done cooperatively with the Bureau of the Census to meet their requirements for the 1990 census. Several other Federal agencies, including the Bureau of Land Management, the Federal Emergency Management Agency, the Nuclear Regulatory Commission, and the Department of Transportation, have also identified requirements for these data. As additional requirements are identified, other data categories will be added after 1987.

Production is being accomplished by raster scanning feature separation plates from the 1:100,000-scale maps, raster editing of the data, raster to vector conversion, and attribute tagging using interactive graphic systems.

Data from the 1:100,000-scale data base will be available beginning in March 1985 in 30-minute blocks (one-half of a 1:100,000-scale map sheet). Data will be distributed in digital line graph standard, graphic, or optional format depending on the software with which they are to be used.

STORING, RETRIEVING AND MAINTAINING INFORMATION ON GEOGRAPHIC STRUCTURES: A GEOGRAPHIC TABULATION UNIT BASE (GTUB) APPROACH

David Meixler and Alan Saalfeld
Bureau of the Census
Washington, DC 20233

ABSTRACT

The various levels and subdivisions of geography and their interrelations may be organized for computer storage, retrieval, and update through the use of multilist structures. In this process several files are created to correspond to the different partitions of the whole space or universe. A directory is used to keep track of entities for each entity type—state, county, congressional district, etc. The "state" directory contains 57 entries, for example, one entry for each state (50) or territory (7).

A fundamental file of unnamed, elementary entities is the basis for storing information on the geography of the space. The records of the fundamental file correspond to elementary geographic areas which are minimal intersections of entities, and which are traditionally called GTUB's for "Geographic Tabulation Unit Base." Each GTUB record contains pointers for several lists. Each list type corresponds to an entity type, say "county". The county list type occupies a unique field in each record, and the value in that field is a pointer to the next GTUB with the same county code. Each list type corresponds to a partition of space and provides a method to store many specific lists of this type. The list elements are the subsequent records of the same entity.

The GTUB's are equivalent to the atomic elements in the complete geographic tabulation lattice. The multilist structure permits the reconstruction of the geographic lattice if desired—hence the properties of lattices may be utilized implicitly with the GTUB.

INTRODUCTION

A topologically sound geometric structure can guarantee mathematically consistent areal classification. The fundamental element of area in a topologically structured map file is the 2-cell; and in the United States, there are over ten million 2-cells. The surface of the country may be theoretically subdivided into those ten million 2-cells with all land area counted (covered) exactly once. All geographic areas are composed of sets of 2-cells. In that sense the 2-cell is the basic unit or fundamental building block for geographic regions. The 2-cells may be unnecessarily small and too numerous for the kind of "building" required to keep track of geographic relationships. Often contiguous 2-cells will behave as a group, remaining together under any geographic partition of the whole space. These groups which behave as single units make up the GTUB's. The 2-cells belonging to a single GTUB will all have the same geography; that is, they are contained in the same regions. GTUB's will be characterized by the property that their constituent 2-cells all have the same geography; and that any 2-cell not belonging to the GTUB must have different geography in at least one category.

A GTUB is an intersection of more familiar geographic regions. A computer data structure called a **multilist** will be used to represent GTUB's. The example in the next section is provided to illustrate the relations among 2-cells, GTUB's, higher-level geographic structures, and the multilist representation of the geographic relations.

AN ILLUSTRATED EXAMPLE OF GTUB/MULTILIST STRUCTURES

Consider the following subdivisions of the same rectangular area into 2-cells, regions, zones, and districts:

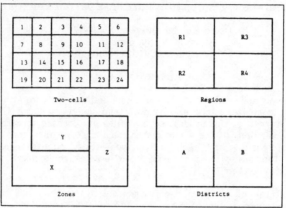

Figure 1. Different Subdivisions of a Rectangular Area.

The 2-cells are automatically a finer physical partition than any of the other subdivisions. All lines used to delineate geographic area boundaries **plus** all street and other linear feature lines go into determining the boundaries or outer limits of the 2-cells. The regions are also a finer partition than the districts in this example; zones are not comparable to regions or to districts in the sense that neither is finer than the other. A finer partition is called a **refinement**. A common refinement may always be obtained for any collection of partitions by taking intersections of areas. The common refinement for the zones, districts, and regions is shown below. This refinement has as its elements the GTUB's for zones, districts, and regions.

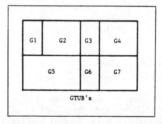

Figure 2. GTUB Sets for Zone-District-Region Geography.

The GTUB's depend on the collection of partitions of geographic areas. Adding other levels of geography will increase the number of GTUB's and make the new GTUB set a refinement of the earlier GTUB set. The GTUB set will always have fewer elements than the 2-cell set because every GTUB will still be composed of 2-cells. In the example above there are 24 2-cells and 7 GTUB's. If every 2-cell were subdivided into 100 2-cells in the example above, there would still be only 7 GTUB's. The number of GTUB's depends on the geography above the GTUB's, not the geometry below (i.e. the 2-cells, which originate from street and other linear feature patterns). The following figure is a Haase diagram showing the inclusion relations of the districts, regions, zones, and GTUB's. A line between entities indicates inclusion. The higher entity contains the lower entity.

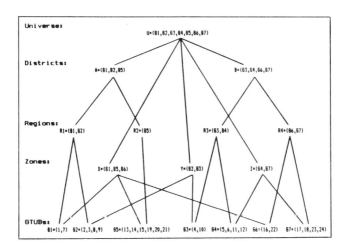

Figure 3. Haase Diagram for Set Inclusion Relations for GTUB's.

In terms of lattice theory, the GTUB's form the set of greatest lower bounds for sets of selected elements from every partitioning set. The GTUB's are also least upper bounds for sets of 2-cells having the same geography. The GTUB's provide a new level of geography between 2-cells and other entities which simplifies the relational scheme, as shown in the figures below.

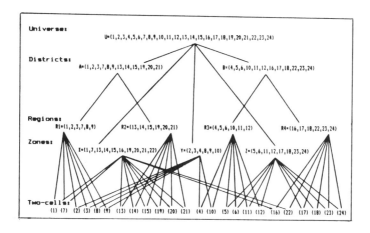

Figure 4. Set Inclusions of 2-cells Without GTUB's.

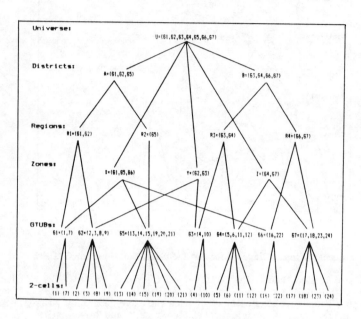

Figure 5. Set Inclusions of 2-cells with GTUB's.

Figure 4 illustrates the structure required to link 2-cells directly to the geography. Without GTUB's, every 2-cell must be joined to every minimal level of geographic entity type (in this case, zone and region). In this particular example there are only two minimal levels. In the more general case, for **m** minimal levels and **n** 2-cells, **mn** links are required. GTUB's provide a unique minimal level of geography on which to anchor all 2-cells.

Figure 5 reveals the clear separation of geography and 2-cells that the GTUB's provide. There are no fundamental inclusions of 2-cells in entities other than GTUB's. This permits the GTUB's themselves to function as the fundamental unit of geography for many applications which do not require the detail of 2-cell information. At the same time, however, the GTUB's are directly linked to the family of 2-cells to permit extracting information on those 2-cells when it is needed.

Figure 5 suggests that different data structures for GTUB's-and-above and for GTUB's-and-below are advisable. The GTUB's simply partition the set of 2-cells, whereas more complex interactions take place above the GTUB's. This will always be the case because the GTUB's are least upper bounds for 2-cell sets and greatest lower bounds for the other geography.

The next section describes the multilist representation for the geography with GTUB's and illustrates the description with the example given above.

Multilist Representation of Geography with GTUB's.

A multilist is a computer data structure capable of providing several simultaneous linkages of data. The data points are regarded as belonging to several lists—each data point belongs to one list for each list type or entity type. GTUB records are the data points, and unique fields on the records identify the containing lists and the list links. Directories support the multilist of GTUB's by keeping track of higher level geography. Higher level geography refers to entity types which have proper refinements between themselves and the GTUB's. An example of higher level geography in the illustration above is the district. The figures below present the collection of data structures required for multilist representation of the GTUB relations presented earlier.

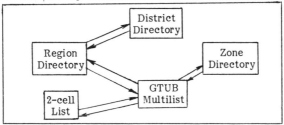

Figure 6. Files and File Links for Multilist GTUB Representation.

More specifically, the files for the example given above will consist of the following set of records:

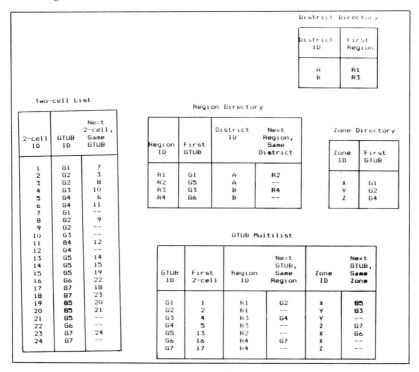

Figure 7. Contents of All Files in the Multilist Database.

Within each file, records correspond to entities or specific areal units. Information is grouped as follows on the records:

Entity identifier: links to lower levels: links to higher levels.

Not every record will contain all three groups of information, although two of the three will always appear. The 2-cells have no lower levels; and, in the example above, districts and zones have no higher levels.

Although the entire collection of geographic entities does not form a hierarchy, some subsets of entity types do form hierarchies. The entity type hierarchies result when one level of geography is a refinement of another level (counties within states, for example). The file organization will reflect this.

Within the GTUB records, the various fields corresponding to the "Next GTUB, same entity" link the records in individual lists that constitute the multilist. In the example above, the "Next GTUB, same entity" field of the GTUB record corresponding to entity type "zone" produces 3 linked lists of GTUB's: (G1, G5, G6,), (G2, G3), and (G4, G7) corresponding to the zones X, Y, and Z, respectively.

Notice that GTUB lists are not generated for higher level geography, such as district. GTUB lists may be built for higher level geography by aggregating lists of lower level entities which constitute the higher level entities. The aggregation procedure imitates the initial list formation routine: references to the "Next lower level entity, same higher level entity" generate lists of lower level entities. Each lowest level entity corresponds to a list of GTUB's. The lists of GTUB's for every lowest level entity in the higher level entity, can be concatenated to produce a list for the higher level entity.

Decomposition procedures of higher level entities of any type into lower level entities amount to list building and concatenation operations for the multilist structure described above.

MULTILIST FILE COMPLEXITY

The example presented in the previous section greatly simplifies the interactions of files in order to illustrate clearly the internal file structures. The next figure from a Census Bureau TIGER specification memo illustrates some, but not all, of the geographic areas which were used for tabulation geography on the 1980 Census of Housing and Population. The entities shown are those 1980 geographic classifications which will again be retained through the 1990 Census. The following diagram represents a subset of both the 1980 and the 1990 geographic areas.

The next figure, figure 8, although far more complex than the illustration of figure 6, represents approximately one-third of the total levels of geography that will be carried in the 1990 Census of Housing and Population. Currently there are close to twenty lowest level geographic entity types immediately above the GTUB level being considered for inclusion in the 1990 geographic database.

In addition to possessing complex interactions, the geographic data base is subject to change. Layers of geography may be added or subtracted. Entities within entity types may be redefined or reorganized. Boundaries may change, necessitating the creation of new units at every level, even down to the 2-cell.

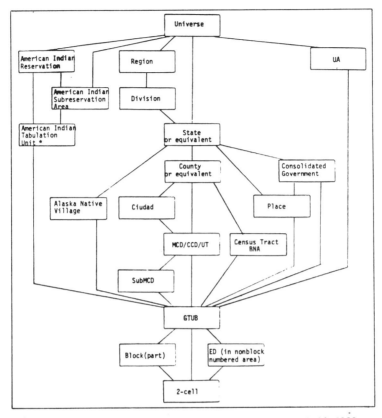

Figure 8. Some 1980 U.S. Geographic Areas to be Included in 1990

In order to organize large dynamic geographic structures such as the structure diagrammed above, the multilist database will require many file management capabilities. Some of those capabilities are summarized in the final section.

ONGOING RESEARCH AND DEVELOPMENT WITH GTUB'S

The Haase diagram in figure 5 illustrated the way that GTUB's can replace 2-cells as the fundamental unit of geography. The efficiency gained in working with GTUB's will depend to a great extent on the relative sizes of the GTUB file and the 2-cell file. As partitioning sets increase, the GTUB's get smaller and their number increases. In theory, the GTUB's could become as numerous as the 2-cells. Some investigation is underway to study the trade-offs and to determine the point at which the GTUB's become inefficient.

The GTUB's, even with their potential inefficiencies, nevertheless provide a minimal set of building blocks from which to build all of the geographic entities. This aspect of GTUB's was not emphasized in the earlier sections; and it is an important consideration for maintaining geographic consistency. Regions may be reconstructed from their GTUB's by means of multilists; and, moreover, the boundaries of those GTUB's may be added using a Boolean sum to reconstruct the boundary of the containing region. The integrated approach of the GTUB and the multilist totally eliminates the polygon overlay inconsistencies that result from storing different types of entities on separate files. The overlay figure below illustrates the manner in which GTUB's behave as jigsaw puzzle pieces which can be used to build any geographic entity.

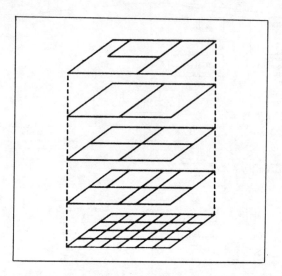

Figure 9. All Overlays are Made Up of GTUB's.

Although the directories in the example presented in this paper were given the same multilist structure as the GTUB file, in practice there may be shortcuts and savings gained by using other structures for higher level directories and files. Census coding of many entities embeds higher level identifiers in the codes of the lower level entities. For example, a Minor Civil Division code carries digits identifying the county and state. This additional structure will be utilized in implementing a multilist database for GTUB's, although it was not specifically acknowledged in the illustrative presentation above.

GTUB file building may be accomplished by adding the layers of geographic entity types one layer at a time. As layers are added, the new GTUB partition becomes a refinement of the old GTUB partition. GTUB elements are split according to way that the new entities partition the 2-cells within the old GTUB. Processing of the splitting of GTUB's by additional layers of geography may be automated to a high degree provided locational information on 2-cells can be retrieved readily, and provided that the new layer's entities are also defined geometrically.

Removing layers of geography results in the consolidation of GTUB's. This process may be accomplished without locational information about the 2-cells.

Several other areas of database management for the GTUB multilist are candidates for study for efficient algorithms. Because of the large file sizes, efficient management is critical.

BIBLIOGRAPHY

Bureau of the Census, 1984, **TIGER Specification Memorandum Series**, GEO-CAT—Phase I, Ch. 2, Doc. 5, Washington, DC.

Wiederhold, G., 1977, Database Design, McGraw-Hill, NY, NY.

AN APPROACH TO MICROCOMPUTER-BASED CARTOGRAPHIC MODELING
Andrew P. Mitchell
Harvard University Graduate School of Design
Cambridge, MA 02138

BIOGRAPHICAL SKETCH

Andrew P. Mitchell received his B.A. degree in Geography-Ecosystems from the University of California, Los Angeles, and his Master in Landscape Architecture degree from the Harvard University Graduate School of Design. He is currently a senior analyst at Criterion, Inc., San Diego, CA, where he is involved in development of cartographic software for planning applications. Andrew is a member of the ASP, APA, and the Southern California Computer Aided Mapping Association.

ABSTRACT

The use of computer-based cartographic modeling techniques in environmental planning has increased dramatically in recent years. However, many users of this technology must rely solely on microcomputers for applications involving large cartographic data bases. Existing microcomputer-based cartographic modeling systems are inadequate for this task due to limitations in size, speed and/or specific operations. An approach has been developed which utilizes a compressed raster data structure and local processing of data to overcome these limitations. This approach has been implemented through the development of a software program which contains rudimentary data input/output and transformation operations. Testing of this program confirms that use of a compressed-raster data structure significantly reduces data storage requirements, while the local processing structure results in acceptable processing time. While this approach does have limitations, it presents a viable method for modeling large data sets using microcomputer technology. However, its greatest potential may be as one component of a system which utilizes vector, raster and compressed-raster structures.

INTRODUCTION

Environmental planners, landscape architects, foresters and others involved in the planning and management of the physical environment are confronted with increasingly complex demands and large amounts of data on which to base decisions. Cartographic modeling is one method available to these managers to utilize the increasing amounts of data in the decision-making process. Tomlin refers to cartographic modeling defined as "the act of formally synthesizing geographic information as part of a decision-making process" (Tomlin, 1983). In practice, this includes methods such as overlay mapping and proximity analysis.

While this technology has become well established in the past decade and continues to advance, many users still do not have access to mainframe or minicomputer-based systems. Many of these users are involved in cartographic modeling with data bases that cover large land areas. This is especially true of developing countries which are involved in the development of resource management plans for large areas but may not have the infrastructure or resources to support large systems.

However, recent advances in microcomputer technology have created the possibility of access to this technology for those users who do not have access to the larger systems (Youngmann, 1981). Microcomputers also present the opportunity to make this technology available in the field where it can be utilized on a daily basis by resource managers, thus reducing reliance on systems operated through a distant regional or national office.

Within the past several years, work has begun on a number of micro-based cartographic modeling systems. Despite recent advances in both hardware and software, these existing systems have shortcomings which prevent their full utilization for cartographic modeling involving large data bases. These problems result mainly from the limited storage capacity of microcomputers, and from increased processing time. Most grid based systems are limited to a maximum map size of 100 by 100 cells, or 10,000 cells total. While vector based systems can accomodate larger data bases, the processing time these systems require limits their use in interactive cartographic modeling.

AN APPROACH TO MICROCOMPUTER-BASED CARTOGRAPHIC MODELING

In order to overcome these limitations, research was undertaken to develop a micro-based system which will allow cartographic modeling of a data base virtually unlimited in size while maintaining acceptable processing speed. The approach developed utilizes a compressed data structure in conjunction with local processing.

Compressed Data Structure
A compressed data structure based on a grid rather than polygon structure was chosen as the basis for the system in order to maintain the cartographic modeling capabilities associated with grid format.

Background Compressed raster format has been used frequently in digital image processing applications, but has been used less often in regard to cartographic data, and rarely in cartographic modeling applications. This structure, the general term for which is run-length encoding, has been discussed in numerous texts on digital image processing. (Gonzalez, 1977; Pratt, 1978; Hall, 1979) As applied to digital image processing, the method involves scanning along a row of an image and comparing gray values of adjacent picture elements (cells). Adjacent cells for which the gray value does not significantly differ are grouped together and termed a 'run'. As described by Gonzalez, "the sequence of image elements along a scan line (row) x1, x2...xn is mapped into a sequence of pairs (gi, li) (gk, lk) where g denotes the gray value and l denotes the run length of the ith run", that is, the number of image elements in the run (Gonzalez, 1977). Pratt distinguishes between run-length encoding and what he terms "run-end" encoding. Rather than specifying the number of image elements in a run, run-end encoding specifies the location of the end of the run in relation to the beginning of the row (Pratt, 1978).

Miller discusses the application of compressed raster structure to the encoding and storage of cartographic data and documents several alternative methods of compressed raster encoding (Miller, 1980). One such

VALUE	LENGTH	ROW	VALUE	COLUMN	ROW	VALUE	COLUMN	VALUE	POINT
0	3	1	0	3	1	0	3	0	3
1	2	1	1	5	1	1	5	1	7
1	2	2	1	2	2	-1		0	9
0	2	2	0	4	2	1	2	1	13
1	1	2	1	5	2	0	4	2	16
1	3	3	1	3	3	1	5	1	19
2	2	3	2	5	3	-1		2	23
2	1	4	2	1	4	1	3	1	24
1	3	4	1	4	4	2	5	2	25
2	1	4	2	5	4	-1			
2	3	5	2	3	5	2	1		
1	1	5	1	4	5	1	4		
2	1	5	2	5	5	2	5		

RUN LENGTH   RUN END   ENDING COLUMN   VALUE-POINT

	1	2	3	4	5
1	0	0	0	1	1
2	1	1	0	0	1
3	1	1	1	2	2
4	2	1	1	1	2
5	2	2	2	1	2

figure 1. Comparison of four compressed-raster data structures. Note that run length and run end notation require three integers per run as compared to two per run for ending-column and value point notation. Note also that value-point notation requires the least number of integers to store the map, but also requires the largest integers.

system records the position of the first element of the run and the last element of the run, both identified by column number, as well as the value of the attribute of the run and the number of the row. An alternative system records the row, beginning column, and attribute value of the run.

A third system, termed "ending-column notation", is similar to the run-end encoding described by Pratt, but utilizes a "-1" in the data field to indicate the end of a row. The position of the run is specified by the number of the column in the x-y matrix in which the run ends. This approach has several distinct advantages over systems which use both beginning and ending column since only two values are stored instead of three. In addition, data searches are facilitated since the location of the required data can easily be determined from the column and row values, which are known. Data search becomes much more difficult using run-length encoding since data must be located by summing the run length values to determine the beginning and ending points of a run (Miller, 1980). Figure 1 compares several compressed raster data structures.

Value-Point Notation A variation of the ending column notation was developed as part of this research. Termed "value-point notation", this approach uses two integers to describe each run: the value of the attribute associated with the run, and the ending point of the run, which is the position of the point in the x-y matrix (figure 1). This method has the advantage of further reducing the total number of runs, and thus the storage required for the map, since a run may encompass several rows. Using this approach, a map of a constant value would be stored as one run. Using ending column notation, the same map, of x rows, would be composed of x runs. For comparison, the same map stored in grid format would be composed of x * y points, each represented by an integer value. Thus an overlay stored in value-point notation can be thought of as a continuous string of runs of varying length and value, with the endpoint of the final run equal to the number of cells in the overlay. While data search is more readily accommodated using this method, operations involving display of data are made somewhat complex. In addition, the largest value stored using value-point notation is the value associated with the last point of the map, which is equal to the total number of points (cells) in the map. This may be a limitation for some microprocessors. In contrast, the largest value stored is equal to the number of columns in the map for ending column notation, and equal to the length of the longest run for run-length notation.

Discussion of compressed-raster structure There are disadvantages associated with compressed raster encoding, such as the complexity of algorithms for cartographic modeling required by compressed raster, and the increased difficulty in data searching. The predictable nature of traditional raster, in which each data element is explicitly located, facilitates both of these tasks. However, when working within the limitations of microcomputer technology the greater storage and processing requirements of traditional raster offset the advantages to be gained. One major advantage of the compressed raster approach is that fine resolution mapping, approaching that associated with traditional cartographic methods, can be achieved. When using a traditional raster format, doubling the resolution of a data set will quadruple the number of cells. The result has been that users decrease resolution in order to decrease the number of data elements. This problem is reduced with a compressed-raster data structure since doubling resolution only doubles the number of runs (figure 2).

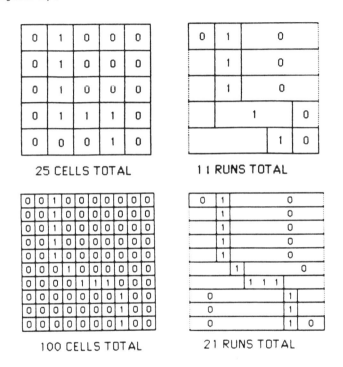

figure 2. Comparison of the amount of additional data generated for traditional grid vs. compressed raster when resolution is increased.

The savings in storage to be gained from use of compressed raster will vary based on the configuration of the overlay being encoded. Hall has noted, in regard to digital image processing, that "The efficiency of run length encoding depends on the number of gray level transitions or edges and could therefore be expected to be most efficient for images with a small number of edges and gray values"

(Hall, 1979). When applied to cartographic data, this means that those maps with few attributes distributed over large areas will require fewer runs than those maps which have many attributes dispersed widely and in small clumps. The extreme example is surficial data, such as elevation, in which each cell can be expected to have a value unlike its neighbors, thus resulting in many runs. The "worst case" example would involve a map in which each cell requires its own run--since a compressed raster structure requires two values per run (value of the attribute and run length or ending column or point), such a map would require twice the space of the same map stored in traditional raster format. However, since for many modeling purposes the surficial map would be reclassified to several ranges of value (slopes of 0-8%, 9-15%, 16-25% and >25% for example), the result would be a map which could efficiently be stored in compressed format.

Local Processing of Data

In order to allow processing of large data sets using microcomputer technology, it is necessary to take into account the limited main memory capacity of existing microcomputers. While microprocessor technology continues to develop, thus increasing the memory capacity of microcomputers, many machines in use are limited to as little as 16k bytes of main memory. One approach to accommodating this limitation is to implement local processing of data. This approach entails retrieving a portion of the data set from secondary storage, processing it, and returning it to secondary storage. This process is repeated such that the data set is processed sequentially from beginning to end, one portion at a time.

The advantage of this approach is that only a small amount of data is contained in main memory at a given time, thus main memory capacity can be quite small. By comparison, many traditional raster systems require that entire maps be stored in main memory at one time, and often two are required for data transformation operations.

The major disadvantage of local processing is that the number of input/output operations is greatly increased and thus processing time is increased. The result is a trade off between space and time: the more main memory available, the larger the portion of the data set that can be processed at a time, and the fewer the input/output operations. Theoretically, the size of the portion to be processed would be specified by the user to be as large as possible with a given microcomputer. To process data sets stored in traditional raster format using local processing would probably result in prohibitive processing time and cost. However, by local processing of compressed raster data, the amount of data to be processed is already greatly decreased, thus the number of input/output operations will be decreased. It is this combination of compressed raster data structure and local processing which allows the capability for processing of large data sets using microcomputers.

IMPLEMENTATION

A software program using the described structure has been written to test the viability of the above approach as a basis for cartographic modeling of large data bases using microcomputer technology.

Structure of the System
The current program, tentatively entitled MAP2, is written in FORTRAN 77 high level programming language and utilizes the value-point notation data structure, described previously, as well as local processing. Runs are represented as two-dimensional arrays and are labeled "VP" (for value/point, the order in which indicator integers are listed in the array). To take advantage of local processing, arrays are packaged in records of a specified length. In the current version, each record contains one array of 256 runs. Since each run is represented by two integers of 4 bytes each, each record requires 8 * 256 bytes or 2k bytes.

Although structured to run on a microcomputer, the program was developed on an IBM 370 mainframe computer to take advantage of the CMS operating system utilized by the IBM. CMS files in binary format are created by the program, stored in CMS, and are called as input by the program as required. The program uses a prompt format for user interface: users are asked to enter input for specific operations (for example, the number of the map file to be processed, the number of the map file to be created, and data to be used in the processing, such as values to be assigned when reclassifying a map).

Operations
A number of data input, output and transformation operations have been developed for this system. Data input operations include the ability to download existing raster format maps to compressed raster format, encode maps directly into compressed raster format and create a map of a constant value. Data output operations include the ability to display a map using alphanumerics to represent the values associated with the map, and to convert an existing compressed raster format map to traditional raster format.

Data transformation operations include the ability to alter the values associated with the attributes of a map, assign a unique user-specified value to each combination of values encountered when two maps are compared (figure 3), and to create a map with values obtained by summing the values on two or more existing maps.

Demonstration of MAP2
To test the efficiency of the compressed raster structure and local processing approach as implemented in the MAP2 program a comparison between MAP2 and a traditional raster based system was undertaken.

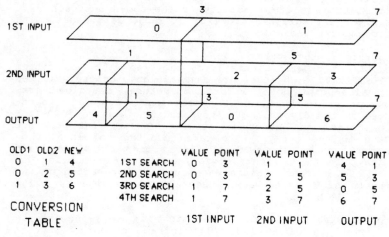

figure 3. The Cross operation. A new run is created each time the endpoint of an existing run is encountered on either input array. Note that on each search the array incremented is the one which contained the shorter run on the previous search.

Hardware and software utilized The comparison utilized an IBM 370 mainframe computer and compared storage and processing time of MAP2 with the Map Analysis Package (MAP) (Tomlin, 1983), a typical, in terms of data structure, traditional raster based system. MAP assigns a 2-byte integer to each cell in a map and requires, for most operations that two entire maps be in main memory for processing. MAP2 defines each run in terms of two 4-byte integers and processes maps in records of 256 runs each. Thus, this comparison is of the efficiency of both the compressed raster structure and the local processing approach combined.

Data bases The two programs were compared using two separate data bases. The first, Petersham, consists of a 36 square mile area in central Massachusetts. The data base is 180 rows by 180 columns for a total of 32,400 cells. The data base consists of approximately ten data maps, three of which, Waterbodies, Soils and Slope, were chosen for use in the comparison. The second data base, Alaska, covers a 54 square mile area of the Tongass National Forest in Southeastern Alaska. The data maps are 250 rows by 90 columns for a total of 22,500 cells. The data base consists of six maps, four of which, Waterbodies, Wildlife, Visual quality, and Land Management Zones, were used in the comparison. All maps were originally encoded in traditional grid format and were downloaded to compressed raster using the Gridin operation of the MAP2 program.

Comparisons When comparisons were made of storage space required, substantial savings were achieved for almost all maps when stored in compressed raster format (Table 1). The exception, as would be expected, was the Petersham slope map which required more space in compressed raster than in traditional grid format. Maps with large continuous areas of a single value, such as the Petersham soils map or the Alaska land management zones map, required the least space. It should be noted that since both of these data bases consist of less than 32,768 cells, the compressed raster arrays could actually have been specified using two 2-byte integers, thus cutting required storage space to half of that shown by the figures in the table for MAP2.

Petersham Data Base	MAP2	MAP
Waterbodies	39kb	64kb
Soils	32kb	64kb
Slope	190kb	64kb
Alaska Data Base		
Waterbodies	10kb	45kb
Wildlife	10kb	45kb
Visual Quality	14kb	45kb
Land Management Zones	12kb	45kb

Table 1. Comparison of storage requirements of several maps utilizing MAP2 and MAP.

An objective comparison of processing time for specific operations, was difficult to achieve since many extraneous factors were involved in running the programs (for example, the time required to type in commands). However, comparisons of processing time seem to indicate no substantial increase in processing for the compressed raster data, this despite the fact that the specified MAP2 record length was 2k bytes. Presumably, increasing the record length would result in more efficient processing for the compressed raster data.

SUMMARY

It is apparent that use of a compressed raster data structure presents a viable approach for most cartographic modeling operations. In addition, when combined with the use of local processing, this structure provides the basis for a microcomputer based system. However, the data structure does make certain operations more complex in development and execution, specifically those which involve sampling of points surrounding a cell. In addition, the efficiency of this structure does not apply to surficial data.

Possible applications of this approach, apart from as a basis for a microcomputer-based system, include high resolution cartographic modeling and image processing.

However, its most valuable application may be as one component of an integrated system which utilizes polygon, raster, and compressed raster data structures as appropriate for the type of data utilized and modeling operations performed.

ACKNOWLEDGEMENTS

The author wishes to thank Dr. Dana Tomlin for his collaboration on some of the concepts presented in this paper, and Dr. Carl Steinitz and Mr. Denis White for their valuable suggestions.

REFERENCES

Gonzales, R.C. and P. Wintz. 1977. Digital Image Processing. Addison-Wesley, Reading, Mass.

Hall, E.L. 1979. Computer Image Processing and Recognition. Academic Press, New York.

Miller, S.W. 1980. A compact raster format for handling spatial data. Technical Papers of the ACSM Fall Technical Meeting, Niagara Falls, NY.

Pratt, W.K. 1978. Digital Image Processing. Wiley, New York.

Tomlin, C.D. 1983. Digital cartographic modeling techniques in environmental planning. PhD. Dissertation, Yale University, New Haven.

Youngmann, C.E. 1981. Portable geographic information systems for on-site environmental data collection. Harvard Library of Computer Graphics/1981 Mapping Collection, vol. 15, Computer mapping of natural resources and the environment. Harvard University. Graduate School of Design.

AN INTERIM PROPOSED STANDARD FOR DIGITAL CARTOGRAPHIC DATA:
BACKGROUND AND REVIEW

Prof. Harold Moellering, Chairman
National Committee for Digital
Cartographic Data Standards
Numerical Cartography Laboratory
158 Derby Hall
Ohio State University
Columbus, Ohio 43210

The Steering Committee of the National Committee for
Digital Cartographic Data Standards is composed of:

Harold Moellering, Chairman	Ohio State University
Lawrence Fritz, Vice Chairman	National Ocean Service
Dennis Franklin	Defense Mapping Agency
Robert Edwards	Oak Ridge National Laboratory
Tim Nyerges	Northwest Cartography
Jack Dangermond	Environmental Systems Research Institute
John Davis	Kansas Geological Survey
Paula Hagan	Wang Laboratories
A.R. Boyle	University of Saskatchewan
Waldo Tobler	University of California
Dean Merchant	Ohio State Univeristy
Hugh Calkins	State University of New York at Buffalo

ABSTRACT

The National Committee for Digital Cartographic Data Standards has been in operation for the past 3 years. The first two years included defining the issues and examining the alternatives available. Several reports have been issued on these topics. The third year of work has involved the definition of an Interim Proposed Standard. Working Group I on Data Organization has developed a data exchange superstructure with the widest applicability in digital cartography. Working Group II on Data Set Quality has proposed a "truth in labeling" approach to data set quality and the categories of quality information that should be made available to a prospective user of a data set. Working Group III on Cartographic Features has proposed an integrated set of feature definitions and a set of examples of the definitions and associated attributes. Working Group IV on Terms and Definitions has proposed a standard set of primitive and simple cartographic objects which can be used with digital cartographic data bases. They have also produced a set of terms and definitions associated with the standards developed by the other three Working Groups. The Committee is now seeking public comments on these proposed standards. The presentation will focus on the background of this overall effort and provide an overview of the current work.

ARC/INFO: A GEO-RELATIONAL MODEL FOR SPATIAL INFORMATION
Scott Morehouse
Environmental Systems Research Institute
380 New York Street
Redlands CA 92373

ABSTRACT

A data model for geographic information is described. Originally designed for thematic mapping and map analysis, the model lends itself to tabular data processing applications as well as automated cartography. The model is a combination of the topological model (to represent feature locations and topology) and the relational model (to represent feature attributes).

GOALS FOR THE DATA MODEL

A geographic information system is a spatial data base together with a set of spatial operators. Any spatial data base is derived from a model of geographic information. The usefulness of a geographic information system depends on having a data model appropriate for geoprocessing. This is particularly true when systems and data bases must serve a variety of purposes. The ARC/INFO data model was designed as the basis for a generalized geographic information system. The overall goal is for a practical data model with as much generality as possible. Specific goals are described in the following paragraphs.

Generality. The data model should support data bases developed at a variety of scales and for a variety of purposes. It should be suitable for applications ranging from thematic mapping to land inventory to topographic mapping to urban base mapping.

Simplicity. The data model should be as simple as possible and still meet its other goals. A simple data model is the key to implementing efficient and reliable geographic data bases and algorithms.

Efficiency. The data model should provide the basic data structure for all geoprocessing functions. It should support efficient geoprocessing functions directly without requiring conversion of data to special "analysis" or "edit" formats. For example, a function such as polygon overlay should be done directly with the data model rather than requiring a grid cell copy of the data base. Functions which should be easily implemented using the data model are: bulk digitizing, high quality map graphics, polygon processes such as overlay and dissolve, non-graphic query and analysis, and network simulation.

Adaptability. It should be possible for both the system user and system programmer to extend or adapt the data model for particular applications. This is especially true for feature attributes. It should be easy to add new attribute information to an existing data base. More importantly, it should be possible to relate information in the geographic data base to existing non-spatial data. Geoprocessing applications often require more than just "map data" -- other non-spatial data are usually involved (e.g., soil interpretation matrices for land planning or property ownership records for urban planning).

Freedom from Restrictions. The data model should contain no inherent limitations on its size or content. It should handle both very large and very small databases well. Any limits that are placed on data volume or content (e.g., maximum 2000 points per polygon) will soon be challenged and require messy adaptation to application programming and system use. The absence of restrictions is particulary important since the data model is intended for very large production applications.

DERIVATION OF THE DATA MODEL

The ARC/INFO data model is based on the idea that geographic data can be represented as a set of features. Each feature has associated locational and thematic data. For example, if our features are cities, then their locational data might be latitude/longitude coordinates; the thematic information might be population, area, etc.

Early in the design of ARC/INFO, it became evident that data structures optimal for the analysis of locational data were not optimal for thematic data. In addition, it was clear that these two views of geography, features in space and features with thematic attributes, are both equally important. It is a mistake to think of thematic data merely as attribute codes tagged onto the end of the coordinate definition of a feature. It is equally incorrect to regard locational data as yet another item in a thematic data base management system.

For this reason, ARC/INFO was designed using a hybrid data model. Locational data are represented using a topological data model (similar to the USGS Digitial Line Graph, USGS 1984). Thematic data are represented using a tabular or relational model. In the name "ARC/INFO", "ARC" refers to the topological data structures and algorithms, "INFO" to the tabular data structures and algorithms, and "ARC/INFO" to the composite data model and associated processes. The data model is a geo-relational as it combines a specialized geographic view of the data with a conventional relational data model.

The topological model was chosen to represent locational data because it has a sound theoretical and practical basis and also met the goals outlined above. It has been studied theoretically (see, for example, Puecker and Chrisman 1975, White 1980, and Corbett 1979) and has served as the basis of a number of successful systems: DIME, GIRAS (Mitchell et al 1977), and ODYSSEY (Morehouse 1982). It has proven useful for a variety of applications, ranging from address matching and network flow simulation to detailed storage of map data.

A number of other well-understood spatial models were rejected because they could not meet the overall system goals. Grid cell encoding and its variants, such as quad trees, were rejected because of their inability to handle large amounts of data with precision. Conventional polygon encoding, as implemented in MOSS (WELUT 1982), was rejected because it is inefficient for many geoprocessing functions (e.g., polygon overlay). Graphic element encoding, used in computer-aided design systems (see, for example, SYNERCOM 1982), was rejected for similar reasons even though this structure is well adapted for interactive graphic editing.

For thematic data processing, the relational (or tabular) model was chosen for its adaptive and simple characteristics. This model is the subject of extensive theoretical investigation. In addition, it has been successfully implemented in a number of systems (SAS, INFO, ORACLE, and others). I prefer to use the term "tabular data model" here rather than the more fashionable "relational data model" because the term "relational" has a more restrictive usage. The conventional statistical matrix as implemented in SAS, for example, illustrates the power and utility of the tabular model. However, SAS is not strictly a relational data base. Another trend which enhances the utility of the tabular model is the emergence of "fourth generation programming languages". These programming languages contain program statements for screen input, report generation, file sorting, and merging and for computation of new record values. The INFO system, used by ARC/INFO for tabular data processing, is one of these.

An underlying strategy in the design of the ARC/INFO data model was the use of existing data models. ARC/INFO can thus benefit from continuing technological advances in these areas.

THE DATA MODEL

The Coverage
The coverage is the basic unit of data storage in ARC/INFO. A coverage is analogous to a single map sheet or separation in conventional cartography. It defines the locational and thematic attributes for map features in a given area.

A coverage is defined as a set of features, where each feature has a location (defined by coordinates and topological pointers to other features) and, possibly, attributes (defined as a set of named items or variables).

Feature Classes
There are several kinds of features that may be present in a coverage. Each of these feature classes may have associated locational and thematic information. Figure 1 shows some of the feature classes that may be present in a coverage.

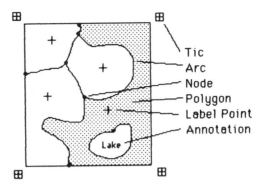

Figure 1: Feature Classes in A Coverage

Tics are registration or geographic control points for the coverage. They allow the coverage to be registered to a common coordinate system (e.g., UTM meters, latitude / longitude, etc.)

Arcs are lines used to represent the location of linear features (e.g., roads) and the borders of area features. Arcs may be topologically linked to the nodes at each end of the arc as well as to the polygons on each side of the arc. Also, an arc attribute table may be created, defining values for thematic attributes (e.g., type of stream, length, etc.).

Nodes are points at the end of arcs. Nodes may be topologically linked to the arcs which meet at the node. A node attribute table may be created, defining values for thematic attributes (e.g., section corner type).

Polygons are areas enclosed by arcs. All polygons are defined by topological pointers to the set of arcs which compose the polygon border and the set of label points inside the polygon. A polygon attribute table may be created, defining values for thematic attributes (e.g., section number, area, etc.).

Label Points are points used to represent information such as wells. Label Points are also used to associate thematic data with polygons and for positioning text within polygons. When used for polygon labelling, label points are topologically linked to the surrounding polygon. A point attribute table may be created, defining values for thematic attributes (e.g., well type).

Annotation is a set of text elements used to label the map on hard copy plots. Place names appearing on a map sheet are recorded in the coverage as annotation elements. Annotation is used solely for display purposes; it is not used in analysis processes such as overlays or section subdivision. Annotation may be topologically linked with features represented.

These feature classes are the basic vocabulary used to define geographic information in a coverage. By varying the types of features contained in a coverage and the thematic attributes associated with each feature class, the coverage can be used to represent many types of map information. For example, the general land office grid can be represented as a set of arcs (section lines), nodes (section corners), and polygons (sections). Thematic attributes can be associated with each of these feature classes.

Feature Attribute Tables
In theory, each of the feature classes described in the preceding sections can have an associated feature attribute table. In practice, feature attribute tables have been implemented for arcs, label points, and polygons. The node attribute table is in the process of being implemented. All attribute tables have the same general structure (see Figure 2).

Poly#	area	type
1	37.5	103
2	18.2	84
3	4.3	161
4	16.2	84

Item ↓ (pointing to "type"); ← Record (pointing to row 2)

Figure 2: Structure of an Attribute Table

In ARC/INFO, the rows of the table are called <u>records</u> and the columns are called <u>items</u>. There may be one table for any feature class in the coverage. Within the feature attribute table, there is one record for each feature of that class. All records in the table have values for the same set of items (or thematic attributes). Items are defined by type and the number of bytes used to store the item.

The feature attribute tables are an integral part of the coverage and are processed by ARC/INFO commands which affect the coverage. For example, when two polygon coverages are overlaid to create a new composite coverage, the polygon attribute tables of the input coverages are merged and written as the polygon attribute table of the output coverage.

In addition to the feature attribute tables, the user can define any number of additional attribute tables. These tables can be related to the feature attribute tables and each other in a variety of ways. For example, the polygon attribute table for a forest stand map could simply contain a single item, the stand identifier. There could be a separate table containing detailed information for each stand in the entire forest. This would allow the stand attribute information to be assembled and maintained independently of the stand maps, but still allow data from both to be related for any application.

The Workspace

A workspace is simply a directory which contains one or more coverages. In addition to the coverage locational data, the workspace has an INFO data base containing the coverage attribute tables and any other, related, attribute tables. Workspaces provide a convenient means for organizing coverages into related groups. They also provide a place for the storage of tabular data not directly tied to a coverage (soil interpretation matrices, for example). Each workspace is completely independent. However, ARC/INFO processing commands allow coverages from different workspaces to be used together. The decentralized organization of coverages in workspaces allows an unlimited number of workspaces (and coverages) to be managed on a single system.

The Map Library

A centralized data structure is useful for the management of very large geographic data bases. The map library is a device which allows coverages to be organized into a large, complex geographic data base. Coverages are organized simultaneously in two dimensions -- by subject or content into layers and by location into tiles (see Figure 3).

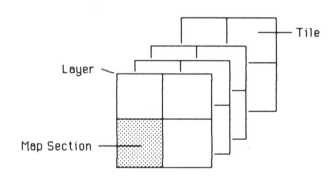

Figure 3: Structure of a Map Library

Tiles. The geographic area represented by a map library is divided into a set of non-overlapping tiles. Although tiles are generally rectangular (e.g., 30° squares), they may be any shape (e.g., counties or forest administration units). Tiles are a digitial analogue for the map sheets of a conventional map series. All geographic information in the map library is partitioned by this tile framework.

Layers. A layer is a coverage type within the library. All data in the same layer have the same coverage features and feature attributes. Examples of layers are land sections, roads, soil types, and wells. It is useful to think of a layer as a coverage which crosses several tiles.

Map Sections. Once a layer has been subdivided into

tiles, it consists of a set of individual units called map sections. A map section is simply a coverage as defined previously.

In terms of the map library, a geographic data base is defined as a set of tiles and layers. The tiles are defined in a special INDEX coverage, where each polygon in the INDEX coverage represents a single tile in the library. Layers are defined in the same way as coverages, by defining the feature classes present and the thematic items associated with each feature class.

CONSTRUCTING A DATA BASE USING THE DATA MODEL

In order to develop a functional data base, geographic information must be adapted to the concepts and vocabulary of the data model. For example, roads are represented by arcs; lakes by polygons; and so on. Feature attributes are recorded in the coverage feature attribute tables. A critical part of data base design using this model is determining which geographic features are stored in the same coverage or layer. For example, should single-line drainage be stored in the same layer as double-line rivers and lakes? These questions can be resolved by considering the use for the data base and by striving to simplify the feature attribute tables associated with various layers. Issues raised in this aspect of data base design are similar to those encountered in the design of relational data bases (see the discussion of normalization in Kent 1983, for example).

As an example of an ARC/INFO data base design, a simplified design for the information contained on a USGS 1:24,000 quad sheet might be:

```
Layer:  General Land Office land net
    arcs:   section lines
            items: line type (e.g.,township border)
   nodes:   section corners
            items: type (surveyed, photo identified)
polygons:   sections
            attributes: township, range, section, meridian

Layer:  Hydrographic Lines and Points
    arcs:   drainage lines
            attributes: type (intermittant, channel),
                        stream hierarchy, river mile index
  points:   wells, gage stations, etc.
            attributes: type

Layer:  Hydrographic Polygons
    arcs:   shorelines
polygons:   lakes, rivers, mudflats,etc.
            attributes: type (lake, flat, river,etc.),area

Layer:  Transportation Network
    arcs: roads, railroads,transmission lines,etc.
            attributes: type,class,DOT segment
```

Layer: Topography
   arcs: contour lines
      attributes: elevation
   points: spot elevations
      attributes: elevation

Several points can be made here. First, by separating hydrographic lines from polygons and by adding channels through the polygons, network analysis models can be used to easily determine upstream/downstream relations and to compute paths between points in the river network. In addition, because both the hydrographic lines and transportation network are simply arcs, the same application code can be used in both cases. This is also true for digitizing, editing, plotting, and all other GIS functions. Because each function simply operates on abstract coverage features (e.g., arcs, nodes, etc.), there is no need for specialized data-dependent logic.

A second point in this example is the inclusion of pointers to non-map data in the geographic data base. Every road arc has a DOT segment number which relates to a statewide transportation data base. This allows information carried in the external data base to be easily related to the cartographic features on the map.

APPLICATION OF THE DATA MODEL

The ARC/INFO data model has been designed to complement and support the geoprocessing functions of the ARC/INFO system. This is in contrast to most other geographic data models, which have been designed either as abstract "models of space" or as standardized repositories for map data. The applications which presently operate using the ARC/INFO data model are outlined in the following table (ESRI 1984).

Input Functions
   interactive digitizing and editing
   read DLG, GIRAS, DIME, and SIF
   build topology from unstructured data
   input screens for tabular data
   interactive edge matching
   coordinate geometry (COGO)
   GLO legal description
   grid cell conversion
   interactive districting

Analysis Functions
   map projections
   polygon overlay
   point in polygon
   line in polygon
   feature selection
   feature removal
   polygon aggregation
   ajust coordinates to control
   section subdivision
   network allocation
   minimum path calculation
   compute buffer zones

relational operators for tabular data

Output Functions
mapping based on feature attributes
- point, line, area symbols
- automatic text placement
- key and legend generation
report writer for tabular data
DLG files
convert to grid cell
interactive map query

Data Management
map sheet split
map sheet merge
map library management

The ARC/INFO geographic information system is presently in use at over 60 sites. In addition to data bases developed by other users, Environmental Systems Research Institute has developed a number of large data bases using the data model during the last three years.

UNEP/FAO World Data Base
area: world
scale: 1: 5,000,000
layers: coastline, FAO soils, FAO agro-ecological zones
size: 1.2 million points; 25,000 polygons

UNEP/FAO Africa Data Base
area: continent of Africa
layers: coastline, elevation, terrain units, rainfall, rainfall, wind velocity, hydrography, roads administrative units, railroads
size: 1.4 million points ; 53,000 polygons

Illinois Lands Unsuitable for Mining Program
area: State of Illinois
scale: 1: 500,000 (areas at larger scales)
layers: terrain units, hydrography, administrative units, oil and gas pipelines, transmission lines, coal resources, land net, roads, surficial deposits
size: 1 million points; 70,000 polygons

Alaska Land and Resource Mapping Program
area: Alaska - 80,000 sq. miles
scale: 1: 250,000
layers: terrain units, hydrography, administrative units, infrastructure, energy and mineral resources, elevation provinces, historic sites
size: 1.8 million points; 62,300 polygons

North Slope Borough Resource Inventory
  area: Alaska - 97,000 sq. miles
  scale: 1: 250,000
  layers: terrain units, hydrography, administrative units,
  infrastructure, energy and mineral resources,
  elevation provinces, historic sites, North
  Slope Borough planning data, subsistence land use
  size: 1.4 million points; 56,600 polygons

REFERENCES

ESRI, 1984, ARC/INFO Users Manual, Environmental Systems Research Institute, Redlands CA

Corbett, J. 1979, Topological Principles in Cartography, U.S. Bureau of the Census, Washington

Kent, W. 1983, "A Simple Guide to Five Normal Forms in Relational Database Theory", Communications of the ACM, Vol. 26, No. 2, pp. 120-125

Mitchell, W. et al 1977, GIRAS: A Geographic Information Retrieval and Analysis System for Handling Land Use and Land Cover Data, U.S. Geological Survey Professional Paper 1059

Morehouse, S. and M. Broekhuysen 1982, ODYSSEY Users Manual, Harvard Graduate School of Design, Boston

Puecker, T. and N. Chrisman 1975, "Cartographic Data Structures", American Cartographer, Vol. 2, No. 1

SYNERCOM, 1982, The Synercom Information and Mapping System INFOMAP, Synercom Technology Inc.

USGS, 1984, USGS Digitial Cartographic Data Standards: Digitial Line Graphs from 1:24000-Scale Maps, U.S. Geological Survey circular 895-C

WELUT, 1982, MOSS Users Manual, U.S. Fish and Wildlife Service, Western Energy Land Use Team, Fort Collins

White, M. 1980, "A Survey of the Mathematics of Maps", Proceedings of AUTO-CARTO IV, Vol. 1, pp. 82-96

AN INTERIM PROPOSED STANDARD
FOR DIGITAL CARTOGRAPHIC DATA ORGANIZATION

Edited by: Timothy L. Nyerges
National Committee for Digital
Cartographic Data Standards
Numerical Cartography Laboratory
158 Derby Hall
Ohio State University
Columbus, Ohio 43210

Working Group I on Data Organization:

Timothy Nyerges (Chair)   Northwest Cartography, Inc.
Donna Peuquet (Vice-Chair)   Univ. of California, Santa Barbara
Frederick Billingsley   Jet Propulsion Laboratory
A. Raymond Boyle   University of Saskatchewan
Hugh Calkins   State University of N. Y., Buffalo
                                 (Working Group IV Representative)
Robert Edwards   Oakridge National Laboratory
Robin Feagas   U. S. Geological Survey
William Liles   Xerox Corporation
David Pendelton   National Ocean Service
Daniel Rusco   Defense Mapping Agency

ABSTRACT

An interim proposed standard for digital cartographic data organization focuses on the topic of data interchange. A conceptual model for data interchange describes the stages of data transformation. Objectives which set the philosophy for a general digital cartographic data interchange format are outlined. The format is defined in terms of data definition records and data records. Data definition records provide information for decoding data records. These data records follow the data definition records in a transmission. Alternative forms of data definition are discussed.

ACES:
A Cartographic Expert System

C. Pfefferkorn, D. Burr, D. Harrison, B. Heckman,
C. Oresky, and J. Rothermel
ESL, Incorporated
495 Java Drive
Sunnyvale, CA 94086

ABSTRACT

ACES is an expert system for labeling maps. In developing ACES an attempt was made to capture the expertise cartographers use in the labeling process. This expertise includes how work is organized, what techniques are applied in specific situations, and how those techniques are selected. To accomplish a labeling task effectively the system must also make use of task specific information which cartographers also evaluate. ACES is a system still under development that can currently solve moderately complex map labeling tasks involving point, line, and areal features. ACES also provides a general planning framework which can be tuned for varied graphic labeling applications.

INTRODUCTION

Cartographic products are used in support of a number of activities: nagivation, operation of various networks and systems, and the exploration for and management of natural resources, to name a few. To a great extent, cartographic composition is performed using conventional manual techniques which are extremely labor intensive, with the process of feature labeling consuming much of the cartographer's time and effort.

To produce quality maps, cartographers make many labeling judgements using techniques which may vary both by product and specific situation. In applying judgement, cartographers may be allowed to generalize, smooth and exaggerate features, move and delete symbols, or modify placement strategy while incorporating various aesthetic considerations.

In order to significantly automate the production of publishable maps, it is necessary to model the expertise of cartographers within the computer. Artificial Intelligence (AI) design structures help accomplish this modelling task by manipulating internally the symbolic representations of those patterns and interrelationships of placement "rules" which are used by cartographers.

To facilitate the development of AI systems, a number of knowledge engineering tools have been developed, including: object oriented programming, rule based systems, automatic storage management, sophisticated programming environments, and powerful man-machine interfaces (raster graphics, window managers, menu packages, extended keyboards and peripherals).

The first section of the paper describes the problems of the cartographer, the second describes how ACES was designed based on AI techniques. An example follows which demonstrates current results, conclusions put the work in perspective, and future development is discussed.

CAPTURING EXPERTISE

"For 150 years, distinct rules concerning type placement spread among topographers and cartographers by word of mouth. The master taught his journeyman and the latter inculcated them in his apprentice..." (Imhof 1975)

The initial step taken to define a working set of map labeling heuristics, and hopefully facilitate the process of inculcation mentioned by Imhof (Imhof 1975), was to identify what expertise was required by cartographers to successfully label a map. How is the work planned? What generic and specialized techniques are considered for use, and why are specific ones chosen over others?

Analysis indicates that cartographic type and label placement is governed by four factors:

(1) Generally accepted cartographic rules
(2) Organizational requirements and standards
(3) Individual style and "rules of thumb"
(4) Mediation by internal review

From map to map, or project to project, the knowledge base (cartographic rules, product dependent specifications, etc.) required to successfully accomplish the labeling task is reorganized into three parts:

(1) General procedural knowledge
(2) Task specific knowledge
(3) Heuristics based on "rules of thumb"

ACES employs a general procedural knowledge base[*] which provides guidance for situations which vary from the most simple problems of point symbol labeling, to the most complex, involving all three major feature label types (point, linear, and areal). Incorporation of a task specific knowledge base helps resolve those problems associated with a defined set of labeling requirements.[**] Yoeli (Yoeli 1972) describes the cartographic-geographic criteria for "easy

---

[*] For example, start with area labels, which are the most constrained.
[**] Product scale dependent symbology, font, text sizes, etc. For example, where should a product standard symbol be placed, and what variations in font, size, and orientation are permissible?

legibility and identification of the map names" as:

(1) Precise graphic relationship between the name and the relevant item.
(2) Minimum of mutually disturbing interference between the names and the other contents of the map.
(3) Application of didactical principles, i.e. the placement of the name in such a way as to amplify, if possible, the characteristics of the items named (e.g., flowing placement of river names, etc.).

What individual techniques, or "didactical principles", cartographers bring to bear for optimal label placement are often transferred informally and are best described as "rules of thumb". The potential to incorporate these informal rules is realized via the ability to include heuristic software modules that can be selectively accessed and modified, depending on varying placement considerations. The ability to combine generally defined, task variable, and floating rule of thumb knowledge bases provides ACES with a symbiotic capability for cartographic labeling tasks.

PROBLEM SOLVING

To capture the expertise of cartographers, it is necessary to represent the potential decisions they make and provide a problem solver for searching through these potential decisions. The ACES problem solver is based on a well known heuristic search method which can be characterized as a process of searching through a tree whose nodes are situations and whose branches are operations on those situations (Newell 1969; Hewell & Ernst 1965; Newell & Simon 1963, 1972; Nilsson 1971; Simon 1971; Slagle 1971; Pfefferkorn 1975).

In ACES the situations are subproblems consisting of a set of map entities or "mapnodes" to be labeled, and the operations are strategies or "actions" that can be applied to the subproblems to accomplish the labeling.

KNOWLEDGE REPRESENTATION

The knowledge representation used by ACES can be divided into three parts: mapnode description, interaction graph, and a decision or design tree.

The mapnodes description represents mapnode attributes such as location, type, icon, label text and font, influence rectangle and possible label positions. This information is used to develop an interaction graph which consists of a set of pointers. Each pointer connects two mapnodes to indicate that their influence rectangles overlap. An influence rectangle is the region which encloses all possible label positions for a mapnode. The generated interaction graph is used for planning and checking of potential label overlap.

The decision tree is used to control the search behavior of the ACES problem solver. It contains a history of what strategies have been tried and where the problem solver can continue its exploration. Figure 1 illustrates an example of an ACES decision tree.

PROCESSING SEQUENCE

Before executing the map labeling actions, the initialization procedure computes the following for each mapnode: possible label positions, influence rectangles, the interaction graph, and a class priority rating. The priority rating is computed for each map object (with associated mapnode) based on a user defined class lattice as illustrated in Figure 2.

All possible mapnode label positions stored within ACES have been chosen from several cartographic labeling studies. (Imhof 1975; Freeman & Ahn 1983; Yoeli & Loon 1972; Hirsch 1982).

Area feature labels are processed first because they tend to be the most constrained (Yoeli, 1972), followed by point, and then linear label placement. *

For areal features, potential label positions are computed based on each feature's particular shape and size. Positions can be determined using various algorithms (e.g. weighted centroid, skeletonization), and if the feature crosses a minimum area threshold, additional positions can be located around the perimeter.

Currently, point features have sixteen potential label positions depending on class priority, with optional provisions for multiple line placement when necessary. Point processing includes a comparison with the area labels which have been previously placed.

Linear (and curvilinear) placement can be computed using parameters based on the feature's shape and size, as well as priority rating and any guides or rules for placement (USGS, 1963).

Once the possible label positions for a mapnode are determined, the system calculates the mapnode's influence rectangle. After all the influence rectangles have been computed, all possible label positions for point feature icons, areal boundaries, and symbolized linear features are checked for overlap. If an overlap is detected, information is gathered concerning which features are overlapped and where the overlap occurs. This information is stored in the interaction graph. Using the interaction graph, the system thus computes the nearest neighbors for each mapnode. A nearest neighbor is defined as any node whose influence rectangle overlaps the influence rectangle of the initial node.

---

* Implementing areal feature labeling in this manner is an example of starting with a simple strategy, and expanding the approach based on emperical results.

Using this stored information, the labelling process begins. Adjustments are performed iteratively until all feature classes have either been placed successfully, or identified in a set for remedial action. Currently, remedial processing is performed automatically after all classes have been processed.

CONCLUSION

The current ACES system has demonstrated that much of the map labeling process can be automated (see Example 1). The current system is experimental and effort is required to produce a production quality system. We are continuing the exploration of applicable AI and graphics processing approaches before any effort is made to produce such a system. The ACES planning framework provides: a capability to focus appropriate strategies that are applicable to different problem sets, the use of techniques in priority order until one is successful, the ability to split problems into subproblems, and an ability to handle the successful and unsuccessful application of various strategies. Specific rules governing action strategies may be developed for each new application.

FUTURE DEVELOPMENT

It is clear to us that many of the AI techniques utilized in ACES are applicable to many other cartographic tasks. The generation and placement of symbolized graphics from multi-scale attribute data bases is one associated application. Other potential directions include map recognition (pattern processing), featurization, and scale dependent or product specific data extraction and generalization.

AI techniques are also applicable to many related tasks involving the differential processing and presentation of graphic and symbolic data. Annotation of any kind of image or graphic data, be it a map, engineering drawing, or business chart, could be performed with application development of the existing ACES system.

REFERENCES

Freeman, H. & Ahn, J. 1983, A Program for Automatic Name Placement: Autocarto Six (Proceedings of the 6th International Symposium on Automated Cartography), 444-453.

Hirsch, S. 1982, An Algorithm For Automatic Name Placement Around Point Data: The American Cartographer, Vol. 9, No. 1:5-17.

Imhof, E. 1975, Positioning Names On Maps: The America Cartographer, Vol. 2, No. 2:128-144.

Monmonier, M. 1982, Computer Assisted Cartography. Englewood Cliffs, NJ: Prentice Hall.

Newell, A. 1969, Heuristic Programming: Ill-Structured Problems. In Aronofsky (Ed.): Progress in Operations Research,, Vol. III, New York: Wiley.

Newell, A. & Ernst, G. 1965, The Search for Generality, IFIPS65, 17-21.

Newell, A. & Simon, H.A. 1963, GPS, A Program That Simulates Human Thought. In Feigenbaum, E. & Feldman, J. (Eds.), Computers and Thought. New York: McGraw-Hill.

Newell, A. & Simon, H.A. 1972, Human Problem Solving. Englewood Cliffs, NJ: Prentice-Hall.

Nilsson, N.J. 1971, Problem-solving Methods in Artificial Intelligence. New York: McGraw-Hill.

Pfefferkorn, C.E. 1975, A Heuristic Problem Solving Design System for Equipment or Funiture LAYOUTS. CACM, VOL.18, NO. 5:286-297.

Sheil, b. 1983, Power-tools for Programmers. In Barstow, D., Shrobe, H. & Sandewall, E. (Eds.), Interactive Programming Environments. New York: McGraw-Hill.

Simon, H.A. 1971, The Theory of Problem Solving, IFIPS71, 249-266.

Slagle, J.R. 1971, Artificial Intelligence; The Heuristic Programm-ing Approach. New York: McGraw-Hill.

U.S.G.S. 1963, Map Editing and Checking. Topographic Instructions of the U.S.G.S., Bk 4, Chaps. 4C1-4C3.

Yoeli, P. 1972, The Logic of Automated Map Lettering. The Cartographic Journal, 9:99-108.

Yoeli, P. & Loon, J. 1972, Map Symbols and Lettering: A Two Part Investigation. Final Technical Report to European Research Office, United States Army, London, England, Contract Number DAJA37-70-C-2659, Technion Research and Development Foundation Ltd., Technion City, Haifa, Israel.

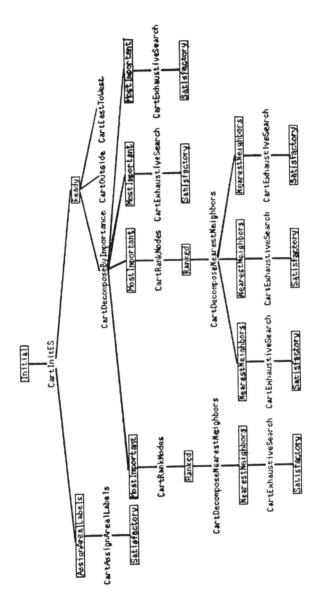

Figure 1: User Defined Decision Tree

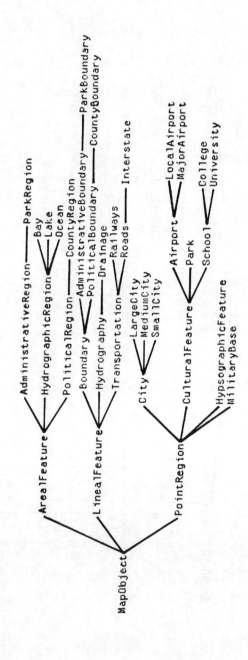

Figure 2: Defined Class Lattice

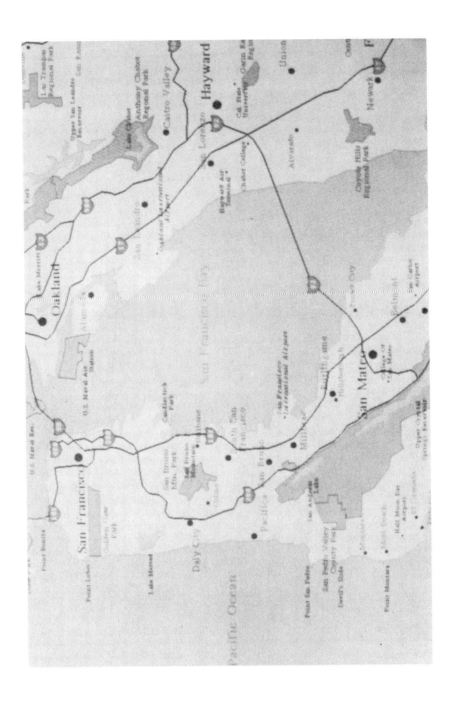

Example 1: Current ACES Product

A STEP TOWARDS INTERACTIVE DISPLAYS OF DIGITAL
ELEVATION MODELS

Thomas K. Poiker
Lori A. Griswold
Simon Fraser University
Canada

BIOGRAPHICAL SKETCH

Thomas K. Poiker (formerly Peucker) studied at several West German Universities, receiving his PhD (Geography, 1966) from the University of Heidelberg. He spent several post-doctoral periods at Harvard University and the Universities of Washington and Maryland. He is presently Professor of Geography and Computing Science at Simon Fraser University where he has taught since 1967. He is also Affiliate Professor at the University of Washington, Editor-in-Chief of Geo-Processing (Elsevier, Amsterdam), Principal of GeoProcessing Associates, Inc. and Senior Associate with Tomlinson Associates.

Lori Ann Griswold received her B.A. in Geography (emphasis cartography) from the State University of New York at Buffalo and is currently M.A. candidate (emphasis Computer Mapping) at Simon Fraser University.

ABSTRACT

Many full-fledged Digital Elevation Models (DEMs) are in existence and the number of programs that display digital surfaces are counted in the hundreds. Surprisingly, however, most of them do nothing more than contour the surfaces or display them by parallel profiles with the hidden portions removed. Few of them allow to add other information (road, houses, etc) and none is known to allow adding or deleting information by pointing at the 3-D display. The major bottleneck for the inclusion of higher sophistication, especially interactiveness, into the display of surfaces is a conceptional one: the transformation from one coordinate system to another is considered one-directional and visibility is a quantity computed during the conversion and lost at the same time. The solution is to make visibility a property of the spatial unit (point, square, triangle). The paper describes the conceptual aspects of this idea and then applies them to the display of three-dimensional surfaces. A few new and potentially powerful ways of computing visibility are demonstrated.

INTRODUCTION

For over two decades, researchers in different disciplines have worked in the area of Digital Elevation Models (DEMs). Many programs have been developed for the contouring of surfaces (Gold, 1984), three-dimensional display of different types (Peucker et al, 1975), slope analysis, highway planning, land-use analysis (Woodham, 1985), etc. Extensive research has been done on the types of data structures that work best for operations on surfaces (Peucker, 1978) and much care has been exercised to make the displays as pleasing as possible (Wolters, 1969).

Considering the extent of research and development in the field of DEMs, it is surprising how unsophisticated the results still are. The level of sophistication that we mean here is related to the graphic complexity of the results, not to their graphic quality.

This allegation has to be explained more. Most of the work in the field of Digital Elevation Models has been for very well defined results of relatively simple structure. In most cases, the computer method is of a variation that looks primitive in comparison to the manual equivalent.

A good example is that of the perspective display of surfaces, or block-diagrams as they are called in the field of manual production. All programs but a few use vertical profiles with the removal of hidden lines. A handful display contours and even less allow the combination of these two graphic primitives or support the display of random features on the surface. If we compare this to the sophisticated and artistically magnificent block-diagrams whose production has almost vanished by now, the computer products can only be called pitiful.

COMPUTER GRAPHICS

Whereas in the early stages of the computer's development, computer cartography pioneered several of the concepts and techniques in computer graphics, it has fallen behind in the pace of development in recent years compared with it's sister discipline. There are several fields in graphics that the mapping community has largely left untouched over the last years.

A good example for this is Computer-Aided Design (CAD). Whereas cartography has participated intensively in the development of the early stages of CAD systems, it has stayed with it's simpler aspects and left the more complex developments to computer graphics.

Like three-dimensional modelling. In computer mapping, the treatment of three-dimensional surfaces has been so far almost entirely for the purpose of display, very little for analysis even though the demand always seemed to be there and many people had promised different capabilities for a long time.

An example may clarify what we mean: Once a three-dimensional display of a surface (a block diagram) has been produced, one might want to perform one or more of the following functions:
(a) Shade all those areas that have steep slopes.
(b) Draw another data set (e.g. roads) on the surface, only showing the visible portions, of course.

(c) Point at the screen (through a tablet, etc) and request the position (i.e. the data-base coordinates) of the point that is indicated.
(d) Draw on the surface (again through a tablet or the like) additional graphic information.
(e) Slowly rotate the surface without recomputing the visibility of every element.

DIGITAL ELEVATION MODELS

With most DEMs that are presently available, the visibility of the elements of a surface is computed during the production of the display and the information is lost once the display is completed. The most frequent method computes one profile after the other, starting with the closest to the viewer, and testing each subsequent profile against the horizon which is the sum of the highest segments of all previous profiles. As the system passes a spatial unit of the surface (a square or a triangle), all information that has come into the picture at that point has to be ready and is displayed right then.

With this method, none of the above functions can be performed. For the first two, the system has to remember the visibility for later use, for the next two the system has to be able to inverse the projection function to find all the points on the surface that have the given coordinates on the screen, and then find the one point that is visible. And the last function will work if we can compute the visibility of an element for more than one viewpoint at a time.

In our continuing research in the area of Digital Elevation Models, we have developed a concept that allows the implementation of all these functions and have developed an algorithm for the visibility of surfaces that supports these functions.

THE CONCEPT

The concept is very simple: Instead of preparing the complete data base for the display of each visible element, make the visibility a property of the database and deal with it as with all other variables in the base. In other words, the computation of the visibility is separated from the production of the display.

Even though the separation of visibility and display into two processes is fairly obvious, it leads to a chain of other considerations. If these two functions can be separated, why not separate the functions into even more processes?

As will be shown later, part of the computation of the visibility of a unit is the computation of the "normal vector", i.e. the vector that stands perpendicular to the surface at the unit. The visibility is then determined by computing the angle between the view axis and the normal vector. If that angle is less than 90°, the unit is visible. If the normal vector is stored as part of the data base, half of the visibility test for any view axis is done.

In the following it will be shown how these ideas were implemented in our DEM for the modelling of surfaces. First, we describe a new algorithm for the computation of visibility on a surface and then we discuss its application for the different functions described above.

THE TIN SYSTEM

The framework for developing the visibility is the Triangular Irregular Network (TIN) (Peucker, et al, 1977). The TIN system is based on irregularly distributed points that are first triangulated into a series of connecting facets. The triangulation algorithm allows the generalisation of the surface by selecting points based on Cheybyshev's approximation criterion.

The facets are always triangles which are defined by their three vertices and three edges. In the main data structure, the vertices are stored with their geometric coordinates and pointers which represent the labels of those points which are at the endpoints of the connecting edges. The sequence of the edges around a point is always clockwise. Triangles can therefore be identified by finding one vertex, determining one of its neighbors (pointers) and the next pointer in the pointer list. A secondary data structure is based on the triangles and contains pointers to the three vertices and the three neighboring triangles.

The DEM was initially implemented on an IBM mainframe and contained, besides the basic structures and the triangulation algorithm, programs for contouring, block-diagrams, visibility charts, relief shading, inclined contours, slope map, and random drawings on the surface.

In 1978, a third generation of TIN was initiated on microcomputers (the first generation was a simple system developed by D. Cochrane (1974) as his Master's Thesis). This new system is termed Micro-TIN (uTIN) and operates under UCSD Pascal (Poiker, 1982).

VISIBILITY

As has been said before, the computation of the visibility takes place in stages. The major steps involve the following computations:
(1) "Exposure".
(2) "View vector".
(3) "Potential visibility".
(4) "Potential horizons"
(5) "Display space".
(6) "Visible horizons".
(7) "Visible triangles".

Several of these levels supply data that can be entered in the original data base. Here is a description of the different stages.

1. The "exposure" is the vector that stands perpendicular to a triangle, with the origin in the center of the triangle. This line is also called the normal vector. In vector notation it is the cross-product of the coordinates of the triangle's three vertices. Stored with the other data for each triangle, it serves the computation of the visibility from any observation point as well as for relief shading, slope analysis, etc.

2. In the case where parallel projection, relief shading, sun intensity, etc., are computed, the "view vector" is given for all triangles as one value. If the observation point is a finite distance from the surface, it has to be computed individually for every triangle. The view vector is computed in vector form.

3. When the view vector and the normal vector are given for a triangle, the "potential visibility" is positive if the angle between the two is less than 90 degrees. That means that the triangle will be visible unless some other triangles lie in front. The cosine of the angle between two vectors is their dot-product. If the cosine is less than one, the triangle is visible.

4. If two triangles are neighbors and one of them is visible and the other is not, the separating edge is part of a "horizon" or a valley. We call them "potential horizons" since all final horizons will come from this set of edges. Valleys can be eliminated by testing the valley-edge against both third vertices or kept for the next step where they would fall out automatically.

5. Up to this point, all computations have been undertaken in what is called "data space", i.e. using the coordinates as they are given in the original data base. Now the coordinates have to be converted from "data space" to "display space", i.e. into the (two-dimensional) coordinates of the final display. But remember, the data set that has to be converted is largely reduced compared to the original data set: only approximately half of all tiangles are potentially visible, all others can be eliminated at this stage. It should be noted here that each triangle has to have a unique label which has to be carried over into the display file. The reason for this extra parameter will become clear later.

6. When an artist draws a landscape with mountains, he/she usually starts with the horizons and then fills in the rest (David H. Douglas has incorporated this function into his block-diagram program and finds the results very appealing. Douglas 1971). To do this in our system, we first have to eliminate those horizons that lie behind and below others. This implies a trivial intersection of all horizons and the deletion of those segments that are hidden by other segments. Of course, all valleys would be eliminated at this stage. We are left with all "visible horizons".

7. Next, all triangles are tested against these horizons. In doing this, the triangles will fall into three groups:
   (a) Triangular facets that are totally visible.
   (b) Triangular facets that are partially visible.
   (c) Triangular facets that are totally non-visible.

If the visibility of a triangle is being stored in the original data structure, each triangle is labelled appropriately but at the same time a display file, i.e. a list of all the visible triangles, is established. The triangles that are partially visible are split into smaller triangles and the visible sub-triangles are added to a special list in the original data base and entered as visible triangles into the display file. This means that the original data base is not enlarged each time a display is computed.

This last process can be done in a more elegant way: Each string of edges that form one continuous horizon throws a shadow that is also a continuous string of lines and the two together create a closed polygon. This method seems to be faster, at least for the determination of shadow areas and the so-called radar maps. However, care has to be taken with shadow areas that have visible islands, i.e., mountains that become visible behind other mountains.

USE OF THE VISIBILITY

After the computation of the visibility of a surface, we are left with two data files: One is the augmented original file, augmented by the visibility information and two extra files, one giving a list of all fully visible triangle and the other containing those triangles that are broken out of partially visible triangles. The other file is the display file which contains the visible triangles in display coordinates. Both files are so structured that for any element (triangle, vertex, edge) in either file its complementary element in the other file can be accessed without long comparisons. The strong relationship between the two files is essential for the success of any further use of these files.

If we now return to the five functions that we want to perform with our DEM, we can summarize them into three different groups.
   (a) Convert from the original data base to the display file.
   (b) Convert from the display file to the original data base.
   (c) Change the external condition slightly.

An interesting aspect of this approach is that most operations can be performed with much less overhead than one might initially expect. For example, there is no need to develop any transformation parameters for the conversion from the display file to the original data file and the slight change of the observation point can be performed with many of the data from the previous view.

ADDITIONS TO THE DISPLAY

Once the basic display file has been developed, any additional computation becomes a local process. If we want to draw a curve on the displayed surface, we do this using the following steps: Given the curve in the original coordinate system, every point along the curve is determined with respect to the triangles of the DEM. In other words, its "local coordinates" are computed. These local coordinates contain the number of the triangle in which the point is located and the x,y coordinates of the point, standardized to the local coordinate system of the triangle. These points are then transferred into the display file and converted from the local coordinate system of the same triangle (this time in display coordinates) into the coordinates of the display system. This might sound more complicated than it is, in most cases the computation is two-dimensional and can use many shortcuts.

Any set of data in the original file can be converted into the display file. Smoothing usually takes place in the display file. Equally, one can draw on a digitizer or tablet (or with a mouse) and the line will be displayed on the display as if it was drawn directly on the surface. Hachuring is also done first in the original data base and the display of the parallel lines takes advantages of the similarity between adjacent lines, a characteristic that is usually called graphic coherence.

MANUAL INTERACTION

When we point at the screen with a stylus, a mouse of a tablet, or digitize a hardcopy of a three-dimensional display, we actually interact with the display file and the program finds for every point that we identify the corresponding triangle in which the point is located.

From there onwards, the process is basically the same as the conversion from the original data base. Since the triangles are known in both data sets, the conversion is again a local problem. Asking for the coordinates of a location that the user points at is therefore a direct process.

Drawing on the screen is a little more complex. Here, the display coordinates have to be converted to data base coordinates and then back to display coordinates. Since the system has to remember everything that the user does with the surface, a direct drawing on the screen (as is done on several graphic systems with a bit-mapped display) is not possible since the results cannot be changed afterwards.

ANIMATION

With this approach the production of several views of a surface from observation points that are changed only slightly (the typical flight over a surface), several time-saving shortcuts can be employed since the change from one image to the next will be minor. Some of these shortcuts are:

(a) Recompute the coordinates of the visible points only every second frame.
(b) Recompute actual visibility only after the observation point has passed through a few degrees (without recomputing the horizons).
(c) Ignore partially visible triangles, treat them as if they were completely visible and draw the horizon as a thicker line.
(d) Recompute the horizons only after the observation point has passed through a few degrees, more than in the previous points.
(e) Maintain the normal vector for each triangle.

CONCLUSION

At the beginning of this paper, a fairly strong statement was made about the state of development of Digital Elevation Models. It is obvious that the presented work is only a very small part of an evolution of DEM studies that eventually will lead to some very powerful tools for surface handling. We believe, however, that we are marching in the right direction, even if very slowly.

REFERENCES

Cochrane, D. 1974, A System for Triangular Representation of Surfaces. Master's Thesis, Simon Fraser University.

Douglas, D.H. 1971, personal communication.

Gold, Christopher M. 1984, Common-Sense Automated Contouring-- Some Generalizations. Cartographica, vol. 21, 2+3, pp. 121-129.

Peucker, T.K. 1978, Data Structures for Digital Terrain Models: Discussion and Comparison. Harvard Papers on Geographic Information Systems, Dutton, G., ed., vol. 5, Cambridge, Mass. 1978.

Peucker, T.K., R.J. Fowler, J.J. Little and D.M. Mark 1978, The Triangulated Irregular Network (TIN). Proceed., Digital Terrain Model Symposium, ASP, St. Louis, May 1978.

Peucker, T.K., M. Tichenor and W.-D. Rase (1975), The Automation of Relief Representation. In: J.C. Davis and M. McCullagh (eds.), Display and Analysis of Spatial Data, New York, pp. 187-197.

Poiker, T.K. (ed.) 1982, The uTIN Programs: User's Guide and Programmers Guide, Simon Fraser University.

Walters, R.F. 1969, Contouring by Machine, a user guide. Amer. Assoc. of Petrol. Geologists, Bull., vol. 53. pp. 2324-2340.

Woodham R.J. and T.K. Lee 1985, Photometric Method for Radiometric Correction of Multi-Spectral Scanner Data Canadian Journal for Remote Sensing, to appear.

Abstract

Title: A non-problematic approach to cartography within the constructs of a Geographic Information System.

Author: Stephen H. Pratt
President
GeoBased Systems, Inc.

Traditionally, cartography has been used to communicate complex information in a simplified, easy-to-read form. With the implementation of computer technologies, a major dichotomy has developed between those people interested in automated cartography and those people interested in using geographic data in an information system context. This separation is exemplified in the orientation of Geographic Information Systems as opposed to that of CAD mapping and drafting systems. Typically, Geographic Information Systems performs information management tasks with limited graphic capabilities. The CAD mapping and drafting systems produce detailed and accurate graphics while being difficult to optimize for information content.

The proponents of Geographic Information Systems often proport that graphics are not of major importance and the CAD proponents claim that a simple retrievial system with detailed drafting and mapping capabilities is a more meaningful approach. GeoBased System's experience has been that these two forms of communications, ie. graphic and tabular, are in fact both supportable in an automated environment.

By implementing a sophisticated but simple parallel data structure design, in combination with a distributed processing environment, one can have high quality cartography within the boundaries of a Geographic Information System.

This paper presents the STRINGS data structure and the distributed processing environment developed by GeoBased Systems, Inc. The examples used in this paper represent systems and projects where the cartographic rendition of maps has been accomplished within a context of a Geographic Decision Support System.

MAP MODELLING AS MORE THAN WEIGHTING AND RATING

Robert Puterski
Sr. Cartographer
State of Colorado
13220 West 60th Avenue
Golden, Colorado 80403
(B) (303)866-3199
(H) (303)279-1824

Geographic information systems have traditionally used simple forms of map compositing techniques involving simple weighting and cumulative adding of each variable to develop some type of spatial analysis. Too often users of these tools violate the simple mathematical rules of levels of measurement. Users also make unintentional mistakes by disguising these fallacies under the name of decision analysis or some similar technique. As geographic information systems and automated mapping techniques become easier to use, less expensive, and diffused throughout the various professions, potential problems exist. Users with minimal training in both cartography and spatial analysis need education in these disciplines. This paper will review some of the common techniques used by most systems, their potential pitfalls, and means of overcoming these problems.

HYPER-ISOMETRY : n DIMENSIONAL MAPPING
IN TWO PROJECTIVE SPACES

Robert H. Raymond, P.L.S.
9631 Evergreen Valley Rd.
Olympia, WA. 98503

ABSTRACT

Mathematical projections have facilitated the expression of complex mapping data in modes convenient to it's distribution and manipulation. The various planimetric displays of geographic data are a common example of this utility. To date, perceptual limitations have restricted most mapping operations to three dimensions. By the sequential application of appropriate transformation equations to the rectangular coordinates of an n dimensional point it is possible to display forms and functions in the two and three dimensional spaces with which we are familiar. The simple relationships between the projection elements and the minimal distortions of the projected figures using the isometric projection make this a suitable schema to discuss here. No attempt will be made to address rotation or translocation of the projected figures; nor will proofs of the operations be discussed, as these considerations are beyond the scope of this paper.

INTRODUCTION

Beginning in 1637 with the Geometry of Descartes, analytic geometry has provided the essential foundation for the mapping of the universe. Descartes limited his discussion to two dimensions; and the technique has since been expanded to visually describe numerical data in three dimensions. However, due to our ability to directly perceive in three dimensions at most (exclusive of our perception of time, in time) three dimensional space has been the effective limit of our mapping abilities.

Various graphic techniques have been developed to allow the display of forms in fewer dimensions than those occupied by the objects displayed. Mathematical projections are an example of such techniques.

This paper will describe the geometry of one such projection. It will illustrate an algorithm for developing isometric transformation equations for mapping n dimensional forms in n-1 spaces. Through a sequential application of these equations it will become possible to graphically display information from any number of dimensional spaces in the two and three dimensional spaces in which cartographers usually communicate.

CONCEPTS AND CONVENTIONS

Isometry and Axonometry
The isometric projections are a special class of those projections known as axonometric. Axonometric projections are

distinguished in that all lines of projection are normal to the space upon which the projection is made. Isometric projections , as used here, will be understood as that set of axonometric projections where the direction angles from all of the n space coordinate axes to the lines of projection are equal. Figure 1 illustrates these ideas as applied to projections from three dimensional space into two dimensional space.

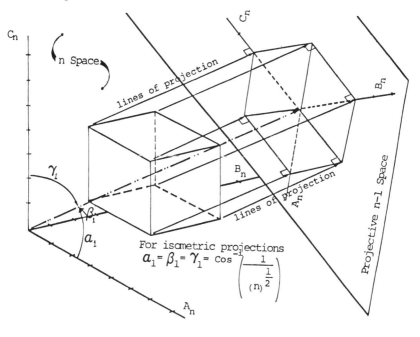

For isometric projections
$$\alpha_1 = \beta_1 = \gamma_1 = \cos^{-1}\left(\frac{1}{(n)^{\frac{1}{2}}}\right)$$

FIGURE 1.

Notation Conventions

Due to the unlimited dimensional spaces to which this process may apply, certain notation conventions for angles, coordinates and axes will be modified. For this discussion the following notation will be observed:

A) The rectangular coordinates of a point will be expressed as $(A,B,C,...)$ rather than as $(X,Y,Z,...)$. In this context the terms $A,B,C,...$ will be refered to as the "coordinate elements." Similarly, the respective axes are termed the A axis, B axis, C axis,... .

B) Points in n space will be denoted as P_n, with coordinates $(A_n, B_n,...)$. Points in n-1 space shall be denoted as P_{n-1}, with coordinates $(A_{n-1}, B_{n-1},...)$. The coordinates of points P_n and P_{n-1} so expressed shall be known as the "coordinate sets".

C) The direction angles from the axes to any point shall be labled $\alpha, \beta, \gamma, ...$

D) The angular distance between the n projected axes in n-1 space are all equal to each other; the angular

419

distance will be expressed by the symbol ℵ .*
In addition, each of the coordinate, axial and angular expressions will be subscripted to reflect the appropriate dimensional space.

These labeling conventions will ultimately prove inadequate for dimensional spaces such that $n > 24$, and may prove inconvenient where the english and greek symbols are assigned to constants and variables by previous custom. I have adopted this notation only to facilitate the present discussion, and will defer a final resolution till the future.

PROJECTION PROCESS

Coordinate Grids

As all coordinate sets will express the rectangular coordinates of a point, the coordinate axes are perpendicular to each other in their appropriate dimensional space. To project the n axes of the coordinate grid of n space it suffices to generate the median and vertices of the unique equilateral simplex polytope in n-1 space.** The median of this figure represents the origin of the n space grid as projected into n-1 space, and the lines extending from this median through the vertices represent the n projected axes. The negative extension of these axes is the reverse extension through the median from the appropriate vertex. The vertices are situated at a distance, h, from the median. The value of h will be determined below.

The angular distance between each of the projected vertices is determined as follows:

(eq. 1) $\aleph_{n-1} = \cos^{-1}\left(\frac{-1}{(n-1)}\right)$

This is illustrated with the projection into two dimensions of the three dimensional coordinate grid using the median and vertices of an equilateral triangle (the equilateral simplex polytope in two dimensions), where:

(eq. 2) $\aleph_2 = \cos^{-1}\left(\frac{-1}{2}\right) = 120°$

The scale along the projected axes, l_{n-1}, is related to the scale in n space, l_n, by a function of the direction angles to the lines of projection in n space ($\alpha_1, \beta_1, ...$), where:

(eq. 3) $\alpha_1 = \beta_1 = ... = \cos^{-1}\left(\frac{1}{(n)^{\frac{1}{2}}}\right)$; and

(eq. 4) $l_{n-1} = l_n \cdot \sin \alpha_1 = l_n \cdot \sin \beta_1 = ...$

If a true scale is desired for the projected figure in n-1

---

* This symbol, especially when subscripted, must not be confused with ℵ (Aleph-sub-_ series) which are used in transfinite number theory as the cardinality of infinite sets.
** Simplex Polytopes are the n dimensional figures having n+1 vertices, expressed by the Shlaffli symbol (3,3,...,3). The tetrahedron, a simplex polytope in 3 space, has the Shlaffli symbol (3,3):3 triangular faces at each vertex.

space let $l_{n-1}=1$. The resulting figure is more properly termed an isometric drawing; but involves no additional distortions than those resulting from the projection. Depending on the intended use of the projection either scale may be employed.

Geometric Relationships Between Grids
To facilitate the remainder of the discussion the following assumptions will be made regarding the relationship between the projected n space grid and the rectangular n-1 space grid:
A) The origin of both grids is coincident, and labeled O.
B) The highest order axis on both grids (those having the largest alphabetic expression where $A < B < ... < Z$) are coextensive. The highest order axial pair are coplanar. The highest ordered n-1 combinations are cospatial in the appropriate number of spaces. (see Fig. 2)

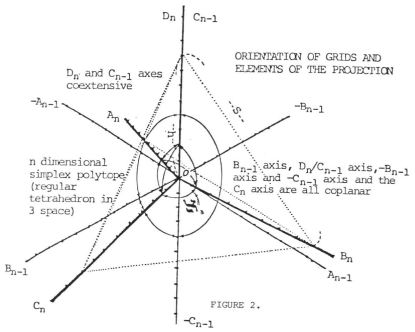

FIGURE 2.

The Projected n Space Grid
The vertices of all equilateral simplex polytopes are equidistant from it's median at a distance, h, which is related to the side length, s, by the following equation:

(eq. 5) $\quad h_n = \dfrac{(n-1) \cdot (s^2 - (h_{n-1})^2)^{\frac{1}{2}}}{n}$

By assigning a coordinate set $(0,0,..,0)$ in n-1 dimensions to the median and orienting the vertices of the simplex polytope for n-1 space as noted above, the n vertices take the coordinate elements as listed in table 1; and the sum of all like elements must equal 0. Table 1 is calculated assuming s=1.

TABLE 1.*
n-1 Space Coordinate Elements

	A_{n-1}	B_{n-1}	C_{n-1}	D_{n-1}	E_{n-1} ...
A_n	-0.500000	-0.288675	-0.204124	-0.158114	-0.129099
B_n	0.500000	-0.288675	-0.204124	-0.158114	-0.129099
C_n	0	0.577350	-0.204124	-0.158114	-0.129099
D_n	0	0	0.612372	-0.158114	-0.129099
E_n	0	0	0	0.632456	-0.129099
	$\Sigma=0$	$\Sigma=0$	$\Sigma=0$	$\Sigma=0$	

(Vertices of n Space Simplex Polytopes)

Thus, to calculate the n-1 dimensional coordinates of the n vertices of the respective simplex polytope, oriented as noted, it is sufficient to calculate the value, h. This value is assigned as the highest order coordinate element in n-1 space to the highest order n dimensional axis. The lower order axes are defined by the remaining vertices, which have a value, c, where:

(eq. 6) $c = -(\frac{h}{n-1})$

The vertices of an equilateral triangle so oriented with s = 1 will then have the coordinates: Vertex$_n$ = (A_{n-1}, B_{n-1}, C_{n-1}, ...).

A_3 = (-0.50000, -0.2886) B_3 = (0.5000, -0.2886)

C_3 = (0, 0.5773)   See Figure 3. & Table 1.

In order to calculate the direction cosines between any n-1 rectangular coordinate axis and the various projected n dimensional projected axes, plane trigonometry yields:

(eq. 7) $\cos \alpha_{n-1} = \frac{A_{n-1}}{h}$; $\cos \beta_{n-1} = \frac{B_{n-1}}{h}$; $\cos \gamma_{n-1} = \frac{C_{n-1}}{h}$; ...

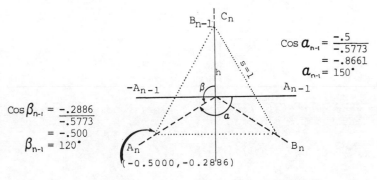

$\cos \alpha_{n-1} = \frac{-.5}{-.5773}$
$= -.8661$
$\alpha_{n-1} = 150°$

$\cos \beta_{n-1} = \frac{-.2886}{-.5773}$
$= -.500$
$\beta_{n-1} = 120°$

FIGURE 3.

*The n-1 coordinate elements for the n vertices are constructed from the appropriate n-1 x n matrix, from table 1, taking the first n-1 elements in each of the first n rows.

This calculation of direction cosines is repeated until all necessary values are generated. Tables 2,3 and 4 list the values for projections from three space, four space and five space respectively:

TABLE 2.
Direction Angles from 2 Space Axes

Projected 3 Space Axes

	α_2	β_2
A_3	150°	240°
B_3	30°	120°
C_3	90°	0°

TABLE 3.
Direction Angles from 3 Space Axes

Projected 4 Space Axes

	α_3	β_3	γ_3
A_4	144°.7356103	118°.1255056	109°.4712206
B_4	35°.2643897	118°.1255056	109°.4712206
C_4	90°	19°.4712206	109°.4712206
D_4	90°	90°	0°

TABLE 4.
Direction Angles From 4 Space Axes

Projected 5 Space Axes

	α_4	β_4	γ_4	δ_4
A_5	142°.2387561	117°.1573328	108°.8292311	104°.4775112
B_5	37°.7612439	117°.1573328	108°.8292311	104°.4775112
C_5	90°	24°.0948425	108°.8292311	104°.4775112
D_5	90°	90°	14°.4775112	104°.4775112
E_5	90°	90°	90°	0°

Transformation Equations
Since the n coordinate elements of a projected point, P_n, will each be located on it's corresponding axis at a distance from the origin specified by the numerical value of each particular element, these points define the vertices of another simplex polytope in n-1 space. This new simplex polytope will not, in most cases, be equilateral. The n-1 dimensional coordinates of these vertices are a function of the direction angles to the appropriate projected n space axis upon which they are situated.

The projected point P_{n-1} is the median of this second simplex polytope. To calculate the n-1 dimensional coordinates for P_{n-1}, the coordinates for each of the n new vertices is separately determined. Next, the median value for each of the n like elements is calculated, and the resulting set of median values is assembled as the coordinate set of the point P_{n-1}.

(eq. 8) $A_{P_{n-1}} = \dfrac{\sum_{m=1}^{n} A_m}{n}$; $B_{P_{n-1}} = \dfrac{\sum_{m=1}^{n} B_m}{n}$; ...

(eq. 9) $P_{n-1} = (A_{P_{n-1}}, B_{P_{n-1}}, C_{P_{n-1}}, ...)$

Thus far we have assumed a scale factor between the two grids of 1:1. This assumption allows for the generation of either isometric projections or isometric drawings. The following set of equations combines the operations described in equations (8) and (9); and provides for the application of the desired scale factor, l_{n-1}. The equations are listed for projections from three space into two space, four space into three space and five space into four space. Further levels of projection may be developed from the preceeding operations and observations.

(eq.10) $A_2 = l_2 \cdot ((A_3 \cdot \cos 150°) + (B_3 \cdot \cos 30°))$

(eq.11) $B_2 = l_2 \cdot (((A_3 + B_3) \cdot \cos 120°) + C_3)$

(eq.12) $A_3 = l_3 \cdot ((A_4 \cdot \cos 144°.7356103) + (B_4 \cdot \cos 35°.2643897))$

(eq.13) $B_3 = l_3 \cdot (((A_4 + B_4) \cdot \cos 118°.1255056) + (C_4 \cdot \cos 19°.4712206))$

(eq.14) $C_3 = l_3 \cdot (((A_4 + B_4 + C_4) \cdot \cos 109°.4712206) + D_4)$

(eq.15) $A_4 = l_4 \cdot ((A_5 \cdot \cos 142°.2387561) + (B_5 \cdot \cos 37°.7612439))$

(eq.16) $B_4 = l_4 \cdot (((A_5 + B_5) \cdot \cos 117°.1573328) + (C_5 \cdot \cos 24°.0948425))$

(eq.17) $C_4 = l_4 \cdot (((A_5 + B_5 + C_5) \cdot \cos 108°.8292311) + (D_5 \cdot \cos 14°.4775112))$

(eq.18) $D_4 = l_4 \cdot (((A_5 + B_5 + C_5 + D_5) \cdot \cos 104°.4775112) + E_5)$

Order of Processing

As the stated intent of this paper is to introduce a method for generating images in two and three dimensions of forms and functions extant in higher dimensional spaces, the equations developed above must be processed in descending order; from n space into n-1 space, then from n-1 space into n-2 space, etc. Once the projection is made into three dimensions any of the existing methods of projection into two dimensions may be employed. However, a certain consistency of results is lost if the projection schema is changed.

APPLICATIONS

As with any mapping operation, it is the responsibility of the cartographer to determine the appropriate projection method for the display of the information. To date, the options for mapping higher dimensional spaces has been quite limited. Hopefully, this is now a temporary state of affairs.

A complete list of possible applications for this type of projection is unnecessary; indeed, it would be impossible to delimit. However a suggestion of the potential for such a system may be gleaned from the following discussion of the relativistic curvature of the universe; "The analogy," visualizing the two dimensional space of a plane curving around the surface of a sphere,"collapses because it is hopeless to imagine what the extra spatial dimension looks like. No one has ever seen it." Callahan (1976) We may still be unable to see in these spaces; but we may now visualize constructs in them, and we may manipulate those constructs to our purposes.

The following set of illustrations are various regular n dimensional polytopes from the indicated number of dimensions projected using the above described processes.

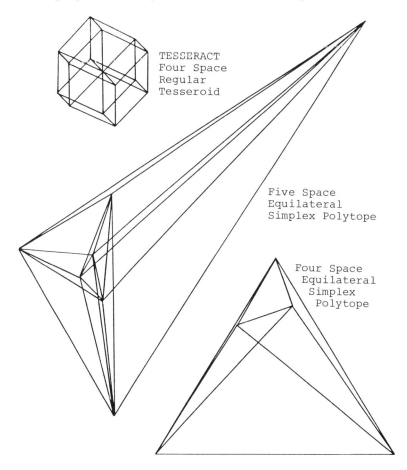

TESSERACT
Four Space
Regular
Tesseroid

Five Space
Equilateral
Simplex Polytope

Four Space
Equilateral
Simplex
Polytope

FIGURE 4.

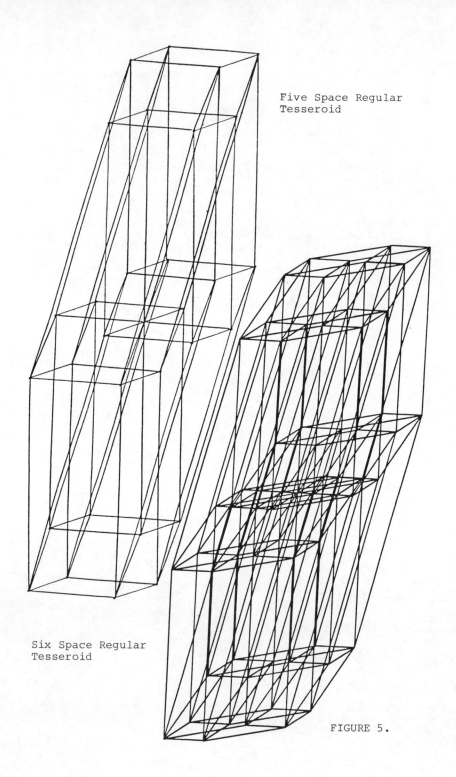

Five Space Regular Tesseroid

Six Space Regular Tesseroid

FIGURE 5.

BIBLIOGRAPHY

Abbott, E., 1963, Flatland, Barnes & Noble, New York

Burger, D., 1965, Sphereland, Crowell, New York

Callahan, J., The Curvature of Space in a Finite Universe: Scientific American, Aug. 1976, pp 90-100

Coxeter, H., 1964, Projective Geometry, University of Toronto Press, Toronto & Buffalo

Descartes, R., 1637, Discourse on the Method for Rightly Conducting the Reason; trans. Smith, D. and Latham, M.: Great Books of the Western World, vol.31, Wm. Benton, Chicago, London, Toronto

Raymond, R., 1983, Hyper-Axonometry, A Demonstration of Four Dimensions in Two Projective Spaces, unpub.

Thurston, W. and Weeks, J., The Mathematics of Three-Dimensional Manifolds: Scientific American, July 1984, pp 108-120

Building a Functional, Integrated GIS/Remote Sensing Resource Analysis and Planning System

Merrill K. Ridd and Douglas J. Wheeler
Center for Remote Sensing and Cartography
University of Utah Research Institute
Salt Lake City, Utah 84108
801-524-3456

Most state governments and many local jurisdictions express a need to encompass the disparate mapping and resource planning activities into a centralized system. Several states have made remarkable progress in establishing an automated geographic information system (GIS) to meet the challenge. Other states, perhaps with more complex infrastructures or smaller budgets, or without a driving force, still look to the day when they might begin.

This paper stresses two points: (1) to be an effective tool for resource analysis and planning, a GIS needs to be integrated with a digital remote sensing capability; and (2) to be truly functional, the paired system needs to be driven by grass-roots local needs. A case study in Utah will be used as a working model.

Several factors have conspired to bring these needs about: our expanded knowledge of resource dynamics, our experience with risks associated with lack of mapping and planning, the emergence of sophisticated remote sensing technologies to map and analyze the resources, and the emergence of computer technology to assimilate vast quantities of spatial and attribute data in GIS systems at controlled scales. The need for "vertically" integrated coordination between agencies from federal to local, and for "horizontally" integrated coordination from the local, to state, to regional level is self evident.

A potential pitfall for a state GIS office is the temptation to go out and digitize every map in sight to add to the library of accessible spatial data in case somebody may be able to use it. The tendency for the office to "sell" the system and recruit digitizing personnel may end in disappointment and/or files of questionably useful maps. To let the system be driven by specific resource questions in particular places might bear more palatable fruit.

Two basic questions are: (1) What data should be acquired to directly serve the resource planning needs and thus to make functional use of the GIS? and (2) How shall that data be acquired and placed into the system? The first question calls for a user-defined set of resource data needs, and the second calls for an integrated remote sensing/GIS system to acquire the data and get it into the GIS for analysis.

The case study presented is couched in a Soil Conservation District (SCD) in northern Utah. A group of agency representatives responsible for the condition and use of various resources in the SCD have assembled a prioritized list of resource information and planning needs. It was determined that the most fundamental data sets to be entered into the GIS analysis system in the first round were:

1. Land use/Land cover
2. Geomorphic/soil unit data
3. Hydrologic unit data
4. Digital terrain

The cheapest way to obtain acceptable land use/cover data in mountain areas was through Landsat digital classification (using a PRIME/ELAS system). The best way to get accurate data for the agricultural and built-up areas of the valleys was through recent photography, manually interpreted. The latter was digitized and merged with the Landsat data as a single "overlay" with integrated legend (using modified ELAS modules on a Tectronics digitizer). The results were segmented digitally into quadrangle chunks. Geomorphic/soil units were determined from B/W stereo photos and field work, coupled with Soil Conservation Service maps. They were digitized and added to the system. In similar fashion the hydrologic units were determined and entered. Relevant attributes per polygon and line were added. Finally, elevation, slope, and aspect (DMA) data were added.

The data were all formatted to enter the state's PRIME/ARC-INFO GIS. The data are now being interrogated for resource management decisions related to such issues as agricultural preservation, urban expansion, soil erosion control, and dam siting. The procedure and results will be illustrated.

EXPERT SYSTEMS IN MAP DESIGN

Gary Robinson
Michael Jackson

Thematic Information Service
Natural Environment Research Council
Holbrook House
Station Road
Swindon SN1 1DE

BIOGRAPHICAL SKETCHES

Dr Gary Robinson leads the digital cartography research section of the Thematic Information Service, NERC and is the convenor of the British Cartographic Society (BCS) special interest group in digital cartography. His interests cover all aspects of digital cartography, computing, expert systems (especially the man-machine interface), Geo Information Systems and error analysis. He graduated from Imperial College, London with BSc and PhD degrees in Physics and Astronomy in 1976 and 1981 respectively.

Dr Mike Jackson is Head of the Thematic Information Services (TIS) NERC. He worked with UK Department of the Environment and Atomic Energy Research Establishment, Harwell before joining NERC as head of the Experimental Cartography Unit (now incorporated within TIS) in 1980. He is Chairman of the European Association of Remote Sensing Laboratories Working Group on Integrated Geo-information Systems and Principal Investigator to NASA for the Landsat 4/5 LIDQA Programme as well as sitting on UK and International Committees related to digital cartography and remote sensing. His research interests relate particularly to the integration of digital cartography and remote sensing data and technology and the development of a knowledge-based integrated GIS.

Both authors are current members of BCS Council.

ABSTRACT

Increasingly spatial data is being collected, analysed, modelled and manipulated using digital computers. The users of such systems include planners, social and environmental scientists, businessmen and engineers. Where they once had recourse only to printed maps and plans they are now able to combine their data sets with spatial data and in the process generate new maps tailored to their specific needs. However, such people rarely have either the cartographic expertise or access to guidance from a professional cartographer. The result can be poorly presented maps and graphic material which fail to impart the information intended, or worse, mislead. What is required therefore is a means whereby the computer can itself act as a friendly cartographic 'advisor'. How can this be achieved? The paper presents one possible solution, using a so-called 'expert', or Intelligent Knowledge-Based System. It outlines the nature of such systems, discusses other potential applications within the field of cartography and describes a collaborative project in the United Kingdom between the Thematic Information Services of the Natural Environment Research Council, Aberdeen and Glasgow Universities and Kingston Polytechnic which involves the investigation and eventual implementation of such a system.

INTRODUCTION

The past decade has seen a rapid rise in the use of computer-based tools such as data-bases and mapping packages by a wide range of people, from scientists and engineers through to planners and administrators. Whereas these people originally had access to fairly limited data sets which they manipulated and displayed by relatively simple programs they are now able to combine disparate data sets such as digital vector and raster data with non-spatial information such as statistics. This all too often results in a product which fails to impart the information intended, or worse still misleads (Carter & Meehan 1984). Some may argue that 'bad workmen blame their tools', but if the workmen don't know any different then perhaps some responsibility should be accepted by the people who develop 'Geo-information Systems' and other mapping packages which are used by cartographically ignorant users.

A typical user rarely has the services of a professional cartographer with the necessary knowledge of map design and production techniques necessary for optimal display and presentation of data. This may be because he may not realise that one is needed or if he does then there may be insufficient cartographers or funds (or both) to go around. A partial solution is to provide cartographic training to all users. In general this is infeasible because of limited resources, cost, aptitude etc. What is therefore essential is that the user realises that a problem exists in the effective display of spatial data.

Probably the user most at risk is the one who produces maps or other graphical output for his own use or for limited circulation. There are several facets to the overall problem. Firstly the user, in designing his output, will often use an interactive graphic facility and therefore he needs to optimise the information appearing on the screen appropriate to his particular expertise. Secondly, the final product may appear on a totally different medium, e.g. paper, which leads to further problems. Thirdly the producer must take into account who the final product is intended for, and for what purpose.

These problems are ones of 'information transfer'. One solution which has been around for several years and is gaining increasing attention, particularly due to the Japanese 'Fifth Generation Project' is the so-called 'expert system' in which the knowledge and skills of one or more experts are encapsulated in the form of 'rules' which are capable of being manipulated by computers.

Do expert systems provide a suitable mechanism for solving the problems encountered in map design? To answer this question it is useful to outline what expert systems are, how they work, which fields they are currently employed in and describe a few example 'rules' involved in map design. Subsequent sections describe a collaborative project in the UK to implement an expert system-based map design system.

EXPERT SYSTEMS

Definition

A formal definition of an expert system, approved by the British Computer Society Specialist Group on Expert Systems (Naylor 1983) is:

'An expert system is regarded as the embodiment within a computer of a knowledge-based component from an expert skill in such a form that the system can offer INTELLIGENT ADVICE or take an INTELLIGENT

DECISION about a processing function. A desirable additional characteristic, which many would consider fundamental, is the capability of the system, on demand, to JUSTIFY ITS OWN LINE OF REASONING in a manner directly intelligible to the enquirer. The style adopted to attain these characteristics is 'RULE-BASED PROGRAMMING'.

How They Work

An expert system or 'rule-based program', contains three basic components:

assertions are statements of fact which are known or unknown and attain logical, numerical or textual (i.e. character string) values. Examples are:

'number_of_different_areal_symbols' (1)

'user' (2)

'map_is_confusing' (3)

rules are used to combine assertions together in such a way that the values of further assertions can be determined. For example:

'if number_of_different_areal_symbols 20 and
user = "tourist" then map_is_confusing = TRUE' (4)

the inference mechanism determines the order in which rules and assertions are processed. There are two fundamental ways of doing this which the preceding examples may be used to illustrate:

(i) To determine the consequences of showing a map containing 22 different areal symbols to a tourist the inference mechanism would apply rule (4) to assertions (1) and (2) to deduce using assertion (3) that it is probably too complicated for his use.

(ii) Conversely, one may want to know under what circumstances the map would be difficult to understand. In this situation the inference mechanism would use rule (4) to decide which assertions ((1) and (2)) are necessary to make assertion (3) (the consequence) true.

In either case the expert system would ask the user for the values of unknown assertions, unless a rule can be used instead. These approaches are known as 'forward chaining' and 'backward chaining' respectively. Which is used depends on whether the application concerned is predominantly 'data-driven' or 'goal-driven'.

Characteristics

The collection of assertions linked by rules resembles a network, or to use a better analogy a 'directed graph'. The way in which the inference mechanism navigates this graph varies from system to system and is the major factor governing the behaviour of individual expert systems.

Expert systems differ from ordinary computer programs in that the rules and assertions are treated as data, rather than being 'hardwired' or built into the code. This enables several different 'topics' to be accessed by a single inference mechanism. Also, most systems are able to dynamically update their rule-bases as new information becomes

available, for example during an inter-active session with a user.
An additional feature of backward chaining systems is their ability to
list the path currently being traversed in the rule-base, giving the
impression of justifying the line of reasoning.

Fields Currently Using Expert Systems

The number of fields in which expert systems are currently employed is
rapidly increasing. The originals chemistry, geology and medicine
are still the major areas though, with the chemical analysis system
DENDRAL reputedly being the oldest, dating back to 1965. Other well
known expert systems are (Michie 1979, Hayes-Roth et al 1983): MYCIN
which is used to diagnose medical ailments (specifically relating to
blood disorders); PUFF (based on MYCIN) is used to diagnose breathing
problems; and PROSPECTOR which assists in the location and evaluation
of potential mineral ore deposits.

Expert Systems in Cartography

Some areas of cartography and digital mapping where expert systems
could be of benefit are:

Manual and Automated Map Design. The knowledge of map design, in the
form of rules and heuristics seems reasonably well suited to expert
systems and forms the basis of this paper. Considerable advances have
been made in the area of automatic name placement (Freeman & Ahn 1984).

Digital Data-base/User Interface. The rapidly increasing complexity
of systems, diversity of data sets, usage and the increasing knowledge
required of their users means that this area is a prime candidate for
exploitation by expert systems. Several workers are already active in
this area (e.g. Peuquet 1983, Bouille 1983).

Cartographic Education and Training. Computers are playing an increasing role as teaching aids at all levels of education. Although
this is looked upon with mixed feelings it is clear that expert systems
will have a major impact in this field, and the teaching of cartography
is no exception.

Spatial Data Error-train Analysis. Systems holding rules about these
could assist a user in overlaying and combining different data sets by
indicating likely sources of error and which statistical techniques
are most appropriate.

Data Capture and Storage Standards. Since expert systems can be used
as computer-based repositories of facts they are ideal for on-line
storage of data standards and digitising conventions, with the possibility of allowing inconsistencies between different standards to be
weeded out.

Data Format and Transfer Standards. Expert systems offer a way of
holding different digital data formats in a self-contained manner and
could therefore provide a consistent mechanism for data transfer and
conversion operations.

Replacing Cartographers. As expert systems are intended to substitute for human experts then conceivably they could replace all
cartographers. However, this is not really practical since:

(i) It is unlikely that any human expert could (or would) formalise all his knowledge in such a form that a computer could take over his job (although there is one recorded instance of this happening, Feigenbaum & McCorduck 1983).

(ii) Cartography, like most subjects is evolving, especially in the light of new technological developments. Cartographers would therefore still be required to undertake research work, even if it was only to up-date cartographic expert systems.

(iii) Most importantly, cartography involves artistic elements which would be (as yet) impossible to capture in an automated system.

MAP-AID: A UK PROJECT

The subject of map design is very complex, particularly in view of the subjective decisions made in determining what constitutes a 'good' map. However, it is essential that an effort is made to solve some of the problems outlined in the introduction, which have arisen from the rapid increase of computing power, cheaper colour output devices and wider access to spatial data banks and other data sets.

Project Origins and Progress

The Thematic Information Services (TIS) of the Natural Environment Research Council (NERC) incorporates the Experimental Cartography Unit (ECU) which was a pioneer in digital cartography. Over the years the Unit acquired several hundreds of megabytes of structured digital map data contained within its Mk 1 database (Jackson et al 1983). In 1981 TIS became the centre for image analysis for NERC and rapidly acquired large quantities of airborne and satellite remotely sensed data. To exploit these data sets powerful interactive image analysis systems were purchased. The equipment includes one I^2S Model 70 and two I^2S Model 75 processors with host computers plus video and digitiser table input and colour raster output. Software was developed to allow the integration of the map and image data (Jackson 1984). In 1984 with the growing acceptance of these facilities as a standard research tool for geologists, ecologists, etc the process of providing local facilities was commenced and the transfer of the facilities into the NERC Computing Services began.

The above developments, whilst welcomed by users, increased the dangers indicated in the introduction, that is, of poor visual representation of data in the interactive and final presentation stages of a project with the consequent loss of information. The impetus was thus provided to transfer the cartographer's map planning, design and production skills to the user through the embodiment of his knowledge in an expert system.

The multi-disciplinary nature of the TIS staff with professional cartographers, geographers, mathematicians, physicists, etc, many with considerable computing experience, provided a sound basis for the research. In addition specific theoretical and practical cartographic expertise was sought from UK universities and the project team now includes J Keates of Glasgow University and M Woods of Aberdeen University.

Finally, independent computing expertise was incorporated through the participation of scientists at Kingston Polytechnic (G Wilkinson and P Fisher).

The Project

The early stages of the project have been concerned with:

Design and Planning. A project plan involving the authors and above identified collaborators in a 3-year programme of work has been prepared. The first significant working test system is planned for mid-1985. The use of the System Designers Limited Poplog program development environment which includes POP11, Prolog and a limited LISP compiler is proposed, supplemented by code written in FORTRAN and C. The GIMMS mapping package (Waugh 1980) is to be used for graphical output. More extensive graphic, map and image options will be added later. The computing environment comprises a 4 Mbyte VAX 11/750 running under VMS supported by an I^2S Model 75 image analysis system, a Britton-Lee IDM relational database machine and two ICL PERQ microcomputers with associated colour raster displays.

Increasing the Group's Knowledge of Existing Expert System Tools. A preliminary evaluation of languages and packages has been carried out to assess their suitability for map design applications. The SAGE package (SPL Int.1984) was selected for an in-depth three month assessment using a trial map data set. The necessarily limited purpose of this study was to identify suitable symbolisation for the different features to be included in the trial map. Conventions were incorporated into the rule base together with rules covering the relative importance of different features and what constituted acceptable combinations of symbols. The user then had the choice of picking from an available range of symbols, with the system providing defaults or warning when conventions were broken or poor combinations selected. The experience gained from this limited exercise suggested that many of the available commercial systems available were going to be too limiting for the complex task of map design. In particular groups of symbols or other features need to be selected together in parallel if frustrating and time consuming iterations are to be avoided. This parallel mode of symbol or feature presentation and selection will require powerful real-time graphics facilities. The hardware is available within low cost systems but software development is necessary. Because many of the decision making processes occur in parallel or near-parellel mode Prolog appears to be the best logic programming language for future developments of the MAP-AID system as it naturally lends itself to implementation on parallel processing systems (Clocksin & Mellish 1979).

Systems Design. The MAP-AID system (figure 1) is composed of three elements linked at the system level: the expert system; data-base system(s) and graphics package(s). Communication between these is via inter-process links (provided by most computer operating systems such as VMS and UNIX) on the same computer system, physical links (possibly over a wide-area network) between different computer systems, or a combination of both.

(i) Expert System. The expert system divides into four logical sections: the 'core' contains the map design rule-base and other information held as rules in a knowledge-base; the user module through which the user controls the entire system and interacts with the knowledge base; a set of data-base system modules (one per data-base) and a set of graphics package modules (also one per package).

(a) The Core. This contains the map design knowledge-base and is independent of the format of the user, data-base system(s) and graphics package(s) modules. To allow communication between the

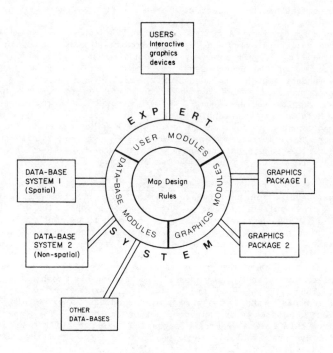

Figure I   STRUCTURE OF THE MAP-AID
            EXPERT-SYSTEM

core and the surrounding modules, a uniform set of standard interface procedures is defined, each of which must have a corresponding implementation in every user, data-base and graphics package module.

For example, suppose a particular design rule needed to know the scale of an input map. The core would invoke a sub-goal of the form:

'return the scale of the data set in the selected data-base'.

The selected data-base would be known (from a previous sub-goal or query) and the next sub-goal to be tested would be triggered in the rule-base of the relevant data-base module.

(b) User Module(s). The user module(s) operates as an intelligent or 'user-friendly' interface by converting what the user types in at a keyboard (or enters via other input devices) into a format suitable for processing by the core. Similarly, when the core requires information to be supplied by the user, because for example the data-base system is incapable of giving it, the user module will convert the query into an English-style question(s).

(c) Data-Base Module(s). To satisfy any queries asked by the core each module must be capable of either:

generating a query in the syntax of the relevant data-base system, issuing it, receiving the response, decoding it and then reacting accordingly.

or returning an 'unknown' response, in which event the system could ask the user to supply the answer via the user module (the user module resembles a data-base module in this respect).

The procedures that create, issue and receive data-base queries are written in 'C' or FORTRAN 77 for efficiency.

(d) <u>Graphics Package Module(s)</u>. In a manner analagous to the data-base modules, the core has a uniform set of defined graphical procedures. Each graphical package module has a corresponding implementation, if applicable.

To enable the map design process to operate efficiently requires the core to interrogate the graphics packages as to what line styles, colours area patterns etc are available. This is similar to the approach used in graphics systems such as Graphical Kernel System (GKS). GKS appears to be a useful starting point for defining the set of uniform procedures necessary in defining the interfaces between the core and the graphics package module(s). A GKS-based graphics package could even be used as one of the mapping systems. A similar approach could be used in the definition of the standard data-base procedures.

Each module may be viewed as comprising: an interface to the external system (the user, data-base or graphics package); internal rules to govern the functions performed by the module; and an interface to allow communication with the core.

(ii) <u>Data Base System</u>. The data-base(s) hold the spatial and associated non-spatial data in the form most suitable for the application in hand, and can be proprietry systems, specialised or house systems or even a combination thereof.
An essential feature of any data-base selected is that it has a well-defined interface at the system level, i.e. what the user would see if it were used interactively.

(iii) <u>Graphics Package</u>. Similar remarks about the data-base system apply to the graphics package. Communication between the data-base and the graphics package (via the expert system) is slightly easier to implement if the latter has some date-base capabilities, as GIMMS has.

<u>Rule Identification</u>. The rule identification stage raises fundamental questions in cartography and constantly tempts one to be sidetracked from the initial objectives. Thus, a simple sounding rule as expressed by the design cartographer such as 'don't use too many strong colours on complicated data' leads one into questions of perception of colour, measurement of colour, spatial interaction of colour etc. In trying to provide guidance in the use of colour the following parameters (and more) may need to be considered:

(a) the number of classes to be depicted (and number of colours to be used).

(b) the total area of the final map.

(c) the amount of overprinting (by lines and text)

(d) the mean size and variance of polygons

(e) the autocorrelation function

(f) the number and ratio of classes to sub-classes

(g) the degree of contortedness of the polygons

(h) whether the map contains polygons of discrete variables (e.g. land use) or continuous variables (e.g. contours)

(i) whether the map is intended for experts

(j) the relationship to conventions, etc

Rather than proliferate rules for the rule-base initial selection of the more general and powerful is sought even where full quantification cannot be achieved. i.e. the system will prompt the user for inputs on a nominal scale based on subjective assessment where precise values cannot yet be calculated.

Model testing. The cartographic model and rule-base are being interactively developed with theoretical evaluation of the model. The model is stepped-through conceptually to test for integrity and shortcomings. This process is useful in the early learning stages and allows more advanced concepts to be tested than in the implementation stage. The process is itself interactive with increasingly complex models being evaluated theoretically before test implementation. Results of the implementation using Poplog will be reported in a forthcoming paper.

CONCLUSIONS

Expert systems can play a useful role in many aspects of cartography and more work should be undertaken in this direction by cartographers and computer scientists. This is particularly important since it is pointed out (Taylor 1984) that unless cartographers get more involved in the 'New Cartography' they will be supplanted by computer graphic designers with inadequate knowledge of map design techniques but who will be responsible for producing the output for systems such as Telidon.

The direction of computer assisted cartography has over the years tended to move away from the original aim of producing 'look alike' versions of manually produced maps and instead focus on the information contained in the data. Many reasons have been presented for this, the most quoted being that computers are inefficient at the simple replication of manual techniques but are better at other tasks. The application of expert systems to map design may lead to this trend coming full circle.

ACKNOWLEDGMENTS

This paper is published with the permission of the Director, Natural Environment Research Council Scientific Services, United Kingdom. Our thanks go to Julie Swann for patiently typing many drafts and to Darrell Smith for producing the figure.

REFERENCES

Boulle F, 1983, A Structured Expert System for Cartography Based on the Hypergraph-based Data Structure, Proc 6th Int Symp on Automated Cartography, Vol II, pp 202-210.

Carter J R and Meeham G B, 1984, Austra Carto One, Perth, pp 259-276.

Clocksin W and Mellish C S, 1981, Programming in Prolog, Berlin, Springer-Verlag.

Feigenbaum E A and McCorduck P, 1983, The Fifth Generation, London, Pan Books.

Freeman H, Ahn J, 1984, Autonap - An Expert System for Automatic Map Name Placement, Proc Int Symp on Spatial Data Handling, Zurich, Vol II, pp 544-569.

Hayes-Roth F, Waterman D A, Lenat D B, 1983, eds, Building Expert Systems, Reading, Massachusetts, Addison-Wesley.

Jackson M J, Bell S M B, Diaz B M, 1983, Geographical Data-base Developments in NERC Scientific Services, Cartographica, Vol 20, No 3. pp 55-68.

Jackson M J, 1984, A Methodology for an Integrated Spatial Data Model, Proc of EARSeL Symposium on Integrative Approaches in Remote Sensing, Guildford UK, 1984.

Michie D, 1979, ed, Expert Systems in the Micro-electronic Age, Edinburgh University Press.

Naylor C M, 1983, Build Your Own Expert System, England, Sigma Technical Press.

Peuquet D, 1983, The Application of Artificial Intelligence Techniques to Very Large Geographical Data-bases. Proc 6th Int Symp on Automated Cartography, Vol I, pp 202-210.

SPL International, 1984, SAGE User Manual, SPL International, Abingdon, England, UK.

Systems Designers, 1983, Poplog User Guide, System Designers Ltd, Camberley, Surrey, England, UK.

Taylor D R F, 1984, Vol I, 12th Int Conf ICA, Perth, pp 455-467.

Waugh T C, 1980, The Development of the GIMMS Computer Mapping System, in Taylor D R F, 1980, ed, The Computer in Contemporary Cartography. New York, Wiley, pp 219-234.

ABOUT DIFFERENT KINDS OF UNCERTAINTY IN COLLECTIONS OF SPATIAL DATA

Vincent B. Robinson
Department of Geology & Geography
Hunter College - CUNY
695 Park Avenue
New York, NY 10021

Andrew U. Frank
Department of Civil Engineering
University of Maine at Orono
Orono, ME 04469

ABSTRACT

We differentiate between two broad types of uncertainty. Type I uncertainty deals with our inability to measure or predict a characteristic or event with uncertainty, where the characteristic or event is inherently exact. In type II uncertainty, there is a situation of intrinsic ambiguity regarding the concept to represented. The many sources of the two types of uncertainty are listed. Methods of handling differing kinds of uncertainties contained in collections of spatial data are suggested. Due to recent advances and its ability to represent and process the vagueness of natural language concepts, fuzzy logic is identified as major area for future research in managing uncertainty in spatial information systems.

INTRODUCTION

Quality of data and information is of great concern to designers and users of all spatial information systems. This paper first defines the 'correctness' of data in a formal way and then attempts to classify different types of uncertainty and indicates methods to deal with them. It is felt that the methodical discussion of terminology may be useful for the current discussion of standards for Geographic Information Systems (Moellering, 1982).

Data accuracy is thought of as the accordance between 'reality' and the information stored in a spatial database using a fixed interpretation (mapping). Inexact data may stem from inaccurate measurements or from differing interpretation of the same data. This leads to the identification of two broad types of uncertainty commonly associated with spatial information systems. The first type of uncertainty is related to our inability to measure or predict a characteristic or event with uncertainty, where the characteristic or event is inherently exact. The second type of uncertainty is derived from the intrinsic ambiguity regarding the concept to represented. It is hoped, that from this presentation a better understanding of the nature of uncertainty will result and lead to development of methods to handle some of the manifest problems.

FUNDAMENTAL HOMOMORPHISM OF INFORMATION SYSTEMS

Data is collected and placed in a database so that that human users can find answers to certain types of questions they need to fullfill their functions in an organisation. It is assumed, that using the database to find the necessary information is simpler, faster and less expensive than if every user has to collect the information directly from reality. In doing so the user relies on the implicit assumption that the answers they retrieve from the database are essentially the same they would find, if they wen out and collected the data themselves.

In order to further formalize this notion, we use the mathematical concept of a homomorphism. Given two sets of objects, original and model, and for each operations on the original objects or model objects. A homomorphism is then a mapping between original and model, such that the result of operations on the original object correspond to the result of performing the corresponding operations on the corresponding model object and vice-a-versa. For an information system (ie. a model of reality) this is to says that information received from the information system (ie. a model operation on model objects) corresponds to information gathered from corresponding operations on reality (Frank, 1982).

SPATIAL INFORMATION SYSTEMS

We use the term **spatial information system** (SIS) to include all sorts of information systems that contain and process data with respect to location in space, especially over the surface of the earth (Frank, 1980). A spatial information system may include databases commonly described as a geographic information system, land information system, geographic expert systems, etc. We assume that the following discussion applies to all such systems insofar as they store and treat data with respect to location in 'reality'. Some of the concepts are even more general and may be applied to other, non-spatial information systems. Each information element in a spatial information system consists of two pieces, namely the description of the nonspatial properties of the object and its location and extension.

In the flow of data-to-information common to spatial information systems the measurement process and the subjective assessment/classification process are attempts to gather data relevant to the purpose of the information system. That is to say that data gathered from reality is filtered as a function of the task domain of the SIS. Furthermore, since SIS's tend to be more general purpose than most information systems, definition of the task domain is itself often vague.

ON THE NATURE OF UNCERTAINTY

The distinction between exact concepts and inexact concepts has important implications in how we view the uncertainty contained in a collection of spatial data. Figure 1 illustrates how the two fundamental elements of information in an SIS may combine to define four possible states of uncertainty. We find areas that are clearly

441

determined, precisely measured, and properties of which are measured
with exactness. On the other hand, we have areas, that are very
imprecisely determined, eg. the Lowlands of South Carolina.
Although we may have at our disposal the means to exactly define and
measure the concept of population as an attribute of a spatial
entity, because of our inability to exactly define the location and
extension of the Lowlands of South Carolina we may not be certain
about the total population reported to be residing in the lowlands.

		Attribute of Spatial Entity	
		Exact	Inexact
Locational Definition of Spatial Entity	Exact	No Uncertainty	Uncertain Attribute
	Inexact	Uncertain Location	Uncertain Attribute Uncertain Location

Figure 1. Combinations of Attribute and Locational Exactness/
Inexactness Possible in a Spatial Information System

To elaborate, lets consider two spatial phenomena. First, a
cadastral land information system where property boundaries define
the boundary of a 'crisp' spatial set. That is, a portion of the
earth surface can either belong to 'property A' or 'property B' but
not both. However, in a landuse/landcover data base there are
several situations where 'crisp' boundaries are not in reality
detectable. Using classical Boolean logic, Robinove (1981) provides
a review of the difficulty in defining 'exact' boundaries. In
effect, he argues for modifying the existing system of
representation so that vague boundaries, such as ecotones, are be
more accurately represented as the vague, or fuzzy, spatial
phenomena that they are. In the former it is a question of how
accurate an 'exact' concept is measured, while in the latter it is
how to 'exactly' represent an intrinsically 'inexact' concept.

Classes of Uncertainty

We can identify with the procedures of data gathering different
imperfections. The user must live with such differences and
with the resulting uncertainty. In the traditional logic system
that underlies the current genre of SIS's, if perfect accuracy were
attained there would be :

1. No imprecision in determining the location or extension of a
certain phenomenon.

2. No error in measuring the essential characteristics that define
the phenomenon measured.

3. No uncertainty regarding the relative location or label of the
phenomenon.

4. No inaccuracy in the assessment of the phenomenon with respect to the interpretation of concepts.

5. No differences between subjects in interpreting the concepts used.

6. No changes in reality that are not immediately reflected in the data stored.

Normally, perfect accuracy can not be achieved for the following reasons:

1. Objects to be measured are often only vaguely defined.

2. Measurements are inherently imprecise (but with additional measurement expensed, nearly arbitrarily levels of precisions can be achieved).

3. In measuring, gross errors not of a statistical nature may slip in.

4. Schemes for classification, or 'labelling', are always imprecise and lead to different attributions depending on subjective judgements.

5. Attributes encoded on a ordinal scale (e.g. dense, medium, sparse), function largely as approximate qualifiers of labels.

6. The subjective and context-sensitive interpretation of 'facts' influences the encoding of facts and affects data during the use of a data collection.

7. Large differences between the intended use of a data collection and the actual use may lead to subtantial differences in the definition of terms and categories, thus leading to semantic error.

8. Facts as represented in the data collection usually represent a past state of reality.

We differentiate between two types of uncertainty. Type I uncertainty deals with our inability to measure or predict a characteristic or event with uncertainty, where the characteristic or event is inherently exact. Error propagation resulting from the distribution of measurement/observation error is an example of Type I uncertainty. The mapping here is from an exact characteristic to an exact representation of an exact concept. Another kind of measuring error may give rise to Type I uncertainty. In the course of a measuring process gross errors not of a statistical nature may slip in and be encoded in the collection of spatial data. The use of certain measuring and computational methods (Baarda, 1973) may limit the possible effects of undetected gross errors on the results (reliability). The underlying assumption for these two aspects of uncertainty in measurements is that the phenomenon being measured does in fact exist without imprecision, and that the sources of our uncertainty are found in the measurement process.

In type II uncertainty, there is a situation of intrinsic ambiguity regarding the concept to represented. The use of 'prototypes' , ie.

examples of 'pure' cases, is a common tool in an attempt to minimize
uncertainty. What is interesting about many examples of this type
of uncertainty is their relation to natural language concepts. It
is common for spatial data to be collected in a database format for
purposes of representing the type and distribution of land use or
land cover over a portion of the earth's surface. In addition, the
use of Landsat data to gather such representations provides an
excellent example of how we move from 'objective' information to
'subjective' labels.

To illustrate that the problems mentioned above are real, consider
the problem of optimizing the labeling of image classes. Aronoff's
(1984) method like others begins from the proposition that a
'location' on a map must belong to one class. That assumption is
couched in the proposition that 'map error' is a yes or no matter.
In his logic framework there is no such thing as a degree of error.
The existence of ambiguity is virtually ignored. However, the
existence of a single ambiguous point illustrates that 'pure'
classes are a construct that maintained for the sake of conceptual
convenience and tradition. It is also significant that the basis
for 'verification' is the interpretation of data by a human analyst.
Thus, classification error, in this case, is itself subject to the
imprecision with which humans manipulate linguistic concepts such as
'forest', 'residential', etc.

Like many other landuse/landcover studies objective data from
Landsat is subjected to unsupervised classifications to obtain image
classes, then the results are subjectively assigned landcover, or
resource, class labels by a human interpreter. The interpreter
typically uses aerial photography to accomplish this task (e.g. see
Pettinger, 1983). Thus, this is an inherently subjective task in
which the interpreter is attempting to match objectively derived
image classes with linguistic concepts that are represented by
linguistic prototypes in the mind of the interpreter. It is not
surprising then that there is variation in interpretation of the
very same data, ie imagery, among interpreters. This particularly
bothersome when the result is stored in a database because at this
point an inherently imprecise concept is stored as an exact
representation. Furthermore, a particularly questionable assumption
of this procedure is that somehow the interpretation of the ground
data is the benchmark against which to measure accuracy.

To illustrate the point that a concept such as a landcover class is
inherently vague, let us consider the problem described by Aronoff
(1984) when developing a data base   to be used to map areas of
Douglas Fir and areas of White Fir. In Aronoff's (1984) study it is
reported that a full third of all pixels verified as Douglas Fir
were classified as White Fir ! Surely there is a reason for this
very large discrepancy between objective methods and the subjective
ones used to 'verify' the results of objective measurements. As an
aside it is interesting as well that a full 41 those pixels
verified as White Fir were classified as Douglas Fir by objective
means. Further contemplation of this situation would lead us to
question the 'purity' of the respective fir stands. For example,
exactly, not approximately, but **exactly** when does an area become
White Fir rather than Douglas Fir ? When does this occur ? What is
it classified if in the pixel there are 40% Douglas Fir, 40% White
Fir, 10% Ponderosa Pine, and 10% Red Fir. The natural ambiguity of

vegetational communities, as defined by human interpretation is one of the issues that led Robinove (1981) to suggest that we begin mapping landuse without 'crisp' boundaries. Let us carry this problem further. When moving from one vegetational community to another, there is usually not a clear boundary. This has been known for centuries, yet we still map landcover as if boundary between forest and grassland can be exactly determined as a 'crisp' boundary.

Treatment of Uncertainty

Statistical Variations. Measurements with their associated statistical errors can be adjusted using statistical theory. Propagation of errors when combining several measurement values to compute a new value is straightforward, according to the law of error propagation:

$$m^2 = \left(\sum (f.i)^2 (m.i)^2 \right)^{\frac{1}{2}} \qquad (1)$$

where m.i is the statistical error on term i and f.i is the influence of this term, ie. the total differential of the function for this term at an approximate location. It is obvious from this formula, that error propagation increases the error, never decreases it.

Ambiguity. Uncertainty arising from ambiguity or subjectivity of the encoding method can be treated with Fuzzy Logic. In this formal system, exact information is viewed as the special case where inexactness has been reduced to zero. Buckles and Petry (1982, 1983) have presented a fuzzy representation of data for relational data bases that satisfies two of problems most often encountered in collections of land use data. The need for a single land use type to associated with a tuple has in the past been due to restrictions placed on data base management by the underlying logic. Using a fuzzy representation of data for relational databases (Buckles and Petry, 1982), Robinson and Strahler (1984) have shown how one can represent seemingly conflicting landuse/landcover classifications in a fuzzy data base. Thus, preserving explicitly the uncertainty inherent in landuse/landcover labeling.

Repeated interactions with a spatial data base often results when a user searches for an accurate representation of an approximate spatial concept using a system that can not represent, much less retrieve, an approximate concept. The user ends up being uncertain about the how well the retrieved data represents the approximate concept, this leads to further interaction with the information system in an attempt to lower that level of uncertainty below some level and in the end may make decisions based on a low confidence in retrieved information. Many of these additional steps would be eliminated if the system were capable of representing and retrieving such approximate information. Robinson and Strahler (1984) have shown how fuzzy representations of data can be incorporated into a landuse/landcover data base management system. Using the results of the work by Buckles and Petry (1982), they show how fuzzy logic can be used to retrieve landcover data stored as linguistic variables.

Another rather subjective class of data found in collections of spatial data are attributes encoded on a ordinal scale, for example - low, medium, high. Terms such as high, low, medium can be treated as linguistic hedges within the context of the above discussion of linguistic information such as landuse/landcover types. Retrieval of land information in the form of ordinal labels leads to a large number of interative operations when an exact system is used to represent and retrieve such linguistic hedges.

UNCERTAINTY AND THE EXPLOITATION OF SPATIAL INFORMATION SYSTEMS

Spatial 'facts' can be represented by one of four fundamental types of data - ratio, interval, ordinal, and nominal (see Figure 2). However, it has been noted with interest that "...the overwhelming majority of GIS applications concern some type of discrete phenomena. Topographic feature codes, place names, geocodes, parcel identifiers, land use types, all fall into the same broad group" (Chrisman, 1984: p. 309). The data Chrisman (1984) referred to is in the main nominal data. On the other hand, data such as elevation, spectral, and planimetric measurements are data of the interval/ratio type. Often the nominal data entered into a GIS derived from interval/ratio data. However, a common practice is then to consider only the nominal data during the retrieval process. If interval/ratio data is stored then the query is structured in such a manner that in concept it is a labeling, or classification, process for retrieving 'labelled' data. This appears to be consistent with recent research in forecasting and man-machine studies suggest that human information processing is geared towards the processing of 'qualitative' information rather than 'quantitative' information (Zimmer, 1984).

Figure 2 depicts the relation we suggest exists between the four fundamental data types, objective information, and meaning. Nominal and ordinal data are generally characteristic of the data type desired as output products by the average 'user' of geographic information systems. These are generally expressed in linguistic terms related to either attributes or relations. For example a nominal attribute might be 'residential', while a relation might be 'places Near Bangor, ME'. These terms are meaning-laden, that is

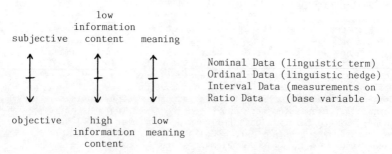

Figure 2. Relationship between information content, meaning, and data types.

to say that they may vary upon accumulated experience of the user and context of query. Ordinal data generally act as linguistic hedges used to place qualifications upon nominal data, often in an attempt to provide a sense of variation in the nominal concept arising out of a measurement process utilizing interval and/or ratio data. Thus, nominal/ordinal data is more meaning-laden to human users than is numerical data of the interval/ratio kind, while interval/ratio data typically contains more 'objective' information.

The above discussion is somewhat related to our previous critique of maps showing crisp boundaries where only fuzzy areas of slow transition exist. It is evident, that maps as products are very helpful sources of information because they present data in a grossly classified way and are therefore easy to use provided the mapmakers and the mapusers concepts agree. One of the main advantages of a map over a aerial photograph, to be classified is its information representation. The map produces less cognitive load on a user, therefore the user is more likely to make correct decisions under pressure (or would you like to use aerial photographs to navigate your automobile around New York City ?). This argument is not true if maps are considered as data collection intended for multiple use. We suggest that the representation of the data should be as near to the source as reasonable feasible and the classification to produce the desired nominal or ordinal data for the user should only be performed when output is produced.

Another subjective source of uncertainty was suggested by Chevallier (1983). The subjective interpretation of facts depending on the background and the goal of the task does not only influence the encoding of facts, but also affects data during the use of a data collection. If multipurpose data collections are built, differences between interpretations during data collection and use of the data must be taken into account.

Large differences between the intended use of a data collection and the actual use may lead to subtantial differences in the definition of terms and categories (semantic error). Discrepancies of this type should not be considered errors in encoding, but are systematic, regular differences in interpretation. Little is known about how such differences can be detected and taken into account. However, whenever they occur the effect may be considerable, especially in influencing decisionmaking process. Thus, much may be learned if the potential of natural language systems capable of representing approximate concepts can be exploited to study semantic error.

CONCLUDING DISCUSSION

Uncertainty in collections of spatial data can arise from a myriad of processes. Nevertheless there are two fundamental types of uncertainty present in most collections of spatial data. Type I uncertainty is related to the error associated with the mapping from reality to the information system model of an exact concept. Historically, this is the only type of uncertainty that has been formally acknowledged as present in collections of spatial data. We consider a second type of uncertainty. Type II uncertainty results from a mapping from reality to the information system model of an inexact concept.

It is evident that there is little activity in devising methods of explicitly incorporating uncertainty in the management and retrieval of information in SIS's. The most common prescription it to attach lineage and measures of Type I uncertainty to maps or digital data files, leaving the processing of this information regarding uncertainty to the individual human user. Furthermore, Type II uncertainty has received little, if any, attention.

Evident from our review of different kinds of uncertainty is that the role of natural language (NL) concepts has been virtually ignored. This is a serious omission since (1) much 'data' in SIS's is of a linguistic nature, (2) linguistic rules are the basis for defining many criteria for data collection, (3) users often demand linguistic terms as the output from a SIS, (4) quality of information, including lineage, is most often described using linguistic terms derived from human judgement. Thus, one of the implications of our review is that the role of natural language concepts in SIS's should receive more rigorous, systematic attention. Another is that there is a clear contribution to be made in this area by man-machine studies. Consideration of both NL concepts and methods of explicitly incorporating uncertainty in the processing data in SIS's suggests that fuzzy logic may be of some considerable utility in the domain of SIS's.

Working within the framework of fuzzy data representation and management allows one to explicitly represent uncertainty in such a manner that it is able to become an integral part of spatial information processing functions. Linguistic variables are able to be represented and retrieved in a manner most consistent with their ambiguous nature. It is also clear that this framework has the potential to store differing characterizations of linguistic data on a user-specific basis, thus allowing the conduct of systematic research on variations in meaning according to user, context, and task.

Finally, we hope that this short overview asks more questions than it answers, but provides a flexible framework for additional, indepth investigations. In particular, we feel that present spatial data processing systems should be revisited in light of the two types of uncertainty identified in this paper.

ACKNOWLEDGEMENTS

Partial support was provided by NSF Grant IST-8412406 and a Research Grant from PSC-CUNY. We would also like to acknowledge the support provided by the City University of New York Computing Center, the University of Maine Computing Center, and the BITNET network, without which this collaboration would not have been possible.

REFERENCES

Aronoff, S. 1984, An approach to optimizing labeling of image classes, Photogrammetric Engineering and Remote Sensing, 50 : 719-727.

Baarda,W. 1973, S-Transformation and criterion matrices, Netherlands Geodetic Commission, New Series, Vol. 5, No. 1, Delft.

Buckles, B.P. and Petry, F.E. 1982, A fuzzy representation of data for relational databases, Fuzzy Sets and Systems, 7 : 213-226.

_____ 1983, Information-theoretical characterization of fuzzy relational databases, IEEE Transactions on Systems, Man, and Cybernetics, SMC-13 : 74-77.

Chevallier, J.J. 1983, Une approche Systemique des Systemes d'Information du Territoire et de leur integrite, Ph.D. Thesis, Swiss Federal Institute of Technology, Lausanne.

Chrisman, N. 1984, The role of quality information in the long-term functioning of a geographic information system, Proceedings AUTO CARTO-6, 303-312.

Frank, A. 1982, Conceptual framework for land information systems - a first approach, paper presented at meeting of Commission 3 of the FIG, Rome, Italy.

_____ 1980, Land Information Systems - An attempt towards a definition. Nachrichten aus dem Karten- und Vermessungswesen, Reihe 1, Vol. 81, Inst. f. angew. Geodaesie, Frankfurt FRD 1980]

Moellering, H. 1982, The goals of the National Committee for Digital Cartographic Data Standards, Proceedings AUTO-CARTO 5, 547-554.

Robinove, C.J. 1981, The logic of multispectral classification and mapping of land, Remote Sensing of Environment, 11: 231-244.

Robinson, V.B. and Strahler, A.H. 1984, Issues in designing geographic information systems under conditions of inexactness, Proceedings of 10th International Symposium on Machine Processing of Remotely Sensed Data, 198-204.

Zimmer A.C. 1984, A model for the interpretation of verbal predictions, Int. Jrnl. Man-Machine Studies, 20: 121-134.

NAVIGATING WITH FULL-COLOR ELECTRONIC CHARTS:
Differential Loran-C in Harbor Navigation

By Mortimer Rogoff
Navigation Sciences Inc.
6900 Wisconsin Avenue
Bethesda, MD 20815

ABSTRACT

Computer technology is ushering in a new generation of navigation equipment. Microprocessor computers together with improving graphics allow advances both in the interpretation and utilization of radionavigation signals, as well as in the integration of those signals with navigation instruments in a new range of useful tools. Systems now being built offer improvement both in positioning accuracy and useful visual displays for navigators.

INTRODUCTION

This paper is an extension of a report submitted by the author at the IEEE Plans 80 Navigation Symposium. In that report the author predicted the advent of new solutions to navigation problems based on advances in microcomputer technology. It is now possible to describe an existing integrated navigation system founded on the principals presented at the previous symposium.

This new navigational system is based on the utilization of Loran-C radionavigation system in combination with a radar interface and electronic charting that results in a full-color, integrated graphic display.

When the differential Loran-C ownship position and the radar images are combined on the electronic chart, the resulting full-color display presents the navigator with an integrated image of ownship, waterborne radar targets, land and bottom contours, and navigation aids. This enhanced presentation is especially useful in harbor navigation where precise maneuvering is essential.

Navigation Sciences Inc. has developed this integrated system and markets it under the name VIEWNAV. The system's reliability and accuracy have been demonstrated by the U.S. Coast Guard, commercial users and a number of pilot associations.

Moreover, the integrated technology has applications far beyond use by vessels in harbors. For example, when signals from the Global Positioning System (GPS) are substituted for Loran-C, the resulting display will have significant land-based applications, too.

VIEWNAV: THE INTEGRATED SYSTEM

Despite sophisticated navigation devices, harbor navigation remains a difficult proposition, especially in foul weather. Accidents, groundings, rammings and down time due to poor operating conditions hamper harbor traffic, delay critical schedules, and result in additional expenses to marine operators.

The bridge of the modern commercial maritime vessel is equipped with an array of electronic systems to aid the pilot and captain, but the task of locating exact position and steering a vessel is a labor intensive job. Plotting fixes, for instance, is still done manually on a paper chart. Computation of waypoint information is another example. Of course, the instinct and experience of a good skipper, pilot or navigator are invaluable, but any help they can get from new technology to aid in quick and correct decision-making and therefore reduce the chance for mistakes and accidents is quickly appreciated and applied.

A good example of a problem that remained unsolved was navigating when floating aids were missing or off station. In northern waterways, ice often pushes buoys off station; or the Coast Guard may completely remove buoys in anticipation of the ice. The navigator is left to find the safe water -- the channel -- without the properly positioned channel markers.

The need here is for an easy to interpret navigational system which will show the decision maker current ownship position relative to the channel. In addition, the same display, by interfacing to the ship's radar can provide current and accurate tactical information -- such as other ship movement -- necessary to ensure safe passage.

The result is a display which serves as the single command center on the bridge. The first of this type of system -- called VIEWNAV -- now exists and it is being used daily in various harbors on the East coast of the United States.

To achieve this visual presentation, which includes a precisely-located image of the vessel (ownship) and waterborne radar returns updated in real time against the background of an electronic chart, three fundamental maritime navigation and maneuvering tools must be combined:

- o Loran-C radionavigation signals are refined through calibration and data processing in a differential mode to yield repeatable accuracies of 15 feet.

- o Paper charts are digitized and converted to accurate, flexible electronic charts.

- o A radar image is superimposed on the electronic chart and through the use of different colors, only waterborne returns are displayed.

ACCURATE POSITIONING WITH DIFFERENTIAL LORAN-C

Most modern means of fixing a vessel's position on water -- whether a ground-based system such as Loran-C or a space-based system such as NAVSTAR -- relies on emission and reception of radio signals. The signals remain constant, but the way they are received and interpreted determines the relative accuracy of a given vessel's position.

Refining the Loran Signal

In the case of Loran-C, all receivers, coordinate converters and plotters available today use Loran time differences (TDs) measured to the nearest tenth of a microsecond. These TDs are then applied to Loran charts based on mathematical predictions rather than actual measurements to locate the Loran lines of position (LOPs). Existing systems allow the potential for large margins of error. When TDs are measured to a tenth of a microsecond and converted to LOPs, the actual position can be off as much as several hundred yards. Given these parameters, few systems have been able to attain steady position accuracies more precise than 100 yards.

The use of differential Loran-C breaks through these limitations. First, TDs are measured and displayed to the nearest nanosecond, which represents approximately one foot. Then, the differential variation is achieved by correcting the existing Loran readings by applying offsets observed at a stationary monitor.

Stabilizing the Loran Grid

Loran signals in a given waterway are surveyed and calibrated. The surveys measure the actual geographic position of the lines of constant time differences rather than their predicted position. Calibration consists of determining the Loran time differences that correspond to carefully surveyed latitudes and longitudes. These correlations are used to convert Lorad TDs to vessel position.

In addition, a land-based Loran monitor is established to identify and correct for any grid shifts that occur during the survey. The result is a highly accurate "snapshot" of the Loran-C grid.

Loran Monitor System

Loran signals received in the same location frequently change by a small amount due to minor variations in radio wave propagation between the transmitters and ownship. These signal fluctuations are due to natural atmospheric phenomena The effect of the changes is to randomly offset position by as much as hundreds of yards during the course of a year and up to tens of yards in one day.

This difficulty is overcome by carefully observing monitor receivers in the local area and broadcasting any shifts to vessels that, in turn, use this data as a correction to the Loran signals being received on board both during the survey period and operational vessels.

It has been demonstrated that when the Loran grid is stabilized and monitored, and the Loran signal is read to the nanosecond, the result is repeatable position accuracy of 15 feet.

The U.S. Coast Guard contracted Navigation Sciences Inc. to survey the Loran-C grid in the Upper Chesapeake region. A lengthy study of observed TDs in this region demonstrated that the correlation of hourly and daily changes between TDs at scattered locations was very strong. If the TDs at monitors within this region were compared, the difference in computed position resulting from using changes at one monitor location to compute position at another resulted in position errors of about 4 yards (90% CEP). In other words, TD variations monitored at one location can be the basis for making corrections at any place within this region of correlation to an acceptable degree of accuracy. In the case of the Upper Chesapeake, this region of correlation extends all the way from the C & D canal to Baltimore, a linear distance of 40 nautical miles. Other regions will vary in their behavior, and have to be measured in order to establish the area over which sufficient correlation exists to be considered a single region for the purpose of differential Loran.

ELECTRONIC CHARTING

Electronic charts are an application of computer technology whereby a digitized version of a marine chart is displayed, in color, on a cathode ray tube (CRT) monitor. Numerous advantages flow from the act of liberating the chart from its classical paper foundation; the electronic medium endows the chart with a degree of flexibility that has been unthinkable in the paper version.

The utility of the electronic chart as both a valuable single advancement and as the background for displaying additional information are recognized in a report assessing the technology's impact on Canadian shipping. A report from the University of New Brunswick says, "Our most significant conclusion is that a major breakthrough in marine navigation is occurring and its success is inevitable. This is because, for the first time since high technology began producing aids such as the gyrocompass, radar, and Loran-C for the mariner, rather than adding to his workload these interactive navigation systems relieve him of the tedious part of his duties and aid him with real-time information on the most important part -- navigating his ship safely. Given that the widespread use of these systems will occur in the near future, it follows that the day-to-day use of the electronic chart will surpass that of the paper chart."

There are a number of methods that can be used to accurately produce an electronic chart. Navigation Sciences' electronic charts are generated by tracing chart contours on a digitizing table. The contour tracing methods possess the advantage of producing a selective degree of detail from the printed original.

The electronic chart allows the addition of a great deal of associated chart data than can be displayed or repressed with the traced contours. This can include water depths, channel boundries, the location and identification of various aids to navigation (buoys, lights, structures, tanks, towers, etc.), as well as place names and other labels.

Chart accuracy is high. The present state-of-the-art of electronic digitizers permits reproducible positioning of the tracing cross hair to within 0.003". A typical nautical harbor chart, drawn to a scale of 1:10,000, will cover approximately six nautical miles on 36 inches of paper. This reduces to one yard in 0.003". Thus, the electronic stylus or cross hair will be able to resolve details on the original printed chart that are one yard apart. This degree of resolution probably equals, or even exceeds (in some places on the chart) the resolution of detail available to the cartographer. Thus, when the electronic chart is well made, it can approximate the accuracy and resolution of a printed chart.

The content of the electronic chart can vary from a complete duplication of the paper original, to a simplified rendition of just a few features contained in the original. However, one important feature of the electronic chart is that detail can be selectively added and subtracted, as in layers, because of the flexibility of the process. Thus, a Mercator grid can be displayed to help locate a particular chart feature, and then eliminated in order to clarify the remaining features. These layers of detail are created by reading from the many individual data files that are the source of features on the chart. There can be individual files on shore contours, depth contours, channel edges, buoys, lights, coordinate grids, etc. These files can be read as required, and then the features associated with each of these files can be removed from the electronic display when it is desirable to simplify the presentation.

The scale of the chart can be changed by a simple manipulation of the data processing constants involved in the electronic display. The actual chart displayed can be thought of as that portion of the digitized area that falls within a "window" of the required dimensions. Changing the chart scale is a matter of changing the height and width of the window.

One of the principal problems associated with paper charts is the need to stay current. Harbor information changes daily; buoys are moved, dredging operations create new or temporary channels, obstacles appear, etc.

One of the principal necessities associated with charts used for navigation is the rapid, timely, accurate updating that allows the user to see any changes that may affect safe navigation in the area covered by each chart. The electronic chart lends itself to this requirement because each individual item of data is accessible to the computer that drives the display. Any element of the chart can be changed by making the appropriate change to the file that contains that particular element. These changes are made electronically, for example, by data transmission via a telephone line, or over a radiotelephone or satellite communications data channel. As another alternative, a memory disk can be introduced that contains the required changes.

Detail concerning these changes is available from various government authorities (NOS, NOAA, USCG, DMA) that normally maintain the aids to navigation that are affected by the changes. To keep abreast of the changes and transmit them to electronic chart users, Navigation Sciences gathers the information on a regular basis and communicates with all the appropriate sources.

TRUE MOTION VESSEL DISPLAY

The first integration involves the computerized differential Loran positioning system and the electronic chart.

With proprietary software, the Loran signals are translated to a constantly changing presentation of ownship on the electronic chart. The vessel appears as a to-scale, hull-shaped image advancing across the electronic chart in true motion matching the ship's path on the water. In operation, when the vessel passes a buoy directly to port, the ship's image on the electronic chart will pass the same buoy at the same time.

As the vessel nears the edge of the currently displayed electronic chart, the screen automatically advances to the next chart.

Radar Interface

The second integration involves radar returns that can be superimposed on the electronic chart.

The radar interface, which is a proprietary raster scan converter, enhances the way radar information is displayed. This device converts the distance/bearing coordinates of radar targets to their corresponding latitude and longitude based on the accurate ownship position determined from Loran-C and the current ship's heading provided by the gyrocompass. Once this conversion is accomplished, the radar image can be added to the electronic chart.

Color can now be assigned to the radar returns, the water, and the land mass. Land masses on the chart are opaque yellow so that the red radar image of the land cannot be seen through the yellow color of the chart. Radar images, colored red and magenta for high and low intensity, are seen only in the blue water areas. Thus, there is an immediate clarification of the display by the simple elimination of the radar's usual land clutter.

As a system safeguard, any ownship position error will result in an alignment error between the red radar shore line and the yellow charted land contours.

The integration of the electronic chart with radar and the true motion display of ownship enhances safety at sea and in harbors by making out-of-position buoys and channel markers clearly visible. The electronic chart shows where the buoy should be; the radar return shows the buoy's actual location. If the buoy is placed correctly, these two images will coincide; if not, the radar return will be separated from the symbol of the buoy on the electronic chart.

The electronic chart presentation also permits a history of radar returns to be retained on the display. If the radar target is a moving vessel, the track indicates its true course and speed.

CONCLUSION

The next generation of computer-aided navigation equipment is entering the market. It involves the use of microprocessors to measure and refine existing radionavigation signals and computer graphics software for displaying a whole range of navigation information. The key feature of this system is an integration of various existing and developing electronic aids into a single screen. The screen can serve as both a graphical and alphanumeric presentation of navigational and tactical information.

The combination allows the mariner to see, at a glance, where he is in relation to the harbor and other vessels, without looking beyond the intgrated electronic display.

REFERENCES

Hamilton, A.C. et al, 1984, "The Expected Impact of the Electronic Chart on the Canadian Hydrographic Service," University of New Brunswick, Fredericton, N.B., Canada.

Newcomer, Kenneth, 1983, "A Survey of the Loran-C Grid on Baltimore Harbor," Navigation Sciences Inc., (under USCG contract).

Rogoff, Mortimer, 1979 "Calculator Navigation," Chapter 5, W. W. Norton & Co., New York.

Rogoff, Mortimer and Peter M. Winkler, 1980 "Integrated Vessel Navigation and Control," paper presented for IEEE Plans '80 Position Location and Navigation Symposium.

MATCH CRITERIA FOR AUTOMATIC ALIGNMENT

Barbara Rosen and Alan Saalfeld
Bureau of the Census
Washington, DC 20233

ABSTRACT

Locally affine rubber-sheeting transformations are being used interactively at the Bureau of the Census to align maps. In trying to align two map representations of the same region, a person will use many visual clues to pair up features on the two maps. Although two individuals will not always agree on the point or line features to be matched, nevertheless, the criteria used to evaluate sameness or difference will probably be quite similar if not identical. The objective of this monograph is to classify the tests, checks, and clues one may employ to determine if a feature on one map is "probably identical" to a feature on another map.

Because exact matching is not always possible (for example, the two maps may be structurally or topologically different), one needs to develop relative or probabilistic measures of matching in order to achieve some degree of comparison of alternative pair-ups. Indeed, each individual's implicit assigning of some unspecified probabilities to possible pair-ups results in his matches being different from those of some other individual. While some "informed assignment of probabilities" and, hence, some arbitrariness is required, nonetheless, many rules may be given quite explicitly and without ambiguity. Those rules deal with relative likelihood or probability in situations and circumstances which are comparable. Without actually assigning an absolute numerical probability (such as 80%) to the event of a match occurring, one may determine that a particular match is more likely than another match because it satisfies all the criteria of the other potential match and then some additional criteria. Or perhaps the first potential match satisfies the same criteria to a higher degree based on some clear relative measures.

INTRODUCTION

Matches are made of corresponding map features: 0-cells of one map are matched to 0-cells of the other, 1-cells are matched to 1-cells, and 2-cells are matched to 2-cells. After a matched pair has been identified, the element in the moving map is repositioned directly on top of the corresponding element in the stationary map. Repositioning brings the moving map into better alignment with the stationary map; and additional matching features are more readily identified.

The approach described in this paper for matching map features through alignment is iterative. First the "most likely" matches are proposed for review by an operator. Then the approved matches are brought into exact alignment by a process of rubber-sheeting the moving map. Again "most likely" matches are proposed (some new candidates may appear due to the rubber-sheeting process) and reviewed by an operator. This alternating match/rubber-sheet process terminates when no new matches are proposed, usually after two or three iterations.

Next independent match criteria are weakened and less likely matches are proposed for review by the operator. At the same time, dependent match criteria (depending on prior matches) may be added to generate potential matches for operator review. Review is always necessary because even strong criteria may result in incorrect matches being proposed. Once again the selection of matched pairs by the operator is followed by a rubber-sheeting of the moving map, and the criteria are reapplied to the rubber-sheeted map until no new potential matches are proposed.

Match criteria may be weakened in small or large steps depending on the experienced reliability of the different criteria to produce accurate matches for the particular map type. For example, urban and rural areas exhibit different success rates with different match criteria. Match criteria may be independent of one another; and the operator may choose to weaken one criterion or another first.

The next section of this paper describes several candidates for match criteria and classifies them as discrete or continuous, topological or geometrical, local or semi-local, and independent or dependent. These criteria have been or are being implemented at the Bureau of the Census as part of a project to match and merge digital map files of the United States Geological Survey and the Bureau of the Census.

CLASSIFICATION OF MATCH CRITERIA

A map may be regarded as a topological object consisting solely of 0-cells, 1-cells, and 2-cells. The 1-skeleton of a map is simply the collection of 0-cells and 1-cells. The 1-skeleton of a map is a mathematical object called a graph. Graph theory studies combinatorial properties of sets of 0-cells and 1-cells or "points and lines," as they are more commonly called. If two maps are topologically the same, then their 1-skeletons will be the same (or isomorphic) graphs. If two maps are similar, their graphs will be similar. One may borrow terminology and results from graph theory to analyze similarity of maps. The graph of a map is also a network; and, hence, one may further borrow network terminology and concepts to analyze maps. A map is also a geometric object, and, as such, may be studied from a geometric point of view as well. Geometry includes the concepts of continuous distance and angle measure in addition to the topological and graph-theoretic notions of adjacency, incidence, orientation, and connectedness.

A 0-cell match is the most elementary match since a match of 1-cells engenders matches of the end point 0-cells of the 1-cells. Similarly, a 2-cell match implies 1-cell and 0-cell matches along the 2-cell boundary. The approach to feature matching in this paper is from smaller to larger: 0-cells are matched first; then 1-cell matches are recognized when corresponding segments have their respective end points already matched as 0-cells; and finally 2-cell pairings are identified as matches after their corresponding total boundaries have been matched. This paper will therefore focus first on 0-cell match criteria.

The first 0-cells to be matched should be special in some fashion. They should possess unusual, unique, or distinguishing characteristics. The 0-cells correspond to intersections of streets; and street intersections may be distinguished by the number and the directions of entering streets. On a map, a place such as "Seven Corners" is quite distinguishable from a typical north-south-east-west city block corner. Seven Corners is often chosen by a map reader in Northern Virginia as a reference point because of its uniqueness. If a map had several intersections with seven corners, however, then having seven corners would not be special and matching would not be so straightforward. This last example is an illustration of dependence—the notion that the likelihood of a match of a particular pair of 0-cells depends on the surrounding 0-cells and how they support or dispute the match.

Consider the following criteria or characteristics of 0-cells:

A. <u>The position of the 0-cell on the map</u>. When two maps are nearly aligned, the corresponding pairs of points should be in approximately the same position. After an initial alignment, pairs of points, one from each map, which form mutually nearest neighbors, become likely candidates for matching. These nearest neighbor pairs are the candidates tested to satisfy other matching criteria before distant pairs are checked. Position is a **geometric** and not a **topological** property because it changes when a topological transformation is applied. Position is a **continuous** variable and not **discrete**. Position of a feature of the rubber-sheeted moving map is **dependent** on the number and kinds of rubber-sheeting iterations performed. Position of a feature **depends** also on matches made of neighboring features. Points near to matched points are moved according to the way the matched points move. Therefore, position is a **semi-local** characteristic or criterion. Semi-local refers to the fact that behavior of nearby 0-cells affects the characteristic of the 0-cell under consideration.

B. <u>The number of 1-cells attached to the 0-cell</u>. This is called the **index** of the 0-cell and is a **discrete** measure because it can assume only positive integer values: 1,2,3, etc. If each of two maps have a unique 0-cell of index 7, then those 0-cells are a likely match. The uniqueness itself is a **dependent** property because it depends on other 0-cells' indices. Having index 7 is an **independent** property because it is not affected by other 0-cells' indices. Having index 7 is also a **local** property because it does not involve characteristics of neighboring 0-cells. Finally, having index 7 is a purely **topological** property and not a **geometric** property because it does not involve the direction or the lengths of the attached 1-cells nor does it involve the position of the original 0-cell.

The figure below illustrates different indices for 0-cells. A 0-cell with an index of 1 corresponds to a dead-end street; and a 0-cell with an index of 2 is called a shape point or inessential 0-cell.

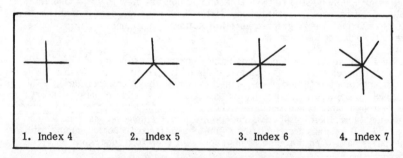

Figure 1. 0-cells of Index 4, 5, 6, and 7.

C. <u>The directions of the 1-cells emanating from a 0-cell</u>. Directions can vary **continuously** and may differ among 0-cells by arbitrarily small amounts. Tolerances and thresholds are required for comparing directions of emanating rays of two 0-cells. If angles are sufficiently similar, then the **local** 1-cell pattern or ray pattern should be regarded as a match. Alternatively one may divide the circle into sectors and thereby consider only a finite number of possible directions. This alternative approach will change direction measure to a discrete measure whose discrete value is the sector in which the 1-cell falls. An eight sector model of the eight principal directions is illustrated below in figure 2.

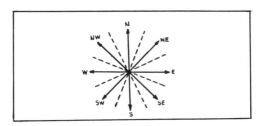

Figure 2. Eight Sectors and Their Defining Principal Directions.

One may also form four-sector or sixteen-sector models. A discrete sector-based measure, while easier to implement than a continuous measure of angle difference, may incorrectly classify as non-matches those intersection patterns whose rays fall near the same sector boundary, but not in the same sector. One may avoid this pitfall by testing for either an eight-sector match or a sixteen-sector match, since the sector boundaries of the two tests do not coincide.

In experiments with real maps at the Bureau of the Census, the intersection pattern was coded as an eight-bit binary number with each bit storing information on the existence (1) or non-existence (0) of a ray emanating in the corresponding sector. The eight bit positions corresponded to the following directions:

$$\frac{SW}{1} \quad \frac{W}{2} \quad \frac{NW}{3} \quad \frac{N}{4} \quad \frac{NE}{5} \quad \frac{E}{6} \quad \frac{SE}{7} \quad \frac{S}{8}$$

The binary numbers corresponding to the intersection patterns in Figure 1 are as follows:

1. (Index 4): 01010101.
2. (Index 5): 11010110.
3. (Index 6): 11011101.
4. (Index 7): 11111011.

The binary coding of intersection patterns as a single number proved to be an effective shortcut method for storing information. Matches were made with an excellent rate of success after an initial map alignment was performed and nearest neighbor pairs were compared.

Two drawbacks of the eight-sector binary representation are that only eight rays may be coded, and two rays in the same 45° sector cannot both be coded. These limitations did not present any problems when working with real maps. Matches that may have been missed because of these constraints were detected by means of other tests.

The binary coding of the total intersection ray pattern as a single number has been descriptively named the **spider function** of the 0-cell. The spider function exhibits some rather useful properties when two 0-cells have similar, but not identical, intersection ray patterns:

(1) For the eight-sector spider function, counterclockwise rotation of the ray pattern of 45° multiplies the spider function value by 2 modulo 255. If one 0-cell has spider function equal to twice the value of the spider function of another 0-cell (mod 255), then the street pattern of the first 0-cell may be a slight rotation of the pattern of the second 0-cell.

(2) If one ray pattern has an extra ray and everything else the same, then the spider functions differ by a power of 2.

(3) If two ray patterns have the same number of rays, and all ray pairs but one in the same sector; and if furthermore the one pair of rays which do not have matching sectors fall into neighboring sectors, then the spider functions will differ by a power of 2.

Properties (2) and (3) above suggest the likelihood of similar ray patterns if spider functions differ by a power of 2. Property (1) suggests another similar possibility when the quotient of spider functions is 2. When one is looking for clues for matching, one may perform various checks on the spider function in addition to testing for equality.

Dependent Match Criteria

The focus on individual 0-cell properties in the previous section could not avoid the discussion of interactions among 0-cells. The context and the map structure force many dependent relations on every 0-cell—an index value will be unique or not unique based on other index values. There is an intrinsic dependence resulting from underlying topological and geometric structure—there cannot be an odd number of 0-cells of odd index, for example. Another kind of dependence which should be examined for clues to matching, however, is not global (i.e., depending on the whole map), but rather semi-local, depending on neighbors of the 0-cell. Semi-local criteria have been proposed for study at the Bureau of the Census, but have not yet been tested to determine their efficacy. This section will therefore merely list some of the possibilities for semi-local matching criteria.

D. Match precipitation. Once a match has been made of a pair of 0-cells, the 1-cells emanating from those 0-cells lead to very likely candidates for additional 0-cell matches. The neighboring 0-cells, that is, those 0-cells which are a graph distance 1 away from the original 0-cells, may be tested with somewhat relaxed criteria for matches (such as one of the weaker spider function criteria). One may iterate this procedure, following the network out from a match in a spider-web-like manner until either everything is matched or one of the networks terminates or changes form. The figure below indicates the network expansion of matches resulting from the match on the unique 0-cell of index 5:

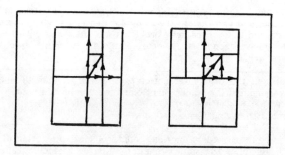

Figure 3. Matches Precipitated by an Index 5 Match.

Matches which can be made by following the network may be made in stages. Each stage involves consideration of all 0-cells at a graph distance 1 from current matches.

Unmatchable criteria as well as match criteria can be defined. The spider function can be applied there as well. The process of matches precipitated through a network may terminate when the sets surrounding the matched set (that is, all those 0-cells at a graph distance 1 from the matched set) are unmatchable.

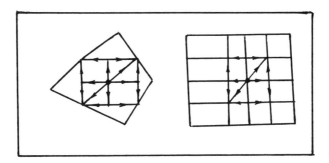

Figure 4. Networks Matched up to Surrounding Unmatchable Set.

E. Triangle or window equivalence (Positional dependence). In the course of the rubber-sheeting process, all of the 0-cells are assigned to enclosing triangles in successively finer triangulations of the space under consideration. On other occasions, quadrant windows are used to partition the space and consequently the 0-cells. When the three 0-cell vertices of a triangle are matched to three image vertices, the triangle interior is assumed to match the interior of image triangle. Naturally it is reasonable to search for 0-cell matches within corresponding triangles. Similarly, after reasonable alignment of the maps, windows should contain corresponding pairs of 0-cells; and it is again reasonable to limit the search for matching pairs to rectangular areas which correspond.

Matching criteria which compare only 0-cells in the corresponding triangles or rectangles are expected to miss a small number of correct matches; however, the search time saved will be considerable in the early stages of matching; and other non-positional criteria may be used later to detect any out-of-triangle or out-of-rectangle correct matches.

If matches are made which fall out of corresponding triangles, the rubber-sheeting may distort the space by producing folds (not one-to-one, but many-to-one affine functions). It is important to detect possible out-of-triangle matches in order to modify the locally affine rubber-sheeting map to recover its one-to-one character.

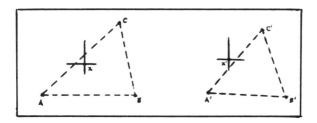

Figure 5. Potentially Dangerous Out-of-Triangle Match.

461

Another globally dependent match criterion is the following:

F. Minimum total "distance" measure. In order to match features on two maps *en masse*, one may define various "distance" measures (which may include infinite "distance" between known unmatchable pairs) for every possible pair of features, one from each map. James Fagan of the Bureau of the Census has begun experimenting with different distance measures as input to the classic Assignment Problem of linear programming. The solution to the map-match Assignment Problem is a pairing of similar cells which minimizes a measure of combined distance between paired cells. The method uses a single combination measure and its output proposed pairs may be tested by further applying other matching criteria. As experimentation continues, other procedures which are shown to discriminate further among the proposed matches may be incorporated into the defining distance measure. The Assignment Problem is an optimization procedure which produces a unique solution based on the distance measure chosen. More experimentation with distance measures will help assess the value of this approach. The technique may be modified to produce successful matches of only a few of the features as a first alignment routine.

CONCLUSION

Several matching criteria for automatic alignment of maps have been proposed and are being tested at the Bureau of the Census as part of a major development program to match and merge map files from the United States Geological Survey with the Bureau of the Census' own digital files. Combinations of the criteria already tested have produced useful iterative matching routines; and additional criteria are scheduled to be tested and eventually used in conjunction with the working routines to improve their efficiency and success rates. Many areas of mathematics have lent themselves to the development of match criteria, including topology, graph and network theory, geometry, statistics, and optimization.

BIBLIOGRAPHY

Corbett, J., 1979, Topological Principles in Cartography, Bureau of the Census Technical Paper 48, Washington, DC.

Harary, F., 1972, Graph Theory, Addison Wesley, Reading, MA.

Lynch, M.P., and A.J. Saalfeld, 1985, "Conflation: Automated Map Compilation—A Video Game Approach," **AUTOCARTO 7 Proceedings.**

Minieka, E., 1978, Optimization Algorithms for Networks and Graphs, Marcel Dekker, Inc., New York, NY.

DETECTING LAND USE CHANGE ON OMAHA'S URBAN FRINGE
USING A GEOGRAPHIC INFORMATION SYSTEM

John Ross
Army Corps of Engineers
Planning Division
Omaha, Nebraska 68105

BIOGRAPHICAL SKETCH

Mr. Ross received a Master of Arts degree in geography from the University of Nebraska at Omaha in December 1984. He is currently with the Planning Division of the Army Corps of Engineers. His responsibilities include developing a project resource geographic information system.

ABSTRACT

Many urban planners find it difficult to inventory changing land uses. A Land Use Change Geographic Information System (LUCGIS) was created to test the feasibility of using an automated approach for monitoring these types of changes. Polygon outlines were entered in a vector format using an X-Y digitizing table. A vector to raster conversion was employed to make digital maps usable on a raster-oriented color graphics terminal. Utilization of a graphics terminal made it possible to view the map creation and overlaying processes, as well as allowing for manual editing. Raster-oriented programs for map overlay and acreage calculation were written. The value of LUCGIS was found in the precision of the resultant 'change' maps. The computerized method of detecting change appears to be more accurate than most conventional methods. By overlaying these land-use maps, many comparsions of land-use categories were produced in the form of dot-matrix printer maps and acreage calculations.

INTRODUCTION

Geographers and cartographers have long been interested in using new methods that make it possible to investigate and depict different aspects of the physical and human/cultural environment. In recent years, Geographic Information Systems (GIS) have proven to be useful in studying spatial relationships and may be useful in monitoring certain types of change in the environment. A GIS is most often a computerized system designed to store, manipulate, analyze, and display large volumes of spatial data. The various applications of geographic information systems have yet to be explored.

One possible application of the GIS concept is in monitoring the rural-urban fringe. This area is a zone of transition between well recognized urban land use and the area devoted to agriculture (Wehrein, 1942, p.53). As the urbanized area spreads outward, so does this fringe area and, as a consequence, the rural-urban fringe area changes rapidly. It is the purpose of this research project to test the feasibility of using a raster-oriented GIS to monitor changes for a portion of the rural-urban fringe of western Omaha.

The Importance of Change Detection

Land-use change information is very important, especially to urban planners. Multi-temporal change analysis may assist the planner in determining a spatial trend. F. Stuart Chapin saw the need for assembling and summarizing land-use data. He also saw the need for construction of systematic procedures for keeping account of changes (Chapin, 1972, p. 298). Even though this is an important aspect of urban planning, the lack of systematic procedures to update and maintain land-use changes has hindered many planning agencies from keeping accurate up-to-date records.

The concept of change detection was a key element in the formulation of the Land Use Change Geographic Information System (LUCGIS) created for this research project. Before change detection was carried out, the maps created for this project had to be encoded and stored in a computerized map file. Automated change detection involves the comparison of two maps that have been encoded into numbers. In this case, two land-use maps were compared to automatically produce a map which displays the areas where change has occurred. A geographic information system is designed to perform such functions.

GIS Data Handling

A fundamental concern in designing a GIS is the encoding of spatial information. Consideration should be given to purpose for the GIS and those who will be using the information. The two most common techniques for encoding spatial data are vector and raster. Vector format refers to single pairs of digital X-Y coordinates that are connected with lines. The vector algorithms used for overlaying maps are particularly complex and consume a great deal of processing time. The other major type of representation for automated cartographic data is raster format. Grid formats, a form of raster organization, were frequently used in the earliest days of automated spatial data handling before specialized graphic output devices were available, since they lent themselves to line printer output. Each grid cell was translated into a character position on a line printer to produce a simple but crude form of graphic output (Peuquet, 1979, p. 132).

Vector format is used more often because of historical predominance of vector-oriented algorithms, and vector encoding is best adapted to retaining the logical map entities familiar to humans. However, new technology has provided better resolution for raster representation, therefore improving the capabilities to depict map features. With this improved output, raster systems may be favored over vector systems because of faster processing, less complex programs, and more efficient computer storage. Improved raster technology will aid immensely in displaying land-use change for the rural-urban fringe.

CREATION OF LUCGIS

The creation of the Land Use Change Geographic Information System involved two major steps: 1) preparation of the map overlays, and 2) digitization of these overlays. The land-use overlays were mapped from aerial photography using a Bausch and Lomb Zoom Transfer Scope (ZTS). The ZTS allows the scale of the photograph to be adjusted to the scale of a base map permitting the conveyance of information from an aerial photograph to the map. This process assures that the geometric distortion inherent in aerial photography is removed, as well as conforming all of the overlays to the same scale.

In order to map land use, the development of a land-use classification system was essential. Normally, five land-use categories are used by the City of Omaha: single-family residential, commercial, public, industrial, and vacant. Two additional categories were used in the LUCGIS land-use classification system: multi-family residential and agriculture.

Data Entry

Upon completion of the map overlays, the task of entering data into a computer at the Remote Sensing Applications Laboratory was initiated. The method of data entry chosen for this project was the ordinary string method, also referred to as the spaghetti method of data entry. The spaghetti method allows the user to simply record X-Y coordinates during the process of digitization. The digitized polygons were later changed to a grid cell representation in the rasterization process. The best aspect of using the spagetti method was the time saved during the digitization process. All land-use maps were digitized within one week's time. The biggest problem with this type of digitizing was the inability to exactly register lines, which became evident when map files were overlayed.

After the land-use maps were digitized, a vector to raster conversion was implemented. The conversion of this information was necessary to use an image processor equipped with raster-oriented software. Rasterization was accomplished by using a pre-existing set of vector-based subroutines to write the vector information to the image processing color graphics terminal. Once the vector information was in the memory of the color graphics terminal, the maps could then be saved and processed in a raster mode.

After the rasterization process, it was necessary to perform a number of cosmetic and semi-automated steps to produce the gray-tone images that would be used in the land-use change analysis. Before polygons could be shaded, the rasterized land-use polygon maps had to be compared to insure that coincidental lines matched exactly, pixel by pixel. To do this, each of the digital land-use polygon overlays was entered into one of the three picture planes on the image processing terminal. In positions where pixels were out of place compared to other overlays, the correct pixel was entered and the out-of-place pixels were erased using a 'paintbrush' program.

Polygons were then assigned gray tones according to the corresponding land-use category. A seed-polygon fill program was used to assign one of the 256 gray tones available on the image processor to each polygon. After all polygons were filled, a program that compared the two maps (this program was used later in the actual overlay process) was employed to find any areas on the maps that were still misregistered. Upon completion of these digital land-use maps, analysis was implemented.

ANALYSIS OF GIS AND LAND USE

The best feature of LUCGIS is the ability to easily overlay maps. In this process, two polygon map files which are expressed in raster format, are compared pixel-by-pixel to produce a change map. The raster method for overlaying was chosen because the programs for raster overlaying are not as complex as programs used in the vector domain. Whereas raster overlaying involves the comparison of one cell to another resulting in the output of another value, the vector method of 'point-in-polygon' compares the lines of both polygons, notes the intersections, and then compares the X-Y locations within that area. The vector overlaying method is therefore more complicated and requires much more processing time.

It was necessary to develop three raster overlaying programs for LUCGIS. The first overlaying program was written for comparison of two land-use maps to determine which chosen value (representing a land use) was in the same or different locations. This was used for the production of the land-use change maps and verification of accurate registration. The second program was developed for the rural to urban land-use map. It compares two land-use maps and displays all values from one map that are in the same position as the chosen value of the other map. The final overlaying program produced for LUCGIS writes all chosen values from one map over another. It was used primarily for overlaying the road map, legend, title, scale, and north arrow onto the change maps.

LUCGIS also includes the programs needed for map construction. A program was written which uses several subroutines to manipulate the vector formatted data entered from an X-Y digitizing table so digital polygon maps can be displayed on the raster-oriented color graphics terminal. To construct the gray-tone maps, a polygon fill program was developed to enter the desired pixel values into each polygon.

Calculation of acreage was an important aspect of LUCGIS. A program was needed that compares two land-use map files and calculates the number of pixels that have changed between the categories of two maps to output a two-dimensional table (see Table 1). The pixels were then multiplied by a factor of .03 which represents the area (in acres) covered by each pixel. Pixel counts from pre-existing programs were used on single maps and multiplied by this factor to determine acreage.

Examples of Land-Use Comparisons

The maps produced for this project display land-use change between 1971 and 1981. These multi-temporal change maps show the land-use category as it exists in three possible time frames: 1) 1971, the land-use changed to another category before 1981, 2) 1971 and 1981, the land-use was the same in both years, or 3) 1981, the land-use was changed from another category since 1971.

Because of the nature of the rural-urban fringe, the growth of single-family residential was particularly evident. The Single-Family Residential Land Use Change Map (Figure 1) exhibits change in the upper left corner of the map, resulting from the growth of two large housing developments. Another housing development is seen in the lower left corner of the map. The remainder of the single-family residential added to the study area between 1971 and 1981 is the result of filling in the 1971 residential areas and of smaller developments.

Multi-family residential (Figure 2) has changed greatly since 1971, with almost a 60% increase. Most of the change has occurred in the upper left corner of the map. Of the 133 acres of multi-family residential added between 1971 and 1981, 95 acres are located here. Taking advantage of the increased population, commercial activities followed residential growth into this area (see Figure 3). However,

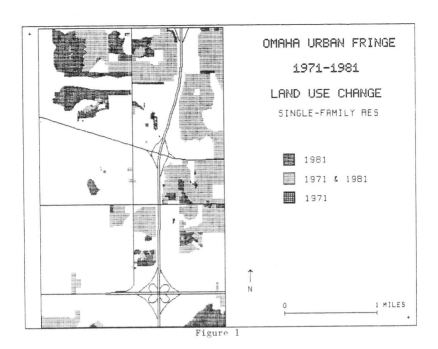
Figure 1

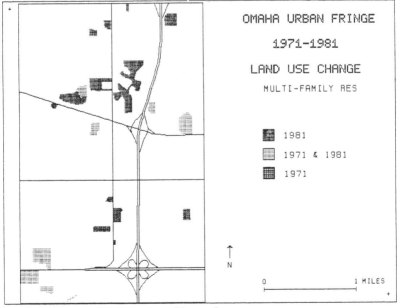
Figure 2

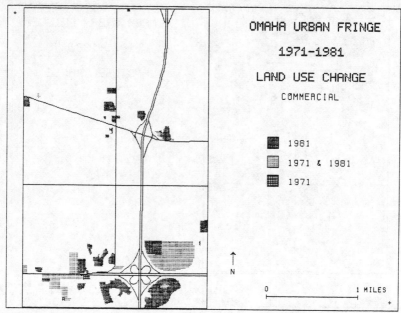

Figure 3

most of the commercial land use is the result of strip development and is located near the intersection of two major roads.

Table 1 also shows land-use comparisons from one year to another. As stated earlier, it was obtained by numerically comparing two map files and counting the number of pixels that have changed from one category to another. This table provides a quick yet efficient way to obtain all acreage comparisons between two maps without having to produce all of the land-use change maps. The table is also very useful when acreage measurements are preferred.

TABLE 1

LAND USE CHANGE ACREAGES

1971 Land Use	(Changed to)	1981 Land Use					
	IND	PUB	COM	SRES	MRES	AG	VAC
IND	---	0	0	0	0	0	0
PUB	0	---	2	0	0	0	2
COM	0	1	---	0	0	0	0
SRES	0	1	1	---	2	0	1
MRES	0	0	0	0	---	0	0
AG	1	177	55	145	103	---	204
VAC	1	106	117	232	28	4	---

Rural to Urban Analysis

One problem that has been studied by urban planners in the recent past is the changing of rural land uses to urban uses. To show rural to urban change by overlaying land-use maps, the LUCGIS classification scheme had to be divided into rural uses and urban uses. Production of the 1971-1981 Rural to Urban Land-Use Change Map involved comparing rural uses of the 1971 land-use map to the urban uses of the 1981 land-use map. To construct this map, vacant land and agriculture were chosen as the rural uses, and single-family residential, multi-family residential, commercial, industrial, and public were chosen for the urban uses.

The 1971-1981 Rural to Urban Land-Use Change Map (Figure 4) identifies land that has changed from rural uses to urban uses in this ten year span. The area with the greatest amount of change to urban uses is evident in the upper left corner of the map. Urbanization resulted primarily from completion of two large single-family residential housing developments, the construction of a golf course, and the addition of several apartment complexes. Conversely, on the right side of the map, very little rural to urban activity has taken place. The reason for the lack of change is that much of this area was already classified as urbanized area as seen on the 1971 Land-Use Map (Figure 5).

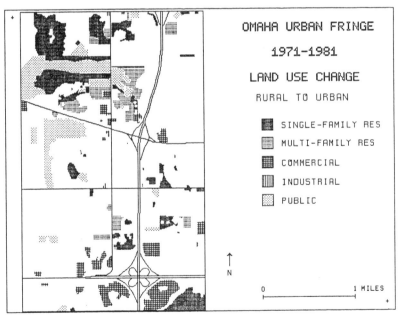

Figure 4

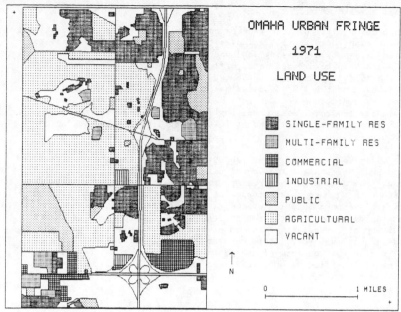

Figure 5

CONCLUSION

Importance of the Raster Approach

A major concern of this project was testing the feasibility of the raster approach for analysis of land-use change. Economy of data storage, minimization of computer time needed for processing, and ease and flexibility in using the data are all increasing concerns as computer utilization increases in cartography (Peuquet, 1979, p. 130). These concerns might be more easily dealt with through raster processing. First, the time it takes to process even a moderate amount of vector-oriented information is much greater than the processing of raster data because of the complex nature of these vector programs. Second, because these vector programs are complex, adding to or altering programs on a vector-oriented system is more difficult than raster programming. Finally, when encoding vast amounts of data, efficient storage is necessary. Data in a raster format can be stored more efficiently than vector data.

The raster method appears to have been a good choice for this particular project. Overall, the raster approach for overlaying polygons to detect change presents a viable alternative to vector manipulation. The only step in this project that might be changed is the way in which data was encoded. The spaghetti method saves time initially, but too much time was spent in the semi-automated registration of the rasterized polygon maps. An arc/node method of data entry would have been preferred, but at the time this project was initiated very little software was available on the system.

With rapid advancements in computer hardware, a raster-oriented geographic information system should be even more valuable to cartographers, geographers, environmentalists, developers, business interests, the military, and others. In the last ten years or so, improvements in such important aspects of raster graphics as data entry devices, the resolution of graphic terminals, and development of the dot-matrix printer, among others have brought great potential to raster technology. It also has brought remote sensing and cartography closer together because of similar computing needs. In fact, LUCGIS was constructed on an image processing system which are most often designed for remote sensing applications. The future may see closer ties between the two disciplines.

Advantages of a Land Use Change GIS

There appear to be several advantages in using an automated Geographic Information System to monitor land-use change:

1) It seems that the resultant change maps would tend to be more precise in detecting change than older methods of visual map comparison or producing hand drawn comparison maps.

2) Data can be stored in a physically compact format such as magnetic tape or disk.

3) Data can be retrieved with much greater speed.

4) Accurate acreage statistics can be accumulated and analyzed.

5) Many comparisons can be extracted from two maps in the form of change maps of acreage calculations.

Most importantly, LUCGIS helps in accounting for land-use change very accurately. Because it took so much time for cartographers to draft these change maps, urban planners seldom used this kind of information. After a system has been created and data base maps digitized, it is only a matter of maintaining the system.

The effectiveness of LUCGIS was found in the precision of the resultant change maps. The computerized method of detecting change appears to have an advantage in speed of map production and accuracy over conventional methods. From the two land-use maps, many comparisons of land-use categories were produced in the form of dot-matrix printer maps and acreage calculations.

REFERENCES

Chapin, F. Stuart, 1972, Urban Land Use Planning, Urbana, Illinois: University of Illinois Press.

Lounsbury, John F., ed., 1981, Land Use: a Spatial Approach, Dubuque, Iowa: Kendall/Hunt Publishing Company.

Peuquet, Donna J., 1979, "Raster Processing: An Alternative Approach to Automated Cartographic Data Handling," The American Cartographer, Vol. 6, pp. 129-139.

Wehrein, George S., 1942, "The Rural-urban Fringe," Economic Geography, Vol. 23, pp. 217-228.

AUTOMATED CONTOUR LABELLING AND THE CONTOUR TREE

Joseph Roubal
Thomas K. Poiker
Simon Fraser University
Burnaby, B.C. V5A 1S6

BIOGRAPHICAL SKETCHES

Joseph Roubal received his B.A. (Cartography/Computer Science) from Western Washington University, and is currently an M.Sc. candidate (emphasis in computer mapping) at Simon Fraser University.

Thomas K. Poiker (formerly Peucker) studied at several West German universities, receiving his PhD (Geography) from the University of Heidelberg. He spent several post-doctoral periods at Harvard University and the Universities of Washington and Maryland. He is presently Professor of Geography and Computing Science at Simon Fraser University. He is also Affiliate Professor at the University of Washington, Editor-in-Chief of Geo-Processing (Elsevier, Amsterdam), principal of GeoProcessing Associates Inc., and an associate of Tomlinson Associates.

ABSTRACT

The contour tree is a construct (a graph) that represents the topological relations of contours on a map. An expansion of the concept allows the logical representation of contours, even if they are incomplete, i.e. have gaps or close outside the map borders. Using certain (usually obvious) assumptions about terrain, an algorithm has been developed that constructs the contour tree structure from line segments, identifies the gaps and labels the contours, given one reference elevation for one of the contour lines. The algorithm works successfully with gaps in up to 15% of the line segments. Wherever there is the possibility for error, the relevant line segments are indicated. Manual intervention, i.e. an operator indicating which segments are peaks, can increase the success rate significantly.

THE PURPOSE OF THE RESEARCH

The creation of digital elevation models and their incorporation into geographic information systems is an important task in many mapping agencies. Although there are many sources for the elevation information, the most complete sources are the contour lines on topographic maps. The ability to use these contours (often in the form of computer-scanned images) to extract the elevation information allows the use of this already existing source, minimizing the need for new collection efforts. Unfortunately, the process of extraction by manual methods is tiresome, painstaking work with much potential for errors of omission. This research is directed towards the partial automation of this task, using as little operator intervention as possible. A prototype computer program

has been developed which can identify the elevations of
most contours in a partially automated system, and promises
to be quite effective. This program recognizes the
relationships between neighboring contours and utilizes
the contour tree structure to establish topological
relationships among the contour lines.

The contour tree
The contour tree is a powerful conceptual structure for
representing the relationships among contour lines on a
topographic map. It can be first found in the literature
by Morse (1965, 1969) and has been expanded to a surface
tree by Mark (1977).

Figure 1 is a simple contour map of Strange Island, B.C.,
and Table 1 is the contour tree representation of the
contours that comprise the map. This example will be used
throughout to illustrate the principles to be discussed.
The root of the tree is the lowest contour line, the one
that encloses all of the other contours. Each link in the
diagram represents a contour line and its enclosing contour
(the lower of the two). Each line may have many upper
neighbors, but may have only one enclosing neighbor.
Each branch in the tree represents a divergence, where
there are two or more contours of the same level that are
enclosed by a common lower neighbor. On the terrain, this
is analagous to a pass between two peaks.

Neighbor relationships and the contour tree
An examination of Table 1 shows that all of the links
represent contour lines that are physically close to each
other on the contour map. This is reasonable, since those
contour lines are also topologically close to each other.
For example, the link between line 3 and line 2 in the
contour tree indicates that line 3 is nested within line 2,
and that these lines are neighbors on the map. This
suggests that the ability to recognize neighbor relationships among the contour lines could lead to the deduction
of their positions within the contour tree.

Defining neighbors from a scanned contour map image
As a human operator examines a series of line segments that
comprise a contour map they are able to recognize the
spatial relationships among the lines, and to make assumptions about the topological relationships. For a computer
program the process is somewhat different, for although the
neighbor relationships can be defined, the topological
significance of each line segment cannot be deduced without
some additional information. Additionally, while all the
topological relationships in the contour tree have their
counterparts in the physical relationships, not all of the
neighbors on the contour map represent links within the
tree structure.

Filtering the neighbor relationships and forming the
contour tree
While the definition of the neighbor relationships within
a contour map is a primarily mechanical process, the recognition of the significant neighbors (those that fit within
the contour tree) is a much tougher conceptual problem.

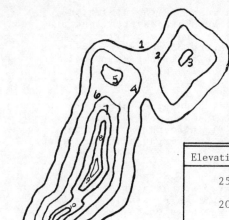

Contour interval 50 m

Elevation of segment 1 is 0 (Sea level)

Elevations	Level	Tree structure
250	6	
200	5	
150	4	
100	3	
50	2	
0	1	

Figure 1. Contour map of Strange Island

Table 1. Contour tree for Strange Island

The approach used here makes certain assumptions about 'normal' terrain and about alternative representations of the contour tree. It consists of a series of rules for manipulating the neighbor relationships to create lists of potentially significant neighbors, and a set of rules for the filtering of these potentially significant neighbors for subsequent assignment into the contour tree.

The tasks of the algorithm are:
- Recognition of the neighbors in a scanned contour image,
- Creating lists of potentially significant neighbors,
- Examining these lists and assigning each segment into the contour tree structure.

Each of these will be discussed more fully below.

DEFINING NEIGHBOR RELATIONSHIPS FROM SCANNED IMAGES

Figure 2 shows a portion of a typical contour map. The position of each of the line segments is captured by scanning the image and producing a computer file that represents each small area on the map as either a line or a blank space (Figure 3). The line segments from the original map can then be reconstructed by connecting those pixels that are next to each other, and assigning an arbitrary identifying number to each. The pixels in the scanned image may be thought of as a grid of graph paper, with each square representing one pixel. In Figure 4 each

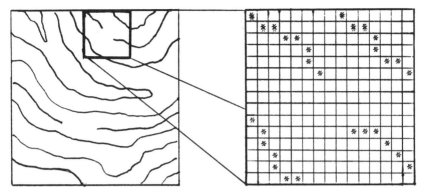

Figure 2. A typical contour map with gaps.

Figure 3. All pixels are 'on' or 'off'.

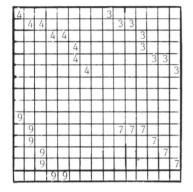

Figure 4. Each 'on' pixel is assigned an identification.

Figure 5. After the first round of propagation.

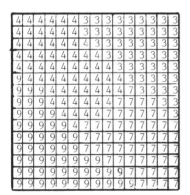

Figure 6. Propagation continues until all space is filled.

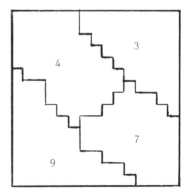

Figure 7. Neighbors and the length of the boundaries can now be determined.

pixel is now assigned a number, rather than being either on or off.

In our example, there are four line segments within the small area, 3, 4, 7, and 9. The method used by the computer program to define the neighbors is to propagate (spread the influence of) each line segment by assigning to each blank space that is adjacent to an assigned pixel the identification number of the assigned pixel (Figure 5). This is done one step at a time until all of the blank space is filled (Figure 6). At this stage it is only necessary to examine the interfaces between the individual line segments and to count the length of each boundary (Figure 7).

After the propagation is complete and the neighbors have been established, the information is compiled into a table that gives the following information for each line segment.
- The number of neighbors
- A list of the neighbors, sorted by length of boundary
- For each neighbor, the absolute length of the boundary

Table 2 is a summary of the neighbor relationship table that corresponds to the Strange Island example. The symbols (+ and -) represent relatively long and short neighbors, compared to the length of the longest neighbor. In practice, these measures are given as a percentage of the longest neighbor, but for the sake of illustration the (+ -) symbols will be sufficient. In general, the long neighbor relationships tend to indicate roughly parallel line segments, while short neighbor relationships tend to indicate knolls, gaps in lines, or other less significant features. An examination of the contour map (Figure 1) and the neighbor relationship table (Table 2) bears out these intuitive guidelines.

CREATING LISTS OF SIGNIFICANT NEIGHBORS

The most significant neighbor relationships are the ones in which one contour is enclosed by its lower neigbor. In an effort to identify these, the program tries to identify local peaks and to create a list of successively lower contours, each one enclosing the line segment above it. These lists are also called traverses.

Local peaks may be identified by the fact that they have only one neighbor, or by operator input. In practice, pits may also appear as one-neighbor segments, so it is important for the operator to identify peaks and pits from the list of one-neighbor segments. Once a traverse has been started, line segments are added to the list by examining the neighbor relationship table and using the rules below. After a segment has been added to the traverse list, its neighbors are considered (excluding the neighbor that already occurs in the tree) for inclusion in the tree. This process is continued until no other segments may be added, then the next traverse is formed.

Table 2. Neighbor relationships and the correct tree structure, by individual line segments

Line segment	Longest neighbor	Other neighbors (by length)	Lower neighbor	Level
1	4	2-	root	1
2	1	3- 4-	1	2
3	2	*	2	3
4	1	6+ 2- 5-	1	2
5	4	6-	4	3
6	4	7+ 11- 5-	4	3
7	6	8- 10-	6	4
8	7	9- 10-	7	5
9	8	*	8	6
10	7	8-	7	5
11	6	*	6	4

* indicates one-neighbor segments
+ indicates a relatively long neighbor
- indicates a relatively short neighbor

Table 3. Traverses derived from the neighbor relationship table (Table 2)

One-neighbor segments				Operator-identified segments		
Col 1	2	3	4	5	6	7
3	9	11	11	10	5	5
2	8	6	6	7	4	4
1	7	4	7	6	1	6
	6	1		4		7
	4			1		
	1					

Rules for forming traverses

Rule 1. Start a traverse at a local peak. Use rules 2-5 to add as many elements as possible.
Rule 2. The traverse is ended if there are no neighbors to consider, or if any entry in the list is a pit.
Rule 3. If there is only one neighbor to consider, that neighbor is added to the traverse, and rules 2-5 are invoked again.
If there are two or more neighbors to consider:
Rule 4. If one of the remaining neighbors is significantly longer than the others, this neighbor is added to the list, and rules 2-5 are invoked again.
Rule 5. If the two longest remaining neighbors are about the same length, the traverse is split into two traverses, one for each of the candidates, and rules 2-4 are invoked.

Using the rules to form traverses (an example)

The use of these rules will be demonstrated with the Strange Island example (Table 2). Examination of the table shows that there are three segments that have only one neighbor (3, 9, 11). These traverses are generated in an arbitrary order, generally in ascending order.

Column 1 of Table 3 shows the traverse generated from segment 3. Segment 3 begins the traverse according to Rule 1. Segment 3 has only one neighbor (segment 2), so segment 2 is added to the list according to Rule 3. Segment 2 has three neighbors, but one of them is segment 3, which has already been added to the list, so that one is not considered. That leaves two neighbors to consider (1 and 4). Since segment 1 is significantly longer than segment 4, segment 1 is added to the list according to Rule 4. Since segment 1 is a pit the traverse is ended according to rule 2.

This explanation may be abbreviated in the following form:

Segment added to traverse (start at peak)	Neighbors to consider (excluding previous traverse element)	Selection	Rule used
3	2	2	3
2	1+ 4-	1	4
1	4	end	2

In a similar manner, the traverse in column 2 is shown:

Segment added	Neighbors to consider	Selection	Rule
9	8	8	3
8	7+ 10-	7	4
7	6+ 11-	6	4
6	4+ 11- 5-	4	4
4	1+ 2- 5-	1	4
1	2	end	2

Columns 3 and 4 are generated from segment 11:

Segment added	Neighbors to consider	Selection	Rule
11	6	6	3
6	4+ 7+ 5-	4,7	5

Rule 5 creates two new traverses.

Column 3

11			
6			
4	1+ 2- 5-	1	4
1	2	end	2

Column 4

11			
6			
7	8- 10-	end	none

In the previous examples, all of the traverses began with
one-neighbor segments. While the program can generate some
correct traverses with almost no operator assistance, the
ability of the program to recognize peaks and pits (even if
they are not one-neighbor segments) improves the performance of the program quite significantly and involves very
little additional operator intervention.

Columns 5, 6, and 7 represent traverses that are possible
if segments 5 and 10 are recognized as local peaks through
operator input. They are generated in the same manner as
the previous examples.

Segment added	Neighbors to consider	Selection	Rule
10	7+ 8-	7	4
7	6+ 8-	6	4
6	4+ 11- 5-	4	4
4	1+ 2- 5-	1	4
1	2	end	2

5	4+ 6-	4	4
4	1+ 6+ 2-	1,6	5

Rule 5 starts two new traverses

Column 6

5			
4			
1	2	end	2

Column 7

5			
4			
6			
7	7+ 11- 5- 8- 10-	7 end	4 none

With the addition of columns 5-7, all of the line segments
from the original map are accounted for and may be assigned
into the tree structure. In the absence of the last three
columns, segments 5 and 10 would have been unaccounted for.

Up to this point, several relationships have been isolated
as potentially significant, but not all of these are valid.
The next step in the solution is to manipulate these
traverses with the aim of extracting the contour tree
structure from them.

ASSIGNING LINE SEGMENTS INTO CONTOUR TREE POSITIONS

The position of a line segment within the contour tree is
defined by the level of the contour and its lower neighbor,
since each element may have only one lower neighbor. Each
level in the contour tree represents an elevation
difference of one contour interval, with the root having
the lowest elevation (Figure 1). The program must examine
the traverses and determine the lower neighbor of each line
segment, and the level of that segment relative to the
lowest pit.

If a traverse that runs from a peak to a pit is defined correctly, it can be assigned positions in the tree, working from the pit and proceeding 'uphill'. As an example, consider the traverse 9-8-7-6-4-1. Starting with segment 1 and going towards segment 9, each element in the list is the lower neighbor of the following segment, and each element is one level higher than its lower neighbor. These can be seen to be the correct relationships in the contour tree (Table 1). If a traverse contains any contradictions in either level or lower neighbor, the program must detect it and ignore it. The following guidelines are followed when assigning traverse elements into the tree structure:

* Reverse the order of all traverses, so that the pits are first and the peaks are last. This facilitates going uphill.

* The correct choice of the first traverse to be processed is important, as it may influence the outcome of the tree. The best choice appears to be to select the longest traverse that begins with the lowest pit and ends in a peak (column 2 of Table 3).

* The first element is added arbitrarily (One has to start somewhere). Thereafter, each element is checked to see if it has already been assigned into the tree. If it is not in the tree already, it is added. The previous element is the segments lower neighbor, and the segments level is one higher than its lower neighbor. If it already exists in the tree, then its level and lower neighbor assignments are checked with the values it would have been assigned. If there are any contradictions, the remainder of the traverse is disregarded. As long as elements are either added to the tree or confirmed in their position, each subsequent element of a traverse may be considered for addition into the tree.

Consider the traverses of the Strange Island example. Start with the longest traverse that qualifies (column 2). Segment 1 is the lowest pit and is assigned to be level 1, with no lower neighbor (it is the root). As each element in the list is considered, it will be found not to be in the tree, so each will be assigned the next higher level, and the appropriate lower neighbor (Table 2).

Generally, the rest of the traverses are processed according to the number of elements (longer ones going first). The next traverse to consider is column 3 (ignoring columns 5-7 for now). Segment 1 exists in the tree, so the program assigns the current level and lower neighbor to match the values alreay assigned. Now segment 4 is considered. It also exists in the tree, so its assigned level and lower neighbor are checked with the expected values (segment 1 as lower neighbor and level 2). These values match, so segment 6 is checked, and it matches also, so the current levels are updated. Now, when segment 11 is considered, it is not found in the tree, so it is added, above segment 6, in its proper place in the tree.

When column 4 of Table 3 is considered, however, some contradictions occur. The first element (segment 7) occurs in the tree, so it is assumed to be correct. Now, when segment 6 is considered, it is also in the tree, so its consistency is examined. If it were correct in this traverse, segment 6 would be one level higher than segment 7. The opposite is true, however, (segment 6 is lower than segment 7 in the original traverse) so the remainder of this traverse is rejected. This is just as it should be, since this traverse represents incorrect relationships. It can be seen now why it is so important to choose an initial traverse that has correct relationships, because if the first traverse has inconsistencies, those will be regarded as the correct relationship.

By processing the traverses, the contour tree structure is recovered from the neighbor relationships, and most of the contours can be assigned their correct elevation. Furthermore, the line segments that the program is unable to assign are pointed out to the operator for manual input.

Operator intervention can be minimized, usually confined to delineating pits and peaks, assigning an elevation to at least one contour (preferrably the lowest), and giving the contour interval. Additionally, at the conclusion of the program the operator can confirm or correct information provided by the program, and can handle the situations that the program is unable to untangle.

In practice, the contour line separation plates for topographic maps are the most common source for contour information. These printing plates routinely have gaps for labelling and other interruptions, and may contain other inconsistencies. The program was designed to consider incomplete data, and research was carried out to determine the programs performance. A small area containing 55 line segments was tested with various degrees of errors introduced. The initial research included very little operator assistance (only the lowest contour and the contour interval) and produced fair results. The program was able to assign more than 80 percent of the lines in the complete contour map, with graceful degradation as increasing percentages of gaps were introduced. With gaps in 20% of the line segments, the program began to fail in the identification of the surface structure.

However, most of these problems are avoided with the minimal operator input described earlier. The program can then be expected to annotate almost all of the segments, thus having great potential for significantly reducing the time and effort of extracting heights from scanned contour maps.

REFERENCES

Mark, D.M. 1977, Topological Randomness of Geomorphic Surfaces. PhD dissertation, Simon Fraser University.
Morse, S.P. 1965, A Mathematical Model for the Analysis of Contour Line Data. Tech Rep 400-124, New York University.
Morse, S.P., 1969, Concepts of Use in Computer Map Processing. Communications, Assoc. Comp. Mach. v12, pp 145-152.

LATTICE STRUCTURES IN GEOGRAPHY

Alan Saalfeld
Statistical Research Division
Bureau of the Census
Washington, DC 20233

ABSTRACT

Many geographic files, such as the Census Bureau's Master Reference File on regional areal subdivisions, are partially-ordered by inclusion, but do not have a purely hierarchical relation. Although the Master Reference File was initially conceived as a hierarchical file, the necessity of examining additional levels of geography forced the structure to lose the hierarchy originally present. The ability to add levels of geography is a necessary feature of a geographic system. A lattice-based file structure permits the addition and deletion of areal references and furthermore adds special elements and algebraic operations to the set of file elements in order to facilitate searches and general file manipulations.

This paper describes the elementary properties of lattices and presents illustrations from an implementation of a computerized lattice file structure.

INTRODUCTION

In 1980 James Corbett and Marvin White described a mathematical model of maps which encoded geometric and topological relations of 0-cells, 1-cells, and 2-cells via incidence matrices and which proposed encoding regional partition relations and subdivision relations via lattices. The theory of lattices that was needed to implement the mathematical model was not delineated at the time for one good reason: the theory was not yet available. While the topological and geometric theory in Corbett and White's mathematical model of maps had been successfully computerized at the Census Bureau and elsewhere, no one had yet developed a computerized structure for representing the elements and relations of a lattice. Indeed, no one knew the nature and properties of the complete geographic lattice of regions described in Corbett and White's model and some suspected that the underlying set would be enormous and the lattice operations inefficient. Dr. Lawrence Cox of the Census Bureau had studied properties of manually constructed lattices of large geographic tabulation zones for his important work on disclosure analysis; however, the creation and manipulation of the necessary lattices was done entirely by hand.

The mystery of Corbett's lattices comes from two principal sources. The first is that lattices themselves constituted a new and generally unfamiliar mathematical construct for the geographer or cartographer. The second source of mystery is more fundamental. A lattice is an augmentation of the usual set of regions dealt with in geography. There are diverse regional entities of geography such as states, counties, urbanized areas, congressional districts, etc.; and each of the individual entities constitutes an element of a partially-ordered set or poset (ordered by inclusion). Additional elements need to be added to the "natural" partially-ordered set in order to create an "unnatural" lattice; and the number and relations of those newly added elements are not immediately obvious even to the experienced mathematician. At least in principle, partially-ordered sets are familiar to geographers. The enclosing lattices are not familiar in name or in principle. Moreover, the enclosing lattices are not even uniquely determined by the underlying partially-ordered sets, all of which adds further to the confusion.

One of the first attempts to remedy the confusion was an internal Census technical document, "Lattice Building," (Saalfeld, 1983), which described for any poset a unique minimal enclosing lattice and which further described an original constructive method for building that lattice. The methods described in the paper were implemented for small sets using relational matrix representations. The matrices, of size n-by-n for posets of size n, limited the size of the posets that could be handled by the new methods.

A second internal Census memo, "Structuring a Poset/Lattice File," (Saalfeld, 1983), presented a data representation strategy that exploited the so-called "bottom-heavy" nature of complete geographic lattices. That strategy was further refined and implemented for small test data sets using B-tree storage routines by Brian Smartt at the Santa Clara County Center for Urban Analysis in 1983.

The purpose of this paper is two-fold: the first is to clear up the mystery of lattices in general with an illustrated overview of basic lattice theory as it applies to geography; and the second is to introduce the foundations research that has taken place at the Census Bureau so that others wishing to implement Corbett and White's model can build upon that research.

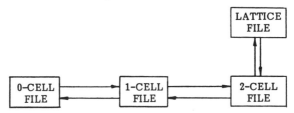

Figure 1. Some Interactions in Corbett and White's Map Model.

The lattice file in Corbett and White's theory permits the geographer to access important combinations of 2-cells in an efficient manner. Internal lattice file operations allow the geographer to add and subtract layers of geography and readily recover the resulting relations of inclusion and intersection of the regions involved.

BASIC CONCEPTS IN LATTICE THEORY

A partially ordered set, or poset, $\{ P, \leq \}$, is a set P with a binary relation $\leq$ between some pairs of elements of P such that:

(1) For all p in P, $p \leq p$. (Reflexivity)
(2) If $p \leq q$ and $q \leq r$, then $p \leq r$. (Transitivity)
(3) If $p \leq q$ and $q \leq p$, then $p = q$. (Antisymmetry)

A lattice is a poset in which every pair of elements posesses a unique least upper bound and a unique greatest lower bound.

A least upper bound is also called a meet and notation used is: lub(x,y) or xVy. A greatest lower bound is also called a join, and notation used is: glb(x,y) or x ∧ y.

A least upper bound is comparable to every upper bound:

(4) $x \leq x \vee y$ and $y \leq x \vee y$.
(5) If $x \leq p$ and $y \leq p$, then $x \vee y \leq p$.

Similarly a greatest lower bound is comparable to every lower bound.

(6) An <u>upper ideal</u> in a poset P is a subset X of P such that if x is in X and $x \leq y$, then y is in X also.

In other words, an upper ideal contains every element that is bigger than any one of its elements. Similarly one may define a lower ideal:

(7) A <u>lower ideal</u> in a poset P is a subset Y of P such that if y is in Y and $z \leq y$, then z is in Y also.

Given any set X in a poset P, the collection of all elements in P simutaneously greater than or equal to all elements of X forms an upper ideal, denoted X^*, where

$$X^* = \{ p \in P \mid \forall x \in X, x \leq p \}.$$

Similarly X_* represents the lower ideal:

$$X_* = \{ p \in P \mid \forall x \in X, p \leq x \}$$

It is easy to show that X is contained in X^* (or X_*) if and only if X is a singleton, $\{x\}$. It is also easy to show that X is always contained in $(X^*)_*$ and in $(X_*)^*$.

Furthermore, the operators upper star, *, and lower star, $_*$, are order reversing. That is, if $Y \subseteq X$ as subsets of a poset, then $X^* \subseteq Y^*$ and $X_* \subseteq Y_*$. The operators * and $_*$ behave as follows with intersections and unions of subsets:

(8) $(A \cup B)^* = A^* \cap B^*$,

(9) $(A \cup B)_* = A_* \cap B_*$,

(10) $(A \cap B)^* \supseteq A^* \cup B^*$,

(11) $(A \cap B)_* \supseteq A_* \cup B_*$.

In general, equality does not hold in (10) and (11).
The following properties of the upper star and lower star operators are easily verified.

(12) For all $X \subseteq P$, $X^* = ((X^*)_*)^*$

(13) For all $Y \subseteq P$, $Y_* = ((Y_*)^*)_*$

(14) For all $X \subseteq P$, $(X^*)_* = (((X)^*)_*)^*)_*$

Property (14) follows from either property (12) or (13). However, it is property (14) that makes the combined operator, $(\ ^*)_*$, a closure-type operator in the sense that the operator is idempotent, i.e. applying the operator twice gives the same result as applying it once.

Property (14) is also useful for limiting the number of generating sets X one must examine in order to study all sets of the form $(X^*)_*$. One may regard property (14) as saying that one need only look at X's which are already lower ideals (and, more specifically, X's which are already lower ideals of the form $(Y^*)_*$ for some Y).

An upper ideal, I, is called <u>principal</u> if there is an element x in P such that

$$I = \{ x \}^* = \{ p \in p \mid x \leq p \}$$

Similarly a lower ideal is principal if it has the form $\{ y \}_*$ for some element y.

In a lattice, L, the following results hold:

$$X^* = \{ \text{lub}(X) \}^* \quad \text{for all } X \subseteq L$$
$$X_* = \{ \text{glb}(X) \}_* \quad \text{for all } X \subseteq L$$

Therefore all ideals arising from upper star, *, and lower star, *, operators are principal in a lattice.

THE NORMAL COMPLETION: A MINIMAL CONTAINING LATTICE

The groundwork has been laid for the following main result:

Given any poset, $\{ P, \leq \}$, consider the family $F = \{ (X^*)_* \mid X \subseteq P \}$ of lower ideals. For each $p \in P$, associate the element $(\{p\}^*)_* = \{ p \}_*$ of F. Let the usual set inclusion, $\subseteq$, give F a partial order. Then $\{ F, \subseteq \}$ is a lattice which contains a copy of $\{ P, \leq \}$ under the injective mapping $i: p \rightarrow (\{p\}^*)_*$.

The mapping i preserves order: $p_1 \leq p_2$ implies $i(p_1) \subseteq i(p_2)$. Moreover if P is already a lattice, then every element of F is principal and i(P) equals all of F. The lattice $\{ F, \subseteq \}$ is called the <u>normal completion</u> of the poset $\{ P, \leq \}$.

The following are properties of the normal completion operation are easily proved. They form a theoretical basis for formal iterative construction procedures used to build a minimal containing lattice.

If the operator, ¯, represents the normal completion operator, one has (up to isomorphism of posets or lattices):

A. For every poset P, $P \subseteq \bar{P}$.

B. For every poset P, $\bar{\bar{P}} = \bar{P}$.

C. For posets $\{ P, \leq_p \}$ and $\{ Q, \leq_q \}$ with $P \subseteq Q$ and $\leq_q$ equal to $\leq_p$ on elements of P, one has $\bar{P} \subseteq \bar{Q}$.

D. For a poset $\{ P, \leq_p \}$ contained in a lattice $\{ L, \leq_l \}$ (where $\leq_l$ restricted to P is equal to $\leq_p$), one has $\bar{P} \subseteq L = \bar{L}$

E. $P = \bar{P}$ if and only if P is a lattice.

EXAMPLE OF A GEOGRAPHIC POSET

The simple figures below will be used to illustrate definition and methods. Consider the following subdivisions of the same region:

485

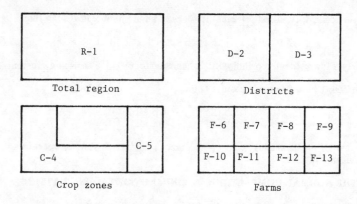

Figure 2. Illustration of Geographic Subdivisions Forming a Poset.

The above geographic relations may be described as a **poset** with the following Haase diagram:

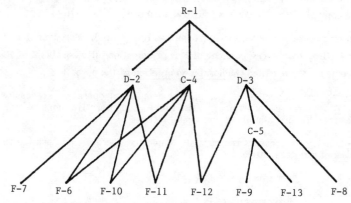

Figure 3. Haase Diagram for Containment Relations of Illustrated Geographic Poset.

The links of a Haase diagram show immediate containment, where there are no intermediate sets. For that reason, no link is drawn between F-9 and D-3 since C-5 lies between Farm 9 and District 3. Notice that the farms are numbered out of order to facilitate the drawing of containment lines with fewer intersections of those lines. At times it is impossible to avoid intersections.

The collection of farms, crop zones, districts and region do not form a lattice, only a poset. Adding the null set as a least element is not sufficient to bring the collection to lattice status. The farms, F-6, F-10, and F-11, have no least upper bound, but rather have two incomparable upper bounds, C-4 and D-2. The addition of the intersection of C-4 and D-2 as a new element will make the collection a lattice.

MATRICES, VECTORS, AND BOOLEAN ARITHMETIC

The above example also serves to illustrate the matrix/vector representation of posets which lend themselves to straightforward arithmetic operations. Consider the above thirteen regions and the inclusions among them; and define the 13-by-13 matrix (a_{ij}) as follows:

$a_{ij} = 1$,   if region i is contained in region j, and

$a_{ij} = 0$,   otherwise.

It is easy to see that (a_{ij}) looks like:

```
1 0 0 0 0 0 0 0 0 0 0 0 0
1 1 0 0 0 0 0 0 0 0 0 0 0
1 0 1 0 0 0 0 0 0 0 0 0 0
1 0 0 1 0 0 0 0 0 0 0 0 0
1 0 1 0 1 0 0 0 0 0 0 0 0
1 1 0 1 0 1 0 0 0 0 0 0 0
1 1 0 0 0 0 1 0 0 0 0 0 0
1 0 1 0 0 0 0 1 0 0 0 0 0
1 0 1 0 1 0 0 0 1 0 0 0 0
1 1 0 1 0 0 0 0 0 1 0 0 0
1 1 0 1 0 0 0 0 0 0 1 0 0
1 0 1 1 0 0 0 0 0 0 0 1 0
1 0 1 0 1 0 0 0 0 0 0 0 1
```

Consider the following Boolean arithmetic tables of 0's and 1's:

Addition:
+	0	1
0	0	1
1	1	1

Multiplication:
*	0	1
0	0	0
1	0	1

By defining a Boolean arithmetic (logical **and/or** operations) on the entries of the square matrix, one may perform matrix multiplications such as $(a_{ij})^2$ or $(a_{ij})^n$. If (a_{ij}) is the matrix of a poset, then $(a_{ij})^n$ will equal (a_{ij}) for all positive integers n because of transitivity and reflexivity. Furthermore, one may check antisymmetry by using the same Boolean arithmetic to verify that $a_{ij}*a_{ji} = 1$ only when $i=j$.

From the above remarks, one sees that a poset's matrix representation can be quite useful. Moreover, a given set of relations may be expanded to include all reflexive relations by adding 1's to the diagonal and all transitively-implied relations by matrix exponentiation as described in the preceding paragraph. The resulting set of relations may then be tested for antisymmetry to see if it is a poset.

THE HAASE DIAGRAM AND THE HAASE MATRIX

The Haase matrix is simply (h_{ij}) where:

$h_{ij} = 1$,   if element i is directly below element j, and

$h_{ij} = 0$,   otherwise.

The Haase matrix for the poset used in the illustration above is:

```
0 0 0 0 0 0 0 0 0 0 0 0 0
1 0 0 0 0 0 0 0 0 0 0 0 0
1 0 0 0 0 0 0 0 0 0 0 0 0
1 0 0 0 0 0 0 0 0 0 0 0 0
0 0 1 0 0 0 0 0 0 0 0 0 0
0 1 0 1 0 0 0 0 0 0 0 0 0
0 1 0 0 0 0 0 0 0 0 0 0 0
0 0 1 0 0 0 0 0 0 0 0 0 0
0 0 0 0 1 0 0 0 0 0 0 0 0
0 1 0 1 0 0 0 0 0 0 0 0 0
0 1 0 1 0 0 0 0 0 0 0 0 0
0 0 1 1 0 0 0 0 0 0 0 0 0
0 0 0 0 1 0 0 0 0 0 0 0 0
```

If (a_{ij}) is the complete relation matrix of a poset, I is the identity matrix, and (h_{ij}) is the Haase matrix; and if (a_{ij})-I is the matrix (a_{ij}) with the diagonal changed to zeros, then:

$$(h_{ij}) = [(a_{ij})-I] - [(a_{ij})-I]^2.$$

The subtractions in the above expression make sense because a **1** entry is never "subtracted" from a **0** entry.

Conversely, to go from (h_{ij}) to (a_{ij}), one has: $(a_{ij}) = [(h_{ij})+I]^m$ for some integer **m**.

The Haase diagram is a directed graph; and the Haase matrix is the usual relational matrix for a directed graph. Many interesting results from graph theory are applicable. In particular, decomposition results on chains and antichains have proven useful in generating ideals efficiently.

VECTOR ARITHMETIC FOR LATTICE ELEMENT CONSTRUCTION

Other meaningful Boolean operations may be performed on entries in vectors and matrices. Some of these are used to construct the normal completion lattice of a poset. A few of those operations are described briefly here which correspond to the "upper-star" and "lower-star" opeators introduced in the section on the normal completion lattice.

The rows and columns of the poset matrix are vectors which represent all principal upper and lower ideals of the poset, respectively. In each case the **n-vectors** are characteristic vectors for the set of **n** elements: each possible combination of **1**'s and **0**'s corresponds to a possible subset of the **n** elements, with **1** signifying that an element is present, **0** that it is not.

For example, the third column has **1**'s in cells 3, 5, 8, 9, 12, and 13, indicating that element D-3 contains the fifth, eighth, ninth, twelfth, and thirteenth elements in addition to itself. Similarly the sixth row has **1**'s in positions 1, 2, 4, and 6, indicating that the element F-6 is contained in the first, second, and fourth elements in addition to being contained in itself. The third column is the characteristic vector for $(D-3)_*$; and the sixth row is the characteristic vector for $(F-6)^*$.

Boolean arithmetic defined termwise on corresponding cells of characteristic vectors amounts to the usual intersection and union of the represented subsets:

Intersection: $(a_i)*(b_j) = (a_k*b_k) = (a_1*b_1, a_2*b_2, \ldots, a_n*b_n)$.

Union: $(a_i) + (b_j) = (a_k + b_k) = (a_1 + b_1, a_2 + b_2, \ldots, a_n + b_n)$.

The vector/matrix/Boolean manipulations required to create new lattice elements (least upper bounds of a collection of poset elements) are sketched below as an illustration of the required arithmetic procedures:

Given poset elements $\{p_1, p_2, \ldots, p_k\}$, find the characteristic vector of $(\{p_1, p_2, \ldots, p_k\})_*$, the upper-lower-star operator of the set.

First, let (v_{ij}) be the row vector of the poset matrix corresponding to p_i^* for $i=1,2,\ldots,k$. Then $(\{p_1, p_2, \ldots, p_k\})$ has characteristic vector:

$$(v_{1j}*v_{2j}*\ldots*v_{kj}) = (v_{11}*v_{21}*\ldots*v_{k1}, v_{12}*v_{22}*\ldots*v_{k2}, \ldots, v_{1n}*v_{2n}*\ldots*v_{kn})$$

since the product * corresponds to the intersection $\cap$; and

$$\{ p_1, p_2, \ldots, p_k \}^* = p_1^* \cap p_2^* \cap \ldots \cap p_k^*.$$

Let $(w_j) = (v_{1j}*v_{2j}*\ldots*v_{kj})$.

Then (w_j) is the characteristic vector for the subset:

$$W = \{ p_1, p_2, \ldots, p_k \}^*,$$

and it remains to find the characteristic vector for W_*.

If (a_{ij}) is the square complete relational matrix of the poset described earlier, then (z_r), the characteristic vector for W_*, is given by:

$$z_r = [\ w_1*a_{r1}+(1-w_1)\ ]*[\ w_2*a_{r2}+(1-w_2)\ ]*\ldots*[\ w_n*a_{rn}+(1-w_n)\ ],$$

for $r = 1,2,\ldots,n$.

The expression $(1-w_i)$ above is just the Boolean complement of w_i; that is, $(1-w_i)$ is 0 when w_i is 1, and 1 when w_i is 0. By adding (in the Boolean sense) $(1-w_i)$ to each factor, one is saying simply to ignore the factor unless w_i is 1.

CONCLUSION

This short paper reviews the theoretical foundations and some of the applications of lattices being explored at the Bureau of the Census. The arithmetic examples given above are presented simply to illustrate the computational possibilities in representing posets and lattices as matrices. Ongoing research in geographic lattice applications at the Bureau of the Census is focusing on efficient storage and retrieval of matrices and sparse matrices, and on additional representation and computational strategies for poset and lattice elements and operations.

BIBLIOGRAPHY

Birkhoff, G., 1967, Lattice Theory, Providence, RI, The American Mathematical Society.

Cohn, P.M., 1965, Universal Algebra, New York, NY, Harper and Row.

Cox, L., 1975, "Applications of Lattice Theory to Automated Coding and Decoding," **AutoCarto II**, Reston, Va., ACSM/Census.

Cox, L., 1980, "Suppression Methodology and Statistical Disclosure Analysis," **Journal of the American Statistical Association**, 75(370), pp. 377-385.

Crawley, P., and R.P. Dilworth, 1973, Algebraic Theory of Lattices, Englewood Cliffs, NJ, Prentice-Hall, Inc.

Preparata, F.P., and R.T. Yeh, 1973, Introduction to Discrete Structures, Reading, MA, Addison-Wesley.

Saalfeld, A., 1983, "Lattice Building," internal Census Bureau technical document.

Saalfeld, A., 1983, "Structuring a Poset/Lattice File," internal Census Bureau document.

White, M.S., 1984, "Technical Requirements and Standards for a Multi-purpose Geographic Data System," **The American Cartographer**, Vol. 11, No. 1, pp. 15-26.

TOPICS IN ADVANCED TOPOLOGY FOR CARTOGRAPHY
Alan Saalfeld
Statistical Research Division
Bureau of the Census
Washington, D.C. 20233
(301) 763-1496

ABSTRACT:

Cartography has embraced the fundamental principles of combinatorial topology with great benefit to the development of the mathematical theory of maps. Many other useful tools available from other areas of topology have gone unrecognized and unused. Several of those areas are illustrated here. Usually the cartographer considers only the most basic topological properties of static, simply-connected, two-dimensional manifolds or surfaces. Here we present some of the advanced topological theory that may be utilized to address the more difficult, higher-dimensional, dynamic problems of cartography. Some of the problems viewed from a new topological perspective include generalization, deformation over time, 3-D representation and 2-D representation of 3-D features, algebraic operations on map features, unified theory of map projections, and orientation. Some topological tools used to examine these problems include homotopy theory, homology and cohomology theory, topological groups and vector spaces, and global analysis. These sophisticated tools are simplified for use by the mathematically adept cartographer:

STREAMLINING CURVE-FITTING ALGORITHMS THAT INVOLVE VECTOR-TO-RASTER CONVERSION

A.J. Saalfeld and R.T. O'Reagan
Statistical Research Division
Bureau of the Census
Washington, DC 20233

ABSTRACT

Curve drawing in automated cartographic applications traditionally consists of the following sequence of operations when raster or dot plotters are used to produce the maps or when the map is reproduced on a raster VDT:

1. Select a curve type or family, such as splines.

2. Select a member of that curve family.

3. Select points on the curve and approximate with a polygonal line consisting of vectors.

4. Convert vectors to a series of dots called the raster image of the vectors and plot those dots or illuminate them as pixels on a screen.

The user rarely has control over the vector-to-raster conversion. In general the conversion routine is automatic and transparent to the user. The user may, however, choose his curve and its evaluation vectors more judiciously if he is aware of the conversion procedure. In this paper common vector-to-raster conversion routines are examined in order to select vectors more efficiently and improve appearance of the eventual raster image and the speed of the computations.

New approaches for generating curves and vectors are presented here. These approaches anticipate eventual raster distortions and try to minimize them. Improved appearance of the final product is the principal objective of these new approaches, although speed in computing and drawing the map curve features will be improved as well with the new procedures.

INTRODUCTION

Automated cartography is moving beyond the "stick maps" of some of the original DIME files to publication quality graphics. The Census Bureau is studying means of improving appearance of its computerized cartographic products; and one area of interest is smooth curve drawing. Many irregular features such as rivers and shorelines are stored as sequences of critical points on digital files. In the past those critical points were merely linked with straight line segments to produce the familiar DIME "stick map." Now, at the time of drawing, the curves are reconstructed from the critical points using algebraic and analytic interpolation; and the curves are then plotted using additional software and hardware approximation and conversion methods. The Census Bureau uses a raster plotter for hard copy and both vector and raster displays for VDT's. When the output is raster or pixel images, the traditional procedure for constructing a curve image includes the following steps:

1. Identify fit point sequence and a family of analytic solutions (curve types, such as splines).

2. Select (i.e. compute) an analytic solution from the family (compute coefficients or other "closed" form).

3. Select points along the analytic solution to use for a piecewise linear approximation with vectors or line segments.

4. Convert the vectors to a raster image.

The fourth step is usually done automatically by the hardware. (Curves other than vectors may also be converted directly to rasters by some hardware routines. If those types of curves can be fitted directly to the critical points, then a step may be eliminated.)

Each of the above steps can affect the appearance of the final image adversely by introducing error due to rounding or approximation or by selecting a final image which the particular display device does not show off well. This paper will examine the way in which each step can compound problems of appearance. It will further explore means of shortcutting the four steps; and, finally, it will explore means of anticipating and reducing or avoiding problems in the final step by judiciously choosing curve families, curves, and vectors in the earlier steps.

The principal emphasis in this examination of curve drawing routines is on appearance. Curve drawing for cartography is not primarily concerned with precision of representation, although a degree of fidelity is required. A city must be on the correct side of a river. Beyond that the cartographer cares about the looks of the wiggles in the river.

For screen graphics on a Video Display Terminal (VDT), speed of computation and appearance are tradeoffs. Several very fast vector-to-raster routines based on incremental arithmetic have been developed for raster displays. These fast conversion routines generally produce considerable aliasing or "jaggies," abrupt pixel pattern interruptions. Other slower routines using partial pixel illumination reduce aliasing by blending edges with gray areas. Still other routines smooth lines by using multiple pixel thicknesses which effectively conceal the incremental shifts of the line center. For all of the above routines, the incoming vector is a given—the routine handles it as best it can. For this paper, however, the vector may also be adjusted (within reason) to influence the final raster image. This paper will focus on single pixel width, single intensity illumination representations in order to simplify discussions and illustrations. Nevertheless, methods introduced here may be extended to drawing with gray scale and drawing with multiple pixel width lines.

PROBLEMS WITH THE FOUR-STEP APPROACH TO CURVE-DRAWING

This section will illustrate some of the difficulties arising at each stage of the four-step approach to curve-drawing given in the introduction. One may work backwards from step 4 and step 1, focusing on appearance difficulties at each step:

Step 4. Convert the vectors to a raster image.

The basic problem here is that some vectors convert to nicer raster images than others. Vertical, horizontal, and 45° slopes are the best, of course. Long, nearly vertical or nearly horizontal vectors for which the raster approximation contains a rare abrupt step are most striking to the eye. More frequent steps seem to be integrated (averaged) more readily, somewhat in the fashion that the eye integrates the gray pixels used for anti-aliasing.

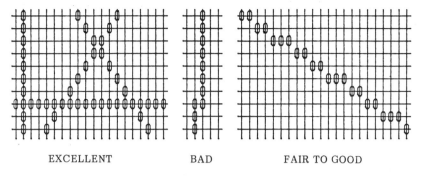

    EXCELLENT   BAD   FAIR TO GOOD

Figure 1. Raster Approximations of Various Vectors.

The above observation suggests that step 3, the selection of vectors (or rather their end points), might include some criteria for avoiding vectors that do not convert well to raster images.

Alternatively one might work with a finite collection of vectors as building blocks, allowing only those vectors which "rasterize" well to enter into the approximating polygonal line. One might assign measures of smoothness to raster images of a small collection of vectors and other measures of smoothness to junctures of such vectors, since the eye also perceives irregularities at such junctures. If this is done, the problem of vector approximation to a curve could be made combinatorial and finite, with the proper solution being that combination of vectors which optimizes the overall smoothness measure.

Step 3. Select points along the analytic solution to use for a piecewise linear approximation with vectors.

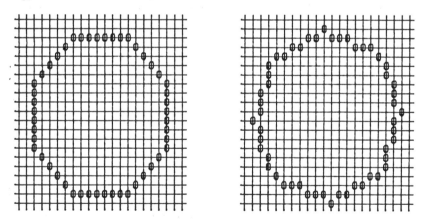

Figure 2. Raster Approximations of Regular Octagons.

For example, if one were to choose 8 points on a circle to approximate the circle by eight vectors, a regular octagon would be a good choice. However, not every regular octagon is equally good for raster conversion. A regular octagon with horizontal and vertical legs is preferable to one with legs sloping slightly off the usual axes because of small angle representation problems seen earlier.

Circle size versus pixel size comes into play in the selection of the approximating octagon as well. The 45° sloping octagon sides can never be **exactly** as long as the horizontal legs, for instance. On discrete grids of rasters or pixels, rounding is an inevitable contributor to distortion.

The following series of illustrations reveals several difficulties which may occur after a smooth analytic solution has been derived for a curve-drawing problem. Additional problems regarding arriving at a smooth analytical solution will be treated in the next section. Here the focus will be on rounding: first from curve to vector end points, then from vector to raster image.

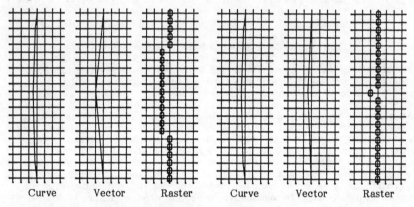

Curve   Vector   Raster   Curve   Vector   Raster

ROUNDING   WITHOUT ROUNDING

Figure 3. Effects of Rounding and Not Rounding Vectors
Before Conversion to Raster Images.

The points which are chosen to fall on a smooth analytic representation of a curve may be rounded to points on a discrete grid before vectors are determined (as shown on the left above) or they may be rounded when the vectors are converted to raster images (right drawing above).

Rounding vector end points to grid points prior to raster conversion has several advantages:

> If the vectors are computed with discrete grid end points, the raster images will be easier to study for irregularities.
>
> Furthermore, raster images of vectors whose end points are grid points may always be endowed with a central point symmetry which guarantees a more regular appearance. That symmetry may be attained by utilizing an incremental pixel selection procedure such as Bresenham's linear algorithm from each end of the vector to the center point. (See figure 4)
>
> A third advantage of rounding vectors to grid points before converting to the raster image is that the initial directional move (of the 8 possible) for an incremental procedure will always be in the correct sector. This property of the initial directional move is a consequence of positioning the vector end point exactly on the grid point, thereby eliminating rounding effects at the first pixel placement. Since this result will be true for all vectors entering a particular juncture, smoothness at junctures will be improved. (See figure 5)

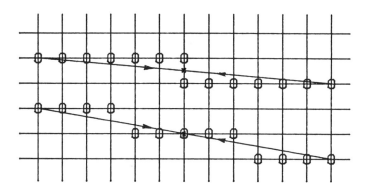

Figure 4. Results of Bresenham's Algorithm Applied from both Ends to the Center.

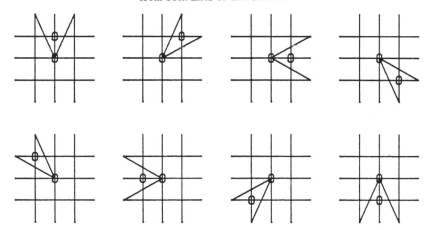

Figure 5. Sector Choices for Initial Directions.

However, the raster images will suffer more total displacement in general if the vectors are rounded prior to raster conversion.

If, on the other hand, the vectors are not rounded to the discrete grid prior to conversion to raster images, then the illuminated pixels will be closer to the continuous polygonal line curve approximation whose points actually lie on the curve. Nearness to the true smooth curve does not guarantee a smoother raster image, as the illustration below indicates.

In addition the raster image may exhibit abrupt atypical directional discontinuities precisely at the junctures of two approximating vectors, further degrading the smooth appearance of the raster image. An angle of $50°$, for example, is represented by a sequence of pixels of steps predominantly $45°$ with some infrequent steps of $90°$. A step of $45°$ is better at a juncture with another vector because it is only $5°$ from the intended direction. If both vectors at a juncture have large and opposite atypical steps, the effect can be a directional change nearly $90°$ greater than the intended directional change. (See figure 6)

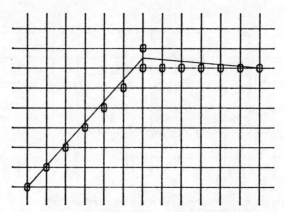

Figure 6. Dangling Pixel Due to Unrounded Vectors.

Vector selection options to approximate a continuous curve (step 3) provide many opportunities for improving ultimate appearance of the raster image. The number of vectors, the placement of vector end points along the curve, and the rounding of those end points prior to converting to raster images all affect the output. The previous step can also influence appearance:

Step 2. Select an analytic solution from the family of curves.

In general this solution will be uniquely determined by the curve family and some additional constraints placed on the curve by the user. Those constraints may involve specifying tangent values or restrictions on higher derivatives. The constraints, while somewhat arbitrary, are added to guarantee uniqueness and replicability of the fitting procedure. The complexity of the solution is directly related to the number of constraints that the user may specify. For example, a polynomial of degree **n** can handle **n+1** constraints.

Instead of picking constraints such as higher derivative values which merely serve to guarantee uniqueness of the analytic solution, the user may seek constraints which facilitate the vector selection procedure in step 3. For example, if the curve passes through grid points, then rounding the vector end points will not be necessary. Because intermediate values on the curve are critical in determining the vectors to be used, a procedure which anticipates later steps is one which chooses the particular analytic solution based on properties of intermediate values.

It is not generally easy to specify intermediate curve values prior to specifying the curve, although some such procedures do exist (see the paper by Fagan and Saalfeld in these **Proceedings**). A greater potential for improving eventual raster appearance lies in the selection of the curve family itself:

Step 1. Select a family of curves.

For some families of curves, the subsequent steps may be more easily implemented or simplified. One example of such a family is the family of parametrized polynomial coordinate functions:

$$X(t) = A_0 + A_1 t + A_2 t^2 + \ldots + A_n t^n.$$
$$Y(t) = B_0 + B_1 t + B_2 t^2 + \ldots + B_n t^n.$$

The theory of finite differences permits any polynomial to be evaluated at regular intervals by means of successive incrementing operations. For example, linear functions have a constant increment, quadratics have a constantly increasing or linear increment, etc.; thus intermediate values of polynomials may be generated using incremental arithmetic only. Methods which exploit the incremental relations are currently being studied at the Census Bureau.

One final possibility which arises in relation to the last example is that of skipping steps in the four-step procedure. An incrementally generated curve for a curve-fitting procedure has great potential for direct raster specification. If one can eliminate the need for vector approximation (steps 2 and 3) and move directly to step 4, potential gains in processing speed will be considerable.

The idea of skipping steps 2 and 3 is not new. Several slow procedures illuminate all pixels near to a curve simply by evaluating the curve on a very dense set of points. The principal deficiency of such an approach is the poor conversion results for some curves. As seen in figure 6, the nearest raster image to an arbitrary curve may contain some dangling pixels which degrade appearance. The new focus proposed here is on appearance. A procedure which selects pixels on a curve successively based on incremental tests can eliminate dangling pixels due to rounding and simultaneously increase computation speed.

BIBLIOGRAPHY

Bresenham, J.E., 1965, "Algorithm for Computer Control of Digital Plotter," IBM Syst. J., 4(1), pp. 25-30.

Bresenham, J.E., 1977, "A Linear Algorithm for Incremental Display of Circular Arcs," Communications of the ACM, 20(2), pp. 100-106.

Foley, J., and A. Van Dam, 1984, Fundamentals of Interactive Computer Graphics, Reading, MA, Addison-Wesley.

Jordan, B.W., and W.J. Lennon and B.C. Holm, 1973, "An Improved Algorithm for the Generation of Non-Parametric Curves," IEEE Transactions on Computers, pp. 1052-1060.

Pavlidis, T., 1982, Algorithms for Graphics and Image Processing, Rockville, MD, Computer Science Press.

Pitteway, M., 1967, "Algorithm for Drawing Ellipses and Hyperbolae with Digital Plotter," Computer Journal, 10(3), pp. 282-289.

Saalfeld, A., 1984, "Building Penplotter Operational Characteristics into Smoother Curve-Drawing Algorithms," Conference Proceedings: Computer Graphics 84, V.2, Fairfax, VA, National Computer Graphics Association, pp. 598-606.

VanAken, J.R., 1984, "An Efficient Ellipse-Drawing Algorithm," IEEE Computer Graphics and Applications, 4(9), pp. 24-35.

AN INTERIM PROPOSED STANDARD

FOR DIGITAL CARTOGRAPHIC FEATURES

Warren Schmidt
National Committee for Digital
Cartographic Data Standards
158 Derby Hall
Ohio State Univerrsity
Columbus, Ohio 43210

Working Group III on Cartographic Features is composed of:

Warren Schmidt (chair)   Digital Mapping Unlimited
Robert Rugg (vice-chair)   Virginia Commonwealth University
Mary Clawson   IIT Research Institute
Beth Driver   Technology Service Corporation
Richard Hogan   National Ocean Survey
Leslie Kemp   Defense Mapping Agency
Joel Morrison   U.S. Geological Survey
Fred Tamm-Daniels   Tennessee Valley Authority
William Hess (observer)   Central Intelligence Agency
Robert Jacober (observer)   U.S. Air Force
Roger Payne (observer)   U.S. Geological Survey

ABSTRACT

The Working Group III - Features of the National Committee on Digital Cartographic Data Standards has since its inception in 1982, studied the issues, devised alternatives, and proposed an interim standard for digital cartographic features. The standard consists of three parts: a descriptive model, definitions, and codes for features. The model includes features, attributes, and attribute values plus two optional categories: feature class and attribute class. Being prepared at Virginia Commonwealth University are the definitions of the features and attributes, samples of which are contained in the Interim Proposed Standard and Supporting Documentation. The latter document discusses relationships, maintenance, and alternatives considered by the Working Group. The final assignment of codes is deferred pending completion and review of the definitions.

AN APPROACH TO EVALUATION AND BENCHMARK TESTING
OF CARTOGRAPHIC DATA PRODUCTION SYSTEMS

David Selden
Battelle Columbus Laboratories
2030 M Street, N.W., Washington, D.C. 20036

BIOGRAPHICAL SKETCH

David Selden is a Research Cartographer with the Battelle Columbus Laboratories, Washington Operations. His current research activities include a study of raster-to-vector conversion processes and development of an analog-to-vector conversion cartographic benchmark testing package for the United States Defense Mapping Agency. He is also a task leader for the design and production of maps of U.S. cancer mortality rates and trends for the U.S. Environmental Protection Agency (EPA). In addition he has recently completed a preliminary investigation of geographic information systems for the EPA Office of Toxic Substances. Mr. Selden holds a B.A. in Geography from Clark University and a M.S. in Cartography from the University of Wisconsin-Madison.

ABSTRACT

The Battelle Columbus Laboratories has developed an approach to evaluating state-of-the-art automated cartographic data/map production systems. Evaluation is performed within the framework of an analog-to-vector (A/V) conversion model, which encompasses the overall cartographic data production process (i.e., source preparation, digitization, feature tagging, spatial coding, and data management). Attention is focused on the unique characteristics of cartographic manuscripts and the limitations of data capture systems. It also addresses editing functions, process speed, procedural trade-offs and production bottlenecks. These are all analyzed in the context of pertinent organizational requirements. A set of benchmark testing materials and procedures have been developed to assist in the evaluation of A/V systems. The paper discusses the A/V conversion model and the development of the benchmark testing package and methodologies.

INTRODUCTION

The need to transform large quantities of analog paper map products into digital cartographic data continues to confront many organizations. The development of sophisticated automated graphic data capture systems has made a significant impact on the map conversion process. Raster scanners and automatic line following devices are replacing manual digitizers. Data editing facilities, built upon state-of-the-art color computer graphics workstations, combine interactive, computer-assisted and automatic functions in raster and/or vector modes to provide advanced capabilities. New algorithms and hardware processors improve raster-to-vector conversion rates. Automatic feature recognition/extraction and spatial/topological encoding software is emerging from research laboratories to support human operators in building feature coded digital cartographic vector data files. These technological advances appear to bode well for the efficient conversion of remaining stores of paper maps and related graphic manuscripts. Significant challenges and difficulties remain, however.

Organizations in search of state-of-the-art capabilities must judge the type and brand of data capture system which will best serve their special needs. Organizations in possession of such capabilities, or less modern facilities, must continually reassess the most effective utilization of their existing systems in light of changing present and future requirements. Those in the market for new systems and those already in production both need to ask similar questions.

The ability to ask meaningful, penetrating questions about automated cartographic data capture systems depends on a solid understanding of current and developing technologies. It also presumes a sound knowledge of map products, digital data specifications, organizational qualities, and the overall cartographic map conversion/data production process. This final layer of knowledge should form the basis for evaluating state-of-the-art automated cartographic data capture systems. Commercial data capture systems are built to convert many different types of analog graphic materials to digital data and are not necessarily specifically designed for cartographic applications. Therefore, it is incumbent upon mapping organizations to integrate such systems into their production process. Conversely, a system's specifications and limitations will often dictate the details of a map/data production process. These often cross-cutting requirements must be reconciled at every juncture in the map conversion process. Thus, Battelle Columbus Labs has developed the analog map to digital cartographic data conversion model (A/V) as a conceptual framework for evaluation of systems in production or those in the marketplace.

THE ANALOG-TO-VECTOR (A/V) CONVERSION MODEL

The conversion of paper map products into digital cartographic data is accomplished through a series of procedures which generically are common to most organizations involved in this activity. The A/V conversion model (Figure 1) identifies the five major components of this process: Preparation, Digitization, Tagging, Spatial Coding, and Data Storage. A sixth optional component, Raster-to-Vector Conversion, is also included where raster scanning has been employed for initial data capture. The remainder of this section will discuss each individual component and highlight critical aspects of the interrelationship between the process and automated data capture systems.

Analog-To-Vector Conversion--Preparation

Preparation of analog cartographic manuscripts for conversion to digital representation addresses two categories of information: analog graphic and analog feature attribute. Analog graphic is any cartographic "paper" product

containing point, line and area elements, derived from
either photogrammetric compilations (e.g., color pencil on
mylar), four color printed maps, color/feature separation
film positives or color proofs. Analog feature attribute
preparation refers to the process of preparing feature code
overlays which annotate codes for each type of cartographic
element to be digitized and processed. Preparation of cartographic manuscripts consists of four phases: (1) Data
Selection, 2) Manuscript Reformat, 3) Manuscript Enhancement, and 4) Manuscript Review Edit.

ANALOG-TO-VECTOR CONVERSION MODEL

Figure 1. Analog-to-Vector Conversion Model

Data Selection. The selection of specific data categories from cartographic manuscripts is the first step in
the preparation stage. In many cases only select categories
are required from an original input (e.g., only contours
from a topographic quadrangle). Alternatively, all information on a compilation manuscript may be submitted for digitization. The results of data selection greatly influence
the proceeding steps in preparation and also challenge the
capabilities of data capture systems to varying degrees.
Its particular impact depends on the data categories chosen,
their cartographic geometries (i.e., non-intersecting lines,
closed polygons, merging networks, crossing networks, point
features, symbolized elements, and labeling - see Figure 2),
material type (e.g., mylar, paper), data representation type
(e.g., ink, scribing, color pencil), material age, data density, data quality (e.g., broken lines), and material-data/
system(s) compatibility (e.g., color input for a black and
white scanner).

Manuscript Reformat. This process concerns the transformation of cartographic manuscipts prior to digitization.
In some cases, analog manuscripts may require recompilation,
photo-reduction, panelling or subdivision into multiple
pieces due to size (i.e., larger than scanning format) or
image density (resulting in too much data for a digitizing
system to efficiently process in a single pass). In many
instances, manuscript reformatting reflects an initial incompatibility between input cartographic documents and a
particular digitization system.

Manuscript Enhancement. The enhancement of analog cartographic manuscripts is done to provide additional data coverage (e.g., intermediate contour lines in sparsely covered areas), assist in topological encoding (e.g., highlighting bridge intersection in a unique color to identify a node location), and generally improve the efficiency of the analog-to-vector conversion (A/V) process.

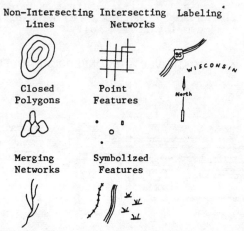

Figure 2. Generic Cartographic Data Types

Manuscript Review/Edit. The final step in the preparation stage is to review the input manuscript prior to digitization to assure data quality (e.g., no gaps in "continuous" lines or merging of "discontinuous" lines such as contours), completeness, and efficient utilization of resources (e.g., analog manuscripts with sparse data coverage may be directed to manual digitization instead of raster scanning).

Analog-To-Vector Conversion--Digitization

The type of digitizing method will dictate to some degree the extent and manner of manuscript preparation. Three basic types of digitization have been identified: raster scanning, automatic line following and manual.

Raster Scanning. This automatic process captures cartographic elements and converts them to digital representation by means of electro-optical, solid state, or laser "cameras" which traverse the manuscript (mounted on a rotating drum or stationary flatbed) in scan lines at prescribed resolutions. Some scanners are color sensitive, capable of capturing and storing up to twelve colors simultaneously, thus offering flexible analog input requirements. Others are solely single color scanners or require multiple pass scanning (through color filters) to achieve full color scanning capabilities. Scanning resolution is variable on some systems and fixed on others. Variable resolution scanners permit the selection of the appropriate setting for maximum "acuity" and minimal data storage. Output of raster scanning typically requires data editing and conversion to centerline vector representation.

Automatic Line Following. This automatic process
(although also operated in "manual" and semi-automated
modes) captures the x,y coordinate positions of cartographic elements through the use of video "cameras" or
deflected laser beams. Two important advantages of this
approach is the capability of concurrent feature tagging and
the elimination of raster-to-vector conversion processing
(although data filtering, editing, and reformatting usually
occurs as a post-process).

Manual Digitization. The manual capturing of x,y coordinate positions of cartographic elements with hand-held
electronic cursors still maintains certain advantages over
automatic procedures. Primarily, it is more effective to
capture map sheets of sparse data coverage in this manner.
Additionally, concurrent feature tagging and no raster-to-vector conversion requirement make this method advantageous
in certain cases.

Analog-To-Vector Conversion--Raster-To-Vector Conversion

This is the process of converting digital cartographic data
from a raster (scan line or grid) data structure to a x,y
vector, centerline (or edge) representation. This is only
required as a result of raster scanning for initial data
capture and reflects typical vector data structure requirements for many digital cartographic data bases or geospatial
analysis packages. The A/V conversion model identifies
three components of raster-to-vector conversion: raw raster
data review/edit, raster-to-vector conversion and raw vector
data review/edit. Individual data capture systems implement
raster-to-vector conversion in unique ways, through alternative algorithms, special hardware processors, or combining
or by-passing entire components (e.g., no raw raster data
review/edit).

Raw Raster Data Review Edit. This is the process where
raw raster data (i.e., direct output of raster scanning) is
reviewed for data quality, completeness, and anomalies or
errors (e.g., gaps, stubs, spikes, sticks, streaks, coalescence, misalignment) commonly associated with the scanning
process or attributable to the analog cartographic manuscript. Three approaches to raw data review and editing
have been identified. Interactive computer graphics techniques permit human operators to visually search through a
file, locate problems and correct them via light pens, cursors, trackballs and other assorted electronic interfaces to
the graphics plane. A second alternative utilizes computer-assisted software functions which automatically identify
errors and facilitate their correction through interactive
computer graphics techniques. A third option relies on
fully automatic error identification and correction software. Typically in this case, certain tolerances (e.g.,
identify and close gaps equal to or smaller than three pixels by two pixels) are set by an operator prior to initiation of a routine.

One evaluation criterion for data capture systems is the
availability of raw raster editing facilities. The provision of one, two or three forms of raster editing support is

a secondary consideration. Finally, the overall effectiveness of any of these functions dictates their true value. A commerical systems lack of raster data editing capabilities in any form may reflect a serious handicap. In recent years, a number of vendors of data capture systems have developed sophisticated capabilities in this area. Certain functions such as "snow" removal (i.e., unwanted, miscellaneous data elements) may be more effectively accomplished in a raster environment. The sole provision of interactive computer graphics for raster data editing limits an organization to costly labor intensive activities in this phase of the A/V conversion process. Computer-assisted and automatic editing functions promise lower manpower requirements and improved editing performance over manual editing techniques. However, this can be very misleading. Successful resolution of data quality problems and the degree to which automated functions create new errors or inconsistencies (e.g., gap closure on contour data can result in the unintended connection of close contour lines) requires scrutiny.

Raster-To-Vector Conversion. This concerns the actual conversion of raster cartographic data to centerline vector representation. One standard approach implements this data conversion in three steps: skeletonization, line extraction, and topology reconstruction. Skeletonization is the reduction of raster elements to one unit of resolution through a process of peeling, ballooning or centerline calculation. A number of data anomalies have been shown to result from the skeletonization process including generation of stubs, gaps, and unthinned data elements. Subsequent correction of these anomalies in raster and/or vector mode is required and should be anticipated for systems applying skeletonization algorithms. Line extraction is the identification of unique line segments and is accomplished in two basic ways, line following or scan line. The line following method tracks vectors until encountering intersections at which point nodes are identified. Scan line processing extracts individual line vectors as they are encountered one scan line at a time. This latter approach requires that topology reconstruction (i.e., the identification of point, line and area geometric, spatial relationships) be explicitly performed as a post-process.

A second approach to raster-to-vector conversion in the CAD/CAM, engineering drawing market has recently been reported.* In this case, boundary or edge chains are generated from line drawings in real time concurrently with raster scanning via a special hardware processor. A subsequent process generates centerline vector data via software calculation of centerline x,y coordinate locations.

The evaluation of raster-to-vector conversion revolves around two basic issues, speed and quality. Processing large amounts of data typically associated with cartographic input often results in production bottlenecks in raster- to-vector conversion. This is complicated by the variability of cartographic geometries and the capability of alternative

* A Defense Mapping Agency (DMA) Raster-To-Vector Analysis, Battelle Columbus Labs - currently unpublished.

conversion algorithms to efficiently process them. The quality of the data resulting from raster-to-vector conversion is of equal concern. Fast conversion with poor quality output is of little value. The extensive efforts and time required for error detection and correction may seriously counteract the perceived advantages of raster scanning data capture (i.e., fast raw data capture with lessened dependence on human operators). Unfortunately, most commercial vendors reveal little or no information about their raster-to-vector software (for proprietary reasons) thus leaving potential customers without critical knowledge.

Raw Vector Data Review/Edit. This process directly parallels raw raster data review/edit. Interactive computer graphics, computer-assisted software and automatic error detection and correction software functions can all be implemented in vector mode. Some commercial systems rely almost solely on the vector editing functions with minimal or no raster editing capabilities.

It should be noted that both manual and automatic line-following digitizing systems also require and usually offer raw vector data review/edit capabilities. Errors such as gaps, spikes, and linear misalignments are quite common as a result of human or machine deficiencies or existing analog manuscript anomalies. The need for the full array of error detection and correction tools similar to those described for raster scanning systems is equally important for vector digitizing systems.

There are two notable trends in the development of editing capabilities. The provision of three levels of editing as previously described is becoming an industry standard. This reflects an editing philosophy based on the "80-20" rule which sets as a goal the detection and correction of eighty percent of all errors via automatic functions and the remaining twenty percent through a combination of interactive computer graphics and computer-assisted routines. (No evidence has been presented which demonstrates attainment of this goal by any commercial vendor. The degree to which a vendor can legitimately claim successful editing automation should be of great concern to prospective customers.) The second trend worthy of note is the development of "parallel" raster and vector data editing capabilities, often within a single computer graphics environment. Both raster and vector data can be edited at a single "station" and in increasing cases vector data can be overlaid on its raster counterpart. This is particularly effective in reviewing certain kinds of anomalies including linear misalignments of centerline vector data.

Analog-To-Vector Conversion--Tagging

This is the process of logically and physically associating vector cartographic data with analog feature attribute information in a digital computer file.

Analog Feature Attribute/Vector Data Merge. This process links vector data with analog attribute information (codes) by means of interactive computer graphics, computer-assisted software, or fully automatic methods. The reported development of automatic contour tagging routines (in commercial vendor research laboratories) is ushering in a new era of automatic feature extraction and tagging capabilities. This presages diminishing labor intensive feature tagging activities which presently require significant organizational personnel and time resources. The evaluation of cartographic data capture systems should consider capabilities in this area and more importantly ascertain a company's commitment to on-going research and development. The successful development of automatic feature tagging capabilities in raster scanning systems will emerge as a significant factor in comparing raster and vector systems.

Tagged Vector Data Review/Edit. This is a quality assurance step which assures the correctness and completeness of the tagging procedure. An array of techniques are available which support this activity. Among them are interactive computer graphics, graphic plotting, and software checking routines.

Analog-To-Vector Conversion--Spatial Coding

Spatial coding is the logical and physical definition of a cartographic data file's internal and external spatial relationships. Three types of spatial relationships are defined: universal referencing, topological encoding, and data structuring.

Universal Referencing. This describes the conversion of a cartographic data file stored in table (cartesian) to a universal reference system (e.g., Universal Transverse Mercator or Latitude/Longitude). Most state of the art cartographic data capture systems provide universal referencing capabilities.

Topological Encoding. This process defines the internal spatial structure of a digital cartographic file by explicitly delineating the adjacency relationships of points, lines and areas. In recent years a number of standard government topological data structures have emerged (e.g., U.S. Geological Survey's DLG-3 or GIRAS and Bureau of the Census GBF/DIME or TIGER). Vendors of commercial cartographic data capture systems currently support production of various types of topological structures and some are reported to be developing software to support the government structures referred to above. As data interchange increases in importance, the need for standard topological data formats will also grow.

Data Structuring. Data structuring is the transformation of a raw data file into a specified data format. All commercial data capture systems maintain their own internal data formats and often support system independent formats to permit conversion to other systems and data structures. Many vendors will develop custom software to support a client's unique data structure requirements.

Analog-To-Vector Conversion--Data Management

Data management is the final component in the analog-to-vector conversion model. It concerns the processing, storage, manipulation and retrieval of digital cartographic data. All data capture systems provide some form of data file management capabilities. Some support attribute storage and retrieval. Generally, these management capabilities are designed to support editing functions and are not intended for comprehensive digital cartographic data and spatial data management applications.

THE DEVELOPMENT OF BENCHMARK TESTING MATERIALS

AND METHODOLOGIES

The discussion of the analog-to-vector conversion model and the interrelationship between cartographic data production and the technology of data capture systems highlights critical production concerns and significant areas of ongoing development. The intention of the previous discussion was to develop a comprehensive understanding of the overall production process in the hope that such knowledge will assist organizations in formulating evaluation criteria of data capture systems which meet their special needs. An integral part of the evaluation process is the actual benchmark testing of systems to produce empirical evidence of system performance. Battelle has developed a conceptual approach to creation of benchmark testing packages for cartographic data production systems.

The benchmark testing capability described below is limited in technical scope to basic analog-to-vector conversion processing technology. Specifically, it addresses the procedures and problems associated with capture and conversion of elemental cartographic geometries found on typical analog cartographic base maps. The geometries consist of point, line and area cartographic elements possessing limited symbology (i.e., line thickness variation and limited dashing, only). This does not imply an ignorance or denial of the emerging importance of advanced data capture system technology (e.g., pattern/feature/character recognition and tagging) and the challenges of processing complex cartographic input. It does emphasize, however, the fundamental role and inherent complexities of processing elemental cartographic data.

Benchmark Testing Materials

Three types of testing materials have been developed: standard products, synthetic cartographic geometries, and a quality/editing capability test sheet. Together these materials provide a comprehensive assessment of a data capture system's level of performance with different types of input and processing.

   Standard Products. A representative set of typical map products which are produced, stored or required by an organization form the basis for a standard products benchmark testing materials package. Selection criteria should be based on representative types of maps, data densities, kinds of materials, method of production (i.e., ink, scribing, etc.) and overall quality variations. The obvious value of such a package is to test a data capture system's capability in assimilating and processing the range of input materials commonly held by a mapping organization. Every organization maintains a unique array of analog map products, thus the creation of a custom set of standard product materials is required.

Synthetic Cartographic Geometries. These benchmark testing materials consist of twelve film positive sheets containing three abstractions of cartographic geometric patterns found on the products of a mapping organization. Each pattern is reproduced four times (thus twelve sheets) with increasing amounts of data. The primary utility of the synthetic cartographic geometry benchmark testing materials is to support analysis of the impact of geometric formation and data density on raster-to-vector conversion throughput. Given a known quantity of linear inches it is possible to calculate raster-to-vector conversion rates for different types of geometries, data densities and alternative algorithms or systems. Thus, without understanding the inner workings of particular algorithms, a prospective customer can empirically test a data capture system for raster-to-vector conversion performance.

Quality/Editing Capability Test Sheet. Given the importance of data quality and the amount of resources currently invested by mapping organizations in quality assurance activities, in addition to commercial development of advanced editing capabilities, there is an obvious requirement to evaluate and test system functions in this area. Figure 3 portrays a small section of the test sheet. It essentially consists of a series of "perfect" cartographic geometries (i.e., abstractions of cartographic geometric patterns with no discernible anomalies or flaws) in the first column. Arrayed to the right each "perfect" geometric sample are replications of the pattern with purposefully created errors (e.g., gaps) with known dimensions. The combination of perfect geometry and geometric degradation samples are repeated two additional times in different lineweights.

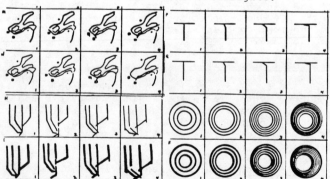

Figure 3. Quality/Editing Capability Test Sheet (portion of sample prototype)

The overall purpose of the quality/editing capability test sheet is twofold. First, as the analog sheet is processed (i.e., digitized, edited, converted, etc.) evidence of system introduced errors can be collected by observation of the "perfect" geometric samples. All errors, anomalies and geometric degradations will be readily apparent and easily identified in the first column of the test sheet after each operation. Secondly, extensive testing of computer-assisted and automatic editing functions is made possible. Tolerances can be set to verify the sensitivity of algorithms to small variations. For example, with known gaps of two, three, five, and seven thousandths of an inch (or a pixel equivalent) in various kinds of geometric patterns, will gap detection and correction routines succeed in resolving only those gaps at the prescribed tolerances? How will different geometric patterns with equal gap sizes be treated? What rate of automatic editing success will be observed. This benchmark testing sheet will provide a prospective customer of any commercial data capture system with the means to more closely, and concretely evaluate system editing functions and overall quality related issues.

Benchmark Testing Methods and Evaluation Criteria

Benchmark testing methods fall into several activities: raster scanning (or alternative digitization methods), automatic raster editing, raster-to-vector conversion, automatic vector editing, and plotting on an automatic vector plotter. Evaluation criteria is based on individual process times, combined process times, virtual image quality assessment, digital plot/analog input "overlay" quality analysis, system integration/user friendliness evaluation, and statistical tests of timing data.

CONCLUSION

A comprehensive understanding of the overall analog map to digital cartographic data conversion process (A/V) is critical to an effective evaluation of cartographic data production systems. Although certain components may appear to wield the greatest influence over production performance (e.g., large format color raster scanners and fast raster-to-vector conversion algorithms), it is the total production process as implemented by a particular system which must be considered. Additionally, the unique characteristics of personnel and products associated with a mapping organization should be factored into the evaluation equation. It is the integration of organizational requirements and system capabilities which ultimately dictates the degree to which production goals are achieved. Thus, development of effective benchmark testing tools and the evaluation of cartographic data production systems depends as much on documentation of "in-house" procedures, capabilities and requirements as it does on an indepth understanding of current technologies.

REFERENCES

Selden, David and Went, Burton, A Defense Mapping Agency (DMA) Raster-to-Vector Analysis, 1984 (currently unpublished)

AUTOMATED MAPPING

B. Shmutter & Y. Doytsher
Technion I.I.T. , Haifa , Israel

Summary

Automated mapping on many occasions makes use of editing stations replacing thereby the labour of the draftsman of the traditional mapping process by the work of the operator of the interactive graphical station. Considering the cost of that equipment and the time consumed to operate it, it becomes apparent that a mapping process which disposes of the editing station can be rather advantageous, particularly for smaller mapping organisations. A brief exposition of such a mapping process is given below.

Regardless of the method by which the mapping is performed, with or without an editing station, the first stage of the mapping is the data collection. Two types of data are collected, topographic and planimetric; the latter comprising all features which are to be presented on the map. During the data collection codes are assigned to the various features. These serve several purposes; indicating which features do not participate in the contour generation, describing the nature of the features and how they have to be plotted (colour, type of line etc.) and determining the mode of processing, the data representing the features are subjected to.

On completion of the collection the data are processed by a series of programs. These execute two major steps, the one - generation of a DTM consisting of a grid of elevations and an array of topographic features, the other - generation of a planimetric data base. The second step includes, among others, the functions usually performed on an editing station, such as "rectification" of buildings, smoothing curved lines, maintaining parallelism between lines, assigning symbols and so on.

At the processing stage, the data collected in overlapping areas between photogrammetric models are adjusted in order to ensure a unique definition of lines passing from model to model.

The data bases so obtained constitute a digital map extending over the entire area under consideration.

Samples of maps produced by the above procedure complete the paper.

DEPARTMENT OF LANDSCAPE ARCHITECTURE • SCHOOL OF NATURAL RESOURCES
COLLEGE OF AGRICULTURAL AND LIFE SCIENCES • UNIVERSITY OF WISCONSIN-MADISON
25 AGRICULTURAL HALL • MADISON, WISCONSIN 53706 • TELEPHONE 608-263-7300

September 17, 1984

TITLE: Dane County Land Records Project

PRINCIPAL INVESTIGATORS: Nicholas R. Chrisman
David F. Mezera
D. David Moyer
Bernard J. Niemann, Jr.
Alan P. Vonderohe
University of Wisconsin-Madison

POSTER ORGANIZER: Joseph Sonza-Novera
Department of Landscape Architecture
25 Ag Hall
University of Wisconsin-Madison 53706
(608) 263-7300

The Dane County Land Records Project, University of Wisconsin-Madison requests display space for the poster session at Auto-Carto 7. We would like to display 8 posters each measuring 28 inches in width and 44 inches in height totaling 20 linear feet. The panels detail the role of the Dane County Land Records Project in the modernization of land records on the local level. A brief description of individual panels follows:

Panel 1: The Dane County Land Records Project: An Overview.

A discussion of the goal and objectives of the project in implementing a modernized multi-purpose land information system. Listed are the federal, state and county agencies, private firms, professional associations, and utilities involved with the Dane County Land Records Project.

Panel 2: Legal and Institutional Requirements for Land Records.

Details of the data requirements mandated by representative Wisconsin State Statutes. Examples include wetland mapping, groundwater protection, soil erosion control and animal waste management legislation. Emphasis is placed on those land/resource records that simultaneously satisfy several legal and institutional data requirements.

Panel 3: Importance of Land Records Modernization:

Costs and beliefs. Results are given from two recent studies detailing the costs to the citizen to maintain the present land information base and te belief disparities among individual producers and users of land records.

Panel 4: Benefits Derived from Modernization of Land Records.

The various types of benefits that can be expected in general from a modernized land records system are outlined. Included are several examples of specific benefits that accrue to local governments, private firms and individual citizens.

Panel 5: Modernization of Land Records: Positioning Technologies.

Results are given from three recent research projects conducted at the University of Wisconsin-Madison as part of the Dane County Land Records Project. Discussed are the technological and economic implications of using Doppler, Global Positioning System (GPS) and inertial surveying systems for geopositioning.

Panel 6: Application of ODYSSEY for Aggregate Resource Planning.

Given are the results of a case study to identify limitations suburban development and zoning restrictions impose on sand and gravel resources. Results are presented in tabular and plotter formats.

Panel 7: Use of a Modernized Land Information System Monitoring Soil Erosion.

Given are the results of a case study to quantify parcel specific soil loss while allowing the capability for generalized county-wide soil loss planning from the same data base. Results are presented in tabular and plotter format.

Panel 8: Conclusions and Recommendations.

General recommendations and conclusions for state and local government are made as derived from the research experiences of the Dane County Project.

THE AUTOMATED REVISION OF TENNESSEE VALLEY TOPOGRAPHIC 1:24,000 QUADRANGLES

Frederick L. Tamm-Daniels
John R. Cooper, Jr.
Major C. R. McCollough
Tennessee Valley Authority
Chattanooga, Tennessee 37402

ABSTRACT

For quite some time, the increasing cost of map maintenance has necessitated a move to automated systems. The Tennessee Valley Authority (TVA) Mapping Services Branch (MSB) began to automate its map production in March, 1983, and now, one-and-one-half years later, has succeeded in automating the revision process for its seven hundred seventy-five 1:24,000 national series topographic maps maintained under an agreement with the U.S. Geological Survey (USGS) National Mapping Program. The first example of this process is the Pulaski, Tennessee quadrangle. This quadrangle, as well as the others in progress, was loaded and revised on TVA's Automated Mapping System (AMS), which consists of an Intergraph VAX 11/780 graphic design system and a Gerber 4177 photohead plotter. The revision process is being implemented in three stages: a combined automated and manual process for the first year, completely digital photorevision for the years following, and completely digital full revision using the AMS stereoplotter workstation. By integrating the map revision process with the necessary digital loading, TVA is speeding up its 1:24,000 quadrangle revision as well as creating the digital data that will be necessary for future revisions, related map products, and geographic analysis.

PROCEDURAL FLOW

The automated interim photorevision of 1:24,000 Valley Topographic Maps (VTM) has been integrated into the existing manual process and parallels many manual color separation procedures.

The major requirement for automated revision is that the current version of a quadrangle be available in digital form. For this reason, only revised data that is normally shown in purple and the woodlands overlay were loaded (digitized) into the system during the first year.

First Year Interim Revision
The Compilation and Edit Unit prepared the revision manuscripts as they have for TVA's manual color separation procedures. Using these manuscripts, the operator created a digital map file using Intergraph's

---

Any use of trade names and trademarks in this publication is for identification purposes only and does not constitute endorsement by the Tennessee Valley Authority.

World Mapping System software. Symbols indicating the four corners of the quadrangle in a polyconic projection were placed in their theoretically correct positions in the graphic file by keying in their locations in latitude and longitude.

After setting up the file, the operators digitized a new woodland overlay and all revisions that were shown on the manuscript, using 59 preassigned logical layers for the various features (See Appendix A). All of the digitized features were symbolized according to USGS specifications for 1:24,000 topographic quadrangles.

Upon completion of the digitizing, a cycle of multi-color pen plots and digitizing edits was conducted by operators on the sheets they had digitized.

Final photoplots were then made on negative film on the Gerber plotter and turned over to the manual Color Separation Unit for registration with the archived separations for each quadrangle.

A particular problem arose in registering the two sets of quadrangle separations: the automated separations did not exactly fit the existing manually produced separations, and registration holes had to be stripped in. The reason for this misregistration was that the separations produced on the AMS were theoretically correct, i.e., all the detail on the quadrangles fit the four corner monuments which were located exactly where they should be in relation to one another, and the scale was exactly 1:24,000. The manually produced separations, however, due to manual production methods, were not as accurate. This made for difficulty in manually deleting content from the old separations and making the exact joins with the purple (revised) data.

After registering the separations, a color proofing and edit cycle was begun, as has been the normal practice.

Present Interim Revision
After the first set of 12 interim revision separations were produced, loading of the next set of quadrangles scheduled for revision was started. Revision of these maps is being accomplished in a similar manner to the partially automated method; however, all of the features, except for text, are available in digital form, so that all revision work, including deletion of features, is being done on the AMS. As in the first phase, symbolization of all features adheres strictly to USGS specifications.

After a digitizing edit, negative separations are plotted for all of the features (except for text) that will appear on the printed map. (The type separation is still prepared manually because of system limitations.) The generation of all of the linework from the AMS has resolved the problem of misregistration with manually prepared materials.

515

A feature of the Gerber photohead plotter that speeds up production is the 28-pin registry system. Pins are available to plot three sizes of separations using a Misomex punch, and to plot four 1:24,000 quadrangle separations during one run when the Gerber multiple plot software is used.

Quadrangle Loading
A quadrangle must be loaded into the system before a fully automated revision can be accomplished. In most cases, an operator obtains the set of negative separations for that quadrangle which has been supplied by the USGS Eastern Mapping Center. When a set of negatives for a quad is not available, the archived paper print is used. Source material is digitized at its original scale of 1:24,000.

A World Mapping System file is set up as in the above revision procedure. (All of the VTM quadrangles are in the polyconic projection.) Following this, the digitizing table surface is activated. Most distortion is resolved through the digitizer transformation (a least squares fit), which causes the actual corners of the quadrangle to match their theoretical positions as shown on the graphics screen. The relationship between the actual and theoretical corners is applied to all data that is digitized.

Three standard files are attached to any VTM graphic file: a cell (symbol) file, a user command index file (which allows the operators easily to access World Mapping System functions), and a color lookup table, which defines the color used for each logical level. The colors have been defined to represent as closely as possible the colors shown on TVA's standard opaque color proof. The background is a light grey to represent the paper and to allow the use of black for map features.

Culture is the first major category to be loaded. Cultural features are separated onto 59 logical levels. Sixty-seven symbols ("cells" in Intergraph terminology) have been created for point, line, and area features. The symbols meet USGS specifications when they are photoplotted and provide accurate representations on the graphic screens for the operator; they are used to represent most point features, such as vertical and horizontal control monuments, school flags, and cemetery crosses. Road casings are created by digitizing either a centerline or one side of a road casing, and then copying the line in a parallel mode as many times as necessary. First and second class road fills are created from digitized centerlines by using a linear symbol as a "pattern" for that line. The outlines of area features are digitized and, depending on the particular feature, the outline is either used to make a peel coat, e.g., for woodlands, or it is defined as a shape and filled with an area pattern on the AMS, e.g., for disturbed surfaces. Drainage is the second feature category digitized. Although this does not present the same problem as culture with respect to symbolization, special treatment is needed for springs and open water areas. For springs, a symbol must be used, whereas for open water, only the outline is digitized and an open window negative is created using the photoplotter and peel coat.

With the exception of depressions, symbolization is not a concern with the third category of features, contours. The sheer volume of data, however, is a major problem. The digitizing of contours is responsible for approximately one-half the time required to load a complete quadrangle. The total time for digitizing a quadrangle in the Tennessee Valley is from four to six weeks. After some experience with contour digitizing using stream mode (where the system is continually being fed data points as long as the cursor button is being held down and moved, provided the new point meets distance and deflection angle criteria), a decision was made to use curve strings to represent contours. This mode of digitizing was easier for the operators, since only one data point is fed into the system each time the cursor data button is pressed. A curve string, as opposed to a line string, adds two extra points to each end of each string. These two points are not visible, but they are used to apply a spline curve to the string of points when "slow curve" display is turned on. Therefore, the number of points needed to represent a contour properly is less than with a line string. On recent quadrangle loads, the elevation text has been added to the "brown text" logical layer.

Text is the one category of information that, for the most part, is not included in the digital file. This is due to the difficulties in creating publication quality type fonts on the AMS.

EVALUATION

Accuracy
The accuracy of the automated revision process may be evaluated at two stages of production: when the data are entered into the system and when separation negatives are plotted. These are the two points where inaccuracies due to digital processing would most likely be introduced.

The hard copy produced by the photohead plotter has been found to be more accurate than manually produced products covering the same geographic area. This has been determined by measuring the distances (using an Invar steel scale) between points of known location, such as grid intersections and quadrangle corners. As previously mentioned, this greater accuracy and precision created a registration problem during the first year interim revision process.

Cartographic data are entered into the system by manual digitizing using a backlighted table with .001 inches resolution and a hand-held cursor with a 2X magnification. A well-trained operator using this equipment can create an accurate digital copy of the features on the source materials. However, due to the

geographic accuracy of the system (through World
Mapping), some slight positional shifts from the original
materials do occur. For the purpose of revising VTM
quadrangles and producing publication quality separation
negatives, TVA has been well satisfied with the accuracy
of loaded quadrangles. Data that are added to the
digital file as a part of the revision process are
digitized from a revision manuscript produced by the
Compilation and Edit Unit This digitizing process is
very much like the existing manual process of scribing
features to be added from a revision manuscript. Any
inaccuracies introduced in this process would be the same
for either the manual or automated revision technique.

Cost Comparison
Cost figures are available only for the first year
interim revisions. The quadrangles being revised this
year in a completely digital mode are not yet far enough
along to permit a cost comparison.

The first year interim revision process costs averaged
$1,500 per quadrangle for the automated operation.
Compilation and edit cost averaged $2,100, and the manual
color separation work cost approximately $3,500. This
made the total average cost per sheet $7,100. The
average cost for a quadrangle revised by manual methods
has been $6,000. Therefore, the first year revision
process cost an extra $1,100 per map.

Two questions should be asked concerning the higher cost
for a partially digital product: why was the cost
higher, considering that "digital mapping" is supposed to
be a money saver, and what extra benefits, if any, were
derived for the extra money spent. In the authors'
opinion, the primary causes of the increased cost were
the need to include the extra step of painstaking manual
matching of the separations, and the difficulty
encountered in properly registering the digitally
produced revision separations with the existing set of
manually produced negatives, and the concomitant
painstaking manual matching of features. The extra labor
and reproduction shop charges were significant. The
benefit that TVA derived, in addition to refining its
revision procedures, was that part of each quadrangle
revised was loaded into the AMS. When those same
quadrangles are fully loaded into the system, that part
of the data will already have been digitized.

Time/Cost For Loads
During this and the previous fiscal year, TVA has been
loading the first set of 27 quadrangles, 12 of which will
be digitally revised in a full production mode this
year. The time required to load a quadrangle has ranged
from four to six weeks depending on the complexity of the
data. This has resulted in an average cost of $4,500 per
quadrangle.

The cost of loading a 1:24,000 quadrangle through manual
digitizing is significant and is being gradually lowered
by the writing of programs designed to speed up specific

tasks. The extra cost of manual digitizing is, to some
degree, lessened by the fact that TVA has tried to avoid
purchasing more equipment than necessary, since capital
depreciation on any such equipment must be paid out of
operating funds.

An evaluation has been presented only for the first phase
of digital revision on TVA's AMS. The production of a
completely digital interim revision prototype, Pulaski,
Tennessee, has shown that publication-quality line work
can be produced digitally. By the end of the 1985
calendar year, enough quadrangles will have been
processed through the digital revision procedure so that
the costs can be evaluated. We fully anticipate that
they will be significantly reduced from the present level
as the implementation phase continues for this revision
method.

FUTURE PRODUCTION

Raster Scanning
TVA's MSB has looked at scanning systems and does not
believe that their cost (in terms of annual depreciation)
can be justified or supported by its operating budget.
However, discussions are underway with other government
agencies for scanning services to be provided on a
reimbursable basis. From the available information, it
seems that this method of data capture would be most
effective for contours and drainage.

Attributes
Attribute ("intelligent") information is not presently
being attached to graphic elements in the AMS VTM files.
Once the graphic part of the automated revision process
is smoothed out, attribute data bases will be created for
these files.

Compilation
In the near future, we will attempt to compile interim
revisions at an AMS workstation to avoid the present
repetitive process of manually creating a revision
overlay and going over the linework a second time in
order to digitize it. This should reduce the revision
cost by removing unnecessary procedures.

Text
The omission of most text from the VTM files is mostly
due to the limited capabilities of the AMS in this area.
Based on inhouse testing, the authors feel that text of
acceptable quality can be generated by the photohead
plotter. However, acceptable font and spacing problems
must still be addressed on the Intergraph part of the
system. Work on text will be continuing inhouse to try
to eliminate the need for a manual type stick-up
operation on VTM quadrangles.

Three Dimensional Files
Although there are advantages to storing cartographic
data in three dimensional files, the difficulties of
digitizing "planimetric" features in 3-D, as well as the

inaccuracy of the elevations, have persuaded us to
continue to use two-dimensional files. For certain
non-VTM related projects that are worked on the
stereoplotter workstation, however, three-dimensional
files will be created and used to generate final map
products.

Contours with Tagged Elevations
Even though the decision has been made to remain with
two-dimensional files, it is still advantageous to
maintain elevation information at least for the contours
on a quadrangle. If contours are logically "tagged" with
their elevations, they can be used as input to digital
terrain modeling packages. Work is underway on a contour
digitizing program that will automatically tag contours
with elevations as they are being digitized.

SUMMARY

TVA's Automated Mapping System was purchased with the
expressed intention of lowering the cost of revising
TVA's map series, one of them being the seven hundred
seventy-five 1:24,000 Valley Topographic quadrangles
maintained under agreement with the USGS National Mapping
Program. Less than two years later, TVA has produced a
digital interim revised quadrangle that meets all USGS
specifications for the printed 1:24,000 topographic map
series. To the best of our knowledge, this is the first
such quadrangle produced in the United States.

Appreciable cost savings are anticipated with the
production of completely digital interim revisions.
Further cost reductions will be achieved through
customized programming and increased use of newer
technologies, e.g., raster scanning and editing.
Benefits beyond a lowered cost for map revisions are
being and will continue to be realized because the
quadrangle is in digital form. The availability of an
accurate digital reference base map is invaluable for use
in the production of project-specific special-purpose
maps, as well as allowing for the creation and use of
digital terrain models.

APPENDIX A

Tennessee Valley Authority
Topographic Mapping (1:24,000)
Level Assignments - Cartographic Files

Level	Feature	Separation Color
1	First class casings	Black
2	Second class casings	Black
3	Red road fills	Red
4	Third class casings	Black
5	Fourth class roads	Black
6	Wagon, jeep, and foot trails	Black
7	------------------------------	
8	Active railroads	Black
9	Inactive railroads	Black
10	Crossings (Bridges, tunnels, etc.)	Black
11	Masonry dams, locks, jetties	Black
12	------------------------------	
13	First class buildings	Black
14	Second class buildings	Black
15	Public service buildings	Black
16	Mines or quarries	Black
17	Transmission lines	Black
18	Pipelines, telephone lines	Black
19	State boundaries	Black
20	County boundaries	Black
21	Corporate boundaries	Black
22	Reservation boundaries	Black
23	Small parks, golf courses	Black
24	Cemeteries	Black
25	Horizontal control points	Black
26	Vertical control points	Black
27	Boundary and miscellaneous monuments	Black
28	Levees and miscellaneous hachuring	Brown
29	State Plane labels	
30	GLO lines and monuments	Red
31	Foreshore and offshore features	Black
32	Index contours	Brown
33	Intermediate contours	Brown
34	Supplemental contours	Brown
35	Neatline, State Plane and Carter ticks, longitude and latitude info	Black
36	UTM	Black
37	Strip Mines and intricate surfaces	Brown
38	Open water (double line drainage)	Blue
39	Single line drainage	Blue
40	Intermittent drainage and ponds	Blue
41	Water wells and springs	Blue
42	Swamps	Blue
43	Channel lines	Blue
44	Lands subject to inundation	Blue
45	Field and fence lines	Red
46	Urban areas	Pink
47	Woods	Green
48	Other vegetation	Green
49	Purple revisions (other than text)	Purple
50	Black miscellaneous separation	Black
51	Blue miscellaneous separation	Blue

52	Brown miscellaneous separation	Brown
53	Red miscellaneous separation	Red
54	Unclassified buildings	Purple
55	Purple text	
56	Black text	
57	Blue text	
58	Brown text	
59	Red text	
60	------------------------------	
61	------------------------------	
62	------------------------------	
63	Photoplot template	

CARTOGRAPHIC IMPLICATIONS OF AUTOMATED MULTILAYER MODIFICATION

Joan R. Terenzetti
Michael T. Wasilenko
Synectics Corporation
111 East Chestnut Street
Rome, New York 13440

ABSTRACT

During cartographic compilation and editing, there may occur an effect of one feature upon several others which intersect and/or coalesce with that feature. These features in traditional production usually exist on separate layers or flaps of the compilation manuscript. The correction or modification of features as they may be affected by some other feature's change, then becomes a multilayered modification task.

As more and more cartographic tasks have become automated to some extent, more of the overall production process is also becoming automated. This includes the multilayered modification task. Synectics Corporation is currently involved in a project for the Rome Air Development Center which is pushing the application and integration of automated cartographic compilation and revision techniques to new limits. This project includes the study, analysis, and implementation of a baseline automated multilayer modification capability.

This paper presents a discussion centering upon the expression of multilayer modification processes as they are performed manually with an evaluation of these manual functions as they pertain to automation. This includes the impact of spatial dimension, scale, and relations of affected features, the expression of vague and subjective actions as specific constructs or rules, and the degree of automation possible, which is unique to each modification environment.

ABSTRACT OF PAPER SUBMITTED FOR AUTO-CARTO 7, MARCH 1985

SPATIAL DATA HANDLING SOFTWARE: A VIEW FROM A UNIVERSITY GEOGRAPHY DEPARTMENT

Derek Thompson
Associate Professor
Department of Geography
University of Maryland
College Park, MD. 20742

Office 'phone: 301/454-2244
Home 'phone: 301/345-4942

The use of the digital form of spatial data has been progressing more rapidly than the ability of many universities to marshall resources to put modern tools in the hands of students. Spatial data handling tools are important as general support for geographers and others undertaking research or participating in learning activities at undergraduate and graduate levels. The instruction about spatial data learning using digital technologies is a specialized part of curricula in many academic institutions.

Today, however, only a few geography departments have extensive support facilities or personnel to achieve both of these goals. Skilled graduates are being attracted to jobs in industry rather than academe; well equipped laboratories for spatial data handling are not numerous; and software representative of the current state-of-the-art is present in only a small proportion of academic geography departments. It is unlikely that business enterprise will create computer packages or libraries in spatial data handling akin to the widely used statistical analysis packages like SPSS, MINITAB, or SAS.

There is here a challenge, then, for universities to collaborate on the development of libraries of algorithms and computer programs that may be useful for educational purposes, general and specialized, in many institutions. Software for supporting student and faculty research could take tha modern conversational form, for example, a MINITAB approach to the older spatial data handling packages like GEOSYS or FLOW. Specialized functionality, for supporting courses like geographic information systems or automated cartography, could take the form of transportable libraries of routines that would demonstrate principles such as line generalization, triangulated irregular networks, or interactive choropleth map design.

A review of spatial data handling software suggests that much remains to be done, but that the advent of lower cost hardware and other technological developments hold promise for the creation of a widely available "GEOPACK". This facility could have tools for handling point, line, area, netwrok, grid cell, and volume data types, and allow for data capture, storage, manipulation, management, and display of spatial features and their attributes, including capabilities for analysis and modelling.

INTERACTIVE CONSTRUCTION OF CONTIGUOUS CARTOGRAMS

by

Prof. W. Tobler
Geography Department
University of California
Santa Barbara CA 93106
(805) 961-3663

ABSTRACT

A cartographer working with an interactive graphics terminal can control the design of a cartogram without having to perform the tedious area calculations or the trial and error drafting. In the environment described the computer algorithm presents suggested drawings which can be modified almost instantaneously by simple graphical intervention. This cycle is repeated as often as necessary until the design is complete, at which time the finished map is routed to a hard copy device.

MAP SYMBOLS FOR USE IN THE THREE DIMENSIONAL GRAPHIC
DISPLAY OF LARGE SCALE DIGITAL TERRAIN MODELS
USING MICROCOMPUTER TECHNOLOGY

Charles T. Traylor
Department of Geography and Planning
Memphis State University
Memphis, TN 38152

James F. Watkins
Department of Art
Memphis State University
Memphis, TN 38152

BIOGRAPHICAL SKETCHES

Charles T. (Tim) Traylor is Associate Professor of Geography at Memphis State University specializing in cartography, air photo interpretation and remote sensing of the environment and is also Director of the Cartographic Services Laboratory. He earned a B.S. degree in Biology and a M.S. degree in Geography at the University of Alabama in Tuscaloosa and the PhD degree from the University of Kansas. He was the Cartographic Director for both the "Atlas of Alabama" and the "Atlas of Mississippi".

James F. Watkins is Professor of Design and Color in the Department of Art at Memphis State University. He is currently coordinating the implementation of digital imaging equipment into the Department's Graphic Design Program. He holds an MFA Degree from the University of Alabama. His articles on kinetic art and visual illusions have been published in LEONARDO, an international journal of visual art. As a winner in an international educational toy design competition, his work has been displayed in the Smithsonian Institute.

ABSTRACT

The need for a better way to display classified LANDSAT images prompted this study. It was found that students had trouble visualizing the look of the ground cover when the information was presented in the traditional form of colored pixels on a video monitor, especially when looking at a small area displayed at a large scale. Converting the image into a three dimensional block diagram that was colored according to the classification scheme helped, but still some troubles persisted. A method is presented whereby the colored pixels of the three dimensional terrain model are substituted with symbols that represent the ground cover. The method includes converting digital terrain data into a gridded block diagram, coding the surface according to the LANDSAT classification and superimposing specially designed symbols as appropriate from a growing library of land cover representations.

The difficulty is enhanced even more when the image is
enlarged to the extent that each pixel is an obvious
rectangle; the perception of looking at the earth's surface
is lost. Figure 2 (previous page) is an example of an 18
pixel by 20 row classified LANDSAT image.

This paper describes a procedure that assists one in the
visualization of such a image by using two dimensional
symbols placed on a psuedo 3-D block diagram.

BACKGROUND

Other ways of viewing this data have been observed: Spectral
Data has an algorithm in their RIPS system that allows the
classified surface to be viewed in perspective.* The Defense
Mapping Agency has a program to generate block diagrams of
digital terrain models in color that appear to the eye to be
plastic relief models of a topographic map.(USAETL 1977)
Small scale LANDSAT images have been portrayed as information
on a relief model.(McKeown & Denlinger 1982) At the other
end of the spectrum, CAD systems have been developed for
designing small environments and to allow one to
interactively take a tour by computer looking at various
components (Atherton 1981), rooms, hallways, side views of
buildings and even the local community. However, the data
for these projects are very precise and specific to a given
project.

Many advances in computer graphics have been made in the last
few years. On large computer graphic systems such attributes
as surface texture, color, light reflectance and others have
been considered. Computer images are being seen daily on
television, some so convincing that the average person does
not realize the part that computers have played. Artists and
designers have become involved with computers that have high
resolution color capability, and have begun to use the
digitizer and monitor as their design medium. The procedure
presented here uses ideas from many of the current
applications put together in a unique way.

EQUIPMENT

The heart of the capability that is presented here is the low
cost add-on graphics board produced by SCION Inc. for the IBM
PC. The PC640 board is a "high resolution" color monitor
driver that produces a 640 by 480 resolution screen with 16
colors displayed out of a possible 4096 colors. The PC640
board appears to the computer as an additional 256 K byte
bank of memory. As such, it can be addressed easily by any
program. The set of 67 primitives built into the firmware
includes such operations as move, draw, box, fill, flood,
copy, pie, text and others. The software used in part of
this project was the EASEL program by TIME-ARTS. Other
commercial software is available that uses the SCION board.
Another section of the project was done on the APPLE II+
microcomputer using the APPLEWORLD software package by United
Software of America.

---
* Personal Communication

INTRODUCTION

As a teacher of remote sensing, I have observed that some students have more difficulty that others in relating to the colors of a classified LANDSAT scene. This is especially true when the display is on a microcomputer which has limited resolution and color capability. Figure 1 shows a typical classified LANDSAT scene as it might appear on a microcomputer.

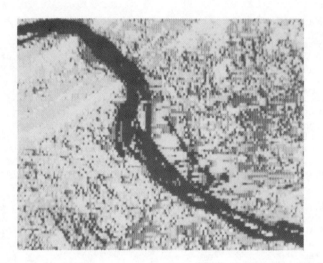

Fig. 1 LANDSAT scene displayed on an APPLE II at a resolution of 140 pixels by 96 rows (B/W photograph of a color monitor).

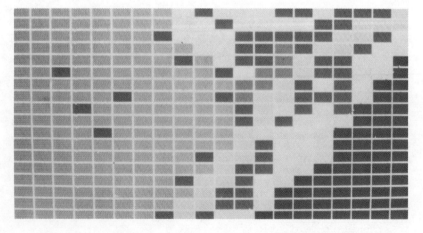

Fig. 2 Large scale LANDSAT classified image

PROCEDURE

The first step in this project was to produce the three dimensional terrain model. This was done by geo-referencing the LANDSAT data to a 1:24000 scale topographic map. Elevations were then sampled at the corners of each pixel in the 18 by 20 pixel study area. The size of the matrix is restricted because of the limited memory in the APPLE II+ microcomputer. One program was written to store the elevation matrix, and another to add the coordinate information and to format the output file. The file was then read by the APPLEWORLD program and converted to the three dimensional model. The model was then digitized as input into the LEADING EDGE (an IBM PC compatable) with the EASEL software. The LANDSAT color classification was transfered to the cells of the digital model. Figure 3 shows the monitor at this stage.

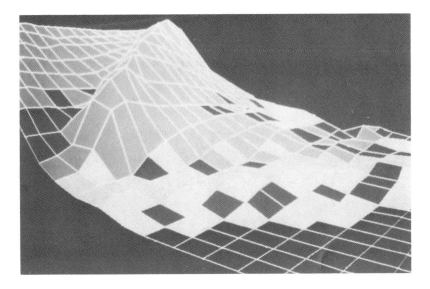

Fig.3   Block diagram of elevation and pixel location with LANDSAT classification

A series of symbols was developed for placing on the surface of the block diagram in place of the color coded classification scheme. It was thought that if the classification was, for example, "Forest," then the viewer's perception would be better served if there were trees on the surface. Symbols were also developed to represent deciduous forest, mixed forest, evergreen forest, scrub forest, pasture, various agricultural crops, residential, urban, commercial and others. The work on the types of symbolic representations is not complete and other types are being added to the list. Figures 4-5 are examples of some of the symbol library.

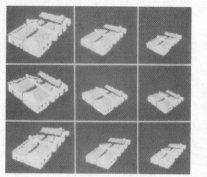 

Fig. 4 Residential area   Fig. 5 Deciduous trees

The symbols are stored on disk and are interactively recalled, scaled and placed on the surface model much as a rubber stamp. Several symbols of the same catagory are available in the cell buffer so that the scene can have some variation of pattern. Figure 6 is an example of a final product. The view is from Fort Hunter, Pa. looking southwest across the Susquehanna River toward the community of Marysville.

Fig. 6 Terrain model with stylized symbols
        representing LANDSAT classification

CONCLUSIONS

The results show that microcomputers can be used to convert LANDSAT imagery into a different format of presentation. Three-dimensional block diagrams with stylized surface symbols representing the classification scheme is a new way of displaying this data. The addition of an artist's interpretation is certainly subjective but the result should be easier for the untrained to visualize what the surface might look like.

REFERENCES

Atherton, Peter R. 1981, A Method of Interactive Visualization of CAD Surface Models on a VIdeo Display: Computer Graphics, Vol. 15, No. 3, pp. 279-287

McKeown, David M., Jr. and Denlinger, Jerry L. 1982, Graphic Tools for Interactive Image Interpretation: Computer Graphics, Vol. 16, No. 3, pp. 189-198

United States Army Engineer Topographic Laboratory (Fort Belvoir, Va.) 1977, Experimental Three Dimensional Map

GENERALIZING CARTOGRAPHIC DATA BASES
Eric K. Van Horn
Geographic Data Technology
P.O. Box 126
Academy Road
Thetford, VT 05074

BIOGRAPHICAL SKETCH

The author received his B.A. from Goddard College in 1973. His initial exposure to computers was as a health planner for Hoffman-LaRoche, Inc., in 1975, but his first serious programming efforts were doing solar system simulations on a microcomputer in 1977. In 1979 he went to work for Creative Computing Magazine as the production manager for the software department, and later became the software development manager. He moved to Vermont and started working for Geographic Data Technology in 1980 as system manager. In 1984 he was promoted to technical director.

ABSTRACT

In applications where the scale and resolution of end use is known, it makes sense to rescale digital maps to this resolution. Maps can be broken up into windows that represent the most likely viewing areas, and each window can be rescaled separately. The algorithm presented by David Douglas and Thomas Peucker in 1973 can then be applied to remove undisplayable detail. Rescaling as a pre-processing step causes more nodes to be removed with less distortion than the Douglas-Peucker algorithm applied alone. Rescaling also aids the deletion of islands and lakes that do not add to the graphic reproduction at the target resolution. In a sample file, the state boundary of Virginia was reduced 73% more than the stand-alone Douglas-Peucker algorithm, yet the amount of detail is greater.

INTRODUCTION

Generalizing, to quote David Douglas and Thomas Peucker, is the "reduction of the number of points used to represent a digitized line". There are three advantages to doing this. First, generalizing reduces the size of a file, requiring less external storage. In microcomputer applications, where external storage is often limited to floppy disks or 10 mb Winchester disks, this is important. Second, the display of maps is a compute-bound process. Ideally, the display of a map should be instantaneous, but on general purpose computers which lack special hardware, this is not the case. On a Zenith Z-100, the display of 4000 segments takes about 6 seconds. Our county boundary file, which has 224758 segments, would take over 5.5 minutes to display at this speed. Certain operations, like continuous zooming and panning, are not possible at these speeds.

Finally, in applications where the map display is done on a low or medium resolution display (up to 400 X 600 pixels), details such as shorelines may compress into blobs when rescaled to fit the resolution of the display. Just as cartographers use less detail when working on large scale maps, computer maps will look better when the amount of detail in the file matches the scale and resolution at which it is displayed.

GENERALIZING METHODS

Generalizing By Point Removal.
The process of generalizing has its origins in applications where maps were stream digitized. In stream digitizing, a map's features are traced and coordinate pairs are sampled at a set rate. A long straight line will potentially be represented by many more points than is necessary for accurate display or computer processing. Generalizing algorithms were developed to reduce the number of points. The most obvious methodology is to eliminate every nth point, effectively reducing the sampling rate. A variation on this is removing a point if it is closer than a given distance from the previous point. Point removal disregards entirely the nature of the terrain. In addition, straight lines are still not necessarily represented by only two points, and square corners may be sliced off at an angle. Examples of reduction by point removal are illustrated in Figure 1.

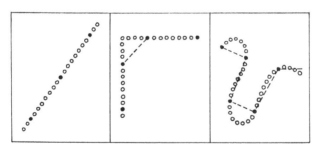

Figure 1. Line reduction by selecting every 8th point.

Feature Analysis.
Attempts have also been made to automate the ways in which a cartographer makes judgments about which features to keep and which ones to ignore. Artificial Intelligence methods can, in theory, be used to teach a program the pattern recognition skills necessary to do this. In reality, this attempts to duplicate a very complex procedure. Other methods try to evaluate the importance of features mathematically. This is also, in fact, a non-trivial solution with many special cases. Neither approach represents a simple generalizing solution.

Line Straightening.
In 1973 David Douglas and Thomas Peucker published a paper called "Algorithms For The Reduction Of The Number Of Points Required To Represent A Digitized Line Or Its Caricature". In their algorithm, a chain of line segments is selected for reduction. A straight line is drawn from the start point to the end point. If the perpendicular distances of all the points in the chain fall within a given tolerance of the straight line, the two end points are selected and all others are discarded. Otherwise, the next point closer to the start point is selected as the end point for the straight line test. The end point continues to move closer to the start point until the test is successful. In the worst case, the start point and end point are adjacent. Figure 2 illustrates how the Douglas-Peucker algorithm works.

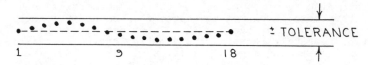

Figure 2. The Douglas-Peucker method. Points 2 through 17 all fall within a given perpendicular distance from the straight line formed by connecting points 1 and 18. In this case only points 1 and 18 would be selected.

This algorithm is simple and useful when used to remove extraneous points along a generally straight line. However, when the tolerance becomes large, features can become distorted and, as stated in their original paper, become a caricature. Figure 3 shows generalizing using the Douglas-Peucker algorithm on the state boundary of Virginia. The original file contains 6519 nodes. The coordinates for this file are in degrees and ten thousandths of a degree of latitude and longitude. At a tolerance of .0001 (one least significant unit), only 4 nodes are removed. Up to a tolerance of .0005, only 45 nodes are remove. When the tolerance is increased to .0010, although the node count is reduced by 2856 nodes to 3663 nodes, the northern border becomes distorted and is no longer a good representation of the boundary. Yet the eastern coastline along the Chesapeake Bay, which is the area most in need of generalization, has been virtually untouched.

RESCALING BEFORE GENERALIZING

Lattices.
All digital devices work in fixed increments. The precision of any digital system depends on the size of this increment. In two dimensional applications like digital mapping, this means there is a lattice or grid which determines the absolute precision of the display. A plotter with a resolution of .0005" may, at a given scale, only differentiate between features that are at least 20 feet

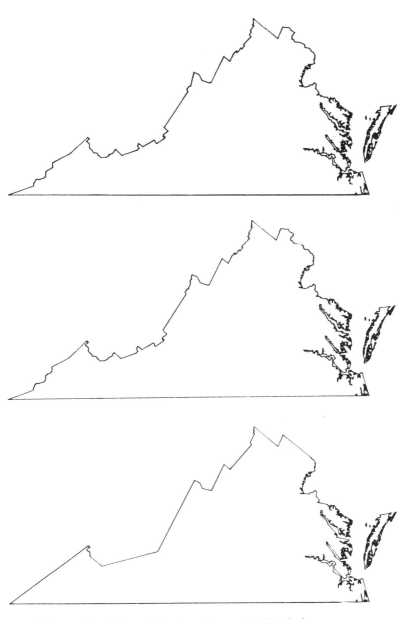

Figure 3. The state boundary of Virginia generalized using the Douglas-Peucker algorithm. From top to bottom, the files were generalized using tolerances of .0001, .0005, and .0010. The largest tolerance reduces the number of nodes from 6519 to 3663, but distorts the northern boundary.

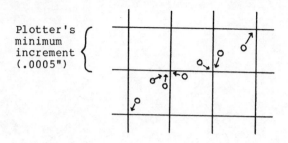

Figure 4. The lattice effect. Points scattered about in space are pulled to the nearest grid location as defined by the digital precision of the device.

apart. Coordinates that fall in between are 'pulled' to the nearest lattice location. Other display devices such as video displays and dot matrix printers have less precision than a plotter. The same map that can be plotted in increments of 20 feet on a plotter will only display to the nearest 734 feet on a 60 pixel/inch graphics screen. Figure 4 shows how measurements in the real world may be mapped onto this lattice.

In the archival storage of maps, the scale or precision of the end use is not known, so the greatest amount of detail possible is necessary. However, in individual applications where the scale and precision of the end use is known, it makes sense to rescale the data to the end use lattice and remove nodes that do not add to the graphic representation at that precision. The Douglas-Peucker algorithm works well when generalizing the rescaled coordinates, because the rescaling, i.e., 'pulling' to a grosser lattice, makes the generalizing a straight line problem at which the algorithm excells.

Figure 5 shows Virginia rescaled to fit a 600 X 400 resolution lattice (the precision of a medium resolution graphics display), and generalized using the Douglas-Peucker algorithm at a tolerance of 1 (1 pixel, or 1 least significant unit). The northern border no longer has the distortion that was present in the non-rescaled .0010 tolerance reduction, even though 5545 nodes have now been removed. The file size has been reduced to 974 nodes, 73% smaller than the .0010 tolerance reduction. Yet, as can be seen, the boundary is a _more_ detailed representation of the boundary.

Window Definition.
In Figure 5, the coordinate extremes for the entire file were determined, and these extremes were used to calculate the factor used in rescaling. In cases where the entire file is likely to be viewed as a single entity, this method suffices. However, if you were to zoom in on a portion of the coastline, it would appear caricatured and unrealistic.

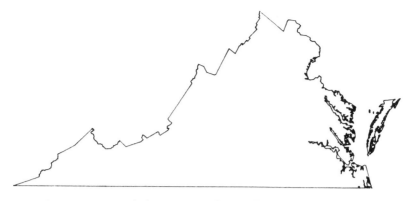

Figure 5. Virginia rescaled to fit a 600 X 400 lattice and generalized at a tolerance of 1 using the Douglas-Peucker method.

In this case it is desirable to specify that portion of the coast to be rescaled as a separate window. Because our primary application is the generalization of boundary files, the software at Geographic Data Technology has been set up to do window definition on the basis of groups of zones (counties, zip codes, etc.).

Figure 6 shows zip code boundaries for Denver. The urban core has been defined as one window, with the remainder of the Denver SMSA defined as another. The zip codes in the urban core were rescaled to the lattice using the coordinate extremes of the urban core. All other zip codes were rescaled using the coordinate extremes of the entire SMSA. In the boundaries that separate the urban and non-urban core, the greater resolution is used.

Window definition can also be done on the basis of geographic heirarchy (where one exists). MCD's, for example, can be rescaled using the coordinate extremes of the containing county. This is much simpler than manually defining zone aggregates.

Rescaling and Islands.
Rescaling and line straightening does not work well in areas where there are many islands. In these cases, small islands must be deleted. Monroe County Florida, which contains the Keys, is shown in Figure 7. In addition to the boundary generalization, islands with an area of less than 25 lattice units (pixels) have been deleted. The size of this file is is 37% smaller, yet the character of the shoreline and the sense of "islandness" have been retained. This deletion also applies to lakes.

CAVEATS

It is possible that using rescaling and straightening will

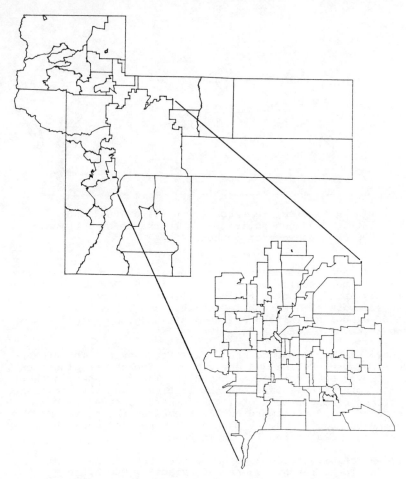

Figure 6. Denver zip code boundaries. The urban core has been generalized as one viewing window, reducing the node count from 1239 to 941. The remainder of the SMSA has been processed as another window, reducing the node count from 1981 to 930. This reflects the way in which the area is most likely to be viewed.

cause lines to cross. There are two reasons this may happen. One is that if the rescaled coordinates are then scaled back into the same units as the original ones, the values may not be the same. The rescaling destroys the precision of the coordinate value. So although the generalizing should be done on the rescaled value, the original value should be retained and used for the final output. This maintains the integrity of the absolute value of the node.

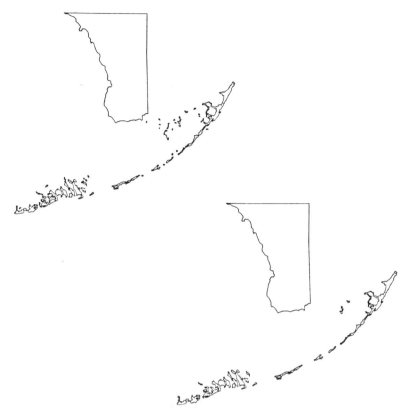

Figure 7. Monroe County Florida shown before and after generalizing. In addition to generalization of the boundaries, small islands have been deleted on the basis of minimum area.

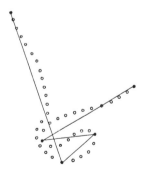

Figure 8. Crossing segments. Files generalized using rescaling and line straightening may cause lines to cross.

Even if the original values are retained, however, some lines may be made to cross. The grid pulling may cause one line to be pulled slightly one way and another line to pulled slightly the other such that in the in-between areas of the lattice the lines may end up crossing. This phenomenon is shown in Figure 8. For this reason, an edit, such as David Douglas's cross algorithm, should be done that checks to see if any lines cross. If they do, a manual procedure can be used to move the necessary nodes.

Where geographic details are extremely jagged, such as the docks in San Francisco or along the Maine coast, line straightening with rescaling will only work if these areas are rescaled using a window set up specifically for them. Otherwise, these areas will still show up as a blob. One way to do this is as part of an interactive process. An operator can zoom in on an area, point to the window, and tell the system to generalize that window at the current viewing scale.

JOB STREAM SUMMARY

Here is a sample job stream to do rescaling and line straightening:

1) Define the normal viewing windows
2) Define the lattice
3) Set the tolerance for the Douglas-Peucker algorithm
4) Rescale the coordinates inside each window
5) Apply the Douglas-Peucker point reduction algorithm
6) Write out the remaining coordinate pairs
7) Load each polygon
8) Determine the area
9) Delete the polygon if the area is too small
10) Check for crossing segments
11) Manually correct segment crossings

CONCLUSION

The Douglas-Peucker line reduction method can be enhanced by rescaling coordinates to fit the lattice of the target environment before it is applied. This causes greater reduction in file size with little or no loss in detail.

REFERENCES

Douglas, David and Peucker, Thomas, 1973, Algorithms For The Reduction Of The Number Of Points Required To Represent A Digitized Line Or Its Caricature, The Canadian Cartographer, Vol. 10, No. 2, pp. 112-122

Douglas, David, It Makes Me So CROSS.

A RELATIONAL APPROACH TO VECTOR DATA STRUCTURE CONVERSION

Jan W. van Roessel[*] and Eugene A. Fosnight[*]
Technicolor Government Services, Inc.
Sioux Falls, South Dakota 57198

BIOGRAPHICAL SKETCH

Jan W. van Roessel is currently employed as a Principal Scientist by Technicolor Government Services, Inc. at the EROS Data Center in Sioux Falls, South Dakota. He received a PhD in Wildland Resource Science and an MS in Engineering Science from the University of California, and an MF from the University of Washington.

BIOGRAPHICAL SKETCH

Eugene A. Fosnight is currently employed as a Senior Scientist by Technicolor Government Services, Inc. at the EROS Data Center in Sioux Falls, South Dakota. He received a diploma in Cartography from the University of Wales, Swansea, United Kingdom.

ABSTRACT

The proliferation of geographic information systems and digital data bases is creating a need for efficient methods to convert data from one spatial data structure to another. One approach is to create ad hoc interfaces, with a potential of N(N-1) interfaces for N data structures. Using an intermediate data structure, at most 2N interfaces are required. An intermediate relational data structure is therefore proposed that takes the form of a set of normalized relations stored in a relational information management system. The advantages of this approach are found in the simplicity of the relational approach, and the availability of relational operators to be used as higher level tools, to convert from and to the relational data structure. The Relational Information Management System (RIM) is used for the ongoing research. In conjunction with the relational data structure, another higher level tool has been developed to cope with linked lists, tree structures, vectors, and matrices, which are not otherwise easily reduced. This data tracking system is programmable at a higher level, in a syntax that allows a concise expression of the desired restructuring. Output from this system can be further operated on by relational operators to arrive at the desired intermediate data structure. This, and other topology verification and checking tools, are visualized as part of a core system dedicated to the conversion and collation of spatial data from diverse origins.

[*]Work performed under U.S. Geological Survey contract no. 14-08-0001-20129.

INTRODUCTION

With a growing number of geographic information systems and digital spatial data sets, a capability to adapt digital spatial data from one system to another has become much sought after. Yet such adaptations are not as easily accomplished as is sometimes believed because of the special characteristics of spatial data. Non-spatial digital data have two main aspects with respect to their organization, physical and logical (Martin, 1977), but spatial data can be thought of as having two additional traits, spatial-logical (topological) and locational (coordinate system, projection, measurement units).

Adaptation is therefore not a simple reformatting problem, as with other data sources, but rather it is a far more complex proposition with spatial implications that cannot be readily deduced by an initial inspection of the data. It may even be impossible to convert certain types of data structures, short of a major processing effort. However, data structures of similiar types, such as an arc-node organization, can be converted from one system to another with a reasonable amount of effort.

One of the obvious first steps for developing a conversion methodology is to obtain a consistent description of the vector data structures involved, and to characterize them in order to anticipate possible conversion problems.

For this purpose a Backus Naur Form (BNF) type of notation was developed to describe the data organization concisely and consistently. The use of BNF for describing the syntax of geographic data was earlier suggested by Cox, Aldred, and Rhind (1980).

A second step in the description process has been to cull the most important characteristics of each data structure. The five major spatial data aspects--physical, logical, logical-spatial (topological), locational, and attribute-- were further reduced to constituent components, and this process was repeated until a satisfactory level of detail was reached. The overall breakdown was then used as a checklist against each BNF description, as well as other documentation, to obtain a set of hierarchical characteristics of the data structure under consideration.

INTERFACING MODELS

One approach to the interfacing of N data structures is to match each one with every other one, thereby creating N(N-1) connections. Alternately, one may implement a "ring" concept whereby each system interfaces to one and only one other system. This yields the minimum number of N interfaces. Its drawback is that data may have to pass through N-1 other systems before a target system is reached, with proportional processing time and potential for errors. A third approach uses an intermediate data structure to which each original structure must be converted before it is transformed to the target configuration. This method requires 2N interfaces and is the one most commonly adopted; it is also the approach used by the authors. Quite often this method is associated with an intermediate tape format. Our concern is to define an intermediate data structure maintained in a database system, which can in turn be readily transferred between systems.

The Workgroup on Data Organization of the National Committee for Digital Cartographic Data Standards (Nyerges, 1984) has proposed a model for digital cartographic data conversion to an intermediate structure as follows:

$$S \xrightarrow{T1} I1 \xrightarrow{T2} I2 \qquad (1)$$

The above model is only for the conversion from source to intermediate; the conversion from intermediate to target is similar. In the above, S represents the source data structure, T1 is a transformation producing an "intermediate" intermediate structure I1, and T2 is a transformation producing the desired standard interchange form. The Workgroup mentions the importance of the data schema for this model. The schema defines the logical aspect of the data structure, and forms a link with the physical representation. A sobering note is the complete absence of any kind of explicit logical or topological schema in any of the existing data structures considered. However, for our approach, the model proves to be useful for considering the different transformation stages in the data structure conversion.

THE RELATIONAL APPROACH

Other investigators have applied the relational model or relational concepts to geographic data handling. At least one prototype geographic information system (GEOQUEL) has been constructed using the relational database management system INGRES (Go, Stonebraker and Williams, 1975). Shapiro and Haralick (1980) have proposed spatial data structures based on relations in relation trees. Garret and Foley (1982) have reported an attempt to build an experimental graphics system using relations, where events trigger continuously evaluated qualified updates, through which the consistency between relations is always maintained. A graphics application (IMPS) based on a relational data base (IMS) was developed by IBM in the United Kingdom (IBM, 1979).

We have selected the relational data model for the intermediate data structures for a number of reasons. One of the main arguments for our choice is the elegance and simplicity of the data representation, resulting from the use of "flat" files and a complete absence of pointers.

A second consideration is the availability of the relational algebra and its unique relational operators. To minimize the effort associated with creating a number of transformation funcitons for different input models, it follows logically that one should attempt to develop higher level tools that can be reused for each conversion. The aim of our study is to investigate whether relational operators fall in this category.

A third reason for considering the relational model is the availability of a number of different software systems, including many operating in the microcomputer environment, as well as hardware implementations in the form of back-end database machines. Research for this paper was performed using the Relational Information Management system (RIM).

Normalization is one of the important considerations in relational database theory. Five normal forms are recognized (Kent, 1983). The first normal form is the most crucial and states that "all occurrences of a record type must contain the same number of fields." This is not an option, but a matter of definition: relational database theory does not deal with variable length records or vector or matrix type attributes. Our proposed intermediate data structure is strictly of first normal form, which allows us to use the full range of relational operators.

Given the simplicity of flat files, the database schema also becomes very simple, and consists of a description of the individual attributes and the composition of the relations in terms of these attributes. Although much can be learned about the contents of a database by inspection of the schemas, the schema is still inadequate for describing the spatial organization of the data.

RELATIONAL DATA STRUCTURE

The type of spatial data structure selected is of the arc-node type. A full complement of spatial elements is represented: regions, polygons, arcs, nodes, and vertex points (line and point type elements are not further considered in this paper). We define a polygon as a closed figure bounded by connected straight line segments, a region as consisting of an outer polygon and zero or more island polygons, and nodes as arc termination points.

The mutual referencing of the elements is resolved as follows (the relational theory excludes direct pointers, so all referencing is in terms of element identifiers). Given the hierarchy obtained by defining more complex elements in terms of their component parts--regions, polygons, arcs, nodes and vertex points--the most austere organization is obtained by having each element reference its subordinated elements. For example, by defining the element identifier attributes regnum, polnum, arcnum, nodenum, strtnode, endnode, and pointnum, as well as the coordinate attributes x, y, we can propose to maintain the following relations:

regpol: regnum, polnum

polarc: polnum, arcnum

arcnode: arcnum, strtnode, endnode   (2)

arcpoint: arcnum, pointnum

nodepoint: nodenum, pointnum

pointxy: pointnum, x, y

where strtnode and endnode refer to nodenum in the nodepoint relation.

The above scheme has downward referencing only. Reverse references can be established with relational operators. For instance, to compute an "arcpol" relation containing polygon numbers by arc, one can isolate the arc column from the polarc relation using the project operator, remove duplicate arc numbers (if not included in the project), and then join this temporary relation with the polarc relation using arcnum as the join item, while maintaining the derived arc ordering.

Spatial references, such as establishing which regions are to the left and which are to the right of the arc, cannot be so easily established in this manner. Considering also the frequency with which this type of information is used, one is led to consider storing some higher level information at the lower levels. In particular, the arcnode relation can be renamed archdr, and be given two additional attributes, lftreg and rgtreg, as follows:

archdr: arcnum, strtnode, endnode, leftreg, rgtreg.   (3)

Also, for network tracking, a nodearc relation is desirable which arcs emanating from each node are stored:

nodearc: nodenum, arcnum.   (4)

The above considerations, and the removal of the pointxy relation, lead to our proposed intermediate data structure I2:

regpol: regnum, polnum

polarc: polnum, arcnum

archdr: arcnum, strtnode, endnode, lftreg, rgtreg   (5)

arcxy: arcnum, x, y

nodearc: nodenum, arcnum

nodexy: nodenum, x, y

One further decision to be made is how to cope with complex islands. The union of the polygons of a complex island constitute the hole in the outer polygon. Not all arcs of the island polygon are therefore island arcs. We have solved this problem by defining the outer boundary of the complex island to be a separate polygon. This gives the advantage of being able to join the regpol relation with polarc and arcxy relations to directly obtain a regxy relation in which each region is explicitly represented in sequence, complete with all its islands, including complex ones. In this way, the topology is defined in a simpler, clearer organization. Thus, our approach is to organize the topological information implicitly by position, rather than explicitly.

Position is critical within most relations. The vertex point coordinates in the arcxy relation are obviously stored in positional sequence. Less obviously, the arcs in the nodearc relation are stored in a clockwise rotation around the node. The arcs in the polarc relation are stored in clockwise order for normal polygons and counterclockwise for island polygons. Finally, in regpol, the first polygon for each region is the outside polygon.

When constructing polygon or region boundary lists, one must keep in mind that arcs must be traversed in different directions for the regions to the left and the right of the arc. The traversal direction is stored implicitly in the archdr relation. If the polygon currently being traversed corresponds to the right region in the archdr relation, the arc is rotating clockwise. The converse is true when it matches the left region.

Other non-spatial relations are defined in addition to the above core of spatial relations. Most important are those for attribute data. They are visualized as a nested set, each consisting of one primary and a number of secondary relations, namely:

$$\text{attprime:} \quad \text{elnum, att1, att2, att3...etc.} \quad (6)$$

This relation links the primary attributes to the spatial elements through the common element identifier. In addition, each attribute in the primary relation can be a key for an entry in secondary relation, in which that attribute is further detailed. For instance, one may have:

$$\text{attsec:} \quad \text{att1, satt1, satt2, satt3...etc.} \quad (7)$$

This allows new attribute relations to be created in which the attribute detail can be varied at will, using join operators.

The overall interfacing effort is reduced not only by the total number of required interfaces, but also by the amount of time required to construct each interface. This time is reduced further when suitable higher level tools are available.

According to the Workgroup model, these tools can be assigned to T1 and T2 categories. The database system used is perhaps an exception; it can be assigned to both categories because it allows the data structures I1 and I2 to exist.

The scope of T1 is the following. The scheme being tested makes use of the fact that the input data structures can be cast in the form of files with fixed- or variable-length records. These files can be translated on a one-to-one basis to unnormalized (with respect to the first degree) records in the RIM system, storing one record per relation row. One would not expect this capability in an RDBMS, but RIM permits records with variable-length attributes, although relational operations cannot be performed on variable-length items. With the transfer, a schema is developed for the unnormalized data structure in the relational database, and I2 represents the normalized intermediate structure presented earlier.

In the T2 category, a breakdown can be made considering tools that allow data to be restructured from T1 to T2, and those to be used to repair errors, perform checks, and make I2 more complete.

RESTRUCTURING TOOLS

The transformation from the unnormalized form to a normalized representation is accomplished by a specially developed program (RIMNET) that allows one to track data

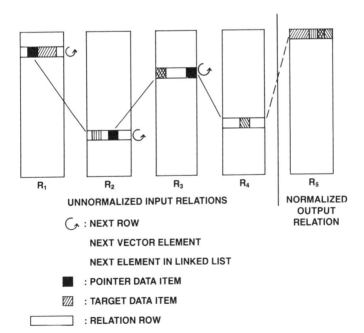

Figure 2.--Schematic of RIMNET operation.

items along predefined access paths in the unnormalized representation. This program is one of the main conversion tools. Using relational operators, the normalized forms are then further reduced to the essential relations in the intermediate relational format (I2).

RIMNET has been designed to serve as a normalization tool, but also to cope with such data organizations as linked lists, which cannot easily be unscrambled with relational operators. The principal idea behind the program is demonstrated in Figure 2.

The desired tracking sequence can be defined by a concise free-format syntax in which the relations, groups, and types of increments are specified. The basic element in the syntax is the group specification. In a group the pointer data item name is separated from the target data item names by a colon. Either pointer or target item names can be omitted. The entire group specification can be enclosed in brackets with following meaning: {} row increment; [] linked list; <> vector increment. Brackets can be nested. The following are examples of valid group and increment specifications:

$$\{POINTER:TARGET\}$$

$$[POINTER:] \qquad (8)$$

$$<:TARGET>$$

$$\{<:TARGET>\}$$

For a relation, group specifications can be concatenated. To indicate the relevant relation, groups are preceded by the relation name, as follows:

$$REL1 = \{POINTER:TARGET\} \qquad (9)$$

When more than one relation is involved, specifications are strung together, as in the following example:

$$REL1 = \{POINTER:TARGET\} = \qquad (10)$$

$$REL2 = <:VECTOR>$$

Here, POINTER points to a row of REL2. The output relation will contain two data items per row: TARGET, and an element of VECTOR.

An example of the programmable tree-traversing that is possible with RIMNET is found in a simple situation in which the rows of the unnormalized relation REL contain a vector attribute, VECTOR, and this relation must be normalized by assigning one vector element to each row of the output relation. The normalizing operation is then specified as follows:

$$REL = \{<:VECTOR>\} \qquad (11)$$

where: VECTOR is the right hand side of the group specification, VECTOR is the target item, and an explicit pointer is not needed because the <> brackets imply a next vector element pointer, while the {} brackets imply a next row pointer. REL is the input relation for which VECTOR is defined as an attribute in the schema.

Execution of the syntax starts at the first row and causes each vector element to be transferred, proceeding then to the second vector of the second row, and so on.

Slightly more complex is an operation in which a variable number of elements must be retrieved, as indicated by another data item TOTEL stored in the rows of the input relation. The syntax is then:

$$REL=\{TOTEL\}<:VECTOR(1-TOTEL)> \qquad (12)$$

If, in addition, the elements may be further spread out over more than one record, and the continuation record is indicated by an item NEXTREC, the syntax becomes:

$$REL=\{:TOTEL\}[NEXTREC:]<:VECTOR(1-TOTEL)> \qquad (13)$$

The square brackets indicate a linked list for which NEXTREC is the pointer item.

RELATIONAL OPERATORS

Once the initial relations (I1) are normalized, relational operators can then be applied to the extent possible to convert them to the desired intermediate structure (I2). Thus, we are testing the concept that these operators can be an important set of higher level tools.

So far, the following RIM operators have been used frequently: JOIN, PROJECT with WHERE clause, INTERSECT, CHANGE with WHERE clause and DELETE DUPLICATES. Especially the INTERSECT command was found to be a powerful operator, with advantages over a JOIN in many instances.

TOPOLOGICAL TOOLS

One of the objectives of the intermediate data structure is to serve as a way-station where improvements and quality checks can be made. Therefore, we have adopted the philosophy of "weak input". This means that we will be able to accept data deficient in topology, because we will make necessary fixes and enhancements with our tools. One of the significant advantages of a relational information management system such as RIM is that the data structure can be inspected, partially copied, changed, and otherwise manipulated in any desirable manner, interactively. One therefore enjoys total access to the data.

In another sense "weak input" means that only certain basic relations need to be extracted from I1; others can be built from these initial relations. The fully defined topology of I2 can actually be created given a "weak" input of two tables. One table must contain the arcnumbers and vertex point coordinates (arcxy), while the other must contain the adjacent regions for each arc (regardless of order -- a weak archdr). A topological restructuring program that uses these two input relations has been written and tested.

It is possible to visualize even weaker input, such as merely a set of connected arcs, or even arbitrary spaghetti. Given a data structure, the question is: how much of its topology should be salvaged. We envision the eventual use of a "generalized vector processor" that will be able to reduce vector data to basic line segment components from which it can recreate the desired topology.

CONVERSION EXPERIENCE TO DATE

The first system for which a data structure has been converted to the relational format is the Geographic Entry System (GES) of Electromagnetic Systems Laboratory's (ESL) Interactive Digital Image Manipulation System (IDIMS). From the relational format GES data have been transferred to the ARC/INFO system of the Environmental Systems Research Institute (ESRI), and AGIS/GRAM of Interactive Systems Corporation (ISC).

A second system for which data structures were converted is the Analytical Mapping System (AMS) used by the U.S. Fish and Wildlife Service. This system is the vector input system to its companion Map Overlay and Statistical System (MOSS).

Subsequent systems for which data were transferred to the relational format were ICARAS (a raster-to-vector conversion system developed by D. Nichols of the Jet Propulsion Laboratory), as well as data from an Intergraph digitizing system (EDCIDS) being developed at the EROS Data Center, and the Unified Cartographic Line Graph Encoding System (UCLGES) of the U.S. Geological Survey (used to structure digital line graph (DLG) data).

Tools such as RIMNET have been successfully used in at least two interfaces (GES and AMS). In the more polished interfaces, the various tools are tied together with the VAX/VMS DCL language.

CONCLUDING REMARKS

Although we are currently at the beginning of our interfacing efforts, the relational approach with the use of higher level tools shows promise for leading towards a conversion model that will be closely aligned with the one proposed by the Workgroup on Data Organization of the National Committee for Digital Cartographic Data Standards.

The simplicity and elegance of the relational approach is reflected in the simple relations of our intermediate spatial data structure, such that the spatial operations can take full advantage of the relational method.

The development of a good set of interfacing and diagnostic and improvement tools will provide us with an excellent capability to tackle diverse interfacing problems.

REFERENCES

Breithart, Y. J., and Hartweg, L. R., 1984, "RIM as an implementation tool for a distributed heterogeneous database", NASA Integrated Program for Aerospace - Vehicle Design Symposium II, Denver, Colorado.

Codd, E. F., 1970, "A relational model of data for large shared data banks," Communications of the ACM, v. 13, no. 6, p. 377-387.

Codd, E. F., 1971, "Relational completeness of data base sublanguages", R. Rustin (Ed.), Data Base Systems (Courant Computer Science Symposia VI), Prentice-Hall, Englewood Cliff, New Jersey.

Cox, N. J., Aldred, B. K., and Rhind, D. W., 1980, "A relational data base system and a proposal for a geographical data type", Geo-Processing, v. 1, no. 3, p. 217-229.

Garrett, M. T., and Foley, J. D., 1982, "Graphics programming using a database system with dependency declarations", ACM Transactions on Graphics, v. 1, no. 2, p. 109-128.

Go, A., Stonebraker M., and Williams, C., 1975, "An approach to implementing a geo-data system". Memorandum No. ERL-M259, Electronics Research Laboratory, College of Engineering, University of California, Berkeley, California.

IBM, 1979, IBM interactive management and planning system under IMS/VS. User guide: SB11-5220. IBM European Publications Center, Uithoorn, The Netherlands.

Kent, W., 1983, "A simple guide to five normal forms in relational database theory". Communications of the ACM, v. 26, no. 2, p. 120-125.

Martin, J., 1977, Computer Data-Base Organization, O. Prentice-Hall, Inc., Englewood Cliffs, New Jersey.

Nyerges, T., 1984, Editor, "Digital Cartographic Data Standards: Alternatives in Data Organization", Progress Report of the Workgroup on Data Organization of the National Committee for Digital Cartographic Data Standards, Report 4, January.

Shapiro, L. G., and Haralick, R. M., 1980, "A spatial data structure", Geo-Processing, v. 1, no. 3,

Shapiro, L. G., 1980, "Design of a spatial information system in map data processing", edited by H. Freeman and G. G. Pieroni, Academic Press.

TESTS TO ESTABLISH THE QUALITY OF DIGITAL CARTOGRAPHIC DATA:
SOME EXAMPLES FROM THE DANE COUNTY LAND RECORDS PROJECT

Alan P. Vonderohe and Nicholas R. Chrisman
University of Wisconsin - Madison
Madison, WI 53706

ABSTRACT

Tests, particularly those based on independent sources of higher accuracy, are crucial to the evaluation of the quality of digital cartographic data. This paper reports interim results of tests applied to positional accuracy of USGS Digital Line Graph data with additional reference to logical consistency. Procedures to carry out tests are discussed.

BACKGROUND: QUALITY STANDARDS

While advances in some aspects of digital cartography have been rapid, studies of data quality have lagged behind. Recently, issues of data quality have begun to attract attention with the work on data standards in a number of countries (Canadian Council on Surveying and Mapping, 1982; National Committee for Digital Cartographic Data Standards [NCDCDS], 1985). This work coincides with the overhaul of US geodetic control standards (Federal Geodetic Control Committee [FGCC], 1984) and a new specification for the accuracy of large-scale maps (American Society of Photogrammetry [ASP], 1985). Although the need for standards is clear, and the professions and agencies have focussed attention on the problem, there has been remarkably little research published that investigates alternatives for quality standards. The published literature particularly lacks examples drawn from prevalent data products. Perhaps agencies or companies perform such tests internally, but the procedures used and the results obtained are not reported.

Before the cartographic standards take on a final form, there should be a fully informed debate on the nature and purpose of quality tests. The overall mission of the NCDCDS covers quality testing, but, as a volunteer committee, it cannot afford to undertake a full testing project. This paper hopes to contribute to a debate on the procedures for testing digital data quality by reporting some interim results of recent research projects.

DANE COUNTY LAND RECORDS PROJECT

The tests were performed as a part of the Dane County Land Records Project (Chrisman and others, 1984), a two year cooperative effort addressing a broad range of issues in modernizing land information. Within the broad goal, the project is directed towards information needs for soil erosion control planning. The cooperators in the project include US Soil Conservation Service, Wisconsin Departments of Natural Resources and Agriculture, Trade and Consumer Protection, Dane County Land Conservation Committee and Land Records and Regulation Department, along with University of Wisconsin faculty. For certain aspects, Wisconsin Department of Transportation, US Geological Survey and Bureau of Land Management have cooperated as well.

COMPONENTS OF QUALITY

The NCDCDS Working Group on Data Set Quality has discerned six interrelated components of quality: lineage, positional accuracy, attribute accuracy, logical consistency, completeness and currency (Chrisman, 1984). Of these, lineage is not testable because it contains a summary of the source material and transformations that lead to the data. Currency can masquerade as each of the other errors, and is difficult to test for separately. The other four components can be tested to different degrees. This paper will focus on positional accuracy, but the intent is to view quality as a whole.

TESTS OF POSITIONAL ACCURACY

The ASP committee and the NCDCDS agree that testing forms a firmer basis for quality assessment than any other method. The NCDCDS interim proposed standard has adopted the draft ASP specification as the basis for conducting a test (Merchant, 1982; ASP, 1985; NCDCDS, 1985). This testing procedure has been applied to digital products of a project preceding the Dane County project (Petersohn and Vonderohe, 1982). In this study we will apply this test to a digital data product of national significance, the Public Land Survey [PLSS] layer in the USGS National Digital Cartographic Data Bank.

USGS Public Land Survey section data

One of the most popular digital data products in the US is the description of the Public Land Survey System. Demand for this information is described in the National Research Council [NRC] report (1982) on the modernization of the PLSS. The states of Minnesota, Illinois and Montana (at least) have digitized this information for state information systems (NRC, 1982, p. 25). Phillips Petroleum (1984) has digitized 30,000 quadrangles in many states. US Geological Survey has also been active. A large proportion of their Digital Line Graph products consist of the PLSS layer, along with related political boundaries. A major recommendation of the NRC report is that USGS complete a comprehensive coverage of the PLSS to contribute to integrated mapping and leading towards a national basis for multipurpose land information systems.

The Dane County Land Records Project is a pilot project for integrating land information, with particular attention to SCS soil surveys and local property records. To help provide a basis for areas of Dane County with limited survey control, we entered into cooperation with USGS National

Mapping Division [NMD]. NMD digitized the PLSS and political boundaries for the 34 seven-minute quadrangles that cover Dane County. On the cost-sharing basis of such projects, costs were competitive with digitizing the data ourselves.

Digitizing was done at the Mid-Continent Mapping Center using graphics workstations followed by conversion into the DLG topological structure (Allder and Elassal, 1983). Unlike some digitizing operations, NMD has access to stable base versions of the maps, precluding the large errors associated with paper maps. This study is not directed at the internal procedures used by NMD, because the proof is in testing the result.

There is a difference between the legal layout of the PLSS and its cartographic representation. PLSS section lines are defined by the section corners (including closing corners) and the quarter section corners. Each quarter corner was supposed to be set on a straight line between two section corners, hence in cartographic representation, particularly at 1:24000, there may be no visible deflection at quarter corners. For this reason, this study will exclude quarter corners. To promote a truly multipurpose system, beyond the cartographic representation, USGS should consider including more explicit recognition of quarter corners.

Comparison Surveys
The quality standards call for tests against an independent source of higher accuracy. Both independence and higher accuracy must be stressed.

Independence. Independent sources would seem to require separate derivation of information, for instance a separate survey. With map information, completely separate information is quite rare because all maps and surveys usually relate to the same geodetic reference network. Independence in statistics refers not to procedures, but to results. Independent processes produce uncorrelated errors. Typically, the use of different technology (for example photogrammetry versus field survey) will suffice to produce uncorrelated errors.

Higher Accuracy. Clearly a test is most useful if one source can be treated as the "truth". The ASP specifies that the "nominal positional accuracy" of the check survey be "three times that required" of the product to be tested (Section 2.4.1). This section also specifies, somewhat redundantly, FGCC second order horizontal control surveys. For the intended large-scale maps the latter may be appropriate, but for the 1:24000 maps the former seems more in order.

Sources of Check Surveys. The digital products tested were dictated by availability of appropriate check surveys. The ASP specification calls for twenty test points distributed in a sheet. The cost of second order surveying for twenty points would be rather high; this project relied on surveys performed for other reasons. One extensive ground survey covers the remonumented section corners of the

Town of Oregon in Dane County. This survey was performed by a registered land surveyor and his measurements have been adjusted as a part of our project (Krohn, 1984). The survey was controlled by three second order stations in the National Geodetic Network, having an approximate six mile spacing. Two of these have no direct connection in the National Network. Coordinate comparisons between this survey and an inertial survey, recently conducted by Bureau of Land Management (BLM) and Dane County Land Records Project personnel, indicate a mean positional discrepancy of 0.433 meters. A great deal of this discrepancy may be attributable to the difference in control for the two surveys. The inertial survey was tied to three second order, class I control stations established by Global Positioning System technology and two stations in the National Network, only one of which was common with the ground survey. In any case, the quality of the ground survey was deemed more than adequate to perform the analysis reported herein. It provides a sufficient sample of check points for one USGS sheet (Oregon) and a smaller sample for the Attica quad.

The other area of testing is in the Town of Westport in the Waunakee quad. For this report, only those surveys performed by University of Wisconsin groups to second and third order standards have been used. All surveys were tied to a second order, class I network established during an earlier project (Crossfield and others, 1983). The number of section corners available in Westport falls below the ASP threshold, but provides additional information to evaluate the Oregon test.

Coordinates. The Digital Line Graph encodes coordinates for section corners as integers. The distance represented by one count is .61 meter on the ground. The integer coordinates are related to UTM through a transformation specified in double precision. On obtaining the DLG data, the coordinates were converted to UTM in double precision (64 bits). Due to the roundoff, there should be at least .61 meter of uncertainty.

The check surveys were computed using the Wisconsin South Zone of the State Plane Coordinate system, in feet. The coordinates in State Plane were converted to UTM using the geodetic software distributed by the National Geodetic Survey in double precision. This software may differ in some details of map projection from that used by NMD, but the results should be identical to a level of precision below .6 meters. A national effort should standardize the software used for these projections to remove any uncertainties. Though coordinates of one map projection may be readily transformed into another, it would seem wise to adopt a national standard, agreed to by the states and the user community, and thereby avoid the process altogether.

Results
The DLG coordinates were subtracted from the survey coordinates to yield the measured discrepancies. A number of summary statistics can be calculated from this raw data. The ASP (1985, Appendix A) suggests bias (mean discrepancy)

and precision (standard deviation) calculated separately for
each coordinate axis. For some purposes it is more
customary to report Root Mean Square Error which measures
absolute deviation, without accounting separately for bias.
As another option, the National Map Accuracy Standards are
specified in terms of straight line positional discrepancy
without regard to coordinate axis. The table below shows
mean distance, and gives the distance of the ninetieth
percentile of the distance distribution. These figures (the
basis of the National Map Accuracy Standard) are difficult
to interpret, because they lack a statistical basis. It has
become common practice to convert the accuracy standard to a
function of standard deviation (Circular Map Accuracy
Standard) (ACIC, 1962). A few of these measures are
reported in Table 1 for the sheets described above. In the
table, eleven of the check points in the Oregon sheet
contained redundancy in the ground survey and twenty-three
were established by sideshots with no redundancy. The row
labelled "all data" includes points with and without
redundancy. In the Attica quad, one of the points was
established with redundancy and the others were sideshots.
All eleven points in the Waunakee sheet contained redundancy
in their ground survey.

Table 1: Summary of positional accuracies
(figures in meters UTM)

Sheet	Points	Bias X	Bias Y	Std. Dev. X	Std. Dev. Y	RMSE X	RMSE Y	Distance ave.	Distance 90%
Oregon									
redundant	11	-1.6	-1.2	6.7	3.6	6.6	3.7	6.2	10.5
all data	34	-1.8	-1.2	5.4	3.5	5.6	3.6	5.7	10.2
Attica	12	1.4	3.0	5.4	3.3	5.3	4.3	5.5	9.7
Waunakee	11	-1.4	6.2	5.7	7.5	5.6	9.2	9.9	16.9

The lack of points for the Attica and Waunakee sheets makes
statistical statements less applicable than in the Oregon
case. The bias detected in the Oregon sheet is
statistically significant, but low on a substantive scale.
The standard deviations for all the sheets are lower than
might have been expected. All but the Waunakee sheet pass
the National Map Accuracy threshold. The Waunakee sheet is
strongly affected by the large Y bias and a few large
outliers. All but one of the eleven points in the Waunakee
sheet showed the survey points on the positive side of the
mapped points. Such errors, if they can be explained, can
be easily removed.

The Oregon sheet not only passes the ASP Class 1 standard at
1:24000, it passes at a scale of about 1:18000. It must be
remembered that the test covers the digitizing and computer
roundoff effects, as well as the inherent error in the map.
The largest error detected, in fact, may be due to a
difference in information available at different times. The
southwest corner of section 13 showed an error of 17.4
meters in the X direction, but the 1960 topographic survey
could not have included the legal position of the section
corner which was discovered a foot below the pavement in

1975. Another large error (13.5 meters) occurs at a section corner where no evidence was found in 1977 and a new monument had to be set. Without more detailed notes on the methods used by USGS to compile the maps, the erroneous assumption has to be made that section corners are perfectly known.

Discussion of Positional Test

The positional accuracy of the USGS Digital Line Graph that we tested was better than the figures usually quoted. In the NRC (1983) report and in other places, the figure of 40 feet (16.2 meters) is quoted freely. This number is derived from the outer tolerance permitted by the National Map Accuracy Standards. Although it proves to be a relatively valid description of the data, the figures on bias and precision take more of the distribution of errors into account.

The NRC (1982) report recommends that a PLSS data bank be constructed by BLM based on the USGS Digital Line Graph. If all the Digital Line Graphs are as accurate as the Oregon sheet, such a data bank might serve some functions in geographic information systems. However, such information systems could support only those purposes having the coarsest accuracy requirements, such as natural resource inventory. If, ultimately, such information systems are intended to meet the needs of all users, and thus be truly "multipurpose", the positions assigned to PLSS corners should not be considered static. The data bank should be improved as new measurements are obtained, through highway construction, property survey, control densification or other processes. We plan to provide a revised copy of the DLG data for placement in the national archive. In the few locations where comprehensive remonumentation and control surveys of the PLSS have been conducted, as in Racine County Wisconsin (Bauer, 1976), the data bank should be constructed directly without digitizing the topographic map. The maintenance of a dynamic geometric framework poses a significant challenge to software development (Chrisman, 1983).

LOGICAL CONSISTENCY

The PLSS DLG's have a topological data structure which encodes relationships between objects on the map. This provides a verifiable structure for each layer of the map, but does not deal directly with relationships between layers. Software can verify the logical consistency of each layer, and USGS is committed to distribute only clean data (so-called DLG-3). Many systems are able to produce logically consistent structures for a single layer, but few are able to handle interrelationships of features.

In this case, the boundary and section data were constructed together. One set of nodes were created for a quadrangle for use with both layers. This procedure implicitly ensures that lines intended to be identical would overlay directly on each other. The geometric relationship is as far as the structure goes. The complex attribute scheme provides for coding a line as coincident with another feature, but it does not permit naming that other feature. Furthermore, this scheme is not always used.

In one case in the Oregon sheet, a section line and a political boundary which should have been the same were separated by a 5.7 meter gap. This created a sliver polygon where a section was placed outside its township. Strangely, the proper line was present, but encoded with the same section on either side.

Another major consistency problem is created by water bodies. Due to public trust doctrines, the PLSS does not cover navigable waters. The topographic maps show section lines running up to water but not across it. With a traditional product, the section lines are in red, and the hydrography is in blue, but the blue lines are needed to make the sections topologically complete. In the digital representation there are lines to represent water boundaries with codes to show that they are intended to coincide with hydrographic features. However, there is no explicit reference possible in the structure to the actual line intended. This structural problem has been noted by others (Dangermond, personal communication, 1984). In our case, the hydrographic layer has not yet been digitized. The PLSS data contains straight lines that connect section lines together, instead of following the shore. This representation robs a rather accurate product of its ability to produce reasonable area estimates and convincing graphics. Furthermore, when the hydrography is encoded, there will be no easy way to determine which pieces to be used. We suggest that a digital line graph requires a new perspective on its purpose. The current PLSS data can be used as a useful source of section corner locations, but not as a description of sections as areas.

ACKNOWLEDGMENTS

The authors wish to express their appreciation to a number of parties. Don Doyle performed the fieldwork and furnished the data for the ground survey in Oregon Township. David Krohn and James LaVerriere performed the least squares adjustment of Doyle's data. Nancy VonMeyer and Mr. Krohn assisted in the inertial survey which provided a check on the Oregon adjustment. The Bureau of Land Management, Cadastral Surveys Division furnished equipment and personnel for the inertial survey. Partial support for this research comes from US Department of Agriculture Hatch and University of Wisconsin University Industry Research funds.

REFERENCES

Aeronautical Chart and Information Center (ACIC) 1966, Principles of Error Theory and Cartographic Applications: Report 96, Washington DC

Allder, W.R. and Elassal, A.A. 1984, Digital Line Graphs from 1:24000-Scale Maps, in USGS Digital Cartographic Data Standards, Geological Survey Circular 895-C, Reston VA

American Society of Photogrammetry (ASP) 1985, Accuracy Specifiations for Large Scale Line Maps, *Photogrammetric Engineering and Remote Sensing* forthcoming

Bauer, K.W. 1976, Integrated Large-scale Mapping and Survey Control Program Completed by Racine Co. Wis., *Surveying and Mapping*, vol. 36, no. 4

Canadian Council on Surveying and Mapping 1982, *Standards for the Quality Evaluation of Digital Topographic Data*, Energy, Mines and Resources, Ottawa

Chrisman, N.R. 1983, The Role of Quality Information in the Long-Term Functioning of a Geographic Information System, *Proc. AUTO-CARTO 7*, vol. 1, p. 303-312

Chrisman, N.R. 1984, Alternatives for Specifying Quality Standards for Digital Cartographic Data, in *Report 4*, National Committee for Digital Cartographic Data Standards, Columbus OH

Chrisman, N.R., and others, 1984, Modernization of Routine Land Records in Dane County, Wisconsin: Implications to Rural Landscape Assessment and Planning, *URISA Professional Papers Series 84-1*

Crossfield, J.K., Mezera, D.F., and Vonderohe, A.P. 1983, The Westport Experience, *Proc. ACSM-ASP Fall Meeting*, p. 18-27

Federal Geodetic Control Committee, 1984, *Standards and Procedures for Geodetic Control Surveys*, Rockville MD

Krohn, D.K. 1984, The Least Squares Adjustment of Oregon Township: An Interim Report for the Dane County Land Records Project, University of Wisconsin-Madison

National Committee for Digital Cartographic Data Standards (NCDCDS), 1985, An Interim Proposed Standard for Digital Cartographic Data Quality, in *Report 6*, Columbus OH

National Research Council, 1982, *Modernization of the Public Land Survey System*, National Academy Press, Washington DC

Petersohn, C. and Vonderohe, A.P. 1982, Site-Specific Accuracy of Digitized Property Maps, *Proc. AUTO-CARTO 5*, p. 607-619

Phillips Petroleum 1984, Digitized Base Map Data Using USGS Quads (marketing brochure)

A GENERALIZED CARTOGRAPHIC DATA PROCESSING PROGRAM

Denis White
Laboratory for Computer Graphics and Spatial Analysis
Graduate School of Design, Harvard University
Cambridge, MA 02138 USA

ABSTRACT

The Harmony program, developed in 1980 and 1981, is a general computer tool for modifying, transforming, and displaying cartographic data. It is a companion program to the Harvard Odyssey system and shares many conventions with that set of programs. As a program managing sequential files of cartographic data, it provides for the creation of new fields in one or more output files by combining one or more input fields with arithmetic, relational, or logical operators in arbitrarily complex formulas. Set membership evaluation, ordinal ranking, numerical recoding, and various geometric transformations including map projections may be included in the formula expressions. One or two input files may be synchronized on one or more key fields in order to aggregate or disaggregate data. Records for output may be selected by an arbitrarily complex formula expression. As a display program, Harmony allows the description of the color, pattern, and size of data symbolism by the same formula mechanism used in data management. An adjunct program called Picture provides a two dimensional figure drawing capability for preparing titles, graphic scales, north arrows, and similar map elements.

INTRODUCTION

As more geographic information is recorded, processed, and mapped with the assistance of computers, programs with a wide range of capabilities are more and more needed. This paper describes a program which acts upon cartographic data like a programmable calculator acts upon numbers. It has a language in which the cartographic elements and their attributes are the objects to be manipulated and various mathematical, logical, and geometric operations are the functions that do the manipulation. Other approaches to cartographic data processing similar to this are Berman and Stonebraker (1977), Burton (1979), Cox and others (1980), Tsurutani and others (1980), and Frank (1982). The program, called Harmony, was developed in conjunction with the Odyssey System (White 1979, Chrisman 1979) with which it shares a number of features. Harmony is built upon a conceptual model of cartographic information, explained in the first section of the paper, using an entity-relationship model of data (Roussopoulos and Yeh 1984).

Subsequent sections of the paper describe the capabilities of the program and the language used in the user interface. The final sections show examples of the use of Harmony and discuss supporting programs.

CONCEPTUAL MODEL

The fundamental structural notion of the cartographic model of Harmony is the decomposition of the landscape into named (or numbered) cartographic elements with the geometric form of points, lines, or areas. As such this program, and all others like it, differ radically from most image processing approaches to cartography which process and map continuous spatial fields of information without identifying individual features. Point, line, or areal features in Harmony are the entities of the cartographic model and they are associated with one another by a set of relationships (or relations). Relations are defined in a general way and include the relations between a feature and its name, and a feature and its location, as well as more sophisticated relations such as that between a feature and its topological neighbors. The model consists of the classes of entities and their relations. Below is a table of some of the relations used in cartographic data processing.

TYPE OF RELATION		
	NAME OF RELATION	DESCRIPTION OF RELATION
Identification Relations		
	Name	Entity's "name is " (e.g., geocode)
	Location	Entity's "location is" (coordinate position)
Spatial Relations		
	Scale	Entity "contains" another entity
	Order	Entity "precedes" another entity
	Boundary	Entity "bounds" another entity
	Interior	Entity "has in its interior" another entity
	Intersection	Entity "intersects" another entity
	Congruence	Entity "is congruent to" another entity
	Dimension	Entity's "spatial dimension is"
	Distance	Entity's "distance from" another entity "is"

Table 1. A Taxonomy of Spatial Relations.

The mathematical background for the relations described in this table are discussed in White (1981). Note that the relations of dimension and distance are only two of a potentially large number of geometric measures which might be used. The spatial relations defined in this table plus other non-spatial relations form the basis for the organization of cartographic information in Harmony. The unit of organization is a collection of like entities, such as a group of point, line, or area features, along with a set of relations that are constant throught the collection. These collections are stored in the computer as files and each individual (record) in the collection has the same set of relations (fields) as every other individual.

Each field in a file has a code indicating the nature of the relation that is stored in the field. One of the fields is designated as the "primary" field and is presumed to contain the names or identifiers of the point, line, or area features that are contained in the file. Following are a list of the numerical codes used to indicate various relations:

Primary Field Types
0 Points
1 Lines
2 Areas
Scale
11 Containers
12 Containees
Order
21 Precedessors
22 Successors
Boundary Topology
31 Nodes clockwise
32 Chains clockwise
33 Polygons clockwise
34 Beginning Nodes
35 Ending Nodes
36 Left Polygons
37 Right Polygons

Interior Topology
42 Adjacent Chains
43 Adjacent Polygons
45 Connected Chains
46 Connected Polygons
47 Separated Nodes
48 Separated Chains
49 Separated Polygons ("holes")
Geometry
50 (X,Y) Coordinate Locations
51 (X,Y) Minima
52 (X,Y) Maxima
61 (X,Y) Centroids
81 Perimeters/Lengths
82 Areas
91 Distance from Point
Non-Spatial Attributes
101 Binary
102 Nominal
103 Ordinal
104 Interval
105 Ratio

Table 2. Types of Relations in Harmony.

When Harmony prints out the field descriptions of a file, the column labeled "Rel" indicates which relation of those above is represented by each field of the file. A special code of -1 is used to indicate the field which contains the number of coordinate pairs in a line or area boundary description (relation number 50). The Odyssey file management system (Morehouse 1978) used by Harmony allows for one variable length field as part of a record. A count field (code -1) always accompanies a variable length field.

HARMONY'S LANGUAGE AND FUNCTIONS

Harmony is a sequential file processing program with three processing "modes". Each mode uses one or more input files and produces one or more output files or a set of output graphics commands. The three modes are sorting, copying, and drawing. Sorting must be explicitly invoked by the user in order to prepare files for further processing such as aggregation. One input file is sorted to one output file. Copying is the basic mode for file transformations and may be performed from one or two input files (two input files must be synchronized, i.e., sorted on the same feature) to any number of output files. Drawing is from any number of input files to a single file of graphics commands or directly to a graphics device. The command for the appropriate action,

> **sort;**
> **copy;**
> **draw;**

is normally given after the relevant files have been opened and the file descriptions have been modified, or after symbolism for drawing has been specified. (See Dougenik (1978), Broekhuysen and Dutton (1983), for a discussion of the Odyssey command language generator used for this program.)

Input files are opened by specifying their internal Harmony file number and their external system file name,

> **open input:file filename:'name';**

Output files are also identified by internal file numbers and external file names, but in addition have optional qualifiers of text, aggregated, or graphics. Text files are written in character format rather than in binary. A file which is opened as aggregated sets up the aggregation mechanism for the subsequent copy operation. The text and aggregation options may coexist. Mutually exclusive from both of these is the graphics option which prepares the file to receive graphics commands. The final qualifier for output files is the input file with which to pattern the output file's fields. Each output file is initially patterned after an existing input file such that it has the same number of fields, each with the same characteristics as the template input file. The statement opening output files is,

> **open output:file filename:'name'**
>   **text**
>   **aggregated**
>   **graphics**
>   **like template:file;**

where the default is binary, non-aggregated, non-graphics, and with a template like input file number one.

The three commands for managing the fields of files are the construct, remove, and modify commands. To construct a new field for an output file the template field from one of the input files must be indicated. The new field is then constructed with the same characteristics as the template field. Any field of an output file (or an entire output file) may be eliminated from processing with the remove statement. The modify statement is one of the most important in Harmony. With it, fields of output files are set up according to the desired contents on output. Most of the functional capabilities of Harmony are in fact initiated by specifying how each output field is to be made. The qualifiers to the modify command are the field's label (which appears in file descriptions as a documentary assist), the data type (which may be integer, real, or character in codes of 1, 2, or 3), the word count (in computer words appropriate to the data type), the relation code (as described above), and, finally, the "formula" for the field. The syntax for these commands are:

```
construct  output:(file,field) like template:(file,field);
remove     output:(file,field);
modify     output:(file,field) label:'label'
                               datatype:code
                               wordcount:count
                               relation:code
                               formula:'formula';
```

FORMULAS

The formula mechanism is one of the distinguishing features of Harmony and provides the program's flexibility and generality. Each field of each output file is built during the copy operation according to its formula. Formulas consist of variables, operators, and functions arranged in a conventional infix notation. The variables are fields of the input files. The operators are the arithmetical, relational, and logical operators found in most algebraic programming languages. The functions are some of the common mathematical functions (trig, log, exp) and plus some special ones. The membership function evaluates the presence or absence of its first argument in the set specified by its second argument. The ranking function returns the position into which its first argument would sort in the ordered table specified by its second argument. The conversion function returns the position in which its first argument is found in the conversion (recoding) table specified by its second argument. The transformation function specifies in its second argument one of twenty coordinate transformations (including ten map projections) to be applied to its first argument. The aggregation function specifies in its second argument one of six (last value, number of values, sum, mean, minimum, maximum) aggregation methods to be applied to its first argument. The conditional function returns its second argument as a value if its first argument is true, otherwise it returns its third argument. The formula language supports any level of nesting of variables, operators, and functions as long as the total length of the formula does not exceed a predeclared maximum (288 characters).

The implementation of the formula mechanism consists of a preparation stage, before file processing, in which all formulas are compiled from their infix representation to a more compact and efficient postfix representation. As each record of each output file is prepared, each field's formula is evaluated in its postfix form to obtain the contents of the output field. The variables (the current contents of the designated input fields) are combined by operators and transformed by functions as specified in the formula. The program performs type conversion (all numerical values are processed as real numbers) but no type checking or word count checking.

The formula mechanism is also used to select records for an output file. Each output file may optionally have a formula which is evaluated for each record to determine whether to include the record in output. Selection formulas must evaluate to a logical value (true or false). In the drawing mode, Harmony also uses formulas to specify the color, pattern, and size of symbolism to represent each entity of each input file to be mapped. With the various operators and the ranking function and the conversion function, any type of value classification which does not require preprocessing all values (such as quantiles or nested means would, for example) can be performed. The relevant statements in Harmony are:

```
select   output:file by formula:'formula';
color    input:file  by formula:'formula';
pattern  input:file  by formula:'formula';
size     input:file  by formula:'formula';
```

Many of the coordinate transformations require parameters which are specified in separate statements in Harmony. The membership sets, and ranking and conversion tables are also specified separately. These statements are:

```
make  center:(xcenter,ycenter)
      radius:distance
      parameters:(one,two)
      resolution:distance
      angle:degrees;
make  membership:number  table:(mem1,mem2,...memN)
      ranking:number     table:(break1,break2,...,breakN+1)
      conversion:number  table:(code1,code2,...,codeN);
```

Further statements are used to specify the graphics device, viewport, and data window for maps, and to specify the fields to be used in sorting. The status of any field or file and the values of various parameters may be displayed; the codes for relations, aggregation types, devices, graphics patterns, and transformations, and the symbols for operators may be reviewed with help commands. Finally, the program supports a scripting or command file feature in which long sequences of Harmony instructions may be prepared with a text editor and subsequently read into the program either interactively or in a batch mode.

AGGREGATION AND DISAGGREGATION

Harmony has a general capability for aggregation and disaggregation. In the aggregation process, data from subsidiary geographic units are grouped together in some way and assigned to the parent unit. The simplest form of aggregation uses one input file sorted on the field containing parent identifiers. The output file normally consists of a primary field of parent units and one or more attribute fields constructed with formulas using aggregation functions. These formulas accumulate the data for one parent unit while that unit's subsidiaries are being processed in the input file, and then store the aggregate data in the output file when a new parent appears in the input. An additional input file may be used if, for example, attributes of the parent units are needed in the aggregation formulas or are to be transferred to the output file from an existing parent level file.

In the disaggregation process, data from a set of parent geographic units are distributed in some way to subsidiary units. For this process two input files are needed: a parent file with aggregate data, and a file of subsidiary units with attributes to be used in the disaggregation formulas (such as the subsidiary units' areas or boundary lengths, for example). Harmony creates the output files as subsidiary unit files with fields of disaggregated data calculated by formulas taking a parent attribute and spreading it to subsidiaries according to some ratio of subsidiary attribute to parent attribute, or, for nominal data, simply copying the parent attribute. As with its other features, Harmony allows any kind of disaggregation which can be expressed in its formula language, but requires the user to "formulate" the result.

FORMULA EXAMPLES

A good way to understand how formulas are used to accomplish cartographic data processing goals is to look at some examples. Many applications using coordinate locations begin with a map projection from spherical coordinates to planar. The first transformation upon a data set of points or lines normally has a formula like:

'TRAN (F(1, -1), 411)'

where this formula would be specified for the output field containing the coordinates. (When the coordinates are contained in a variable length field, as they would in a "chain" file of polygon boundaries, the field number is always designated as -1). The formula specifies that the transformation function indicated by the second argument (an Albers projection #411) is to be applied to the first argument, the variable length field (of coordinates) of the first file. Subsequent processing might calculate areas of polygons, lengths of lines, or, for points, distance from a prescribed point with similar formulas, changing the second argument of the transformation function to specify the appropriate transformation.

A class of operations often using long and nested formulas is the selection of individual geographic units based on one or more criteria. Simple selection can be made from membership tables by first assigning members to the table and then using a formula like:

'MEM (F(1, 1), 1)'

to chose only those input records for output processing whose first field value is a member of table number one. Such a formula can be arbitrarily complex including the logical operators of AND, OR, and NOT to refine the selection. A similar use of formulas is assigning suitability classes to geographic units based on a number of criteria. Nested conditional functions can create a quite complicated formula, for example:

'IF (F(1, 1) > 1, IF (F(1, 2) = F(1, 3), 2, 1), 0)'

where the meaning is "if (file 1, field 1) is greater than one then if (file 1, field 2) is equal to (file 1, field 3) then assign the value of 2, otherwise if condition one is true but two is false assign the value of 1, otherwise if condition one is false assign the value of 0." Any degree of arithmetical, relational, or logical complexity can be handled this way.

Mapping with equal interval value classifications is done with a formula like:

'(F(1, 1) DIV 10) + 1'

that subdivides the range of values in (file 1, field 1) by intervals of 10. Classification by arbitrary intervals can be done by the ranking function or by nested conditionals. Each input file has its own formulas for determining the color, pattern, and size of symbolism of its values.

SUPPORTING PROGRAMS

Two additional programs support the activities of Harmony. The Picture program (White 1983) provides a two dimensional figure drawing capability that allows points, lines, and areas to be drawn as symbols, in outline form, or to be filled with colors and patterns. Text annotation can be created in several fonts. Titles, keys, north arrows, and other map elements are constructed with this program to accompany the symbolized geographic units drawn by Harmony. A program called Global converts data into the correct format for Harmony. Odyssey files can be read directly into Harmony.

CONCLUSION

The Harmony program uses a "higher level" language for cartographic data processing that provides for the specification of the results of cartographic operations in a mostly non-procedural manner. The success of this abstraction of cartographic processes is based on the validity, generality, elegance, and completeness of the underlying conceptual model of cartographic entities and their relationships.

REFERENCES

Berman, R. R., and Stonebraker, M. 1977. GEO-QUEL: A system for the manipulation and display of geographic data. Computer Graphics 11(2):186-91.

Broekhuysen, M. and Dutton, G. 1983. Conversations with Odyssey. Proceedings, Auto-Carto VI, ed. B. S. Wellar. Ottawa.

Burton, W. 1979. Logical and physical data types in geographic information systems. Geo-Processing 1(2):167-81.

Chrisman, N. 1979. A many dimensional projection of Odyssey. Laboratory for Computer Graphics and Spatial Analysis, Graduate School of Design, Harvard University.

Cox, N. J., Aldred, B. K., and Rhind, D. W. 1980. A relational data base system and a proposal for a geographic data type. Geo-Processing 1(3):217-29.

Dougenik, J. A. 1978. LINGUIST: A table driven language module for Odyssey. Proceedings of the First International Symposium on Topological Data Structures for Geographic Information Systems, ed. G. H. Dutton. Laboratory for Computer Graphics and Spatial Analysis, Graduate School of Design, Harvard University.

Frank, A. 1982. MAPQUERY: Data base query language for retrieval of geometric data and their graphical representation. Computer Graphics 16(3):199-207.

Morehouse, S. 1978. The Odyssey file system. Proceedings of the First

International Symposium on Topological Data Structures for Geographic Information Systems, ed. G. H. Dutton. Laboratory for Computer Graphics and Spatial Analysis, Graduate School of Design, Harvard University.

Roussopoulos, N., and Yeh, R. T. 1984. An adaptable methodology for database design. IEEE Computer 17(5):64-80.

Tsurutani, T., Kasahara, Y., and Naniwada, M. 1980. ATLAS: A geograhic database system. Computer Graphics 14(3):71-7.

White, D. 1983. A graphics system for instruction in computer graphics. Laboratory for Computer Graphics and Spatial Analysis, Graduate School of Design, Harvard University.

White, D. 1981. A taxonomy of space-time data relations. Presented at the Princeton Conference on Computer Graphics and Transportation Planning, Princeton, N. J.

White, D. 1979. Odyssey design structure. Harvard Library of Computer Graphics, 1979 Mapping Collection, II:207-15.

Short Biography
Denis White
January 1985

Denis White is a Research Associate in the Laboratory for Computer Graphics and Spatial Analysis, and a Lecturer in Landscape Architecture, at the Harvard University Graduate School of Design where he has been for ten years. His research and teaching interests are computer assisted cartography, the design and implementation of geographic information systems, and computer modeling for environmental design and planning. He has a BA in computer science from the University of Wisconsin, graduate credit in geography from the University of Oregon, and a MA in geography from Boston University.

Building A Digital Map of the Nation for Automated Vehicle Navigation

by
Marvin White
Etak, Inc.
1287 Lawrence Station Rd.
Sunnyvale, CA 94089
(408) 747-1903 (Work)
(415) 858-0877 (Home)

ABSTRACT

A navigation computer with a moving map display is now commercially available at less than $ 2000 for cars and trucks, and this places extraordinary demands on digitial cartography. First, the equivalent of two paper maps must be squeezed into a single digital cassette. Then, the map data must be placed on the cassette tape arranged so the computer can stay ahead of the driver, regardless of speed. Also, coordinates must have a relative accuracy of a car-length. Finally, vast coverage --- the entire U.S. --- is needed by the end of 1986.

To meet this challenge, we have embarked on the largest digital mapping project ever undertaken. A key element of the project, perhaps the most important, is the mathematical basis of the system for digitizing maps and constructing cassettes. This basis facilitates chaining for data reduction, makes organizing cassettes by neighborhood possible, and helps to control the tremendous digitizing effort.

The presentation of this paper will include a short film of the Etak Navigator in operation.

Suitability of Ada for Automated Cartography

Donald L. Williams
Department of Geography and Area Development
University of Southern Mississippi
Southern Station Box 5051
Hattiesburg, MS. 39406
(601) 266-4733 (Office)
(601) 544-2936 (Home)

Map construction algorithms currently exist in a number of computer programming languages and in several different versions of such languages as Fortran and Basic. Like other relatively sophisticated software, mapping programs often depend heavily on special features of the operating or hardware system on which each program was originally written. This dependence stems from the system dependence of most high level programming languages and has the important consequence of making most mapping programs difficult to transport from one computer to another.

Ada is a general purpose programming language sponsored by the U.S. Department of Defense and developed with the specific objective of producing highly portable programs. If Ada is a suitable language for writing cartographic programs, its use could alleviate current portability problems. To determine whether Ada is a practical language for cartographic applications, several common mapping algorithms originally programmed in Fortran and Basic have been rewritten in JANUS/Ada. Results of this effort demonstrate that basic mapping and map data processing programs may be readily written in Ada. Conversion of Fortran programs to Ada, in particular requires changes which have proven more cosmetic than substantive even though the program code appears very different.

ENQUIRY SYSTEMS FOR THE INTERROGATION OF INFRASTRUCTURE

R.J.Williams
University of Wisconsin-Madison
Madison, WI 53706

BIOGRAPHICAL SKETCH

Robert Williams is a Captain in the Australian Army and is currently a graduate student at the University of Wisconsin - Madison. He has eighteen years experience in topographic mapping and aeronautical charting, at both technician and managerial levels. His experience extends through many phases of mapping including geodetic surveying, aerial triangulation, map compilation, cartography, and lithography and includes eight years in computer-assisted mapping. He received his BA in Computing Studies from the Canberra College of Advanced Education in 1979 and is a member of the AIC and ASP.

Note: The views expressed are not necessarily those of the Royal Australian Survey Corps. The paper stems from the results of personal experience and research by the author.

ABSTRACT

Up until now the major effort by organizations which encode data covering large geographic areas has been in the data base creation phase with relatively little effort on the use, or interrogation, of that data, particularly with respect to establishing enquiry systems of infrastructure. It seems that the next stage in development of systems will be in specialist enquiry systems, or expert systems - an expert system being defined as "a set or arrangement of things so related or connected as to form a unity or whole and being skillful and having training and knowledge in some special field". One important application of an expert system is the interrogation of infrastructure which is required for relief operations for natural disasters, search and rescue operations, and also for route planning and charting.

INTRODUCTION

The degree of automation in cartography has increased gradually over the past ten years. Many mapping organizations are now using computer-assisted procedures to produce their cartographic products at a quality that is now "acceptable for reproduction". Up until now the major effort by organizations which encode data covering large geographic areas has been in the data base creation phase with relatively little effort on the use, or interrogation, of that data, particularly which respect to establishing enquiry systems of infrastructure - infrastructure being the substructure or underlying foundation, and especially the basic installations, on which the continuance and growth of a community, state and country depends. It seems that the next stage in development of systems will be in specialist enquiry systems, or expert systems. The need for such systems is reflected in both the public and private sectors. Often quotes referring to the "need to have an integrated emergency management system which takes into account procedures, communications and transportation aspects" appear in newspapers following natural disasters (Governor Anthony Earl (State of Wisconsin) in The Capital Times of April 12, 1984). Clearly the Governor may not have in mind an expert system of the type discussed in this paper, but the concept is recognized, albeit at an elementary level. The private sector, when referring to expert systems are generally more precise as addressed by Bereisa and Baker (1983) when discussing the state of development in automotive navigation systems at Buick Motor Division, General Motors Corporation (Abstract only submitted to AUTO-CARTO 6).

This paper discusses a basic conceptual view of possible expert systems, discusses some characteristics of systems oriented toward infrastructure applications, examines components of an expert system, and considers relationships between various types of data, structure of information and implementation of algorithms by using a case study.

A CONCEPTUAL VIEW OF AN EXPERT SYSTEM

One of the dysfunctions in the development of automated cartography and geographic information systems up until now has been to draw a prospective user and the system designer closer together than, say, the relationship that existed between the user and the cartographer in traditional mapmaking. While this might seem an obvious benefit for any system, the dysfunction occurs in that all too frequently the "system designer" becomes the controlling operator with the "user" taking a subservient role - a point expressed by Bie (1983) who suggested that "autocartography has been technology-driven rather than resulting from user needs". There are innumerable articles and papers which support this notion - papers which describe in detail data input, editing, validating of data, and processing (often just to create valid data) procedures, and then briefly mention possible future applications of their systems. While such work has contributed greatly to our discipline, by addressing techniques, standards, and the like, as well as creating data bases from local project area to global coverage - a fundamental requirement for geographic information systems - the design of expert systems should be approached more directly from a user's perspective.

The development of expert systems should proceed with
specific purposes and scopes clearly determined and defined,
and at a level of sophistication that will serve the user
adequately and efficiently. I am, however, concerned of
uncertainty within the field of geographic information system
development. While some authors have favoured research into
expert systems (Smith 1984), others have been more
pesimistic, noting that "few geographical problems command
such attention" and query "on what topic do we really know
what we are talking about in the sense of expert systems?"
and "where is there a need for a daily (or at least frequent)
use?" (Nystuen 1984 page 359).

Perhaps the problem of conceptualizing expert systems has
been caused by a lack of understanding of the role of
cartography and an inadequacy of suitable definitions. It is
not my intention to attempt to provide any sort of historical
analysis of cartography, but rather to observe the direction
the discipline has taken in recent years. This direction has
been generally to add more "information" to cartographic
products both directly and indirectly. Map symbolization and
specification has been refined to enable more information to
be printed directly onto the map. For example, tourist road
maps contain road distances, rest areas, bus depots, highway
interchange numbers, schematics of major roads with time, as
well as distance, provided, and so on. Indirectly, much more
information is provided in the form of leaflets, books, and
so on, designed to accompany maps.

This direction should be maintained - that is to provide more
information - but should be more selective with respects to
specific needs. For example, a person travelling along
Interstate Highway I70 through Colorado is probably only
interested in accomodation within close proximity to I70 and
not all accomodation in the State of Colorado.

CHARACTERISTICS OF INFRASTRUCTURE ORIENTED SYSTEMS

The scope of infrastructure is extensive and in one way or
another is used on a daily basis. Infrastructure can be
examined from a place or city perspective and from an area or
region perspective. Infrastructure includes information on
population and administration, medical facilities, water
supply, power supply, airfields, ports, railways, roads and
telecommunications.

Analysis of this information is equally diverse. Applications
might include service functions such as the supply of road
maps, flight routes and times, and accomodation as provided
by tourist information centres; county and state functions
such as management of service facilities, planning logistics
for natural disaster relief operations, or rerouting traffic
for highway construction ; or national and international
functions such as aeronautical and nautical route charting
and offshore area determination. It would be an ambitious

attempt to try to list all such facilities, functions and
applications and the intention is not to do so, but to merely
highlight the vast number of applications and to indicate
some parts or roles that expert systems might play in the
future management of these resources. More specifically, some
of the roles might be to plan routes through road networks to
provide tourists with route information or to determine the
best path to route vehicles to provide relief and aid
operations (Figure 1); or to plan aeronautical routes and to
locate navigation aids; or to determine buffer or protection
zones along a coastline for navigation purposes (Figure 2);
or to plot a route around barriers of features (Williams
1980).

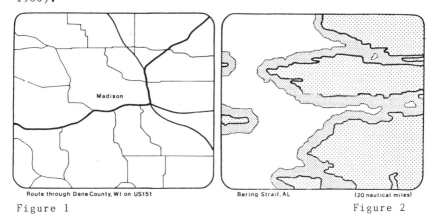

Figure 1   Figure 2

In addition to type of infrastructure and the purpose of
application, the geographic area of coverage has to be
considered. Again this can be extensive in scope, ranging
from local project area to global coverage. Thus a key
characteristic of an expert system would be to manage
information rationally with respect to area of coverage and
application.

COMPONENTS OF AN EXPERT SYSTEM

Communication

The success of an enquiry (expert) system will depend upon
its ability to provide timely and reliable information. That
is, a detailed and accurate response provided in two days
time is of _no_ value if the information is required by
tomorrow. Likewise, too much information is often as bad as
too little information. Thus an expert system should have an
input system, or more accurately, a user communication module
which should be able to interact during the information
gathering process in order to increase or decrease the amount
of information being provided. This interaction may be
provided in a number of ways including by the use of menus,
by question and answer, or by declarative statements

(Williams 1980). In any case, one of the functions of the
communication module should be to validate the response,
advise on the availability of certain information and to
record a historical account of the task.

System response
In most discussions on systems, the processes performed by
the system would be discussed at this stage and that followed
by analysis of output. But with an expert system, it should
be the final products that drive the intermediate stage and
so this component should be examined next. The system output
may be a visual display on a screen, a printed graphic or a
text description and associated tables, with the precision of
system response related to the intended use or user. For
example, a person requiring general information on the route
between Madison, Wisconsin and Green Bay, Wisconsin might be
satisfied by a response which said to take Highway US151 and
State Highway 26 to Oshkosh, a distance of 87 miles, and then
Highway US41 to Green Bay, a further 56 miles, while another
user requiring more detailed planning information, might need
more specific details regarding road identification,
intermediate distances, location of refuelling places and
selected areas for accomodation, and yet another user, say a
construction engineer, might require a detailed drawing and
description of a particular road intersection. Therefore, the
first user would probably be satisfied with a (text)
statement, the second might require a route map annotated
with selected information and accompanying guides, and the
third might require high quality graphics and detailed
information on terrain characteristics.

Information processing
In order to provide information as discussed above, adequate
processors and related data bases are required. The number of
"requests", and therefore algorithms to be analysed is a
function of the actual requirements of an implemented system.
For the analysis of infrastructure for search applications,
algorithms are required for the determination of shortest, or
best, paths in complex, or multi-level, networks such as road
transportation systems; the determination of shortest paths
between unrestricted nodes, those having no "physical links"
but constrained by distance as the case of air navigation
routes; and the determination of "proximity", "closest
location", and associated features.

These algorithms are required to be processed in a
multi-level environment and so the structure of the
information and management of the information has to be
designed accordingly. However, there are usually constraints
on the type of structures that can be represented by various
data base management systems. Most data base management
systems will only support structures that satisfy certain
properties required by the data base management system. The
most common manner of characterizing structures is either as
hierarchical or network structures. In a hierarchical
structure each record type can at most have one owner. With a

network structure more than one owner is allowed for each record type (Hawrysczkiewycz 1976). The interrogation of infrastructure requires retrieval and processing of information at both local and global areas of coverage and, so, a hybrid system of list, hierarchical, and network data structures incorporated into an appropriate relational model is required.

System knowledge
Some of the processes are deterministic while others may only provide estimates and so a portion of the data base and some algorithms could constitute a form of "system knowledge". This introduces the concepts of "knowledge" and "experience". Knowledge can be viewed as data including relationships, and deterministic procedures and techniques for providing finite answers. Experience can be viewed as those procedures and estimates that are "likely" to provide "reasonable" responses based on experience; for example, it is likely that the route between two cities is likely to be shorter using the Interstate Highway system than, say, the County road system.

In order to examine these components more closely, a case study is used. The study is concerned with the analysis of road networks in a multi-level configuration.

CASE STUDY

A case study is used to demonstrate the notions of knowledge and experience and the type of data structures required to perform queries on infrastructure type information. The case study specifically addresses the "shortest route through road networks" problem and processes data across regions down to the level of local roads, by using complex node structures, and a hydrid system of hierarchical, network and relational data structures.

Suppose one wishes to determine the shortest route in a road network between two terminal places or nodes. Then an algorithm (Figure 3) permits the analysis of a graph to produce a path in a network. Raphael (1976) suggests that, with heuristic algorithms, the success of the operation depends upon the "estimator"; that is the ability to efficiently determine the most likely distance to the terminal node from the present position.

As this process contains an "experience" operation, the "estimator", or factor by which the direct distance between a node and the terminal point is multiplied, can be modified by observing the current relationship, for example the class of link (road) currently being processed. Futher, because of the irregularity of road patterns, a route determined between an origin node and a terminal node may not necessarily be the same as a route determined from the terminal node and the origin node, and so an "experience" operation would be to determine both routes and select the shorter.

```
GRAPH PATH (Origin, Destination)
IF    <Origin = Destination>
THEN  <trace path to goal off "closed" list>
ELSE  <generate successors to Origin>
      <determine estimated distance to Destination as the
       summation of distance travelled and direct distance>
      <place node on "open" list>
      <select node from "open" list with lowest value>
      <place node on "closed" list>
      <GRAPH PATH (selected node, Destination)>
```

Figure 3

However, if one wishes to determine the route between a local road junction in the Township of Arena, County of Iowa and State of Wisconsin to a road junction in the County of Winnebago, State of Wisconsin, then the determination is required through a multi-level network. In this study the following relationships have been established: (1) <u>same base unit</u>, where origin and terminal nodes are in the same network, whether it be Town, County, or State; (2) <u>adjoining</u> units, for example adjacent Counties; and (3) <u>hierarchical areas</u>, for example a Township within a County. Figure 4 is a recursive algorithm to determine a route through a multi-level network and Figures 5, 6, 7 and 8 show a graphical representation of a solution.

```
PLAN PATH (Origin, Terminus)
IF    <same base unit>
THEN  <process unit>
ELSE
      IF    <unit on same level>
      THEN
            IF    <adjoining units>
            THEN  <determine transition point>
                  <process adjoining units>
            ELSE  <locate exit from each unit>
                  <process lower order units>
                  <PLAN PATH (with exit points)>
      ELSE
            IF    <hierarchical areas>
            THEN  <determine transition point>
                  <process lower order unit>
                  <PLAN PATH (point to terminal)>
            ELSE  <locate exit from each unit>
                  <process lower order units>
                  <PLAN PATH (with exit points)>
```

Figure 4

It can be seen that with this hierarchical approach, it is possible that important parts of a route will be processed at too high a level for practical use. For example, the City of Madison appears as a single node on the state level data base. This situation can be remedied by defining certain key nodes as <u>complex</u>, or <u>special</u>, nodes whereby directories (part of the knowledge base) permit the evaluation of a node with input and output links to the determination on a lower order data level. This principle can be extended recursively to

include such features as highway interchanges, such as
I90/I94. Similarly, links may become <u>complex links</u> at a lower
order as is the case for a divided highway and so may be
<u>directed</u>. Further, links may be temporarily "non-operational"
due to, say, flooding and so may be <u>obstructed</u>.

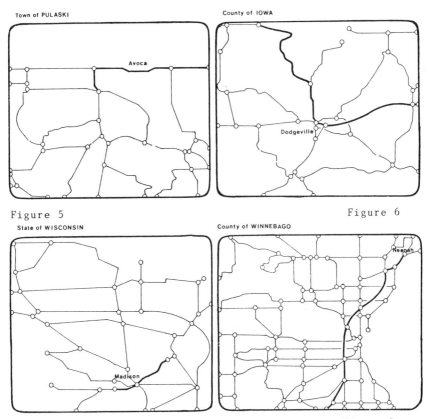

Figure 5

Figure 6

Figure 7

Figure 8

Measurement and reference
While it would be desirable to have a homogeneous data base
with respect to area of coverage and coordinate system, this
is neither practical nor possible - practical in the sense of
having to transfer innumerable maps plans, and documents
currently available to a common reference system; and
possible in the sense that while data up to County level
could be on planar system, State and Country level data
bases should be on a spherical system. Thus an expert system
should be able to detect deficiencies in the data base as
well as transforming between data sets.

Data base structure

The term, data structure, has been used by many authors, often with slightly different connotations. Generally the term is used to describe data format types, for example vector, raster, string, polygon and so on and associated relationships such as topological structure. Information is data that has been processed to obtain specific results of relationships and increases knowledge of the recipient (Burch and Strater 1974). With respect to expert systems, the design phase should be viewed from the higher perceptual level and so the analysis should be of data base structure, or information structure. The case study emphasized the relationships of information. Figure 9 describes the data base structure.

```
DATA BASE STRUCTURE FOR ROUTE SELECTION SYSTEM
------------------------------------------------
The following definition describes the structure of the data base using
Backus notation:

<file title>       ::= <system address> <base area code> {<area code>} <file type>

<system address>   ::= {system disk drive} | {network address}

<base area code>   ::= {US} | <other>
<other>            ::= {CA} | {UK} | {AU} | ....

<area code>        ::= {null} | <sub-unit code> <area code>
<sub-unit code>    ::= <state> | <county> | <town> | <special>         [#1
<state>            ::= {AL} | ... | {CA} | ... | {WI} | ... | {WY}
<county>           ::= ... | {DA} | ... | {IO} | ... | {WI} | ...
<town>             ::= ... | {AR} | ... | {DO} | ... | {PU} | ...
<special>          ::= {town subdivision} | {complex interchange} | ...

<file type>        ::= <directory code> | <data code>
<directory code>   ::= <@> | <&> | <+> | <%>
<@>                ::= {directory of data available}
<&>                ::= {directory file of adjacent areas}
<+>                ::= {directory file of inter-level nodes}
<%>                ::= {directory file of intra-level nodes}
<data code>        ::= <n> | <n#> | <n!>
<n>                ::= {feature e.g. 1=boundaries, 2=roads, etc}
<n#>               ::= {data file of network nodes for feature "n"}
<n!>               ::= {data file of network links for feature "n"}

Note [#1           ::= sub-unit codes are in hierarchical order
```

Figure 9

CONCLUSION

The case study provided evidence that processing of multi-level networks is feasible. However, the study also highlighted the need for further research into <u>experience</u> and estimation operations, and techniques for converting <u>experience</u> information into <u>knowledge</u> information, although intuitively if particular routes are used regularly then this <u>knowledge</u> could be incorporated into directories.

This research indicates that enquiry systems for the interrogation of infrastructure are feasible and such systems will be demanded as digital data becomes freely available.

NOTE

All figures were redrawn and simplified for reproduction purposes from output produced by author-developed software.

REFERENCES

Bereisa, J and K.Baker (1983), "Automotive Navigation Systems and Support Infrastructures" (Abstract), Proceedings AUTO-CARTO 6, Ottawa.

Bie, Stien W. (1983), "Organizational Needs for Technological Advancement - The Parenthood of Autocartography Revisited", Paper presented to AUTO-CARTO 6, Ottawa.

Burch, John G.Jr and Felix R.Strater,Jr. (1974), Information Systems: Theory and Practice, Wiley, Santa Barbara.

Hawryszkiewycz, I T (1976), Computer-Based Forms of Information Systems, Vol 1, Canberra College of Advanced Education Monograph Series No M4.

Nystuen, John D (1984), "Comment on 'Artificial Intelligence and Its Applicability to Geographical Problem Solving'", Professional Geographer, 36(3).

Raphael, B (1976), The Thinking Computer: Mind Inside Matter, Freeman, San Fransisco.

Smith, T.R (1984), "Artificial Intelligence and Its Applicability to Geographical Problem Solving", Professional Geographer, 36(2).

Williams, R.J (1980), "Automated Cartography with Navigational Applications", Paper presented to Fourth Australian Cartographic Conference, Hobart.

Williams, R.J (1984), "Who or What is DES?", Cartography, 13(3).

ABSTRACT FOR AUTO-CARTO 7

The Portable Map

Paul M. Wilson
Diane S. McQuarie
Geogroup Corporation
39 Canyon Road
Berkeley, CA 94704
O-(415)845-8044
H-(415)549-1756

Development of a fully portable graphics system is the logical extension of current trends in information display and computer graphics. Convergence of these technologies will inevitably produce a portable system capable of complex data manipulation and graphic display functions -- a portable electronic map.

This paper examines technological and market trends supporting the evolution of a portable graphics system and discusses design issues surrounding a specialized type of portable system aimed at mapping applications. Demand for this type of system has preceded its commercial availability. Potential users in a variety of fields are currently attempting to assemble integrated systems that partially fulfill their need for a portable unit.

Potential applications emphasized in the paper include utility right-of-way mapping, surveying, local government land assessment, and transportation routing. To support these applications, a portable graphics system must combine the physical characteristics of today's emerging lap computers with the power and sophistication of a desktop graphics workstation.

Electronic design requirements for this type of small, lightweight portable hinge on large mass storage, graphics display capabilities, and a communications component. Necessary elements include: memory capable of supporting complex graphics software; a high resolution flat panel display; a mass storage device (either magnetic or optical); a communications link; and a data management structure enabling quick access to map information.

Mapping, because of its need for large data bases and demanding display requirements, is among the most difficult applications for any computer. Although micro-based mapping systems are becoming available, most large applications are still in the province of mainframes or superminicomputers. Moving this level of capability to a portable system will allow few compromises in hardware and software design.

POTENTIAL CONTRIBUTIONS OF DIGITAL CARTOGRAPHY AND
SPATIAL ANALYSIS IN ASSESSING IMPACTS OF ACID DEPOSITION

James R. Wray
National Mapping Division
U.S. Geological Survey
Reston, Virginia 22092

ABSTRACT

Industrial nations in northern middle latitudes have become acutely aware of adverse effects of acid rain, or acid deposition. One strategy to achieve better understanding of the problem, and to win support for a mitigation program, is to estimate tangible costs to building materials. A key facet of this strategy is to compile a region-by-region inventory of the amount, kind, and location of building materials susceptible to acid deposition. A spatial approach to this problem assumes that the presence, form, and function of buildings—and of the materials of which buildings are constructed—are related to the patterns of land use. The presence or absence of buildings and other infrastructures is a primary clue in the mapping of land use and land cover. Prototype metropolitan and regional building materials inventories are already underway, and automated cartographic and geographic information system techniques are used to perform several related tasks: (1) to adapt an existing digitized land use and land cover inventory, in combination with other data sets, in a problem-solving context; (2) to relate land use data to statistical areas used in the Census of Population and Housing; (3) to stratify sample areas and choose sample buildings for determining the amount of selected building materials present per unit area of each land use class and area sampling frame; and (4) to generate prototype computerized thematic maps, graphics, and area statistics. In addition to the ongoing building materials inventory, the same approach is being considered to assess the impact of acid deposition on other broad land use categories already existing in the data base such as forest land, agricultural lands, and areas of inland water.

Publication authorized by the Director, U.S. Geological Survey, on January 2, 1985.

AN INTERDEPARTMENTAL APPLICATION
OF SPATIAL DATA BASES-BUILDING A STATISTICAL
NETWORK FILE FROM A TOPOGRAPHIC FEATURE FILE

Joel Z. Yan
Rennie Molnar
Jean-Pierre Parker
Statistics Canada
Ottawa, K1A 0T6, Canada
J. Glen Gibbons
Energy, Mines and Resources Canada
Ottawa, K1A 0E9

BIOGRAPHICAL SKETCHES

Joel Yan is currently Chief of Methodology in Geocartographics Subdivision of Statistics Canada where he has been employed as a specialist in geographic information processing and system development for ten years. Rennie Molnar is Head of the Software Development Unit in the Geocartographics Subdivision. Geocartographics is a service bureau to Statistics Canada and other agencies in computer-assisted geography, cartography and graphics. Jean-Pierre Parker is Head of Basefile in the Geocoding Unit of Geography Division. The Geocoding Unit is responsible for creation and maintenance of spatial base files including the Area Master Files for Statistics Canada. Glen Gibbons is currently Scientific Advisor to the Director General of Surveys and Mapping Branch of EMR. Glen has eight years of experience developing digital mapping applications at EMR.

ABSTRACT

For 15 years, Statistics Canada has been building and maintaining digital street network files as a base for micro-area retrieval of census data. The total coverage now includes approximately 150 municipalities in 41 metropolitan areas containing more than half of the Canadian population. This base file is also used throughout Canada by municipal government and emergency dispatch agencies. In order to expand coverage of these files while increasing their application and compatibility with traditional map bases, digital data exchange experiments have recently been undertaken with Surveys and Mapping Branch of Energy, Mines and Resources nada (EMR) and other agencies. This paper describes the result of one experiment which involves building the network Area Master File for the City of Lethbridge, Alberta from a digital topographic file. The topographic file had been extracted from the National Topographic Data Base which is being created for a variety of uses, the primary application being the production of the 1:50,000 NTS map series. This feature file was converted to network form at EMR, and then attribute information such as street names and addresses was added at Statistics Canada. The resulting process, scheduled to be complete by March 1985, holds promise of building a composite digital file which is both compatible with topographic maps and also has potential applications to transportation planning. Furthermore, it is hoped that AMF production costs will be reduced to permit an expansion of the AMF programme for Canada. The potential benefits of this type of standardization as well as related problems of transfer and manipulation of spatial data bases are discussed in this paper.

1. INTRODUCTION

This paper first examines the digital base file and mapping programs at the two agencies involved, and events leading up to the decision to conduct an inter-departmental pilot test using data for the City of Lethbridge, Alberta. Section two looks at the methods used and development required in conducting the test. Finally, Section three looks at the results, provides guidelines for future interdepartmental spatial data base work and raises a number of questions.

1.1 The Area Master File

Statistics Canada initiated the creation of digital street network files prior to the 1971 Census as a tool first for micro-retrieval of census data by user specified area, and later for automated collection maps (Yan and Bradley, 1983). These Area Master Files contain a logical representation of all city streets and other geographic features such as railroad tracks, rivers, and municipal boundaries in machine readable form. The AMF is maintained by the Geocoding Unit of Geography Division. It corresponds in function to the GBF/DIME file created in the U.S. during the same period, and with the TIGER file (Broome, 1984) being prepared for the 1990 Census. There are, however, differences in structure, since the DIME file is based on the block, and the AMF is based on the block-face.

Large urban areas (population 50,000 or more) are divided into block faces. A block-face consists of one side of a street between two successive intersections. These block-face spatial units are small enough that when aggregated they become a good approximation for a user identified query area. Each block face is assigned central x-y UTM coordinates, to which files of households, or persons can be coded i.e., geocoded. A user needing information from a geocoded file, outlines the area of interest on a map. This area is digitized and becomes a special "query area". All block face centroids falling within this area are aggregated and statistical data from the Census are tabulated for those block faces. This process is described in the booklet "Facts by Small Areas" (Statistics Canada, 1972). New applications of the AMF have been described by Boisvenue and Parenteau (1982). Area Master Files now exist for virtually all urbanized portions of the 38 tracted centres in Canada and three other centres. As of June 1981, this constituted coverage of 52% of the Canadian population.

1.2 Other Digital Data Sources

During the past few years there has been a significant increase in the volume of digital spatial data available in Canada, specifically from federal, provincial, and municipal mapping agencies (Tomlinson, 1984). In some cases, digital data is now available for areas within the AMF coverage program. To avoid duplication and reduce overall AMF creation and maintenance costs, Statistics Canada is beginning to look to digital data sources and updating programs of sister mapping agencies. This division of responsibilities compares with the working relationship of U.S. Geological Survey and the U.S. Census in the joint TIGER/U project described by Marx (1983) and Callahan (1984). A number of experiments with other agencies are underway or under discussion. This paper highlights one such experiment conducted with EMR. Other joint projects are listed in Table 1.

Table 1. Agencies Interested in Joint AMF Maintenance Agreements with Statistics Canada

Agency	Agreement	Status
City of Winnipeg	File Maintenance by the City of Winnipeg. Plotting by Statistics Canada.	In production.
Corporation of the District of Burnaby	Burnaby is building a network file with AMF compatible node identifiers. Detailed negotiation is beginning.	Burnaby has submitted a proposal for discussion.
Province of Ontario, Ministry of Natural Resources	Ontario wants to evaluate AMF as a source of attributes and network structure for their topographic data base.	Joint pilot project is underway for the city of Cambridge.
Four Regional Police Forces in Ontario	Coding by clients. Digitizing and processing by Statistics Canada.	AMF creation is progressing well.
Metropolitan Toronto	Metro is maintaining a link to AMF for their planning network file.	A meeting is planned.
C.R.A.R. Inc., Quebec	Network Files for 700 municipalities have been created by C.R.A.R. Incorporated.	Preliminary negotiation is underway.

1.3 The Digital Mapping Programme of Energy, Mines and Resources Canada

The Surveys and Mapping Branch of Energy Mines and Resources Canada has been involved in the automation of the processes of mapping and charting for over twenty-five years. Early activity involved the development of a system for the manual digitizing of 1:50,000 scale map manuscripts and methods for storage, retrieval and automated reproduction of this cartographic data. Parallel to this development, research and development was undertaken into the digitizing of information directly from aerial photography using a hybrid system of photogrammetric instrumentation and in-house developed video-digitizer and software. In both cases it was evident that the collection and manipulation of a complex data set required use of interactive graphics techniques.

Subsequent developments, starting in late 1976 were concentrated in the photogrammetric digitizing approach to data collection with the reproduction capability retained from the manual digitizing developments. This development made use of the interactive graphics capability of Intergraph systems and by late 1978 staff training and engineering development was being undertaken on a production level system.

The production system, which continues to evolve (Gibbons, 1982 and 1984), is predicated on the concept of a digital data file which contains the measured ground positions of terrain features rather than the more usual approach of digitized map coordinates. While this concept requires some additional effort in order to produce a cartographic map, it possesses numerous advantages because it does not suffer the usual distortions, omissions and generalizations of a cartographic product.

Since the beginning of production in the first quarter of 1979, the system has been used to acquire the digital topographic data and produce 1:50,000 NTS map products over some 180,000 km^2. The majority of the effort has been in the more heavily populated area of the province of Ontario following the arc of Lake Ontario from Ottawa - Cornwall to Sarnia.

Based on the results of competitive bids from the surveying and mapping industry, this process of acquiring a digital topographic data base and producing a cartographic map product from this source is about 25% more expensive than producing a cartographic map product by traditional analogue and graphic methods. There is, however, an additional product in the topographic data base acquired.

The utility of the digital topographic data as acquired by the Branch has often been demonstrated by the production of alternate mapping forms. This has included cost effective production of cartographic products for map scales from

1:20,000 to 1:250,000, aeronautical chart bases at 1:250,000 and 1:500,000 and so on. The greater benefit for the approach is, however, expected to be in the non-graphic applications whether as input to mathematical models of the environmental scientist or as a structure to which data sets of spatially distributed themes can be referenced. These non-graphic applications have been slower to develop for several reasons. A significant barrier is the difficulty of transferring digital information without the benefit of established standards for information models, data definitions, data formats, etc. Also the nature of the communication between the producer and the user is changed when exchanging digital data. This is not immediately recognized and imposes certain delays which may frustrate the exploitation of digital data.

There was an identified need to improve the communication with users of traditional map products in respect to digital data. Hence, Surveys and Mapping Branch established a small project team to review with users potential applications for digital data and the needs of these users in relationship to the plans and capabilities of the Branch. The mandate of the project team also included the demonstration of the utility of the topographic data by undertaking small projects with direct assistance to potential users.

Several federal agencies and departments were contacted in the review. However, for various reasons, including the relatively modest amount of data available, only a few demonstration projects were undertaken. A pilot test with Statistics Canada (STC) was selected because of this agency's substantial experience with handling digital graphic files. Furthermore, STC had defined requirements which would provide not only a test of the utility of the data but also an illustration of the characteristics of the data which would be required to meet non-graphic applications.

1.4 The Pre-Pilot Test
Early tests conducted in the summer of 1982 using EMR topographic data for the City of Belleville, Ontario indicated that a number of enhancements were required before application to AMF creation, specifically:
- overshoots and undershoots at intersections of up to 5 meters;
- absence of nodal points at all intersections of 2 or more features; and
- dispersion of logical features (e.g. a street) into several smaller distinct segments not logically connected in any way.

In December of 1982, EMR agreed to create an enhanced file to meet these generic needs for network data as expressed by Statistics Canada. At the same time, Statistics Canada selected the City of Lethbridge for a production pilot trial, since the population of Lethbridge had surpassed 50,000 in the 1981 Census, and an AMF was needed for the

1986 Census. In February 1983, EMR agreed to complete the sheet in question and then to supply Statistics Canada with digital data to agreed upon specifications. Digital files for Lethbridge were received in the summer of 1984, after discussions and software development at both agencies.

1.5 Objectives of the Pilot Test

With the Lethbridge pilot test, Statistics Canada hoped to seriously evaluate the potential benefits and costs, and to identify problems in working together with EMR on digital data projects. The team also hoped to assess potential solutions to these problems and explore them fully with a wide audience before making any permanent changes in procedures.

Statistics Canada recognized a number of potential long term benefits in looking to EMR as a digital data source:
- improved accuracy in existing files;
- improved standardization and compatibility;
- ability to share software development;
- potential unit cost reductions in the long run for new areas; and
- potential compatibile source for elements not currently captured by Statistics Canada such as buildings.

EMR's objectives in conducting the Lethbridge pilot test were as follows:
- demonstrate the utility of its current digital topographic data;
- invite suggestions on improvements to the content and structure of the database;
- assess Statistics Canada as a source of cartographic data; and
- develop an appreciation of the needs and approaches to digital data interchange.

2. METHODS

A methodology was developed which involved development and processing at both departments. The manipulation of the raw topographic data at EMR is described in Section 2.1. The labelling of features and addition of attributes at Statistics Canada is described in Sections 2.2 to 2.4.

2.1 Processing at EMR

The acquisition of the digital topographic data at EMR utilizes techniques and procedures designed to facilitate data collection and subsequent cartographic map production. In consideration of the photogrammetric environment, the data acquisition is mostly unrestricted by factors such as feature sequence and direction. However arbitrary directions or sequences are not readily acceptable as input to the AMF structured file. The critical criteria established for topographic data were the need for completeness, accuracy and a general purpose topographic feature coding scheme which describes the 'real world' characteristics of a feature.

The three major problems associated with the transformation of the basic topographic data to a form more amenable to the requirements of the AMF are: the arbritrary sequences and segmentation of features, absence of coordinates within a string to define intersecting nodes, and overshoots or undershoots. It was also necessary to consider a major difference in the nature of a network approach to data collection from the approach required to ensure an accurate portrayal of features in a multi colour map. This latter point is manifest in the number of vertices used to describe linear features and brings with it the need for appropriate generalization of linear features.

First, the requirement to reorder and reformat data for a variety of applications had already been recognized at EMR. A process was developed to reorder, logically connect and generalize linear features. The process is table-driven with respect to the feature coding scheme and permits the application of distinct criteria for different feature types. The control parameters which can be specified for this process are (i) a truncation value, which defines the minimum permissible distance between successive points in a string, (ii) a curving tolerance which defines the permissible rate of curvature of a string, and (iii) a closure tolerance which defines the maximum distance between successive elements in a chain for the chain to be considered continuous.

The second requirement of intersecting nodes involves three operations: determining the actual intersection of features, rebuilding the line string or coordinate chain to include the determined intersection coordinates and finally, flagging of the intersections or nodes. The process for building the node file for Statistics Canada was coded to operate in a PDP 11/70 environment and has a flow as follows:
- sort all selected features segments by the min-max of the end point coordinates;
- string the segments together into a complete feature after application of the truncation, curving tolerance, and closure parameters;
- starting from the feature with a minimum x-y, determine intersections with successive features, including the cases for undershoot, and build an index of intersections;
- using the index of intersections, process each curving feature to incude new intersection points and handle the cases for overshoots and multiple intersections;
- reprocess the index of intersections as modified for multiple intersections into a single list.

The resulting file should now consist of a number of coded features with intersection points appropriately inserted in each chain and the nodes separately identified as point features. The separation of linear features from the so-called "node features" was necessary to retain a graphics capability for plotting the file at EMR, although a separate data structure combining the coordinate chain and the node flags would be a preferred end product.

The full processing of the pilot test includes extraction from the topographic data base, according to the geographical extent and the feature selection made. This extraction process was only performed once and the results retained as an interim file. The time required to process this interim file for noding was approximately 25 minutes. The transfer to magnetic tape interchange format is very straightforward. The package transferred to Statistics Canada included the formatted tape, and summaries of the features transformed, number of vertices and the like.

2.2 Overview of Processing at Statistics Canada

The primary function of the Interactive AMF Creation System (IACS) is to build an AMF from an EMR topographic feature file and various source documents.

An EMR topographic feature file describes a network of features for a designated area. Each feature is defined by a series of points linked together. One element of descriptive data is also supplied for each feature, a 4-digit feature code that identifies its feature-type according to an EMR classification scheme.

Source documents consist of assorted street lists and maps obtained from the municipality and various other agencies. These documents contain the descriptive data, or attributes, associated with each feature such as NAME, DIRECTION and ADDRESS-RANGES. Since this data is not always available on one map, some attributes must often be obtained from other sources. With Lethbridge for example, address-ranges were not available but postal codes were. Therefore, an extra manual processing step was required to translate postal codes into address ranges. These source documents are also used to supplement the EMR feature network by digitizing additional features such as the municipal boundary. Streets may also be digitized if the source document contains more recent data than the EMR file. The output AMF describes the street network and other physical features for a designated area. Each feature is defined by a series of nodes linked together and is described by such attributes as NAME, FEATURE-TYPE, SUB-FEATURE TYPE, DIRECTION, etc.

The IACS must, therefore, produce an AMF by merging the feature-network received from EMR, with the descriptive data obtained from source documents.

2.3 Detailed Processing at Statistics Canada

The IACS consists of 7 major processing steps. Each process is described below.

A. LOAD - NETWORK - FILE
- the feature network is loaded by a special interface program into a model on an interactive mapping system (AUTOMAP).
- in the process, EMR feature- types are translated into STC internal feature-types and EMR ground co-ordinates are converted to UTM.
- the model is validated by an online spot check to ensure that there are no "overlaps" or "undershoots" and that a node exists in both features where they intersect.
- network plots are produced at BASE-MAP scale for final validation and for transcription of feature-names and other attributes.
- at the end of this step, a model exists containing a description of the feature network without any attribute data.

B. ANNOTATE - NETWORK - PLOTS (Map compilation)
- all feature attributes, including feature name, street type, start point, direction and feature type are coded on the NETWORK-PLOTS from the most recent municipal source documents in preparation for input to the feature network model.
- address ranges are excluded because they are node-attributes and including them interactively would make inefficient use of the cartographic workstation.
- missing features, such as the municipal boundary and new streets, are also drawn onto the network plots for input to the feature network model through digitizing.
- features such as small ponds and trails, which were contained on the EMR file but not required for the AMF, are highlighted in yellow.
- at the end of this step, the network plots are annotated with all feature attributes in a formatsuitable for input to the network model.

C. CREATE COMPOSITE NETWORK MODEL
- a composite feature is created for each desired feature in the model, containing the feature attributes and all of its nodes.
- customized macros were developed to prompt operators for all required information and to provide instructions on procedures.
- missing features are digitized to complete the model.
- some features must also be split if the name changes.
- at the end of this step, the COMPOSITE NETWORK MODEL contains nodes and attributes for all required features.

D. PRODUCE MINI - AMF
- the COMPOSITE NETWORK MODEL is transformed into AMF format and transported from the HP 1000 mini-computer to the the AMDAHL V8 mainframe for further processing.
- the transformation includes creating an AMF FEATURE for each COMOSITE FEATURE and translating internal AUTOMAP feature types into AMF feature types.
- some generated fields such as feature code and sequence number are also calculated.
- some validation is also done and errors are corrected using the interactive graphic editing system.
- at the end of this step, an AMF exists, although some required fields are missing.

E. COMPLETE MINI - AMF
Three major operations are performed on the MINI AMF in this phase by batch processing.
1) LABEL NODES
- node and section numbers are calculated for each detail record based on x and y coordinate UTM values
- section numbers are assigned so that eachsection corresponds to an NTS map sheetof scale 1:5000.
- node numbers are then assigned sequentially within each section, in ascending order based on their x - y location
2) RESEQUENCE FEATURES IN ALPHABETICAL ORDER
- each AMF feature has a unique 6 digit feature code associated with it.
- the features are sorted by feature - name and feature codes are reassigned sequentially in ascending order i.e. alphabetical order
3) INSERTION OF CROSS-FEATURE IDENTIFICATION
- each detail record is linked with other detail records that have the same node and section numbers.
- these represent intersections and the feature name, feature code and feature type of the intersecting feature are recorded in the cross reference fields of each detail record.

At the end of this step, the AMF is complete,except for address ranges and block-face centroid values.

F. INSERT ADDRESSES
Address ranges must be added to the detail records of all addressable features that represent BEGIN, END or INTERSECTION nodes. This process involves several steps as follows:
- if a base-map with address ranges does not exist, then one is compiled from available data.
- a special format printout of the AMF is produced for coding address values.

- coded addresses are then keyed to create a file of AMF update transactions.
- these transactions are processed by standard AMF update software which places address ranges in the AMF and calculates block-face centroid values.

At the end of this step, the AMF is complete except for final validation.

G. VALIDATE AMF

Several steps are taken to validate and if necessary correct the AMF:
- plots are produced for visual verification.
- special programs are run that perform various quality checks.
- final verification is done by a special program that creates ADD-TRANSACTIONS for all AMF features. These transactions are then run through the standard AMF update procedure that contains all the validation rules for the AMF.
- any errors identified are corrected using standard AMF update procedures.

At the end of this step the AMF is complete and available for production use.

2.4 Development at Statistics Canada

One of the attractive features of the IACS was that most of the major components already existed in some form and needed only minor modification to implement IACS function.

The interactive graphic editing system at Statistics Canada, AUTOMAP, had already been used to edit and plot AMFs in both the Computer Assisted Collection Mapping project (Yan, 1982) and the Interactive AMF Update System (IAUS).

For Phase A, loading the feature-network, an existing program developed for the Belleville pre-pilot test was used. The EMR to AUTOMAP program (EA001) needed only minor adjustments to the feature table. Improved run statistics and more informative diagnostic messages were also added.

Phase B, annotating the network plots, is a clerical procedure performed by the Geocoding Unit who have a great deal of experience with similar tasks. Consequently little additional training was required and the only development needed was the documentation of a few procedures.

The third phase, creating the composite network model, was implemented using the facilitities available with the AUTOMAP system. The AMF file structure had already been implemented under AUTOMAP in the IAUS project. AUTOMAP macros were developed to prompt operators for required data from the network plot and to instruct them on how to proceed. Procedures were also documented to augment the online facilities.

Phase D, producing the Mini-AMF, was implemented using the Mini to AMF program (MA001), developed in 1981 for the updating of the St. Catharines AMF as part of the IAUS.

However the MA001 program had to be modified extensively to process an entire AMF and to generate missing fields which were normally present in update mode.

Completing the MINI-AMF in PHASE E was implemented as a mix of old and new. The generation of Node and Section numbers, based on X Y co-ordinates, was an experiment with possible far-reaching impact on the future of the AMF file structure. In the traditional AMF creation method, node numbers are arbitrarily assigned and the AMF is split into sections irrespective of any regular geographic pattern. The section and node numbering algorithm implemented in the IACS ensures that each section corresponds to a fixed 1:5,000 scale map sheet, and that each node number could be calculated based on its x-ycoordinates. If successful, this experiment would demonstrate that it is feasible to create and maintain AMFs by sections representing stable geographic areas as opposed to areas affected by dynamic political boundaries. This method was implemented as a new program and was based on work at the Centre de Recherche en Amenagement Regionale (CRAR) at the University of Sherbrooke.

Resequencing and chaining were implemented using existing AMF programs with virtually no new development.

Phase F, inserting addresses, was implemented based on existing AMF update procedures. A few small utilities were developed to produce the special format print for coding addresses and for capturing and processing the address values. The captured data are formatted into standard AMF update transactions which are processed by existing AMF updates program. Procedures for coding the address ranges were also developed.

Validating and correcting the AMF in Phase G, is done using existing programs. Virtually no development was required because the existing sub-system contains sufficient quality checks. This process was however augmented by various acceptance tests at the end of each phase.

In summary, the IACS is functionally similar in many respects to the traditional AMF creation system with the primary difference being the source and format of the input data. The major difference in implementation is the use of interactive graphic hardware and software to replace the traditional manual coding method.

3. RESULTS

3.1 Lethbridge Test Results

To this date, approximately 40% of the Lethbridge data (240 of 600 features) have been processed up to and including final validation of the AMF. The quality of the end product is acceptable except for nodes missing at many intersections. This was caused by a recently detected software bug at EMR and highlights the need for improved quality checking of digital data received from other agencies. Other minor problems have been identified, all of which can be corrected with small software modification.

Results to date demonstrate that the fundamental approach, the functional flow of the process and the underlying algorithms developed are effective and viable in creating a data structure suitable for network applications. Nevertheless, further testing and quality assurance is required.

Table 2 compares the estimated cost of producing the test AMF using traditional methods with actual timings achieved so far. The interactive data exchange method seems to offer savings in terms of cost and elapsed time in AMF creation. Furthermore, bottlenecks have been identified and fine tuning of the software will result in further savings, particularly in the interactive labelling of features.

Table 2. Comparison of Time for Area Master File Creation – Normal Process vs. IACS Data Interchange Process Based on Preliminary Results

Traditional Process Activities	Planned Person Days	Actual Person Days	IACS Data Interchange Process Activities	Planned Person Days	Actual Person Days
Create address document	6.25	11.0	Load network file	2.0	3.0
Base map compilation	6.25	5.0	Annotate network plots	6.25	5.0
Node symbolization	5.5		Create composite network model	15.0	*25.0
Node numbering	5.5		Produce Mini AMF	2.0	2.0
Coding	22.0		Complete Mini AMF	3.0	2.0
Assign central point	.75				
			Insert Addresses	20.0	**20.0
Digitizing	2.75		Validate AMF	5.0	5.0
Analyse Run 01 results	.15			53.25	62.0
Analyse and correct section plot	11.0				
Analyse and correct node plot	5.5				
Analyse and correct centroid plot	2.1				
Finalize AMF	2.5				
	70.25	72.75			

* 25.0 days based on completing 600 features using the rate achieved during the last week of production. Training and development time is not included.

** Estimated cost for address insertion.

3.2 Differences Encountered
As well as demonstrating the feasibility of digital data exchange between EMR and STC, the pilot study also

identified several areas of incompatibility that must be addressed in the long term.

Many problems were encountered reconciling EMR and AMF feature classifications. For example, EMR recognizes over 20 road-types, based primarily on relative size, whereas AMF recognizes only 5. On the other hand, since EMR has established very few classes for hydrographic features, STC could not be precise enough in requesting the specific features types required. Also, the criteria for classification were not defined sufficiently clearly by either agency, thus further complicating the task.

Feature representation is another area of difference. Whereas EMR represents a divided road or highway as 2 separate features, AMF stores only 1. EMR represents a road with a bridge on it as a continuous feature. AMF, on the other hand, stops the road at the bridge and continues on the other side, effectively splitting the feature. The AMF also maintains some features not maintained by EMR such as municipal boundaries. These features must be added to the AMF model as a separate step.

Requirements for currency of the data also vary for each agency. The AMF is tied strongly to the Census and therefore must be as up to date as possible every 5 years for Census day. New subdivisions identified since the EMR data was captured must be included in the AMF to ensure it is as complete as possible for Census uses.

Finally, the basic unit of work is different. EMR bases their work on standard NTS map sheets, while AMF uses a politically defined municipality as a unit of work.

3.3 Guidelines for Future Work

In the long term, a number of points seem critical to facilitate this type of joint base file maintenance program
(1) Move towards a compatible classification scheme for features at both agencies.
(2) Agree on a common coordinate encoding scheme for feature e.g. boulevard streets, creeks, railway yards, shoreline, bridges.
(3) Develop a standardized transfer format which includes not only topography (coordinates and feature codes) but also topology (nodes, links and areas) and attributes (feature names for example).
(4) Develop a common set of conventions regarding treatment of networks, including nodes, behaviour at intersections, what can cross without a junction e.g. powerline and road, bridges and roads.
(5) Level of generalization should be agreed upon (e.g. in railway yards, how many lines should be coded?)
(6) The final division of responsibility needs to be agreed upon, but it appears that EMR should be the primary agent for distributing basic topography, with other agencies such as Statistics Canada responsible

for recording attribute data and some layers including geo-statistical boundaries.
(7) Planning and scheduling of the complete process beginning with aerial photography should involve significant consultation to optimize currency of critical data elements.
(8) Ongoing consultation by all participating agencies including Statistics Canada is required on a number of aspects including:
- delineation and classification
- transfer format
- standards
- updating requirements

3.4 Action Taken In Implementing Guidelines
There is already progress on a few of the items listed above.
(1) Statistics Canada's AMF staff are planning to review the AMF feature classification against the EMR standard.
(2) A federal-provincial committee has been formed to examine a standard digital transfer format with consideration of topology, attributes and topography based on extending the proposed standard of the Canadian Council on Surveying and Mapping (1982).

3.5 Open Questions
A number of questions remain that must be seriously examined. Recognizing their importance, these questions are provided here to give an indication of future directions.
- Do users always want data by map sheet?
- Should data be available by administrative boundary as well e.g. county or township?
- What should the links be across map sheets?
- What should the links be across administrative boundaries?
- How should digital data be packaged geographically?
- How should digital data be packaged in terms of level of feature?
- How should digital data be packaged in terms of topology?
- Who are the primary users of this digital data?
- How much are they willing to pay?
- How should updating of shared data be handled?

4. CONCLUSIONS

Early results from a pilot test indicate probable benefits to both agencies from continuing joint efforts in digital capture, application and transfer of street network data. A list of suggested procedures and important issues for inter-agency base file maintenance projects has been provided for consideration.

ACKNOWLEDGEMENTS

The authors wish to thank Catherine Gourley, Suzanne Enright, Rejean Fontaine, Andre Landreville and Mark Doherty for assistance with the programming and Carl Sadler and Mila Cardenas for production. The support of Ross Bradley, Andre Boisvenue and J.M. Zarzycki in this interdepartmental venture is also acknowledged.

REFERENCES

Boisvenue, A. and Parenteau, R., 1982, "The Geocoding System in Canada and the Area Master File", Papers from the Annual URISA Conference, pp 226-232.

Broome, F.R., 1984, "TIGER Preliminary Design and Structure Overview: The Core of the Geographic Support System for 1990", presented at 1984 Annual Meeting of the Association of American Geographers , Washington, D.C.

Callahan, George M. and Broome, Frederick R., 1984, "The Joint Development of a National 1:100,000-Scale Digital Cartographic Data Base", Proceedings of the American Congress on Surveying and Mapping, Washington, D.C.

Canadian Council on Surveying and Mapping, 1982, National Standards for the Exchange of Digital Topographic Data, Energy Mines and Resources Canada.

Gibbons, J.G. 1982, "Digital Techniques for 1:50000 Scale Mapping and Revision", Photogrammetric Record, 10(60) pp. 645-652, October 1982.

Gibbons, J.G. 1984, "An Operational System for Digital Topographic Data Acquisition and Digital Cartography", 15th ISPRS Congress, Rio de Janiero.

Marx, Robert W. "Automating Census Geography for 1990", American Demographics, June 983.

Statistics Canada, 1972, GRDSR: Facts by Small Areas.

Tomlinson Associates, 1984, Investigation of Digital Cartographic Status and Developments in Canada, 5 Volumes, DSS Contract No. 03SQ.23246-4-5508 for Energy, Mines and Resources Canada.

Yan, J.Z., and Bradley, D.R., 1982, "Computer Assisted Cartography for Census Collection: Canadian Achievements and Challenges", Proceedings of the Sixth International Symposium on Automated Cartography, Volume 2, pp 135-146, Ottawa.